PROGRAMMING PRECALCULUS WITH PYTHON

FIRST EDITION

LISA SAVY KAUFFMAN

College of Southern Nevada

cognella

SAN DIEGO

Bassim Hamadeh, CEO and Publisher
Susana Christie, Senior Developmental Editor
Alia Bales, Production Manager
Emely Villavicencio, Senior Graphic Designer
Alexa Lucido, Licensing Manager
Natalie Piccotti, Director of Marketing
Kassie Graves, Senior Vice President, Editorial
Jamie Giganti, Director of Academic Publishing

ISBN: 9798823339483

cognella® | CUSTOM
3970 Sorrento Valley Blvd., Ste. 500, San Diego, CA 92121

Brief Contents

Detailed Contents

Foreword

Programming and mathematics go hand-in-hand. The first computer scientists were mathematicians. The first mainframe computers were developed and employed to do complex mathematical calculations. It therefore is a wonderment to me that there has never been a textbook written that teaches mathematics and computer programming as an integrated whole (that I know of). I have seen mathematics textbooks that include sections on integrating technology such as Texas Instruments graphing calculators or software packages like Mathematica. But these are usually afterthoughts, tucked away after the "application problems" or thrown into a sidebar. The technology doesn't *teach* the mathematics and vice versa.

They say the best way to learn a topic is by teaching it. I beg to differ. The best way to learn a topic is to *program* it. You need to know the meaning of the variables (inputs), the formulas to manipulate the variables, how to format the answers (output), how to put together a robust collection of test data, and, most of all, how to interpret and correct errors. In the realm of Bloom's Taxonomy of Learning, traditional mathematics education can take a student as far up the pyramid as "Analysis"—understanding the content and structure of material. Throw in programming, and now you've moved up to the level of "Synthesis"—formulating new structures from existing knowledge and skills.

It is my hope that this textbook will be used to teach both Precalculus I and Python programming in the mathematics department, the computer science department, or a brand new twenty-first-century department that integrates both.

My undergraduate degree was in fact called mathematics–computer science. This was quite forward thinking of UCLA in the 1980s. Yet the courses were not integrated. Half of my upper division courses were in the mathematics department and the other half in the computer science department. I typed my first programs on keypunch cards that were fed into a mainframe using a hopper. (This was decidedly *not* forward thinking of UCLA.) I was thrilled when after graduation, I was finally able to type programs using an actual computer keyboard.

The goals of this book are twofold.

(1) The student learns the fundamentals of Python programming. Python is a high-level object-oriented programming language with many existing libraries (for statistics, plotting, matrix manipulation, algebraic manipulation including calculus and differential equations, an extended math function library, game development, etc.). Best of all, it's easy to learn and free! Python is open-source and can be run on computers, tablets, cell phones, even graphing calculators, making the language even more attractive and accessible to high school integration.

(2) The student learns the fundamentals of Precalculus I within the context of Python programming. The concepts are presented and discussed informally, avoiding the formidable definitions that turn many students off of mathematics.

The sample programs in this book are not necessarily the most succinct, efficient, or elegant. Rather, they are written so that they are easy for a beginning programmer to follow and understand. At the very least, they invite optimization efforts from students. The programs were written primarily using Python v3.3. With future versions, some of the features described in this textbook may become deprecated.

If used in a high school or college course, the instructor can pick and choose which objectives (precalculus and/or programming) they wish to cover. I have tried to present a robust selection of topics—some mainstream and some off the beaten path—to highlight the many applications Python has to mathematics.

This book is by no means an attempt to teach all the nuances and techniques Python offers. The matplotlib library alone could easily fill a book of its own (in fact those books already exist). However, this book does present the basic elements of the language—control structures (if-then-else, for and while loops), list and set operations, built-in math functions, user-defined functions, string manipulations, logical structures, dictionaries, list comprehension, classes, and 2D and 3D plotting.

I hope that instructors and students find this textbook useful and stimulating. If you find an error (there are undoubtedly many), please contact me at lisa.kauffman@csn.edu. I would also love to hear any suggestions for future editions of this textbook. The most difficult part of outlining this project was not what to include but what had to be left out due to time and length limits. Let's make this an evolving communal project!

My third hope is that the reader will be inspired to pursue future programs, mathematics, and projects. Happy programming!

Lisa Savy Kauffman

College of Southern Nevada

December 2022

P.S. If you'd like a free copy of the Answer Key, please drop me a line at lisa.kauffman@csn.edu. Thanks for your purchase of this text!

CHAPTER 1

Sequences and Series

1.1 Introduction to Sequences

A **sequence** is a list of numbers that progress in some sort of pattern, usually written with the numbers separated by commas. Some examples are:

1, 3, 5, 7, 9, ... (odd numbers)

2, 4, 6, 8, 10, ... (even numbers)

1, 2, 4, 8, 16, 32, ... (powers of two)

0, 3, 6, 9, 12, 15, 18, 21, ... (multiples of three)

1, 2, 4, 7, 11, 16, 22, ... (add the next highest integer)

Sequences can be either finite or infinite.

Let's dive into our first program, which will print the first 5 integers: 1, 2, 3, 4, 5.

*Refer to **Appendix A** for instructions on how to enter and run Python Programs.*

For the remainder of this book, comments are preceded by the # symbol and italicized; Python commands, functions, and reserved words are bolded.

PROGRAM 1.1.1

```
print(1)                                  # output command
print(2)
print(3)
print(4)
print(5)
```

Output

1
2
3
4
5

This is possibly the least elegant program in history, but it introduces you to the basic concept of outputting a constant. Let's try making this program more efficient by throwing in the concepts of a numeric variable ***n*** and a **for** loop.

PROGRAM 1.1.2

Image 1.1

Same output as before, but more efficient. So what does the **for** statement do? It repeats the block of indented instruction(s) five times. (Let's call those indented instruction(s) the "body" of the loop.) It starts by setting the value of variable *n* to 0, and then each time it executes the body of the loop, it adds 1 to the value of *n*. It stops when *n* reaches the last number in the range {0, 1, 2, 3, 4}.

Note that the "body" of the loop (instructions to be repeated) are indented by a tab, or four spaces.

Let's say you don't want to have to remember to add 1 every time you use **range**(). You can also specify a range with a set:

PROGRAM 1.1.3

```
for n in {1,2,3,4,5}:
    print(n)
```

However, this is not practical for enumerating long lists. Here's another option for printing 1, 2, 3, 4, 5:

PROGRAM 1.1.4

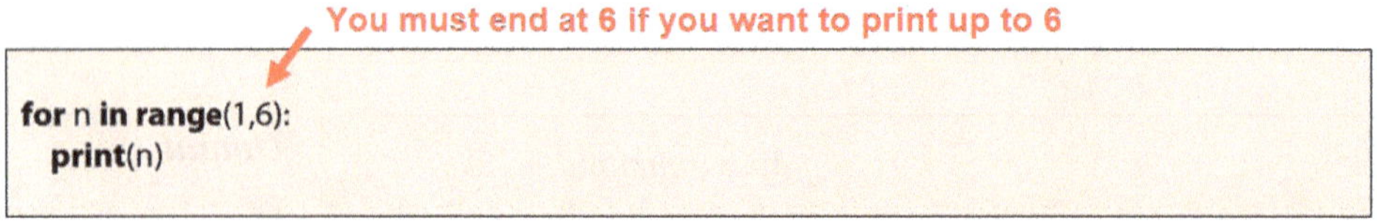

Image 1.2

You may have noticed by now that Python is picky about upper and lower case for variables (such as *n*) and reserved words (such as **for**, **in**, **range**, **print**).

If we want to print the even numbers between 2 and 20, we can change the **for** statement as follows.

PROGRAM 1.1.5

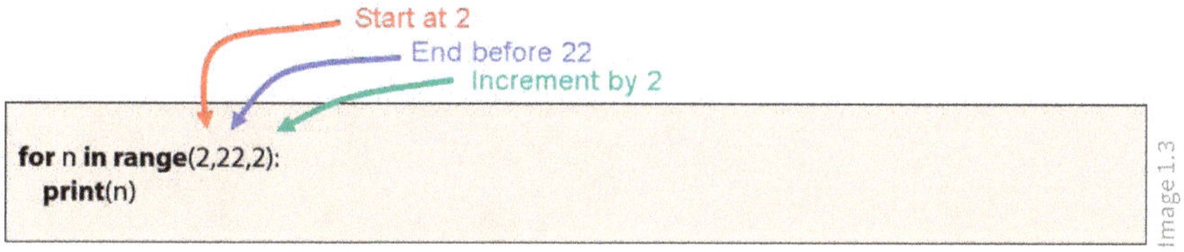

Output

2
4
6
8
10
12
14
16
18
20

Note **range**(2,21,2) would have had the same results as **range**(2,22,2). In general:

---

range(a,b,d) will start at a, abort at b, and increment by d (also called the step size). The step size can be negative.

EXERCISE SET 1.1.1

For each exercise below, modify the **for** statement so that the output will be the given sequence.

1. 11, 12, 13, … 19, 20
2. 3, 7, 11, 15, 19, 23, 27
3. 1, 3, 5, … 15
4. 5, 4, 3, 2, 1
5. 100, 90, 80, … 10

These are called **arithmetic sequences** because each new number is obtained by adding the same constant to the previous number. Arithmetic sequences are often infinite and are described by the first term a_1 and the increment d.

Example

Identify the first term a_1 and the increment d in the following arithmetic sequence:
6, 11, 16, 21, 26, …

Solution

$a_1 = 6, d = 6$

EXERCISE SET 1.1.2

Identify the first term a_1 and the increment d in the sequences in Exercise Set 1.1.1 (repeated below).

1. 11, 12, 13, ... 19, 20
2. 3, 7, 11, 15, 19, 23, 27
3. 1, 3, 5, ... 15
4. 5, 4, 3, 2, 1
5. 100, 90, 80, ... 10

a_1 and d are not affected by whether the sequence is finite or infinite.

We are now going to express sequences using formulas. Let's reset all our **for** statements to **for** *n* in **range**(1,6) so that it counts from 1 to 5.

Is it possible to print the even numbers 2, 4, 6, 8, 10 using **for** *n* in **range**(1,6)? Yes, if we introduce a mathematical expression inside the **for** loop.

PROGRAM 1.1.6

```
for n in range(1,6):
    print(2*n)
```

Output

```
2
4
6
8
10
```

Note: In Python, exponents are expressed using ** (double asterisk). For example, n^2 would be *n***2 and 2^{n-1} would be 2**(*n*–1).

EXERCISE SET 1.1.3

For each exercise below, modify the **print()** function (NOT the **for**) in program 1.1.6 so that the output will be the given sequences.

1. 3, 6, 9, 12, 15
2. 2, 3, 4, 5, 6
3. 3, 5, 7, 9, 11
4. −5, −4, −3, −2, −1
5. 2, 4, 8, 16, 32
6. 1, 4, 9, 16, 25
7. Which of the above sequence(s) is an arithmetic sequence?

SEQUENCE NOTATION

A sequence is typically named a_n, b_n, etc. In other words, a letter followed by a subscript n.

A sequence can be defined using a general term such as $\{a_n\} = \{\text{a formula in terms of } n\}$.

For example, in Exercise Set 1.1.3 above, Exercise 1, the general term is $\{a_n\} = \{3n\}$. Notice the general term also corresponds to the **print()** function in each of the exercises in Exercise Set 1.1.3.

EXERCISE SET 1.1.4

For each of Exercises 1–6 in Exercise Set 1.1.3 (repeated below), write the corresponding general term. The first is already done for you as an example.

1. 3, 6, 9, 12, 15

   Answer: $\{a_n\} = \{3n\}$
2. 2, 3, 4, 5, 6
3. 3, 5, 7, 9, 11
4. −5, −4, −3, −2, −1
5. 2, 4, 8, 16, 32
6. 1, 4, 9, 16, 25

Let's modify the **print()** function in program 1.1.6 to print out the following sequence of fractions:

$$1, \frac{1}{2}, \frac{1}{3}, \frac{1}{4}, \frac{1}{5}$$

Easy, you say.

PROGRAM 1.1.7

```
for n in range(1,6):
    print(1/n)
```

Output

```
1.0
0.5
0.333333333333
0.25
0.2
```

Not what we expected. This is because Python assumes $1/n$ is what is called a **float** type numeric variable. The default is to represent it with a decimal number. If you would like to output the number in the form of a fraction (numerator / denominator), then modify the program as follows.

PROGRAM 1.1.7 REVISED

```
from fractions import Fraction
for n in range(1,6):
    print(Fraction(1,n))
```

Output

```
1
1/2
1/3
1/4
1/5
```

This program introduces us to a few new concepts:

from fractions import Fraction indicates that we are importing a *function* called **Fraction()** from a *module* called **fractions**. A module is a code library that supplements the built-in functions in Python.

Let's try a more interesting sequence of fractions: $\frac{1}{2}, \frac{2}{3}, \frac{3}{4}, \frac{4}{5}, \frac{5}{6}$. First identify the general term of the sequence which is $\{a_n\} = \left\{\frac{n}{n+1}\right\}$. Recall, this corresponds to the **print()** function in the body of the **for** loop.

PROGRAM 1.1.8

```
from fractions import Fraction
for n in range(1,6):
    print(Fraction(n,n+1))
```

Output

```
1/2
2/3
3/4
4/5
5/6
```

EXERCISE SET 1.1.5

For each exercise below, rewrite the **print()** function in program 1.1.8 to output the following sequences.

1. 2, 3/2, 4/3, 5/4, 6/5
2. 1/2, 3/5, 2/3, 5/7, 3/4 *(Hint: 1/2 is the same as 2/4, 2/3 is the same as 4/6, etc.)*
3. 1/2, 1/4, 1/6, 1/8, 1/10
4. 1, 1/2, 1/4, 1/8, 1/16
5. 1, 1/3, 1/9, 1/27, 1/81
6. In program 1.1.7 Revised, what happens if you change (1,*n*) to (1/*n*)? Is the **Fraction()** function very accurate in converting 1/*n* to a rational number?

Our new goal is to print out just a single term of the sequence rather than the first five. We can do this by omitting the **for** statement and directly assigning a value to n. The following program will print out the 10th term of the sequence $\{a_n\} = \left\{\frac{n}{2^n}\right\}$.

PROGRAM 1.1.9

```
from fractions import Fraction   # from now on, module statements will be
                                 # grayed out to distinguish from flow of main
                                 # program
n = 10                           # This is an assignment statement
print(Fraction(n,2**n)           # Use ** for exponentiation
```

Output

5/512

Note that the **Fraction()** function automatically simplifies the expression 10/1024 to 5/512.

Now let's say we want to know the 20th term. Do we need to edit the program and change $n = 10$ to $n = 20$? Not necessarily; there is a way to make your program more versatile using an **input** statement.

PROGRAM 1.1.10

```
from fractions import Fraction
n = int(input("n? "))            # include a space after the question mark
print(Fraction(n,2**n)
```

When you run the program, enter the input in the "console" section of the screen, which is where the output usually appears. Here is an example using https://repl.it/languages/python3.

This is the program section	This is the console section
from fractions import Fraction	n? 20
n = int(input("n? "))	6/262144
print(Fraction(n,2**n))	>

The more important takeaway from this example is the introduction of the **input** statement. We now have an interactive program! Note the effect is to print **n?** and then to wait for your input.

Because the **input** statement always assumes the input value is a string (alphanumeric) type variable, it is necessary to apply the **int**() function to convert the input value to an integer.

Operations on Sequences

It is possible to generate new sequences by adding, subtracting, multiplying, or dividing corresponding entries of sequences. In other words:

If $\{a_n\}$ and $\{b_n\}$ are two sequences with the same number of entries (or both infinite), then a new sequence $\{c_n\}$ can be defined in any of the following ways:

Sum of a Sequence $\{c_n\} = \{a_n\} + \{b_n\}$

Difference of a Sequence $\{c_n\} = \{a_n\} - \{b_n\}$

Product of a Sequence $\{c_n\} = \{a_n\} \cdot \{b_n\}$

Quotient of a Sequence $\{c_n\} = \{a_n\}/\{b_n\}$ provided none of the entries of $\{b_n\}$ are zero

Example

Given $\{a_n\} = \{4n\} = 4, 8, 12, 16, 20, \ldots$ and $\{b_n\} = \{n^3\} = \{1, 8, 64, 125, 216, \ldots\}$ find $\{a_n\} + \{b_n\}$.

Solution

$\{a_n\} + \{b_n\} = \{4n + n^3\} = \{5, 16, 80, 141, 236, \ldots\}$

EXERCISE SET 1.1.6

1. Let $\{a_n\}$ and $\{b_n\}$ be the two sequences in the previous example; i.e., $\{a_n\} = \{4n\}$ and $\{b_n\} = \{n^3\}$. Write a **print()** function that generates the first 5 terms of the product $\{a_n\} \cdot \{b_n\}$. (Assuming it's in a **for** *n* **in range(1,6):** loop.)
2. Modify the program you wrote in the previous exercise so that it asks the user to input a value of *n*, and then print the product for just that subscripted value.

Sample Output

```
n? 10                    # Input any value
1040
```

1.2 Arithmetic Sequences

As mentioned previously, an arithmetic sequence is one in which you just keeping adding the same number, for example:

5, 8, 11, 14, 17, ...

Recall, it is characterized by the first term a_1 and the increment d. In the above sequence, $a_1 = 5$ and $d = 3$.

The most direct way to generate the first 5 terms of this sequence is by using **for** *n* **in range**(5,18,3). This starts with $n = 5$ and keeps adding 3 until it reaches 17.

But the mathematical convention is to generate sequences starting with $n = 1$ and increment by 1, i.e., using **for** *n* **in range**(1,6) or **for** *n* **in range**(1, *whatever*) as we did in Exercise Set 1.1.3. This enables us to think of each sequence in terms of a table:

n	a_n
1	5
2	8
3	11
4	14
5	17

One way to define this sequence is with a general term. In the above sequence, the general term is: $\{a_n\} = \{2 + 3n\}$.

Verify this inside a **for** loop:

PROGRAM 1.2.1

```
for n in range(1,6):
    print(2+3*n)
```

Output

5
8
11
14
17

But where did $2 + 3n$ come from? Is there an easy way to figure that out if you know a_1 and d? As a matter of fact, there is.

GENERAL TERM OF AN ARITHMETIC SEQUENCE

The general term of an arithmetic sequence with first term a_1 and increment d is:

$a_n = a_1 + (n - 1)d$

You can verify on your own that substituting $a_1 = 5$ and $d = 3$ into $a_1 + (n - 1)d$ will simplify to $2 + 3n$.

EXERCISE SET 1.2.1

Each of the following exercises describes an arithmetic sequence. Find the formula for the general term $\{a_n\}$ and then find the first five terms. You can either do this algebraically or by writing a program.

1. $a_1 = 17, d = 13$
2. $a_1 = 200, d = -10$
3. $a_1 = -25, d = 5$

Let's say we are given $a_3 = 8$, $a_7 = 32$ with the same instructions: Find the first five terms.

The first term could be done with a bit of guess-and-check. There are various formulas as well. One way is to recognize that you would need to add the increment d four times to get from a_3 to a_7.

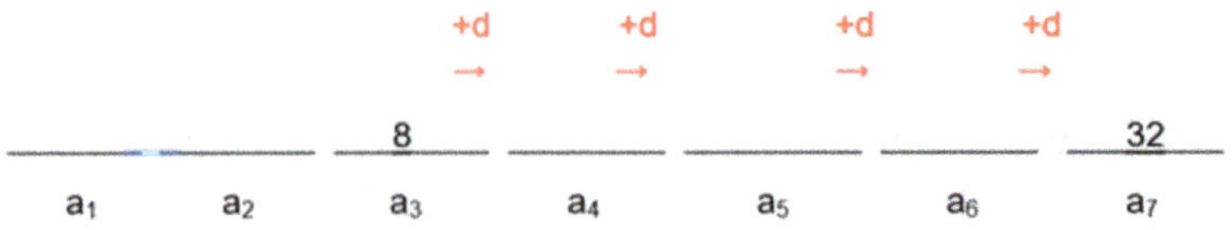

Image 1.4

Algebraically, that would be:

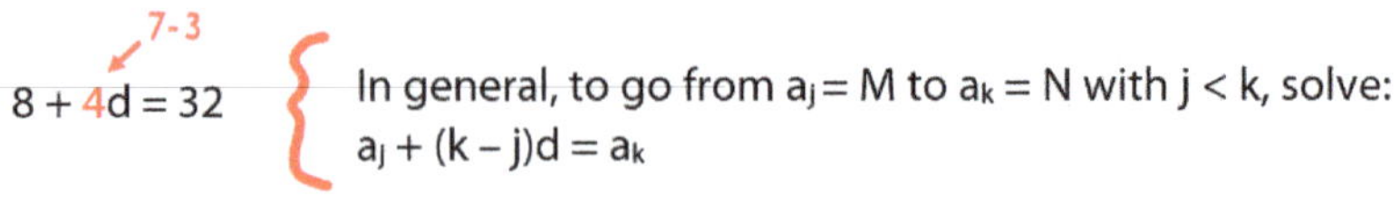

Image 1.5

Solving, you get $d = 6$. Now we can fill in a_4, a_5, a_6 by successively adding 6:

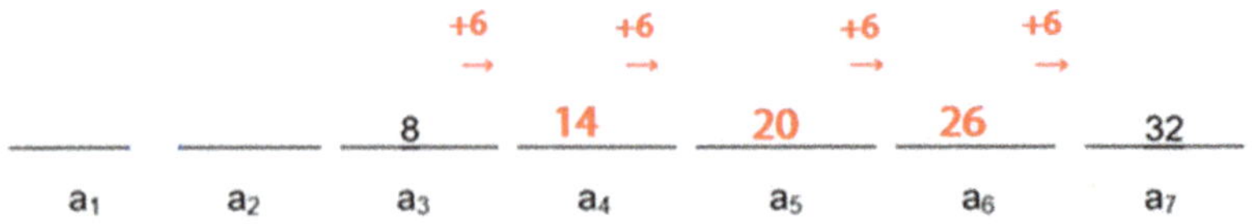

Image 1.6

If you were asked to find a_{100}, it would take a long time to keep adding 6. This is why you need a general formula for $\{a_n\}$, and for that you need to know a_1. For the above sequence, going backward from a_3, you simply subtract 6 instead of add 6.

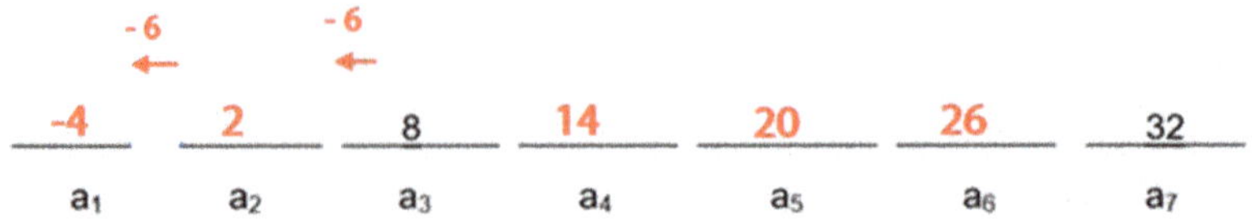

Image 1.7

Now that we know a_1 and d, we can substitute into the general formula: $a_n = a_1 + (n - 1)d$ to get:

$$a_n = -4 + (n - 1)6 = -4 + 6n - 6 = -10 + 6n$$

Let's verify that this formula indeed generates the first 5 terms of the above sequence.

PROGRAM 1.2.2

```
for n in range(1,6):
    print(-10+6*n)
```

Output

```
-4
2
8
14
20
```

EXERCISE SET 1.2.2

Each of the following exercises describes an arithmetic sequence in some manner. For each, find (a) the formula for the general term a_n, (b) the first five terms, and (c) a_{100}.

1. $a_5 = 11, d = 4$
2. $a_4 = 8, d = -6$
3. $a_8 = 8, a_{20} = 44$
4. $a_9 = -5, a_{15} = 31$

Now let's see if we can write a generic program that prompts the user for the first term a_1, the increment d, and then calculates a_{100}.

PROGRAM 1.2.3

```
a1 = int(input("a1? "))
d = int(input("d? "))
print(a1 + 99*d)
```

Output

```
a1? 11
d? 4
407
```

Let's make the output a little more explanatory by concatenating a string in front of the numeric answer.

PROGRAM 1.2.4

```
a1 = int(input("a1? "))
d = int(input("d? "))
print("a100 =",a1 + 99*d)
```

You can string together (no pun intended) as many expressions in a **print()** function as you'd like as long as they are separated by commas.

print(expr1,expr2,expr3, ...)

An expression can be either a string constant (such as "Hello") or a variable. Python will automatically insert a space between each expression. Another example that you can type directly into the console area:

```
>>name="Lisa"
>>age=29
>>print("My name is",name,"and I am",age,"years old.")
```

EXERCISE SET 1.2.3

1. In program 1.2.4, replace **print**("a100 =",...) with **print**("\n a100=",...)

   What is the effect of \n ?

2. Modify program 1.2.4 so that a third input n is requested, and the program outputs a_n using the formula for the general term.

Sample Output

*a*1? 11

d? 4

n? 47

*a*47 = 195

3. Modify program 1.2.4 so that instead of printing only the nth term, it outputs the first n terms.

Sample Output

*a*1? 11

d? 4

n? 3

The 1st 3 terms are:

11

15

19

4. CHALLENGE: Write a new program that will ask the user to input any two terms in the sequence. (Referring back to the example at the beginning of this section, we were given $a_3 = 8$ and $a_7 = 32$.) Calculate and output the values of a_1 and d.

Sample Output

1st Term? 8

Subscript of 1st Term? 3

2nd Term? 32

Subscript of 2nd Term? 7

$a1 = -4$

$d = 6$

Recursively Defined Arithmetic Sequences

Another way to consider an arithmetic sequence is to define each new term as the previous term plus the increment. In other words:

RECURSIVE DEFINITION OF ARITHMETIC SEQUENCE

The arithmetic sequence with first term a_1 and difference d can be defined as:

$a_{n+1} = a_n + \mathrm{d}$

Example 1

Write out the first 5 terms of each of the following recursively defined sequences. (Not all of them are arithmetic.)

A. $a_1 = 10, a_{n+1} = a_n + 5$

B. $a_1 = 0, a_{n+1} = a_n - 2$

C. $a_1 = 64, a_{n+1} = \frac{1}{2} a_n$

Solutions

A. 10, 15, 20, 25, 30

B. 0, −2, −4, −6, −8

C. 64, 32, 16, 8, 4

Example 2

For each sequence, provide the first term a_1 and a recursive definition of a_{n+1}.

A. The *even numbers 2, 4, 6, 8, 10, ...*

B. *The odd numbers 1, 3, 5, 7, 9, ...*

C. *Multiples of* three starting with 3

Solutions

A. $a_1 = 2, a_{n+1} = a_n + 2$

B. $a_1 = 1, a_{n+1} = a_n + 2$

C. $a_1 = 3, a_{n+1} = a_n + 3$

EXERCISE SET 1.2.4

#1–4: Write out the first 5 terms of the following recursively defined sequences. (Not all of them are arithmetic.)

1. $a_1 = 11, a_n = a_{n-1} + 9$
2. $a_1 = 100, a_n = \frac{a_{n-1}}{5}$
3. $a_1 = 2, a_n = (a_{n-1})^2$
4. $a_1 = 1\, a_2 = 1, a_n = a_{n-2} + a_{n-1}$

#5–7: Identify the first term a_1 and a recursive definition of a_n.

5. 7, 9, 11, 13, ...
6. −20, −40, −60, −80, ...
7. 14, 8, 2, −4, ...

The limitation of this definition ($a_{n+1} = a_n + d$), as compared to the general term definition ($a_n = a_1 + (n - 1) \cdot d$), is that it is often difficult or impossible to calculate a_n without first calculating all the terms that precede it. This would be very time consuming with large indices such as a_{100}. Fortunately, that's what computer programs are best at—tedious, time-consuming tasks such as generating a long sequence of numbers.

Let's print out the following sequence:

10, 15, 20, 25, 30, 35, 40, 45, 50

In other words, $a_1 = 10$ and $d = 5$.

PROGRAM 1.2.5

```
a=10
d=5
print(a)                        # print the 1st term
for n in range(1,9):
    a=a+d                       # This is an assignment statement, not an equation
    print(a)                    # print out the next term
```

The output is as previously mentioned: 10, 15, 20, 25, 30, 35, 40, 45, 50.

Note we are using the variable a to represent each successive term a_1, a_2, ... a_9. Because $a_1 = 10$ and $d = 5$, our first assignment statements are $a = 10$ and $d = 5$. This is called **initializing** the variables.

We then enter the loop, which is executed 9 times. Inside the loop, we output the current value of a and then increment it by the value of d.

EXERCISE SET 1.2.5

1. Modify program 1.2.5 so that instead of assigning 10 to *a* and 5 to *d*, the user is prompted to enter values for *a* and *d*. Then print out only the first 5 terms, not the first 9 terms.

Sample Output

*a*1? 11

d? 4

The 1st 5 terms are:

11

15

19

23

27

(Continued)

2. Modify the program you wrote in #1 above so that it also asks the user to input how many terms they want (rather than 5).

Sample Output

*a*1? 11

d? 4

n? 6

The 1st 6 terms are:

11

15

19

23

27

31

We will later see that arithmetic sequences are related to the concept of a linear function.

1.3 Geometric Sequences

Recall an arithmetic sequence is one in which you keep adding the same number (d) over and over again. We now define a geometric sequence as one in which you keep multiplying the same number (r) over and over again, for example, 3, 6, 12, 24, 72, ...

Let's compare arithmetic and geometric sequences:

	Arithmetic Sequence	Geometric Sequence
Example	If $a_1 = 1$ and $d = 2$: 1, 3, 5, 7, 9, 11, 13, 15, …	If $a_1 = 1$ and $r = 2$: 1, 2, 4, 8, 16, 32, 64, …
General Term	$a_n = a_1 + (n - 1)d$	$a_n = a_1 r^{n-1}$

Verify the formula for general term of a geometric sequence ($a_n = a_1 \cdot r^{n-1}$) with the following program.

PROGRAM 1.3.1

```
a1 = int(input("a1? "))
r = int(input("r? "))
print("/n The 1st 5 terms are:")
for n in range(1,6):
    print(a1*r**(n-1))
```

Output

a1? 1

r? 2

The 1st 5 terms are:

1

2

4

8

16

EXERCISE SET 1.3.1

1. Modify program 1.3.1 above so that it makes an additional input request for how many terms.

Sample Output

*a*1? 1

r? 2

How many terms? 10

The 1st 10 terms are:

1

2

4

8

16

32

64

128

256

512

Knowing the formula for the general term of a geometric sequence makes it easy to generate any term(s) in the sequence.

Example 1

For the following geometric sequence, identify the first term a_1, the ratio r, and the formula for the general term a_n:

A. 81, 27, 9, 3, ...

B. 1, −2, 4, −8, 16, ...

Solutions

A. $a_1 = 81,\ r = \frac{1}{3},\ a_n = 81\left(\frac{1}{3}\right)^{n-1}$

B. $a_1 = 1,\ r = -2,\ a_n = (-2)^{n-1}$

Example 2

For each of the following geometric sequences, identify a_1 and r and write out the first five terms.

A. $a_n = 2(5)^{n-1}$

B. $b_n = 4^n$

C. $a_n = \frac{5}{2^n}$

Solutions

1. $a_1 = 2$

   $r = 5$
   First five terms: 2, 10, 50, 250, 1250

2. First rewrite in the form $b_1 \cdot r^{n-1}$:

   $b_n = 4^n = 4 \cdot 4^{n-1}$
   Thus $b_1 = 4,\ r = 4$
   First five terms: 4, 16, 64, 256, 1024

3. First rewrite in the form $a_1 \cdot r^{n-1}$:

$$a_n = \frac{5}{2^n} = 5\left(\frac{1}{2^n}\right) = 5\left(\frac{1}{2}\right)^n = 5\left(\frac{1}{2}\right)\left(\frac{1}{2}\right)^{n-1} = \left(\frac{5}{2}\right)\left(\frac{1}{2}\right)^{n-1}$$

   Thus $a_1 = \frac{5}{2},\ r = \frac{1}{2}$

   First five terms: $\frac{5}{2}, \frac{5}{4}, \frac{5}{8}, \frac{5}{16}, \frac{5}{32}$

Alternately:

$$a_1 = \frac{5}{2^1} = \frac{5}{2}$$

$r =$ the ratio of any two subsequent terms such as $\frac{a_2}{a_1}$.

$$a_2 = \frac{5}{2^2} = \frac{5}{4}$$

Thus $r = \frac{a_2}{a_1} = \frac{5/4}{5/2} = \frac{1}{2}$

Or: r is whatever number is being raised to the variable exponent.

Example 3

Two terms of the same geometric sequence are provided. Find the formula for the general term.

$a_4 = 27, a_7 = 8$

Solution

To go from $a_4 = 27$ to $a_7 = 8$, we have:

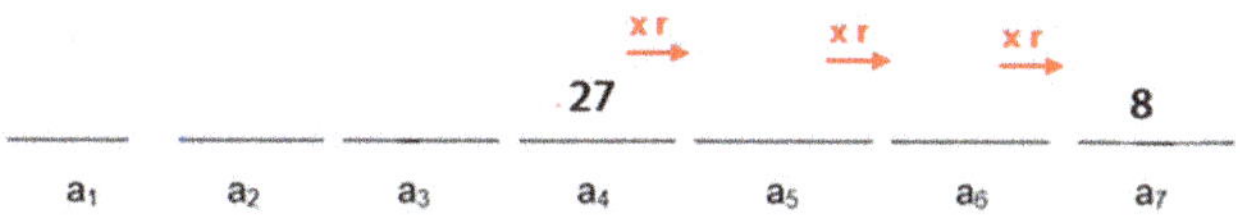

Image 1.8

Multiplying 27 by r three times:

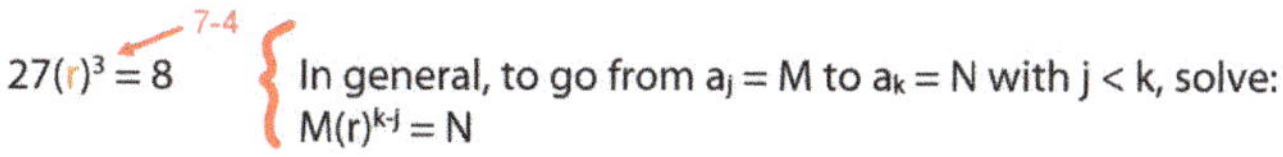

Image 1.9

$$r^3 = \frac{8}{27}$$

$$r = \sqrt[3]{\frac{8}{27}} = \frac{2}{3}$$

To go from $a_4 = 27$ backwards to a_1:

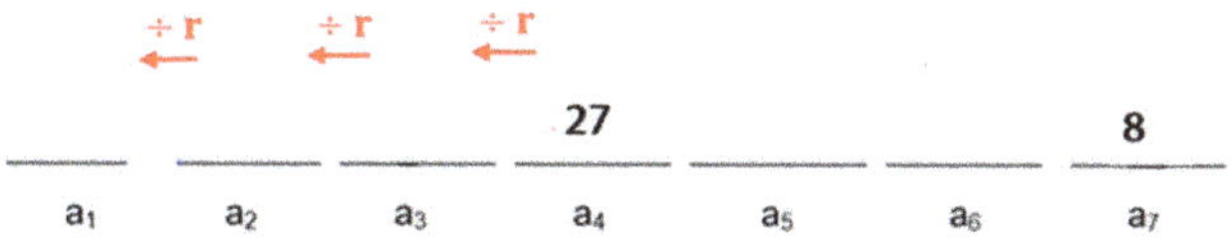

Image 1.10

Dividing 27 by r three times is the same as multiplying by $\frac{1}{r}$ three times:

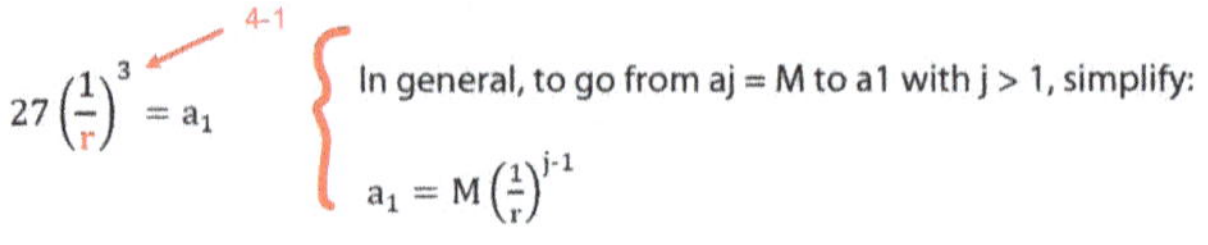

Image 1.11

If $r = \frac{2}{3}$ then $\frac{1}{r} = \frac{3}{2}$.

$$27\left(\frac{3}{2}\right)^3 = a_1$$

Solve for a_1:

$$27\left(\frac{27}{8}\right) = a_1$$

$$a_1 = \frac{729}{8}$$

The formula for the general term is thus $a_n = \frac{729}{8}\left(\frac{2}{3}\right)^{n-1}$

EXERCISE SET 1.3.2

For the following geometric sequences, identify the first term a_1, the ratio r, and the formula for the general term a_n:

1. 1, 10, 100, 1,000, 10,000, ...
2. 1/10, 1/100, 1/1,000, 1/10,000, ...
3. 2, 6, 18, 54, 162, ...
4. 1, −1, 1, −1, 1, ...

For the geometric sequences in #5–7, identify a_1 and r, and write out the first five terms. You can use program 1.3.1 to check your answers.

5. $a_n = \left(\frac{1}{2}\right)^{n-1}$
6. $a^n = 4 \cdot 3^n$
7. $a_n = \frac{3^n}{9}$

Find the formula for the general term.

8. $a_6 = 243, r = 3$
9. $a_2 = 7, a_4 = 1{,}575$

Just as arithmetic sequences can be defined recursively with $a_{n+1} = a_n + d$, we can define geometric sequences recursively as well:

RECURSIVE DEFINITION OF A GEOMETRIC SEQUENCE

The geometric sequence with first term a_1 and common ratio r can be defined as:

$a_{n+1} = a_n \cdot r$

EXERCISE SET 1.3.3

For the following geometric sequences, identify a_1 and r and then rewrite using a recursive definition. The first one is done as an example.

1. $a_n = 10 \cdot 3^{n-1}$

   [Answer: $a_1 = 10 \cdot 3^{1-1} = 10 \cdot 3^0 = 10 \cdot 1 = 10$; $r = 3$; $a_{n+1} = 3a_n$]

2. $1, \frac{3}{4}, \frac{9}{16}, \frac{27}{64}, \ldots$

3. $a_n = \frac{2^{n-1}}{3^n}$

4. $1, -2, 4, -8, 16, -32, \ldots$

As with recursively defined arithmetic sequences, the limitation of this definition ($a_{n+1} = a_n \cdot r$) as compared to the general term definition ($a_n = a_1 \cdot r^{n-1}$), is that it is often difficult or impossible to calculate a_n without first calculating all the terms that precede it. This would be very time consuming with large indices such as a_{100}.

EXERCISE SET 1.3.4

1. Modify the program you wrote in Exercise Set 1.2.5 #2 (which generated an arithmetic sequence using the recursive definition) so that it generates a geometric sequence using the recursive definition.

Sample Output

*a*1? 3

r? 2

n? 5

The 1st 5 terms are:

3

6

12

24

48

Inputting Fractions

What if we wanted a program to generate the sequence 128, 64, 32, 16, 8, ...? In that case, the ratio is a fraction, and an input statement using **int(input**()) will respond with an error if the user tries to enter ½ or 0.5. This is an easy fix: Import the **fractions** module and replace each **int(input**()) with **Fraction(input**()).

EXERCISE SET 1.3.5

1. Modify the program you wrote in Exercise Set 1.3.4 #1 so that it accepts fraction inputs. Does that preclude inputting integers and decimal numbers?

Sample Output

a? 1/2

b? 1/5

n? 5

The first 5 terms are:

1/2

1/10

1/50

1/250

1/1250

1.4 More on Recursively Defined Sequences

Recall, the recursive definitions for:

Arithmetic Sequence: $a_{n+1} = a_n + d$

Geometric Sequence: $a_{n+1} = r \cdot a_n$

These are just two types of recursively defined sequences. In general, a **recursively defined sequence** is one in which each new term is based on one or more terms that precede it. At least one initial term must be provided.

EXERCISE SET 1.4.1

1. The **factorial sequence** is defined by $a_1 = 1, a_n = n \cdot a_{n-1}$. Write out the first 10 terms in the table below. The first few terms are filled in for you as an example.

n	a_n
1	1 *(given)*
2	$2(1) = 2$
3	$3(2) = 6$
4	$4(6) = 24$
5	
6	
7	
8	
9	
10	

2. Verify your results in Exercise 1 by writing a program that outputs the first 10 terms of the factorial sequence. No inputs are necessary.

   We will find out later the important role that factorial numbers play in the field of probability.

3. The **Fibonacci sequence** is defined by $a_1 = 1, a_2 = 1, a_n = a_{n-2} + a_{n-1}$. Write out the first 10 terms in the table below. The first few terms are filled in for you as an example.

n	a_n
1	1 *(given)*
2	1 *(given)*
3	$a_1 + a_2 = 1 + 1 = 2$
4	$a_2 + a_3 = 1 + 2 = 3$
5	$a_3 + a_4 = ?$
6	
7	
8	
9	
10	

Now, if we want to write a program to output the first 10 terms of the Fibonacci sequence, it requires a bit of juggling to keep track of the variables.

PROGRAM 1.4.1

```
        # Initialize and print the first 2 terms
a = 1
b = 1
print(a)
print(b)
        # Calculate and print the next 8 terms
for n in range(3,11):
    c = a + b
    print(c)
    a = b
    b = c
```

Output

```
1
1
2
3
5
8
13
21
34
55
```

Did you notice the "variable juggling" with $a = b$ and $b = c$? Sometimes it helps to make a table to keep track of the value of the variables as you move through each iteration of the loop.

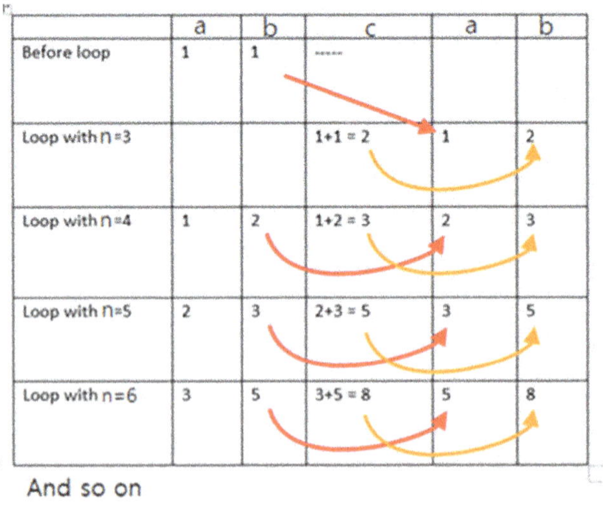

Image 1.12

The Fibonacci sequence is in fact an amazing sequence that is found in art, architecture, and even nature. Many books, websites, and videos are devoted exclusively to this sequence, and I encourage you to explore them.

Most recursively defined functions cannot be represented using general terms (although we have seen that arithmetic and geometric sequences can be defined both ways). The Fibonacci sequence can be approximated with the following general term:

$$a_n = \frac{\left(1+\sqrt{5}\right)^n - \left(1-\sqrt{5}\right)^n}{2^n \cdot \sqrt{5}}$$

Square Root Function

Note the above function involves three square roots. To form a square root in Python, you must first import the math library:

```
from math import *
```

and then take the square root of a number x as follows:

```
sqrt(x)
```

The asterisk is called a wildcard, meaning from math import * will import all the functions in the library.

What if you want to literally print out √2 (using the radical symbol) rather than its decimal equivalent 1.414213562 ...? There is no radical sign √ on a traditional computer keyboard. The workaround is to use the **unicode** "\u221A":

```
print("\u221A",2)
```

will output:

√2

Rounding

You can round a float (decimal) value to a given number of digits using the **round()** function. For example, **round**(3.14159,2) results in 3.14.

round(n,0) will round the value of n to an integer. You can also use **int**(n) to convert a float to an integer.

EXERCISE SET 1.4.2

1. Write a program that generates the first 10 terms of the sequence with general term $\sqrt{a_n}$. Output the sequence using radicals √1, √2, √3, ... √10 rather than decimals.
2. Write a program that outputs the first 10 terms of the Fibonacci sequence using the general term $a_n = \frac{\left(1+\sqrt{5}\right)^n - \left(1-\sqrt{5}\right)^n}{2^n \cdot \sqrt{5}}$. Round to the nearest whole number. How do the results compare to the actual sequence?

(Continued)

Hint: You can represent the above general term as:

((1+ **sqrt**(5))**n-(1- **sqrt**(5))**n)/((2**n)*(**sqrt**(5)))

Notice every open parenthesis "(" must be paired at some point with a closed parenthesis ")". Also remember that it is a good practice to put the entire numerator in parentheses and the entire denominator in parentheses.

3. A method for approximating square roots can be traced back to the Babylonians. The formula to approximate $\sqrt{p}$ is given by the recursively defined sequence:

$$a_1 = k,\ a_n = \frac{1}{2}\left(a_{n-1} + \frac{P}{a_{n-1}}\right)$$ where k is an initial guess.

Write a program that asks the user to input p and k, and then generate the first 10 terms of the sequence. Experiment with the input to evaluate its accuracy.

Sample Output

p? 75

initial guess of square root of *p*? 8

8.6875

8.660296762589928

8.660254037949777

8.660254037844386

8.660254037844386

8.660254037844386

8.660254037844386

8.660254037844386

8.660254037844386

8.660254037844386

Actual square root of 75 is 8.660254037844387

4. CHALLENGE!

The ratio of any two consecutive terms in the Fibonacci sequence is $\frac{a_{n+1}}{a_n}$. This is a special number called the **golden ratio** and itself forms a new sequence. As n gets large, the ratio *converges to* (approaches) a certain constant. Write a program that calculates at least the first 10 terms $\frac{a_2}{a_1}, \frac{a_3}{a_2}, \frac{a_4}{a_3}$, etc. What is the constant (rounded to 3 decimal places)?

1.5 Series

A **series** is the sum of a sequence. For example:

Sequence: $1, \frac{1}{2}, \frac{1}{4}, \frac{1}{8}, \frac{1}{16}, \ldots$

Series: $1 + \frac{1}{2} + \frac{1}{4} + \frac{1}{8} + \frac{1}{16} + \cdots$

Just as with sequences, a series can be finite or infinite. The real fun begins when we tackle infinite series: these provide a definition of π, e, and the foundation of calculus.

Let's start with a simple program that adds the finite sequence 1, 3, 5, 7, 9, 11, ... Recall the formula for general term of an arithmetic sequence is $a_n = a_1 + (n - 1)d$.

PROGRAM 1.5.1

```
a1=int(input("a1? "))
d = int(input("d? "))
n = int(input("How many terms? "))
s = 0                                  # initialize the sum
for i in range(1,n+1):
    t = a1 + (i-1)*d
    s = s + t                          # t is the ith term of the sequence
print("Sum of 1st", int(n), "terms is", s)
```

Sample Output

a1? 1
d? 2
How many terms? 5
Sum of 1st 5 terms is 25

This program introduces a few new concepts. The first is that of an *accumulator,* or running total, variable s. Note we initialize the value of s to 0 at the beginning of the program. Then, each time we go through the loop, we generate the next term of the sequence t and add that to the running total s. After we complete looping, we print the accumulator s, which holds the final sum.

Also notice we made this program generic in that it can add any number of terms of any arithmetic sequence.

EXERCISE SET 1.5.1

1. What happens if you forget to initialize s to 0 at the beginning of the program?
2. What happens when you replace $s = s + t$ with $s += t$?

In exercises 3–5, use program 1.5.1, or any other means, to calculate the sums of the following arithmetic sequences:

3. 1, 2, 3, 4, 5, 6, 7, … , 99, 100
4. $a_1 = 11, d = 5$ (first 5 terms)

(Continued)

5. $\{a_n\} = \{2n + 1\}$ (first 10 terms)
6. Modify program 1.5.1 so that it outputs a "running total" each time it traverses the loop. These are called **partial sums**.

Sample Output

*a*1? 1

d? 2

n? 5

Partial Sums:

1

4

9

16

25

The sum of 1st 5 terms is 25.

7. Modify program 1.5.1 so that it calculates the sum of a geometric sequence instead of an arithmetic sequence. Use it to find the sum of the first 8 terms of $a_n = 3(2)^{n-1}$.
8. Modify your answer to #7 so that it handles fractions and then finds the sum of the geometric sequence $1+\frac{1}{2}+\frac{1}{4}+\frac{1}{8}+\frac{1}{16}+\frac{1}{32}$. Hint: Import the **fractions** library and replace **int(input**()) with **Fraction(input**()).

Desired Output

*a*1? 1

r? 1/2

n? 6

Sum of 1st 6 terms is 63/32

SUMMATION (AKA SIGMA) NOTATION FOR A SERIES

$$\sum_{k=1}^{m} a_k = a_1 + a_2 + \cdots + a_m$$

This simply means to add the first m terms of the sequence $\{a_n\}$.

Example 1

Expand and simplify the series:

$$\sum_{k=1}^{5} \frac{2k}{5}$$

Solution

$$\frac{2(1)}{5} + \frac{2(2)}{5} + \frac{2(3)}{5} + \frac{2(4)}{5} + \frac{2(5)}{5}$$

$$= \frac{2}{5} + \frac{4}{5} + \frac{6}{5} + \frac{8}{5} + \frac{10}{5}$$

$$= \frac{30}{5} = 6$$

Example 2

Express using sigma notation.

$1 + 2 + 3 + 4 + 5 + 6 + 7 + \ldots, + 99 + 100$

Solution

This is an arithmetic sequence in which $a_k = k$.

We are adding 100 terms, so $m = 100$. Substitute into $\sum_{k=1}^{m} a_k$ to get $\sum_{k=1}^{100} k$.

Example 3

Express using sigma notation.

$a_1 = 11$, $d = 5$ (first 5 terms)

Solution

Again, this is an arithmetic sequence, so substitute a_1 and d into the general term:

$a_n = 11 + (n - 1)(5)$

$= 11 + 5n - 5$

$= 6 + 5n$

We are adding 5 terms, so $m = 5$. Substitute into $\sum_{k=1}^{m} a_k$ to get $\sum_{k=1}^{5} 6 + 5k$.

Example 4

Express using sigma notation.

$$1 + \frac{1}{2} + \frac{1}{4} + \frac{1}{8} + \frac{1}{16} + \frac{1}{32}$$

Solution

This is a geometric series with $a_1 = 1$ and $r = ½$. Substitute into the general term for a geometric sequence:

$$a_n = a_1 r^{n-1}$$
$$= 1(½)^{n-1}$$
$$= (½)^{n-1}$$

Substitute into $\sum_{k=1}^{m} a_k$ to get $\sum_{k=1}^{6} \left(\frac{1}{2}\right)^{k-1}$.

There is often more than one way to express a sequence using sigma notation.

For example, $\sum_{k=1}^{6} \left(\frac{1}{2}\right)^{k-1}$ could also be expressed as $\sum_{k=0}^{5} \left(\frac{1}{2}\right)^{k}$.

EXERCISE SET 1.5.2

1. Expand and simplify: $\sum_{k=1}^{7} (k+1)^2$
2. Express using sigma notation: $\frac{1}{2} + \frac{2}{3} + \frac{3}{4} + \cdots + \frac{13}{14}$
3. Express using sigma notation: $1 + \sqrt{2} + \sqrt{3} + 2 + \sqrt{5} + \sqrt{6} + \sqrt{7} + \sqrt{8} + 3$
4. Express using sigma notation: $\frac{2}{3} + \frac{4}{9} + \frac{8}{27} + \frac{16}{81} + \frac{32}{243} + \frac{64}{729}$

Let's go back to Exercise Set 1.5.1 #3:

$$1 + 2 + 3 + 4 + \ldots + 99 + 100 \text{ (or } \sum_{k=1}^{100} k\text{)}$$

This is the type of problem that makes you thankful for computers, as it is very tedious and time-consuming. However, there is a story about a famous mathematician named **Carl Friedrich Gauss** (1777–1855) related to this problem. Legend has it that as a schoolboy he was quite precocious, to the point that his tutor gave him what he thought was a tedious and time-consuming task: Add the integers between 1 and 100. He thought this would get little Carl out of his hair for a good while (and recall this was before calculators). However, Carl was able to do the sum in a matter of minutes, to his tutor's chagrin.

You should know from Exercise Set 1.5.1 #3 that the solution to this sum is 5,050. How did Carl do this calculation so quickly? He noticed the following pattern:

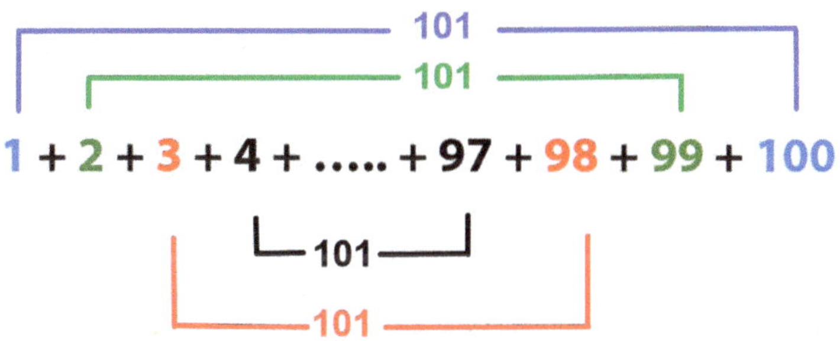

Image 1.13

There are 50 pairs of numbers, and each pair adds up to 101. Thus 50(101) = 5,050.

We can generalize this to a formula. If there are m numbers in the sequence, then there are $\frac{m}{2}$ pairs of numbers. Each pair adds up to the first term a_1 plus the last term a_m.

SUM OF A FINITE ARITHMETIC SEQUENCE

$$S_m = \sum_{k=1}^{m} a_k = \frac{m}{2}(a_1 + a_m)$$

Where a_k represents the general term of an arithmetic sequence, S_m simply means the sum of the first m terms.

The great thing is that this formula works for any finite arithmetic sequence, not just a sequence of consecutive integers.

Example 4

Find the sum:

$-5 + -3 + -1 + 1 + 3 + 5$

Solution

On inspection, this is an arithmetic series with $a_1 = -5$, $a_6 = 5$, and number of terms $m = 6$. That's all we need to know to substitute into $S_m = \frac{m}{2}(a_1 + a_m)$:

$$S_6 = \frac{6}{2}(-5 + 5) = 3(0) = 0$$

Kudos if you were able to recognize immediately that the sum was 0 because all the negative terms cancelled out all the positive terms.

Sometimes the last term is not explicitly given, and you need to figure that out as a first step.

Example 5

Find the sum of the first 20 terms of the arithmetic sequence with general term:

$a_n = -3 + 5n$

Solution

Remember we need two values for the formula: a_1 and a_m. In this example, $m = 20$.

$$a_1 = -3 + 5(1) = 2$$

$$a_{20} = -3 + 5(20) = 97$$

$$S_{20} = \frac{20}{2}(2 + 97) = 10(99) = 990$$

Verify this with program 1.5.1. (You will need to figure out what d is.)

Example 6

Find the sum:

$2 + 4 + 6 + ... + 70$

Solution

We can see that $a_1 = 2$ and $a_m = 70$. But what is m? In other words, how many terms in the series? Set the formula for general term $a_1 + (m - 1)d$ equal to 70 and solve for m:

$$2 + (m - 1)(2) = 70$$
$$2 + 2m - 2 = 70$$
$$2m = 70$$
$$m = 35$$

This is your value of m. You can now substitute $a_1 = 2$, $a_{35} = 70$, and $m = 35$ into S_m:

$$S_m = \frac{m}{2}(a_1 + a_m)$$

$$= \frac{35}{2}(2 + 70)$$

$$= 1{,}260$$

EXERCISE SET 1.5.3

Find the sum of the arithmetic sequences using the formula $S_m = \frac{m}{2}(a_1 + a_m)$.

Check your answers with program 1.5.1.

1. The sum of the first 46 terms of the sequence $2, -1, -4, -7, \ldots$
2. $\sum_{k=1}^{90}(3 - 2k)$
3. $73 + 78 + 83 + 88 + \ldots + 558$
4. $7 + 1 - 5 - 11 - \ldots - 299$
5. Modify program 1.5.1 so that it uses the formula for S_m to calculate the sum rather than a **for** loop. It should be much shorter! Let's refer to this as **Program 1.5.1. Enhanced**. Verify your answers for Exercises 1 and 2 above.

Notice that Exercises 3 and 4 (and Example 6) required two steps:

1. Given a_1, a_m, d: Figure out the number of terms, m.
2. Given a_1, d, m: Find the sum of the first m terms.

The program you wrote in Exercise 5 above (or program 1.5.1) will perform step 2. We need to modify it to also perform Step 1.

We'll first change the input statement requesting n (which we don't know) to request a_n (which we do know). We also need a formula to calculate the value of n.

PROGRAM 1.5.2

```
a1=int(input("a1? "))
d=int(input("d? "))
an=int(input("an? "))
n = [formula]
s= (n/2)*(a1 + an)
print("Sum of 1st", int(n), "terms is", int(s))
```

So what formula do we assign to n? We need to do a little bit of algebra to form this assignment statement. Recall our formula for the general term of an arithmetic sequence:

$$a_n = a_1 + (n - 1)d$$

We need to manipulate this equation so that it isolates m:

$$a_n - a_1 = (n - 1)d \quad \leftarrow \text{Subtract } a_1 \text{ from both sides}$$

$$\frac{a_n - a_1}{d} = n - 1 \quad \leftarrow \text{Divide both sides by } d$$

$$\frac{a_n - a_1}{d} + 1 = n \quad \leftarrow \text{Add 1 to both sides}$$

This is our formula for n. Alternately, you can use $\frac{a_n - a_1 + d}{d} = n$. We can now replace this into the formula above:

PROGRAM 1.5.2

```
a1=int(input("a1? "))
d=int(input("d? "))
an=int(input("an? "))
n = (an − a1 + d)/d
s= (n/2)*(a1 + an)
print("Sum of 1st", int(n), "terms is", int(s))
```

Use program 1.5.2 to check your answers in Exercise Set 1.5.3 #3 and #4.

EXERCISE SET 1.5.4

1. Modify program 1.5.2 so that it handles sequences with fractions.

Sample Output

a1? 2/3

d? 1/6

an? 7/3

Sum of 1st 11 terms is 33/2

Mixed practice. Use one of your programs, or pencil and paper, to answer the following.

2. How many terms must be added in an arithmetic sequence whose first term is 11 and whose increment is 3 to obtain a sum of 1,092?
3. The Drury Lane Theater in Illinois has 25 seats in the first row and 30 rows in all. Each successive row contains one additional seat. How many seats are in the theater?
4. You've accepted a job with an annual salary of $42,550 and a guaranteed annual cost of living raise of $2,200. How many years until your annual salary is $68,950?
5. You've put a new mountain bike on layaway. You make a down payment of $107 and want to pay it off over 24 months. The dealer requires that the monthly payment be increased by $2.50 every month. What is the price of the bike?

Geometric Series

A **geometric series** is the sum of a geometric sequence, for example, $3 + 9 + 27 + 81 + \ldots$

EXERCISE SET 1.5.5

1. Modify program 1.5.1 so that it calculates the sum of a geometric rather than arithmetic sequence.

Sample Output

a1? 3

r? 3

How many terms? 10

The sum of the 1st 10 terms is 88572.

Use your new program to find the following geometric sums:

2. $a_1 = 6, r = -2$ (1st 10 terms)
3. $\frac{3}{9} + \frac{3^2}{9} + \frac{3^3}{9} + \cdots + \frac{729}{9}$
4. $\sum_{k=1}^{11} 2\left(\frac{3}{5}\right)^{k-1}$

Here is a summary of formulas we've learned so far.

	Arithmetic	Geometric
General Term	$a_n = a_1 + (n - 1)d$	$a_n = a_1 r^{n-1}$
Sum of First *m* Terms	$S_m = \frac{m}{2}(a_1 + a_m)$	?

You can see what's next on our agenda.

SUM OF A FINITE GEOMETRIC SEQUENCE

$$S_n = \sum_{k=1}^{n} a_k = a_1 \cdot \frac{1-r^n}{1-r}$$

Where a_k represents the general term of a geometric sequence, S_n simply means the sum of the first n terms.

Example 1

Find the sum.

First 6 terms of the sequence with $a_1 = -20$, $r = 5$.

Solution

$$s_6 = a_1 \cdot \frac{1-r^n}{1-r} = (-20) \cdot \frac{1-5^6}{1-5} = (-20) \cdot \frac{-15{,}624}{-4} = -78{,}120$$

Example 2

Find the sum.

$$\sum_{k=1}^{7} 3 \cdot \left(\frac{1}{2}\right)^n$$

Solution

$$a_1 = 3\left(\frac{1}{2}\right)^1 = \frac{3}{2}$$

$$r = \frac{1}{2}$$

$$S_7 = \left(\frac{3}{2}\right) \cdot \frac{1-\left(\frac{1}{2}\right)^7}{1-\frac{1}{2}} = \left(\frac{3}{2}\right)\frac{\frac{127}{128}}{\frac{1}{2}} = \left(\frac{3}{2}\right)\frac{127}{64} = \frac{381}{128}$$

EXERCISE SET 1.5.6

1. Recalculate your answers to Exercise Set 1.5.5 #2, 3, and 4 using the formula above.
2. Modify program 1.5.1 Enhanced (your solution to Exercise Set 1.5.3 #5) so that it calculates the sum of a geometric rather than arithmetic sequence.
3. Use your new program to calculate the following: $-1 - 2 - 4 - 8 - \ldots -128$.

Time for some mixed practice.

EXERCISE SET 1.5.7

Answer the following questions using whichever method you wish.

1. You have been hired at an annual salary of $42,000 and expect to receive annual increases of 3%. What will your salary be when you begin your fifth year?
2. A new piece of equipment cost a company $15,000. Each year, for tax purposes, the company depreciates the value by 15%. What value should the company give the equipment after 5 years?
3. Initially, a pendulum swings through an arc of 2 ft. On each successive swing, the length of the arc is 90% of the previous length.
   (a) What is the length of the arc of the 10th swing?
   (b) On which swing is the length of the arc first less than 1 foot?
   (c) After 15 swings, what total length will the pendulum have swung (combining all swings)
   (d) When it stops, what total length will the pendulum have swung?
4. An old fable tells the story of a king who decided to reward one of his commoners. The king told him he could have any reward. The commoner replied, "A simple request, sire. Place one grain of rice on the first square of the chessboard, two grains on the second, four grains on the third, and continue doubling until you fill the whole board (64 squares). (a) How many grains of rice need to fit into the 64th square? (b) Compute the total number of grains (over all 64 squares) needed.

   Was the commoner's request a reasonable one?

Image 1.14

5. Let's say you win the lottery and you have a choice of the following two prizes: (1) one million dollars, or (2) one penny the first day of the month, two pennies the second day, four pennies the third day, and so on, until the 30th day of the month. Which choice do you take?
6. In a certain game, you try to avoid having your playing piece "captured." One hundred points is deducted from your score the first time you are captured, 200 the second time, 300 the third time, and so on. How many TOTAL points have been deducted if you've been captured 8 times?

Zeno's Dichotomy Paradox

Let's say you want to travel two miles. In the first hour, you walk one mile. But now you're tired, so in the next hour, you only walk half that distance, which is a half mile. Now you're really tired, so in the third hour, you only walk half the previous distance which is a quarter mile. Will you ever reach your destination?

Realistically, yes, because at some point the distance remaining to your destination is smaller than your shoe size. But let's say you're of infinitesimal size (as long as we're dealing with hypotheticals). Would you ever reach your destination?

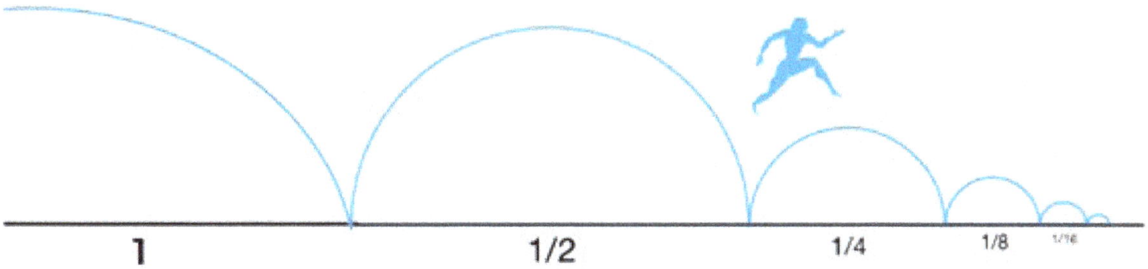

Image 1.15

Hopefully you recognize that this is a geometric sequence with $a_0 = 1$ and $r = ½$. You can easily run one of your programs or use the formula for sum of a sequence:

$$S_n = a_1 \cdot \frac{1-r^n}{1-r}$$

but you'd need to guess a value for n (how many hours).

Let's try $n = 10$:

$$S_{10} = \frac{1-\left(\frac{1}{2}\right)^{10}}{1-\frac{1}{2}} = \frac{1,023}{512} = 1.998046875$$

That's really close to 2 miles, but we're not quite there. We need to try higher numbers. Remember, that's what computer programs are great at—tedious, repetitive tasks.

Presumably your answer to Exercise Set 1.5.6 #2 (find the sum of a finite geometric sequence) looks something like the following. (If you haven't done the problem yet, here's the answer.)

PROGRAM 1.5.3

```
from fractions import Fraction
a1 = Fraction(input("a1? "))
r = Fraction(input("r? "))
n = Fraction(input("n? "))
s = (a1*(1-r**n) / (1 - r)
print("Sum of 1st", n, "terms is", float(s))
```

use float(s) so that s is output as a decimal rather than a fraction

Image 1.16

Sample Output

a1? 1

r? 1/2

n? 10

Sum of the 1st 10 terms is 1.998046875.

We could run this program over and over again, guessing larger and larger numbers for n. But isn't that itself a repetitive task? Let's modify the problem to do our guessing for us.

PROGRAM 1.5.3 REVISED

```
from fractions import Fraction
a1 = Fraction(input("a1? "))
r = Fraction(input("r? "))
n = int(input("n? "))
print("\n First 20 partial sums are:")
for n in range(1,21):
    s = (a1*(1 – r**n) / (1 – r)
    print(float(s))
```

Sample Output

```
a1? 1
r? 1/2
First 20 partial sums are:
1.0
1.5
1.75
1.875
1.9375
:
:
1.9999980926513672
```

We haven't reached two miles yet, but I bet we're really close.

Let's go beyond 20 terms. Let's go to "infinity and beyond." We are going to do this by deliberately creating an infinite loop. I say "deliberately" because usually infinite loops are dreaded and accidental.

PROGRAM 1.5.4

```
from fractions import Fraction
a1 = Fraction(input("a1? "))
r = Fraction(input("r? "))
n = 1                                  # initialize value of n to 1
while 1:                               # loop forever
    s = (a1*(1 - r**n)/(1 – r)
    print(float(s))
    n+= 1                              # increment n by 1
```

Don't run the program yet. First let's make sure you understand what's going on. We have replaced the **for** loop with a **while** loop. What they have in common is that the "body" of each loop (the indented lines) is repeated.

The difference is that the **for** loop did several things for us automatically. Let's say we have **for** n **in range**(1,6). The **for** loop knows to initialize the value of n to 1, add 1 each time it traverses the loop, and repeat the loop while n is less than 6.

The **while** [condition] loop will repeat as long as the condition remains true. Some examples:

while *x* >= 5:	(greater than or equal to)
while *a* != b:	(not equal to)
while response=="YES":	(equal to)

If we want the loop to keep going forever, we use:

while 1:

Because 1 in this context means "true."

But the **while** loop does not automatically initialize or increment a variable for us, so we had to do this ourselves.

Try running the program.

OUTPUT

*a*1? 1

r? 1/2

1.0

1.5

1.75

1.875

1.9375

:

:

2

2

2

Forever

Takeaway #1: Since this program will run forever—or until your device runs out of power—you need to stop it yourself. Find the [x] button in your console, which will abort the program while it is running.

Next notice that eventually (actually fairly quickly) our answer becomes 2. So that's it then, we verified that you will eventually reach two miles.

Actually you didn't. The program just rounded the answers up to 2. They were infinitesimally smaller, but they were still smaller.

Takeaway #2: The program does not always show you the most accurate answer.

Takeaway #3: You will never, theoretically, reach your destination.

Infinite Geometric Series

This should be a fairly quick topic, right? When you add infinitely many numbers, the answer should be infinity? This would be a winning bet at a party of nonmathematical people because the answer is actually NOT ALL THE TIME.

I'm going to introduce the concept of **convergence**, which means if the sum gets really really close to a certain value, we're just going to call it that value. ("Really really close" is discussed more rigorously in calculus.)

The previous example, $1+\frac{1}{2}+\frac{1}{4}+\frac{1}{8}+\frac{1}{16}+\cdots$, is a good example. Can we legitimately call it 2?

Yes. Here's the condition: If $|r|<1$, then the series will converge to a finite value.

SUM OF AN INFINITE GEOMETRIC SEQUENCE

If $|r|<1$, then

$$S=\sum_{k=1}^{\infty} a_1 r^{k-1}=\frac{a_1}{1-r}$$

Example

For each infinite geometric series, (a) identify a_1, r. (b) Does it converge? (c) If so, what is the sum?

A. $10+1+1/10+1/100+\ldots$

B. $2+4+8+16+32+\ldots$

C. $\sum_{k=1}^{\infty}(-0.2)^k$

Solutions

A. $a_1=10$, $r=1/10$ or 0.1. It converges because $|r|<1$.

The sum is $\frac{10}{1-1/10}=\frac{10}{9/10}=10\cdot\frac{10}{9}=\frac{100}{9}$

B. $a_1=2$, $r=2$. It does not converge because $|r|>1$.

C. $a_1=(-0.2)^1=-0.2$. $r=-0.2$. The series converges because $|r|<1$.

The sum is $\frac{-0.2}{1-(-0.2)}=\frac{-0.2}{1.2}=\frac{-2/10}{12/10}=-\frac{2}{12}=-\frac{1}{6}$

EXERCISE SET 1.5.8

For each of the following infinite geometric series in 1–4, (a) identify a_1, r. (b) Does it converge? (c) If so, what is the sum?

1. $1 + ½ + ¼ + 1/8 + 1/16 + 1/32 + \ldots$
2. $8 + 12 + 18 + 27 + \ldots$
3. $\sum_{k=1}^{\infty} 8\left(\frac{1}{3}\right)^{k-1}$
4. $\sum_{k=1}^{\infty} 4\left(-\frac{1}{2}\right)^{k-1}$

It seems like it would be fairly simple to write a program to calculate this sum $\frac{a_1}{1-r}$. Two inputs, one output, no loop.

PROGRAM 1.5.5

```
from fractions import Fraction
a1 = Fraction(input("a1? "))
r = Fraction(input("r? "))
s = a1/(1 – r)
print(s)
```

Note that we are outputting s as a fraction rather than a decimal. It is not necessary to use **print(Fraction**(s)) because s was defined in terms of other fractions.

But this program will only give the correct answer if $|r| < 1$. Thus we are going to check to make sure that the user entered a valid value for r. We will do this with an **if** block. Here is the general syntax:

if [condition is true]:

 Statement 1

 Statement 2

 (etc.)

It's quite simple: The indented line(s) are executed if the condition is true. (The condition does not need to be enclosed in brackets.) Otherwise, the indented lines are skipped.

Let's incorporate this structure into our program.

PROGRAM 1.5.5 REVISED

```
from fractions import Fraction
a1 = Fraction(input("a1? "))
r = Fraction(input("r? "))
if abs(r) < 1:                  # abs is the absolute value function
    s = a1/(1 – r)              # remember to indent this line and next
    print(s)
```

Try running the program with a series that does converge ($a_1 = 1$, $r = ½$) and a series that does not ($a_1 = 1$, $r = 2$). Notice the output for the second case:

OUTPUT

*a*1? 1

r? 2

This is rather sad, as there is no answer and the user has no idea why. So we will enhance our program even more with an **else** clause.

PROGRAM 1.5.5 REVISED AGAIN

```
from fractions import Fraction
a1 = Fraction(input("a1? "))
r = Fraction(input("r? "))
if abs(r) < 1:                  # abs is the absolute value function
    s = a1/(1 – r)              # remember to indent these 2 lines
    print(s)
else:
    print("Series does not converge.")
```

Try running the program again with a series that does converge ($a_1 = 1$, $r = ½$) and a series that does not ($a_1 = 1$, $r = 2$). There is now an informative message in the second case.

OUTPUT

*a*1? 1

r? 2

Series does not converge.

Here is the general syntax of the **if-else** block:

```
if [condition is true]:
    Statement 1
    Statement 2
     (etc.)
else:
    do something else 1
    do something else 2
     (etc.)
```

Note that when comparing if two expressions are equal, you must use == rather than =. Also, your computer keyboard does not have keys for $\leq, \geq$ or $\neq$, so you must use the following comparison operators.

Complete List of Comparison Operators

Equal to	==
Not Equal to	!=
Less than	<
Less than or Equal to	<=
Greater than	>
Greater than or Equal to	>=

EXERCISE SET 1.5.9

1. Prove that .99999 … = 1

   Hint: Think of .99999 … as:

$$.9 + .09 + .009 + \ldots = \frac{9}{10} + \frac{9}{100} + \frac{9}{1000} + \cdots$$

2. Prove that .8888888 = 8/9 using an infinite geometric series.
3. What is the output of the following code snippet:

```
x = int(9.5)
if x == 9:
    print('round down')
else:
    print('round up')
```

(Continued)

4. What is the output of the following code snippet:

```
x = 1/3
y = round(x,2)
if x == y:
   print('one third equals',y)
else:
   print('one third does not equal',y)
```

5. Write a program that asks the user to input a nonzero number. Determine and output whether the number is positive or negative.
6. Write a program that asks the user to input their age. If they are 18 or older, output "You can drink legally." Otherwise, output "You cannot drink legally."

Not all choices in life are binary (on or off, up or down, pass/no pass, etc.). Life is in fact a cornucopia of choices that are sometimes overwhelming. Let's start with three choices.

```
if [condition 1 is true]:
    Statement 1
    Statement 2
    (etc.)
elif [condition 2 is true]:
    do something else 1
    do something else 2
    (etc.)
else # neither condition 1 nor condition 2 is true
    do something else 3
    do something else 4
    (etc.)
```

You can have as many **elif** clauses as you want but only one **else**.
Let's enhance Exercise Set 1.5.9 #5 so that the user can input *any* integer, not just a nonzero integer.

```
n = int(input("n? "))
if n > 0:
    print("positive")
elif n < 0:
    print("negative")
else:
    print("zero")
```

EXERCISE SET 1.5.10

1. Given the following code snippet:

```
price = float(input('price? '))
percent = 0.10*price
discount = 10
if discount < percent:
    print('$10 off is a better deal')
elif percent < discount:
    print('10% off is a better deal')
else:
    print('$10 off is the same as 10% off')
```

   A. What is the output when you input $90?
   B. What is the output when you input $110?
   C. What price will result in “$10 off is the same as 10% off”?

2. Write a program that asks the user to input the weight of their first-class letter in ounces (oz.). Calculate and output the cost of postage using this table (accurate as of 2020):

Weight Not Over (oz.)	Cost
1	$0.55
2	0.70
3	0.85

(Continued)

Sample Output

weight in oz.? 2

0.70

weight in oz.? 5

I don't know how much it is for more than 3 oz.

3. Write a program that asks the user to input the total point value of three playing cards—a number between 3 (highly unlikely) and 30. Output the following message based on the point value.

Total Points	Message
Less than or equal to 16	"Hit"
Between 17 and 21	"Stay"
Over 21	"Bust"

Sample Output

sum of 3 cards? 14

hit

sum of 3 cards? 19

stay

Is it possible to write a program with more than 3 choices? Yes, but this will probably involve "nested" **if-else** blocks, which we will get to soon enough.

1.6 Alternating Sequences and Series

So far we have focused on sequences of all positive numbers—e.g., 10, 20, 30, 40, ...—or sequences of all negative numbers—e.g., −1, −7, −49, −343, ...

We can also have a combination of positive and negative numbers, which we call an **alternating sequence**.

Examples:

1, −2, 3, −4, 5, −6, ...

2, −4, 8, −16, 32, ...

1, −1/3, 1/9, −1/27, 1/81, ...

How do we indicate in the general term $\{a_n\}$ that the sign alternates?

First, consider each term being multiplied by either 1 or −1:

$(1)a_1, (-1)a_2, (1)a_3, (-1)a_4, \ldots$

Then notice that each 1 or −1 can be considered a power of −1:

$(-1)^2\, a_1, (-1)^3\, a_2, (-1)^4\, a_3, (-1)^5\, a_{4,} \ldots$

Thus each term can be considered $(-1)^{n+1}a_n$ if the sequence starts with a positive number and $(-1)^n a_n$ if it starts with a negative.

Example

Write the general term for each alternating sequence.

A. 2, −4, 8, −16, 32, ...

B. 1, −1/3, 1/9, −1/27, 1/81, ...

C. −1, 4, −9, 16, −24, ...

Solutions

A. $a_n = (-1)^{n+1}(2)^n$

B. $a_n = (-1)^{n+1}(1/3)^{n-1}$

C. $a_n = (-1)^n(n)^2$

It's probably time to note that some textbooks use a starting index of n = 0*; in other words, the first term is* a_0 *rather than* a_1*. For simplicity's sake, we are starting all our sequences at* n = 1 *unless otherwise noted.*

EXERCISE SET 1.6.1

For #1–3, write out the first five terms of each sequence.

1. $(-1)^{n+1}\left(\frac{n^2}{2}\right)$

2. $(-1)^n\left(\frac{1}{2n}\right)$

3. $(-1)^{n+1}\left(\sqrt{n}\right)$

(Express your answers in terms of radicals, not decimals.)

For #4–6, find the general term of each sequence.

4. 2, −5, 8, −11, 14, −17, ...

5. $\frac{2}{3}, -\frac{4}{9}, \frac{8}{27}, \ldots$

6. $-\frac{1}{2}+\frac{1}{4}-\frac{1}{6}+\frac{1}{8}-\ldots$

7. 1, −1, 1, −1, 1, −1, ...

8. Write a program that outputs the first 10 terms of the sequence in question 6 above:

$$-\frac{1}{2}, \frac{1}{4}, -\frac{1}{6}, \frac{1}{8}\ldots$$

Now that we've mastered various loops, conditional statements, and the like, we're ready to take on counting and probability in Chapter 2 ... the heart of computer games.

Chapter Projects

I. Triangular and Pentagonal Numbers

A. The following figures represent the "triangular numbers":

Triangular:

Image 1.17

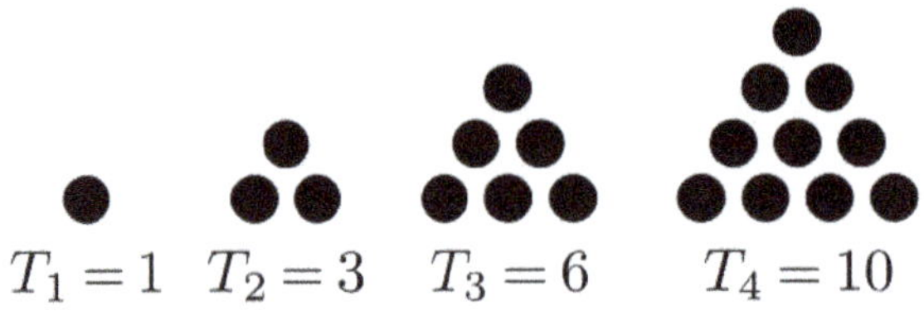

Write a program that asks the user to input the number of triangular numbers they would like generated, then generate and output the sequence.

B. The following figures represent the "pentagonal numbers":

Pentagonal:

Image 1.18

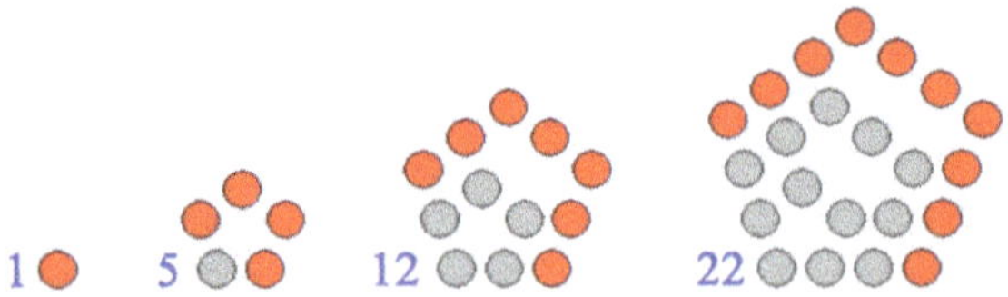

Write a program that asks the user to input the number of pentagonal numbers they would like generated, then generate and output the sequence.

II. Golden Ratio

The **Golden Ratio** is the ratio of pairs of consecutive numbers in the Fibonacci sequence (1, 1, 2, 3, 5, 8, 13, ...):

$$\frac{1}{1}, \frac{2}{1}, \frac{3}{2}, \frac{5}{3}, \ldots$$

A. Write a program that generates the first 50 terms of this sequence. The terms will converge to a constant called ∅ (phi), the "Golden Ratio." What is the value of ∅, as both a radical expression and rounded to 5 decimal places?

B. You can use phi to compute the *n*th number in the Fibonacci series:

$$F_n = \frac{\varnothing^n}{\sqrt{5}}$$

Extend your program to generate the first 10 Fibonacci numbers using this formula. Use the precise definition of $\varnothing$ as a radical expression.

III. Goldbach's Theorem

Goldbach's theorem states that $1 = \frac{1}{3} + \frac{1}{7} + \frac{1}{8} + \frac{1}{15} + \frac{1}{24} + \frac{1}{31} + \frac{1}{35} + \cdots = \sum \frac{1}{p-1}$, where p is called a "perfect power." The denominators are thus:

$3 = 2^2 - 1$

$7 = 2^3 - 1$

$8 = 3^2 - 1$

$15 = 4^2 - 1$

$24 = 5^2 - 1$

$26 = 3^3 - 1$

$31 = 2^5 - 1$

$35 = 6^2 - 1$

etc.

Write a program that prints out the first 50 partial sums. The partial sums for this series are as follows:

$$s_1 = \frac{1}{3}$$

$$s_2 = \frac{1}{3} + \frac{1}{7}$$

$$s_3 = \frac{1}{3} + \frac{1}{7} + \frac{1}{8}$$

$$s_4 = \frac{1}{3} + \frac{1}{7} + \frac{1}{8} + \frac{1}{15}$$

etc.

Are the partial sums converging to 1?

IV. Power Series

In advanced calculus, the following infinite sum is developed:

$$\frac{1}{1-z} = 1 + z + z^2 + z^3 + z^4 + \cdots$$

The series diverges if $|z| < 1$. Write a program that asks the user to input a value of z (between -1 and 1). Output the first 50 partial sums.

CHAPTER SUMMARY

PYTHON CONCEPTS

Output	**print(expr1, expr2, ...)** *Ex:* ***print(****"The answer is",x)*
Set of consecutive integers from 0 to $n-1$	**range**(n) *Ex:* ***range****(5) = { 0, 1, 2, 3, 4}*
Set of consecutive integers from m **to** $n-1$	**range**(m,n) *Ex:* ***range****(1,5) = {1, 2, 3, 4}*
Set of integers starting with *m* and ending before reaching *n*, with increment *k*	**range**(m,n,k) *Ex:* ***range****(2,10,2) = {2,4,6,8}* ***range****(10,1,-2) = {10,8,6,4,2}*
for loop **(repeats indented statements for each *i* in the range)**	**for** i **in range**(m,n,k): statement 1 statement 2 : *Ex:* **for** *n* ***in range****(0,3,10):* ***print****(n)*
Arithmetic Operators	Add + Subtract - Multiply * Divide / Exponent ** *Ex: 2+3*4**2* = $2 + 3 \cdot 4^2$
Types of Variables	Integer **int** Decimal **float** Alphanumeric **string** Rational Expression **Fraction** (must import fractions library) *Ex:* ***int****(2.5) +* ***Fraction****(3,4) = 11/4* ***str****(5) = "5"*

Input	n = **input**(prompt) *Ex:* *today =* ***input****("Day of week? ")* *x =* ***float(input****("Amount? ")*
if block	**if** [condition is true]: Statement 1 Statement 2 (etc.) *Ex:* ***if*** *x < 0:* ***print****("negative")* ***print****("try again")*
If-else block	**if** [condition is true]: Statement 1 Statement 2 (etc.) **else:** do something else 1 do something else 2 (etc.) *Ex:* ***if*** *x > 0:* ***print****("positive")* ***else:*** ***print****("negative")*

(*Continued*)

If-elif-else block	**if** [condition 1 is true]:
	    Statement 1
	    Statement 2
	    (etc.)
	elif [condition 2 is true]:
	    do something else 1
	    do something else 2
	    (etc.)
	else: *# neither condition 1 nor condition 2 is true*
	    do something else 3
	    do something else 4
	Ex:
	if *x>0:*
	    print(*"positive"*)
	elif *x < 0:*
	    print(*"negative"*)
	else:
	    print(*"zero"*)

while loop	**while** [condition is true]:
	    Statement 1
	    Statement 2
	    (etc.)
	Ex:
	answer = 10
	guess = ***int(input***(*"Guess? "*))
	while *guess != answer:*
	    print(*"Bad guess!"*)
	    guess = ***int(input***(*"Guess again! "*))
	print(*"Good job!"*)

Comparison Operators	Equal to: == Note Equal to: != Less than: < Less than or Equal to: <= Greater than: > Greater than or Equal to: >=

Math Functions and Constants	π — *pi* — *(math library)* e — *e* — *(math library)* $\sqrt{n}$ — **sqrt**(n) — *(math library)* $\lvert n \rvert$ — **abs**(n) $n!$ — **factorial**(n) — *(math library)*
Round a float to *n* decimal places	**round**(x,n) *Ex: **round**(1.414159,3) = 1.414*
Convert a float value to an integer	**round**(x) or **int**(x) *Ex: **round**(3.14159) = 3*

PRECALCULUS CONCEPTS

Arithmetic Sequence	A set of numbers (finite or infinite) in which you keep adding the same number. Often identified by first term a_1 and increment d. *Ex: 10, 13, 16, 19, 22, ...*
General Term of an Arithmetic Sequence	$a_n = a_1 + (n - 1)d$ in which a_1 is the first term and d is the increment. *Ex:* *Given 1, 5, 9, 13, ...* $a_n = 1 + (n-1)(4) = -3 + 4n$
Recursive Definition of an Arithmetic Sequence	a_1 = first term $a_{n+1} = a_n + d$ *Ex:* *Given 1, 5, 9, 13, ...* $a_1 = 1$ $a_{n+1} = a_n + 4$
Arithmetic Series: **Sum of the first *n* terms of an Arithmetic Sequence**	$S_n = \frac{n}{2}(a_1 + a_n)$ *Ex:* *Given 1, 5, 9, 13, ...* $S_4 = \frac{4}{2}(1 + 13) = 28$
Geometric Sequence	A set of numbers (finite or infinite) in which you keep multiplying by the same number. Often identified by first term a_1 and ratio r. *Ex: 1, 2, 4, 8, 16, ...*

(Continued)

General Term of a Geometric Sequence	$a_n = a_1 r^{n-1}$ *Ex:* *Given 2, 6, 18, 54, ...* $a_n = 2 \cdot 3^{n-1}$
Recursive Definition of a Geometric Sequence	a_1 = first term $a_{n+1} = r \cdot a_n$ *Ex:* *Given 2, 6, 18, 54, ...* $a_1 = 2$ $a_{n+1} = 3 \cdot a_n$
Finite Geometric Series (Partial Sum): ***Sum of first n terms of a Geometric Sequence***	$S_n = a_1 \cdot \frac{1-r^n}{1-r}$ *Ex:* *Given 2, 6, 18, 54, ...* $S_4 = 2 \cdot \frac{1-3^4}{1-3} = 80$
Infinite Geometric Series: ***Sum of all terms of an infinite Geometric Sequence***	$S = \frac{a_1}{1-r}$, if $\lvert r \rvert < 1$ *Ex:* *Given 1/2, 1/4, 1/8, 1/16, ...* $r = ½$ $S = \frac{1/2}{1-1/2} = 1$
Sigma Notation: **Represents a series**	$\sum_{k=1}^{n} a_k = a_1 + a_2 + \cdots + a_n$ *Ex:* $\sum_{k=1}^{4} k^2 = 1^2 + 2^2 + 3^2 + 4^2 = 30$
Factorial	$n! = n \cdot (n-1) \cdot (n-2) \cdots 2 \cdot 1$ *Ex:* $5! = 5 \cdot 4 \cdot 3 \cdot 2 \cdot 1 = 120$ $0! = 1$ *(special case)*

Credits

IMG 1.14: Source: https://commons.wikimedia.org/wiki/File:Chess_and_backgammon_board_MET_ES4611.jpg, 1830.

IMG 1.15: Copyright © by Martin Grandjean (CC BY-SA 4.0) at https://commons.wikimedia.org/wiki/File:Zeno_Dichotomy_Paradox.png.

IMG 1.17: Copyright © by Melchoir (CC BY-SA 3.0) at https://commons.wikimedia.org/wiki/File:First_six_triangular_numbers.svg.

IMG 1.18: Unknown, https://commons.wikimedia.org/wiki/File:Meyers_b13_s0207_b1.png, 1885.

CHAPTER 2

Counting and Probability

2.1 Sets

You may be wondering, when do we get to start programming games? Most games involve probability. What's the probability of rolling a 7 or 11 when tossing two die? What's the probability of being dealt an ace and a king from a 52-card deck? And so on.

Before we can tackle probability, we must learn the calculation tools of probability, which are permutations and combinations. And that begins with an elementary discussion of sets and counting.

A **set** is a collection of elements such as:

Odd numbers between 1 and 10 = {1, 3, 5, 7, 9}

Multiples of 5 = {5, 10, 15, 20, ...}

Breeds of dogs = {Australian Shepherd, beagle, border collie, bulldog, corgi, etc.}

People who can fly with no assistance = { }

A set can have infinitely many elements, finite, or even none. A set is unordered, as opposed to a sequence, which is ordered. It is not necessary to repeat duplicate elements. For example, {4, 1, 0, 1, 4} can simply be described as {0, 1, 4}.

A set with no elements is called the **null set** or **empty set** and has a special symbol: ∅. (There are NO set braces around the ∅.)

Sets are usually named with an uppercase letter. The notation for number of elements in set *A* is **n(A)**. Let's name and count the elements in the examples above.

Description	Set	Number of Elements
Odd numbers between 1 and 10	$D = \{1, 3, 5, 7, 9\}$	$n(D) = 4$
Multiples of 5	$M = \{5, 10, 15, 20, \ldots\}$	$n(M) = \infty$
Breeds of dogs	B = {Australian Shepherd, beagle, border collie, bulldog, corgi, etc.}	$n(B) = 360$ (in 2019, according to World Canine Organization)
People who can fly with no assistance	$F = \emptyset$	$n(F) = 0$

EXERCISE SET 2.1.1

For each of the sets described below, list up to 5 elements and indicate whether it is finite or infinite.

1. Female U.S. presidents (as of 2021)
2. Prime numbers
3. Numbers that are divisible by both 2 and 3.
4. U.S. states in which the minimum drinking age is 18 (as of 2021)

Comparing and Combining Sets

If A and B are sets that are part of a universal set U ...

Notation	Description	Definition
$A = B$	A equals B	A and B have the exact same elements.
$A \subseteq B$	A is a subset of B	Every element of A is also in B.
$A \subset B$	A is a proper subset of B	A is a subset of B but $A \neq B$.
$A \bigcup B$	A union B	The set of all elements in A or B or both.
$A \bigcap B$	A intersect B	The set of all elements in both A and B (what they have in common).
$B \setminus A$ (or B-A)	Relative complement of A in B	All the elements in B that are not in A.
A' (or A^c)	A complement	All the elements in U that are not in A. *(U, the universal set, is described below.)*
$x \in A$	x is an element of A	x is in the set A.
$x \notin A$	x is not an element of A	x is not in the set A.
$n(A)$	The number of elements in set A.	The number of elements in set A.

The **universal set, U,** is defined within the context of the situation. For example, if D = {dogs} and C = {cats}, then U could be {domestic animals} or {mammals} but probably not {fighter jets}. If Q = {rational numbers} and Q' = {irrational numbers}, then U could be {real numbers}.

Example

Let O = the set of odd numbers between 1 and 20; P = the set of prime numbers between 1 and 20; and E = the set of even numbers between 1 and 20.

A. Find $O \cap P$

B. Is $P \subseteq O$?

C. Find $O \cap E$

D. Find $O \cup E$

E. Find $P \cap E$

F. What would be a logical choice for U?

Solutions

A. The set of odd primes between 1 and 20 = {1, 3, 5, 7, 11, 13, 17, 19}.

B. No, $2 \in P$ but $2 \notin O$.

C. The set of numbers that are both even and odd = Ø.

D. The set of numbers that are either odd or even between 1 and 20 = {1, 2, 3, 4, ..., 19, 20}.

E. The set of even primes between 1 and 20 = {2}.

F. U = {integers between 1 and 20} or just U = {integers}.

Notice there is a distinction between 2 and {2}. The former is a number that can be an element of a set. The latter is a set consisting of the single element 2.

There is also a distinction between Ø and {Ø}. The former is the null set. The latter is a set consisting of a single element that is the null set. (Yes, you can have a set of set(s).)

EXERCISE SET 2.1.2

Let U = the set of positive integers, A = the set of positive integer factors of 24, B = the set of multiples of 3 between 3 and 24; and C = {1, 3, 6}. Identify the following.

1. Is $C \subseteq A$?
2. Is $C \subseteq B$?
3. Is $A \subseteq B$?
4. Find $A \cup B$.
5. Find $A \cap B$.
6. Find $B \cap C$.
7. Is $6 \in A$?
8. Is $6 \subseteq A$?
9. Is $\{6\} \subseteq A$?
10. Is $\emptyset \subseteq A$?
11. Find A'.
12. Find $B \setminus A$.
13. Find $A \setminus C$.

Believe it or not, the answer to #10 is yes: the null set is a subset of every set. In symbols:

$\varnothing \subseteq A$ for any set A

Using Sets in Python

Now that you know a bit of set theory, we can explore the built-in **set** functions in Python. A set in Python is just like the mathematical sets we've discussed so far: they're enclosed in curly braces { }, can consist of any data type, and the order of the elements of the set is not important.

We say that a set is **immutable** in that once defined, the set cannot be changed. You cannot add or remove elements from the set.

TABLE 2.1 Sets and Their Functions

To create and initialize set A with no elements: $A = \{\,\}$	
To create and initialize set A with given elements: $A = \{1, 2, 3\}$ A = {"alpha", "beta", "gamma"}	
To form $A \cup B$: A.**union**(B)	
To form $A \cap B$: A.**intersection**(B)	
To determine if $A \subseteq$ B: A.**issubset**(B)	*(returns True or False)*
To form the difference $A \setminus B$: A.**difference**(B)	
To determine if element x is in set A: x **in** A	*(returns True or False)*
To find the size (number of elements) in set A: **len**(A)	
To determine if two sets are disjoint (*i.e.*, $A \cap B = \varnothing$): A.**isdisjoint**(B)	

"Dot" Notation

Note the syntax *name*.**property()** (for example A.**issubset**(B).) A **property** is a characteristic associated with *name*. We also have *name*.**method()** (for example A.**union**(B).) A **method** is an operation (function) performed on *name*.)

Practice with Set Properties and Methods

Let's try a few quick **set** operations. We will enter these directly at the Python interactive console prompt >>> rather than writing an entire program. The console prompt executes code instantaneously, for example:

```
>>>set1 = {1, 2, 3}
>>>print(set1)
{1, 2, 3}
```

Examples

Given the following sets:

$A = \{1, 2, 3, 4, 6, 8, 12, 24\}$

$B = \{3, 6, 9, 12, 15, 18, 21, 24\}$

$C = \{1, 3, 6\}$

$D = \{5, 7\}$

For the following, give the Python command and the response.

A. Is $A = B$?

Solution

```
>>>A== B
False
```

B. Is $C \subseteq A$?

Solution

```
>>>C.issubset(A)
True
```

C. Find $A \backslash B$

Solution

```
>>>A.difference(B)
{8, 1, 2, 4}
```

(The elements are printed in random order)

D. Are A and D disjoint?

Solution

>>>A.**isdisjoint**(D)

OR

>>>**len**(A.**intersection**(D)) = 0

True

You may be wondering how to use **range**() to quickly assign a sequence to a set. Simply precede **range**() with **set**(). For example:

>>>A = **set(range**(1, 10, 2))

>>>**print**(A)

{1, 3, 5, 7, 9}

EXERCISE SET 2.1.3

Let $U =$ the set of positive integers between 1 and 30 inclusive; $A =$ the set of positive even integers in U; $B =$ the set of integers in U that are multiples of 10; and $C =$ the set of integers in U that are prime. Write the Python command and response for each of the following.

1. Is $B \subseteq A$?
2. Is $A \subseteq B$?
3. Find $A \cap C$.
4. Find $A \setminus B$.
5. Find $A \cap B$.
6. Find $B \cap C$.
7. Find $B \setminus A$.
8. Find $A \cap (B \cup C)$.
9. Find $(A \cup B) \cap C$.

Counting Subsets

Let's list all the subsets of the following sets.

Set A	# of elements in set A	Subsets of A	# of Subsets of A
$\varnothing$	$n(A) = 0$	$\varnothing$	1
$\{a\}$	$n(A) = 1$	$\varnothing$, $\{a\}$	2
$\{a, b\}$	$n(A) = 2$	$\varnothing$, $\{a\}$, $\{b\}$, $\{a, b\}$	4
$\{a, b, c\}$	$n(A) = 3$	$\varnothing$, $\{a\}$, $\{b\}$, $\{c\}$, $\{a, b\}$, $\{a, c\}$, $\{b, c\}$, $\{a, b, c\}$	8
$\{a, b, c, d\}$	$n(A) = 4$	?	?

As you can see, by the time we get to a set with 4 elements, it becomes a little overwhelming to have to list and count the number of subsets. Like all lazy mathematicians (a redundancy), we look for a pattern. Let's consider just the 2nd and 4th columns:

# of elements in set A	# of Subsets of A
$n(A) = 0$	1
$n(A) = 1$	2
$n(A) = 2$	4
$n(A) = 3$	8
$n(A) = 4$	?

It looks like the # of subsets are powers of 2:

# of elements in set A	# of Subsets of A
$n(A) = 0$	2^0
$n(A) = 1$	2^1
$n(A) = 2$	2^2
$n(A) = 3$	2^3
$n(A) = 4$	2^4

So let's place our money on $2^4 = 16$ subsets of $\{a, b, c, d\}$. Check:

# of subsets with no elements	1	Ø
# of subsets with 1 element	4	$\{a\}$, $\{b\}$, $\{c\}$, $\{d\}$
# of subsets with 2 elements	6	$\{a, b\}$, $\{a, c\}$, $\{a, d\}$, $\{b, c\}$, $\{b, d\}$ $\{c, d\}$
# of subsets with 3 elements	4	$\{a, b, c\}$, $\{a, b, d\}$, $\{a, c, d\}$ $\{b, c, d\}$
# of subsets with 4 elements	1	$\{a, b, c, d\}$
Total Subsets	**16**	

Yes, there are indeed $2^4 = 16$ subsets of a set with 4 elements. We can generalize:

If A is a set with n elements, then the number of subsets of A is 2^n.

Our discussion of set theory serves as a precursor to counting and probability theory. For many of our examples involving probability, we will refer to a standard 52-card deck of playing cards. The deck is partitioned into 4 suits (hearts (red), diamonds (red), clubs (black), and spades(black)). Each suit consists of 13 cards: ace, 2, 3, 4, ... 9, 10, jack, queen, king.

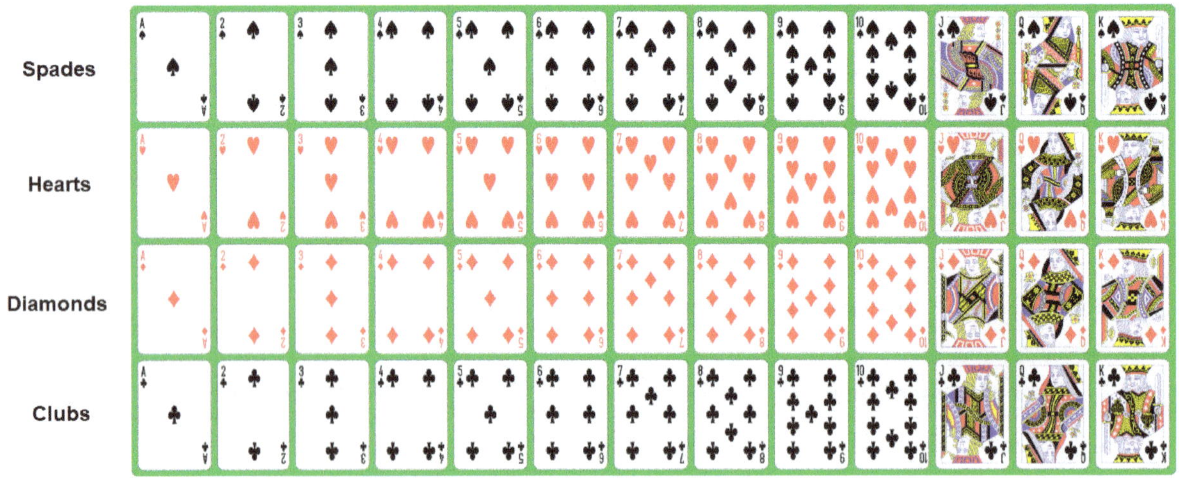

Image 2.1

EXERCISE SET 2.1.4

Let U = deck of 52 cards, H = set of hearts, D = set of diamonds, C = set of clubs, S = set of spades, and Q = set of queens.

1. Find $n(H \cup Q)$.
2. Find $n(H \cap C)$.

3. Find $n(H \cup C)$.
4. Find $n(D \cap Q)$.
5. Find $n(S')$.
6. Find $n(S \setminus Q)$.
7. Find $n(Q')$.

Hopefully your answer to #1 was 16, not 17, because you do not want to count the Queen of Hearts twice:

Image 2.2

... which leads to the following formula, which is essentially "Don't Count Things Twice."

COUNTING FORMULA

If A and B are finite sets, then

$$n(A \cup B) = n(A) + n(B) - n(A \cap B)$$

Special Case

If $A \cap B = \emptyset$ (i.e., disjoint sets), then $n(A \cup B) = n(A) + n(B)$

More General Case

If $A_1, A_2, \ldots. A_n$ are disjoint sets (meaning they have nothing in common), then
$n(A_1 \cup A_2 \cup A_3 \cup \ldots \cup A_n) = n(A_1) + n(A_2) + \ldots. + n(A_n)$

EXERCISE SET 2.1.5

Let H = set of hearts, D = set of diamonds, C = set of clubs, S = set of spades, and Q = set of queens.

1. Find $n(D \cup C \cup S \cup H)$.
2. Find $n(D \cup C \cup Q)$.
3. Find a formula for $n(A_1 \cup A_2 + A_3 \cup \ldots \cup A_n)$ if the sets are NOT disjoint.

Venn Diagrams

Often, **Venn diagrams** are used to represent relationships between sets. Each set is represented as a circle, and the rectangle surrounding them represents the universal set U.

Union of Two Sets

$A \cup B$

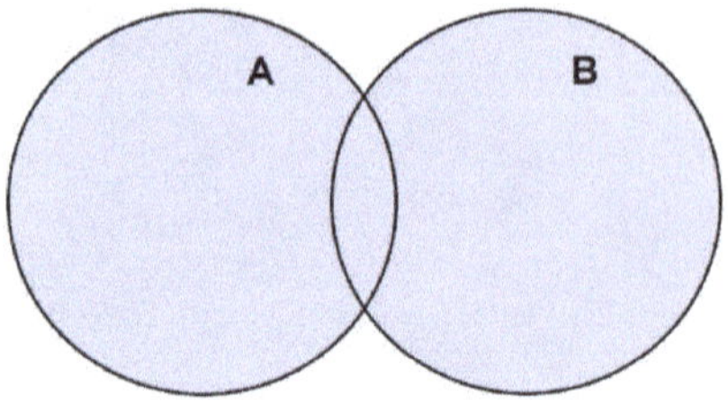

Image 2.3

Intersection of Two Sets

$A \cap B$

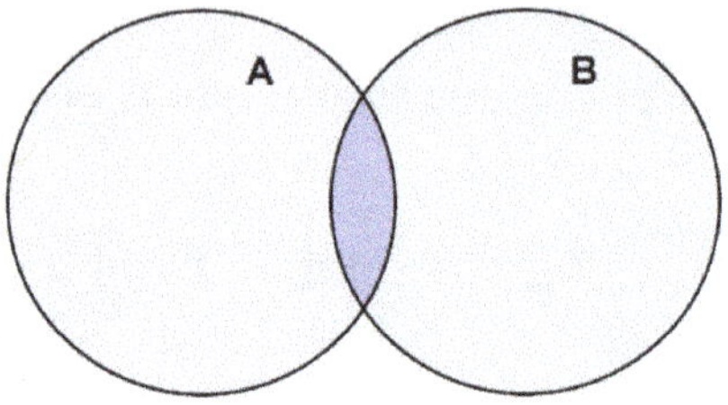

Image 2.4

Complement of a Set

A' or $\overline{A}$

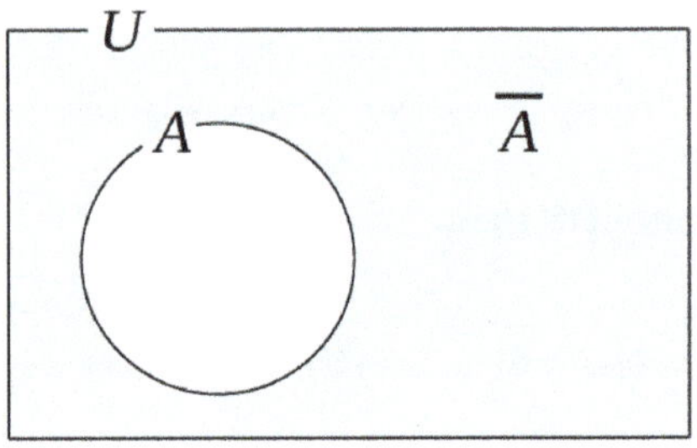

Image 2.5

EXERCISE SET 2.1.6

Draw Venn diagrams (each with two sets A and B) to represent the following:

1. $(A \cup B)'$
2. $A' \cap B'$
3. $B \setminus A$
4. $(A \cap B)'$
5. $A' \cup B'$
6. $U \setminus A$
7. $A \cap B'$
8. $A' \cup B$
9. $A \setminus B'$
10. Rewrite $A \setminus B$ as an equivalent intersection statement.

Venn diagrams are useful in figuring out how many elements are in a set.

Example

Thirty students were interviewed about their favorite subject(s). Twenty said English and 15 said math. Students responded with at least one favorite.

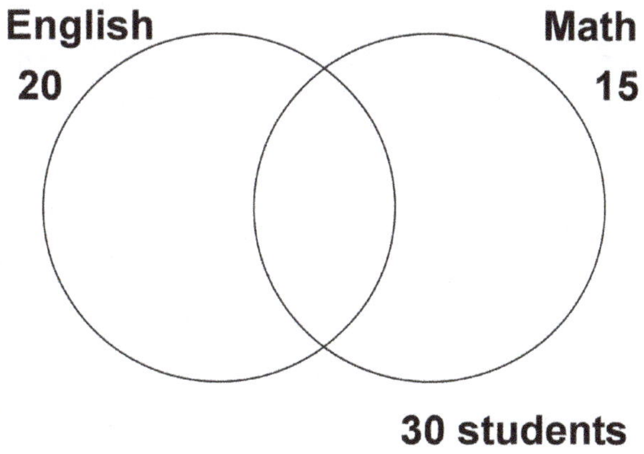

Image 2.6

A. How many students said they liked both English and math?

B. How many students like English only?

C. How many students like math only?

Solutions

This completed Venn diagram might help:

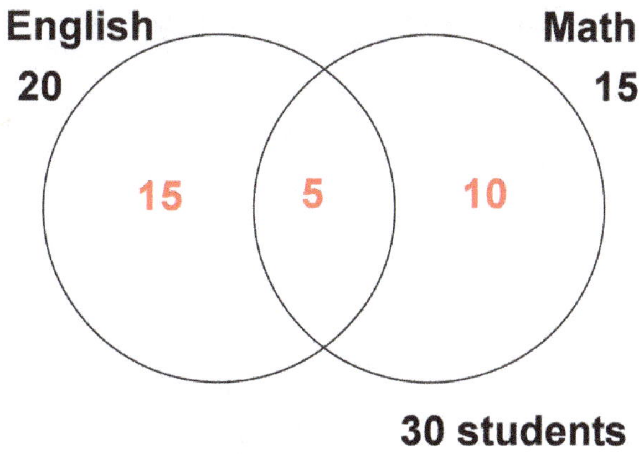

Image 2.7

A. $n(E \cap M) = 5$

B. $n(\text{English only}) = n(E) - n(E \cap \text{M}) = 20 - 5 = 15$

C. $n(\text{Math only}) = n(M) - n(E \cap M) = 15 - 5 = 10$

EXERCISE SET 2.1.7

1. A survey of 85 college students asked if they enjoyed soccer, baseball, or both. (Students who liked neither were excluded from the survey). If 63 students enjoy soccer and 41 enjoy baseball, how many enjoy both?
2. A group of 143 consumers were surveyed; 112 indicated they would be buying a major appliance within the next year, and 27 indicated they would buy both an appliance and a car. Assuming the consumers would be buying at least one of the items, how many indicated they would be buying a car but not an appliance?
3. CHALLENGE: Ninety students went to a school carnival.

   3 had a hamburger, soft drink, and ice cream.

   24 had hamburgers.

   5 had a hamburger and a soft drink.

   33 had soft drinks.

   10 had a soft drink and ice cream.

   38 had ice cream.

   8 had a hamburger and ice cream.

   Fill in the Venn diagram to figure out how many students chose NONE of hamburger, soft drink, or ice cream.

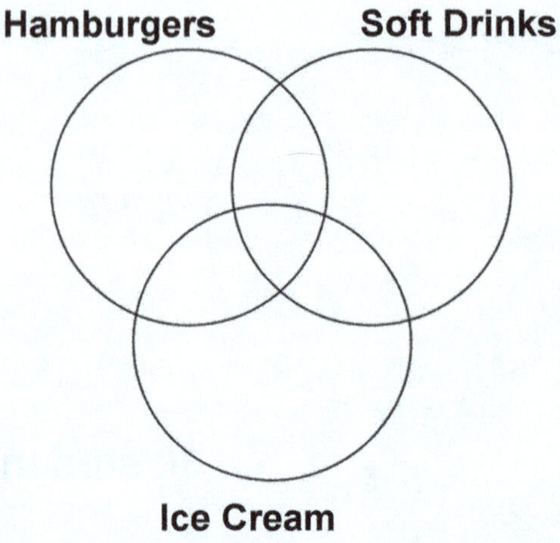

Image 2.8

Program Goal: Write a program that:

1. Asks the user to input $n(A)$, $n(B)$, and $n(A \cap B)$.
2. Calculates and outputs $n(A \cup B)$, $n(A \setminus B)$ and $n(B \setminus A)$.

PROGRAM 2.1.1

```
A = int(input("n(A)? "))
B = int(input("n(B)? "))
A_and_B = int(input("n(A ^ B)? "))
A_U_B = A + B – A_and_B
print("n(A U B) = ",A_U_B)
[ Calculate and output n(A \ B) ]
Calculate and output n(B \ A) ]
```

Sample Output

$n(A)$? 20
$n(B)$? 10
$n(A \wedge B)$? 5
$n(A \text{ U } B) = 25$
$n(A \backslash B) = 15$
$n(B \backslash A) = 5$

EXERCISE SET 2.1.8

1. Fill in the yellow highlighted code that calculates and outputs $n(A \setminus B)$.
2. Fill in the blue highlighted code that calculates and outputs $n(B \setminus A)$.

A Menu of Possibilities

Let's say a certain pizza restaurant offers the following options:
3 types of crust: thin, thick, flatbread
3 sizes: small, medium, large
5 toppings: pepperoni, mushrooms, onions, sausage, bacon

If a customer chooses exactly 1 item from each category (sorry only 1 topping), how many different types of pizza can he/she order?

We are going to write a program to help us generate all the possibilities. In order to do this, we are going to introduce the concept of **lists**. A **list** is basically a **set** in which the order is important, and the individual elements can be indexed.

The elements of a set do not all have to be the same type. For example, ['ace', 2, 3] is a valid set.

Note we use square brackets [ ] for lists instead of curly braces { }, which were used for sets.

TABLE 2.2 Lists and Their Functions

To create and initialize list L with no elements: $L = [\,]$
To create and initialize list L with a set of elements: $L = [1, 2, 3]$ L = ["alpha", "beta", "gamma"]
The elements of list L are referenced as: $L[0], L[1], L[2], \ldots$
To append an element to the end of a list (even an empty list): **L.append**(x)
The size (number of elements) of list L: **len**(L)
To clear out all the elements in list L: **L.clear**()
To "pop off" (remove) the last element in list L: **L.pop()**

We'll learn more list functions as we encounter a need for them, but this is enough to get us started.

Back to the Pizza Problem

Crust: thin (1), thick (2), flatbread (3)

Size: small (4), medium (5) , large (6)

Topping: pepperoni (7), mushrooms (8), onions (9), sausage (10), bacon (11)

Program Goal: Determine all the possible pizza orders given 3 choices of crust, 3 sizes, and 5 toppings. Our strategy is to do three "nested" **for** loops (loop inside of a loop), with the innermost loop (toppings) printing the combination order.

PROGRAM 2.1.2

```
crust = ["thick", "thin", "flatbread"]
size = ["small", "medium", "large"]
topping = ["pepperoni", "mushroom", "onion", "sausage", "bacon"]
for c in range(0,3):
    for s in range(0,3):
            for t in range(0,5):
                print(crust[c],size[s],topping[t])
```

Output

thick small pepperoni
thick small mushroom
thick small onion
thick small sausage
thick small bacon
thick medium pepperoni
:
:
flatbread large sausage
flatbread large bacon

Another possible method of traversing the loops is as follows.

```
crust = ["thick", "thin", "flatbread"]
size = ["small", "medium", "large"]
topping = ["pepperoni", "mushroom", "onion", "sausage", "bacon"]
for c in crust:
    for s in size:
            for t in topping:
                print(c,s,t)
```

Notice in this method that the variables *c*, *s*, and *t* are now strings rather than numeric. How many possible pizza orders is this? Let's have the program count for us.

EXERCISE SET 2.1.9

1. Modify program 2.1.2 (either version) so that a counter n counts how many times you output a line. Remember to initialize n to 0 at the start of the program. At the end of the program, output "Total Pizzas:" followed by the value of n.
2. What is the result of your modified program in #1 ... how many possible pizzas are there?
3. Modify the program so that an additional menu item is added:

   Cheese: regular or extra

   Run your modified program. Now how many possible pizzas are there?
4. Do you think there is a mathematical formula to quickly calculate the result? If so, what is it?

The answer to #4 above is yes:

MULTIPLICATION PRINCIPLE OF COUNTING

If a "menu" consists of n categories, and the number of choices for these categories are $p_1, p_2, p_3, \ldots, p_n$ respectively, then the total number of possibilities in which you choose one option from each category on the menu is:

$$p_1 \cdot p_2 \cdot p_3 \cdots\cdots \cdot p_n$$

This is a very simple formula. Here's an example. Let's say a ballot lists 4 candidates for governor, 4 candidates for lieutenant governor, 5 candidates for senator, 3 candidates for controller, and 6 candidates for secretary of state. You are allowed to select one for each office. The total number of ways to vote is:

$$4 \cdot 4 \cdot 5 \cdot 3 \cdot 6 = 1{,}440$$

EXERCISE SET 2.1.10

1. At a restaurant, you have a choice of 6 appetizers, 10 meals, 8 beverages, and 4 desserts. Assuming you must choose exactly 1 item from each category, how many different meals can be ordered?
2. In a certain state, auto license plates must be 3 letters followed by 3 digits. Repetition of letters and digits is allowed. How many possible license plates are there?
3. The International Airline Transportation Association (IATA) assigns 3-letter codes to represent airport locations such as FLL (Ft. Lauderdale, FL). Repetition of letters is allowed. How many different codes are there?
4. A website specifies that passwords must contain exactly 6 alphanumeric characters. (Alphanumeric means a letter, number, or punctuation mark. For this exercise we will just consider letters and numbers.) Assume the letters must be upper case and repetition is allowed. How many possible passwords are there?
5. Same question as #4, but repetition is NOT allowed. All the characters must be unique.
6. How many 4-character alphanumeric PINs can be generated using digits or upper- or lowercase letters? Repetition is allowed.

2.2 Permutations

In Chapter 1, we defined a mathematical function called the **factorial**:

$$n! = n(n-1)(n-2)\ldots\ldots 2 \cdot 1$$

As we already know how to write a program to calculate a factorial, we can take advantage of Python's built-in factorial function **math.factorial()** in the math library.

We will see how this function lends itself perfectly to counting techniques.

Example 1

In how many ways can you choose a batting order for a baseball team of 9 players?

Solution

You have 9 choices for the first position, 8 choices for the second position, and so on. Thus, the total number of possible batting orders is:

$$9 \cdot 8 \cdot 7 \cdot 6 \cdot 5 \cdot 4 \cdot 3 \cdot 2 \cdot 1 = 362{,}880$$

Which is 9! (The ! is not a punctuation symbol in this context. It is a mathematical notation.)

Example 2

In how many ways can a police officer arrange 6 people for a police line-up? (If you prefer: In how many ways can a photographer line up a family of 6 people for a family portrait?)

Solution

The total possibilities is $6! = 720$.

Example 3

A family of 5 wants to take a picture in front of a waterfall but there is only room for 3 people in the picture. How many ways are there to select and position the 3 family members in front of the waterfall?

Solution

Image 2.9

As you can see, there are $5 \cdot 4 \cdot 3 = 60$ possibilities.

These are all examples of **permutations**.

A permutation is an ordered arrangement of k objects chosen from n objects. There are two notations depending on which textbook you pick up. One of them is $P(n, k)$. The other is ${}_nP_k$. In the above examples, we would say:

1. The number of possible police lineups in which 6 people are chosen from 6 people:

$${}_6P_6 = 6 \cdot 5 \cdot 4 \cdot 3 \cdot 2 \cdot 1 = 720$$

2. Number of possible family photos in which 3 people are chosen from 5 people:

$${}_5P_3 = 5 \cdot 4 \cdot 3 = 60$$

So, one way to calculate ${}_nP_k$ is to start with n and multiply k descending numbers:

Image 2.10

$$\text{nPk} = \underbrace{\text{n(n-1)(n-2) ... (n - k + 1)}}_{\text{k numbers}}$$

For example, let's say we want to choose 4 teams from 32 teams in a basketball playoff. How many ways to order them 1st through 4th place?

Image 2.11

$$\text{32P4} = \underbrace{\text{(32)(31)(30)(29)}}_{\text{4 numbers}} = \text{863,040}$$

There is an alternate formula involving factorials:

The number of **permutations** of k objects selected from a set of n objects is

$${}_nP_k = \frac{n!}{(n-k)!}$$

$$n > 0, k \geq 0, n \geq k$$

Let's recalculate our examples:

1. The number of possible police line-ups in which 6 people are chosen from 6 people:

$${}_6P_6 = \frac{6!}{(6-6)!} = \frac{6!}{0!} = \frac{720}{1} = 720$$

(This is where you can see why we defined 0! = 1 in the previous chapter.)

2. The number of possible family photos in which 3 people are chosen from 5 people:

$${}_5P_3 = \frac{5!}{(5-3)!} = \frac{5!}{2!} = \frac{120}{2} = 60$$

3. The number of ways to select 4 winning teams (in order) out of 32 teams:

Image 2.12

$${}_{32}P_4 = \frac{32!}{(32-4)!} = \frac{32!}{28!} = \frac{32 \cdot 31 \cdot 30 \cdot 29 \cdot \cancel{28!}}{\cancel{28!}} = 863{,}040$$

EXERCISE SET 2.2.1

1. You're playing a word game in which you have 4 letter tiles. How many different words can be formed using those 4 tiles?
2. A student is taking a multiple-choice test with 5 questions. Each question has 4 possible answers. How many ways are there to answer the test?
3. Purchasing a new car, you have 5 exterior color choices, 2 interior color choices, 3 configurations, and 2 choices of radio. How many different models do you have to choose from?
4. A bank asks you to choose a 4-digit PIN. Repetition is not allowed. How many different PINs are there?
5. At a horse race, you'd like to select 3 horses for win, place, and show respectively. If there are 10 horses in the race, in how many different ways can you place your bet?
6. Three people in a room compare birthdays. How many possible combinations of birthdays are there? (Assume no one was born in a leap year.)
7. You're dealt two cards from a blackjack dealer off a new 52-card deck. How many possible hands are there?
8. Write a program that asks the user to input n and k, then outputs ${}_nP_k$. Use it to check your answers to #1, 4, 5, and 7 above.

2.3 Combinations

A key feature of permutations is that the order of the selection is important (like **lists**). This can also be considered as each item being designated a specific position, such as a club selecting a president, vice president, and secretary; or books being ordered left to right on a bookshelf.

We now look at selecting k out of n objects, but the order is **NOT** important (much like a **set**). An example would be a club selecting a committee of 3 members; or choosing 3 out of 100 raffle tickets to win the same prize.

The notation for a combination of k objects selected from n objects is either $C(n,k)$ or ${}_nC_k$ or $\binom{n}{k}$, depending on the textbook. For now, we are going to use ${}_nC_k$.

Let's look at the difference between permutations of a set $\{A, B, C\}$ and combinations of the same set. We want our subsets to each have 2 elements.

Permutations	Combinations
AB	AB
AC	AC
BC	BC
BA	Total $= 3$
BC	
CA	
CB	
Total $= {}_3P_2 = \dfrac{3!}{(3-2)!} = \dfrac{3!}{1!} = \dfrac{6}{1} = 6$	

As you can see, because order is not important, there are fewer combinations than permutations. In the above example, notice that ${}_3C_2 = \frac{3P_2}{2}$, which happens to be $\frac{3P_2}{2!}$.

More generally:

The number of **combinations** of k objects selected from a set of n objects is:

$$ {}_nC_k = \frac{nP_k}{k!} = \frac{n!}{(n-k)!\cdot k!} $$

$$ n > 0,\ k \geq 0,\ n \geq k $$

Example 1

In how many ways can a college math department form a committee of 4 people out of 26 professors?

Solution

There are ${}_{26}C_4$ ways to do this:

Image 2.13

$$ {}_{26}C_4 = \frac{{}_{26}P_4}{4!} = \frac{26!}{(26-4)!4!} = \frac{26\cdot 25\cdot 24\cdot 23\cdot \cancel{22!}}{\cancel{22!}4!} = \frac{26\cdot 25\cdot \cancel{24}\cdot 23}{\cancel{24}} = 26\cdot 25\cdot 23 = 14{,}950 $$

Example 2

A congressional committee is to be formed with 2 senators (out of 100 senators) and 3 representatives (out of 437). How many possible committees are there?

Solution

We first find the total number of senator choices, then the total number of representative choices, and then multiply those two answers.

$$ {}_{100}C_2 \bullet {}_{437}C_3 = 4{,}950 \bullet 13{,}813{,}570 = 68{,}377{,}171{,}500 $$

EXERCISE SET 2.3.1

1. How many ways are there to draw 5 cards from a deck of 52 cards, if the card is replaced and the deck reshuffled each time?
2. How many ways are there to draw 5 cards from a deck of 52 cards if none of the cards are replaced? The order is not important.
3. How many ways are there to draw 5 cards from a deck of 52 cards if none of the cards are replaced, and they are laid in sequence on the table in front of you?

4. The Human Resources Department of a large corporation needs to hire an equal number of men and women to fill ten positions. If 40 men apply and 15 women apply, how many ways are there to choose the new employees?
5. A store owner sells 10 different types of stuffed animals and 15 different types of outfits for them. A very spoiled child comes in with her parents. In how many ways can she pick two animals and two outfits?
6. You'd like to select a dozen bagels from your favorite bakery. There are 16 bagels to choose from. How many selections are possible if you select exactly one of each type?
7. Consider the number of line segments that connect *n* points that lie on the circumference of each circle. How many line segments are there when there are 7 points? What is the formula for how many line segments are drawn when connecting *n* points on a circle?

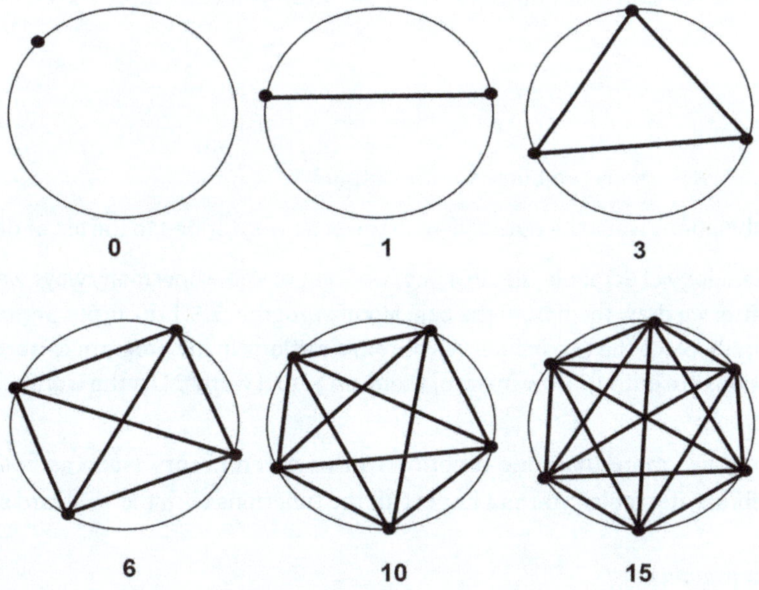

Image 2.14

8. Modify your program in Exercise Set 2.2.1 #8 to calculate a combination instead of a permutation.

Python has many useful libraries that can be imported. The **itertools** module has functions to generate both permutations and combinations of a given set. Let's say we have a bag of 5 Scrabble tiles: {A, E, R, S, T}. We want to know how many different combinations of 3 letters we can choose.

PROGRAM 2.3.1

```
from itertools import combinations
tiles = ['A', 'E', 'R', 'S', 'T']
subsets = combinations(tiles,3)
n = 0
for k in list(subsets):
    print(k)
    n += 1
print(n)
```

Output

```
('A', 'E', 'R')
('A', 'E', 'S')
('A', 'E', 'T')
:
:
('E', 'S', 'T')
('R', 'S', 'T')
```

The **combinations**(set, n) function takes 2 arguments: the set and the number of desired elements in each subset.

Note: the subsets are enclosed in parenthesis ( ) rather than brackets [ ]. This is because they are of the type **tuple,** which is basically an immutable list, meaning you can't alter the elements in the list once they are assigned.

EXERCISE SET 2.3.2

1. Run program 2.3.1. How many combinations are output?
2. How many combinations would be output if an extra letter were added to the list of tiles?
3. Continuing the analogy of Scrabble tiles in a bag, we want to know how many ways we can rearrange the three tiles after we draw them from the bag. Modify program 2.3.1 to output permutations instead of combinations. (Replace the two incidences of **combinations** in the program to **permutations**.) How many permutations are output? How many of them are actual words? List the words.

If you think you'll use more than one function from a given library (such as both permutations and combinations from library **itertools**), you can import all the functions with the wildcard symbol "*" as follows:

```
from [module name] import *
```

2.4 Introduction to Probability

You probably already have an intuitive understanding of probability. What is the probability of rolling a "5" with a 6-sided die? What is the probability of flipping a coin and getting heads? How about winning the grand prize in a lottery if 500,000 people buy tickets? (The actual probability, not wishful thinking.)

When it comes to questions such as the probability of two people having the same birthday in a room full of 30 people, it becomes a bit more interesting.

Let's start with some basic facts about probability.

Probability is expressed as a fraction between 0 and 1, or a percent between 0 and 100%. We'll stick with fractions for now.

Something with a probability of zero means it's never going to happen, like rolling a "7" with a 6-sided die.

Something with a probability of 1 means it's a sure thing, like the sun coming up every day.

Example 1

A 6-sided die is tossed. What is the probability that it is an even number?

Solution

There are 3 even numbers {2, 4, 6}, so the probability is $\frac{3}{6}$.

Example 2

Two 6-sided dice are tossed, as in the game craps. (Thus, there are 36 possible outcomes.)

Image 2.15

A. What is the probability of rolling doubles?

B. What is the probability of rolling a pair of numbers that form a sum of 7?

Solutions

A. There are 6 pairs of doubles, so the probability is $\frac{6}{36}$, which simplifies to $\frac{1}{6}$.

B. There are 6 pairs of numbers that form a sum of 7: 1-6, 2-5, 3-4, 4-3, 5-2, and 6-1. The probability is $\frac{6}{36}$, which simplifies to $\frac{1}{6}$.

Example 3

The American roulette wheel has 18 red, 18 black, and 2 green slots. What is the probability of spinning the wheel and the ball landing in a red slot?

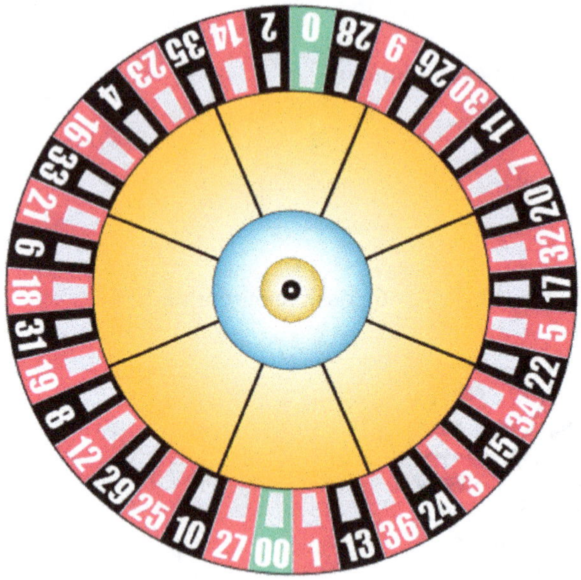

Image 2.16

Solution:

$$\frac{18 \text{ red}}{38 \text{ slots}} = \frac{9}{19}$$

Before we get into more calculations, let's write a program that generates a die roll. For this we will need to import the **random** module and use the **randint**(a,b) function, which generates a random integer between a and b.

PROGRAM 2.4.1

```
from random import randint
ready = input("Press any key to roll the die")
die_roll = randint(1,6)
print(die_roll)
```

Next we'll write a program that generates and outputs a randomly chosen weekday. This will involve using the random integer as the index to a list of strings.

PROGRAM 2.4.2

```
from random import randint
days = ['Mon', 'Tue', 'Wed', 'Thu', 'Fri']
ready = input("Press any key to choose a day")
day = randint(1,5)
print(days[day])
```

EXERCISE SET 2.4.1

1. What is the probability of randomly selecting a day of the week that is a weekday?
2. You randomly select a month of the year. What is the probability that it has exactly 30 days?
3. 68 college basketball teams compete in the annual March Madness tournament. What is the probability of randomly choosing a team that will make it to the Final Four round?
4. If you run program 2.4.2 enough times, you will eventually get an error. What is the error and how can you fix it?
5. Modify program 2.4.1 so that it generates and outputs two die rolls instead of one. (User only needs to press a key once.)
6. Modify program 2.4.2 so that it generates and outputs a randomly chosen month of the year. (Use 3-letter abbreviations for the months.)
7. Write a program that generates and outputs five lottery numbers, each between 1 and 49.
8. Write a program that generates and outputs a random playing card (from our standard deck of 52 cards). You will need two random numbers. The first will determine the suit (hearts, diamonds, clubs, spades) and the second will determine the value (ace, 2, 3, ... 10, jack, queen, king). The output will look something like this:

   You were dealt the queen of hearts.
9. Modify the program you wrote in #5 above so that it deals 2 random cards instead of one. (You will calculate later the probability of being dealt the identical card twice.)

Now let's write a more interactive program—the ubiquitous guessing game.

PROGRAM 2.4.3

```
from random import randint
num = randint(1,5)
guess = int(input("Guess a number between 1 and 5: "))
if guess == num:
    print("Correct!")
else:
    print("Sorry, the number was", num)
```

The two equal signs in *guess==num* are not a typo. Use two equal signs when doing a comparison and one equal sign when assigning a value.

Try out the program a few times. How many times did you run the program until you guessed the right answer? Of course the probability is 1/5 that you will guess correctly each time you play the game. That means that for every 100 times you play the game, you can expect to win about 20 games on average, meaning this is not a very fun game.

EXERCISE SET 2.4.2

1. Modify program 2.4.3 so that you can guess a number between 1 and 10 rather than between 1 and 5.
2. Modify program 2.4.3 again so that if the user guesses a number outside the range (less than 1 or greater than 10), they will get an error message and the program will end.
3. Modify the program you wrote in question 1 above by allowing the user to keep guessing until they get it right. The program will give hints ("too high" or "too low"). Hint: This will probably involve a **while** loop.

Let's return to the blackjack game. In this game, points are associated with the cards as follows:

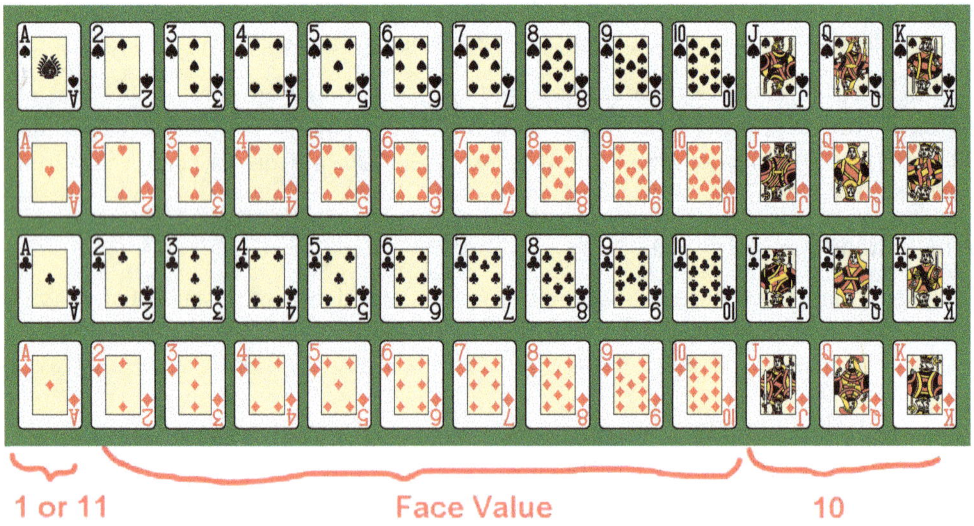

Image 2.17

Let's say we want to write a program that generates and outputs two random playing cards. (We only care about the values A, 2, 3, ... J, Q, K, not the suits).

Assume each card is dealt from a full deck of 52 cards. (This is called ***sampling with replacement****.)*

First, we need to decide on a data structure.

We will use two lists: one for the "name" of the card ['Ace', 2, 3, ..., 10, 'Jack', 'Queen', 'King'] and another for the associated points [11, 2, 3, ... 10, 10, 10, 10].

Each card name will be card_name[*n*] and its associated value is card_value[*n*]*;* for example, card_name[0] = 'Ace' and card_value[0] = 11.

Because we're going to be referring to this program later on, let's give it a filename: blackjack.py.

PROGRAM 2.4.4 BLACKJACK.PY V1

```
from random import randint
        # assign point values to each type of card
card_name=['Ace',2,3,4,5,6,7,8,9,10,'jack','Queen','King']
card_value=[11, 2, 3, 4, 5, 6, 7, 8, 9, 10, 10, 10, 10]
total = 0                                            # initialize the total points
card1 = randint(0,12)
card2 = randint(0,12)
print("You are dealt")
print(card_name[card1], ",", card_name[card2])
total += card_value[card1] + card_value[card2]       # increment total points
print("\nYour total is", total)                      # \n inserts a blank line
```

Sample Output

You are dealt
ace, king
Your total is 21

EXERCISE SET 2.4.3

1. Referring to program **blackjack**:
   A. What is the value of card_name[0]?
   B. What is the value of card_value[8]?
   C. What is the variable type of card_name[0]?
   D. What is the variable type of card_value[8]?
2. If card1 is assigned random number 6 and card2 is assigned random number 4, what is the output?
3. Write a program that simulates one round of the children's card game War. Deal one card to Player 1 and another card to Player 2. If the value of Player 1's card is higher than that of Player 2's, output "Player 1 won" and vice versa. If the values are the same, output "Tie."

Sample Output

Player 1: 10

Player 2: Jack

Tie

Let's make the **blackjack** program more interesting as follows:

(1) Ask the player if he/she wants a third card. If yes, generate a third random card.

(2) Calculate and output the point values of the cards (whether 2 or 3 cards).

SAMPLE OUTPUT

You are dealt: 5, J Do you want another card? Y/N N Your total is 15.	You are dealt: 9, 3 Do you want another card? Y/N Y You are dealt: J Your total is 22.	You are dealt: 2,9 Do you want another card? Y/N Y You are dealt: Q Your total is 21.

The changes to **blackjack v1** are shown in blue.

PROGRAM 2.4.4 BLACKJACK.PY V2

```
from random import randint
# assign point values to each type of card
card_name=['Ace',2,3,4,5,6,7,8,9,10,'Jack','Queen','King']
card_value=[11, 2, 3, 4, 5, 6, 7, 8, 9, 10, 10, 10, 10]
total = 0                                        # initialize the total points
card1 = randint(0,12)
card2 = randint(0,12)
print("You are dealt")
print(card_name[card1],",",card_name[card2])
total += card_value[card1] + card_value[card2]   # increment total points
hit_me = input("Do you want another card? Y/N ")
if hit_me=="Y":
   card3 = randint(0,12)
   print("You are dealt",card_name[card3])
   total += card_value[card3]
print("\nYour total is",total)                   # \n inserts a blank line
```

You might recognize this game as a rudimentary form of blackJack (aka 21). In the real game: (1) the ace can be worth 1 point or 11 points, depending on which gives the player the best score; (2) the dealer deals themself two cards and the player's goal is to get a higher score than the dealer without "busting" (going over 21).

EXERCISE SET 2.4.4

1. Modify program **blackjack** so that if the player's total is higher than 21, output "YOU BUSTED." If the total is exactly 21, output "You win!" If the total is less than 21, output "Good job."
2. Modify program **blackjack** so that it also deals two cards to the dealer. Output the dealer's cards and total score.

Sample Output

You are dealt:

6, K

Do you want another card? Y/N

N

Your total is 16.

Dealer's hand:

Q, 9

Dealer's total is 19.

3. Improve the program by comparing the player's score to the dealer's score and outputting the winner of the round. Use the following rules:

Player gets exactly 21	"Blackjack!"
Player busts (score > 21)	Dealer wins
Player doesn't bust but dealer busts	Player wins
Neither player busts	Higher score wins
Tie	"Push"

Sample Output

You are dealt:

6, K

Do you want another card? Y/N

N

Your total is 16.

Dealer's hand:

Q, 9

Dealer's total is 19.

Dealer wins

To improve this game, we want to give the player the option to request more than one card.

SAMPLE OUTPUT

You are dealt:

9, 3

Do you want another card? Y/N

Y

You are dealt:

2

Do you want another card? Y/N

Y

You are dealt:

5

Do you want another card? Y/N

N

Your total is 20.

We could add another variable card4, but we want a more elegant way to keep track of the cards. Thus we will replace card1, card2, card3, ... with a list: card[0], card[1], card[2], ...

Note the changes to the original program below in blue.

PROGRAM 2.4.4 BLACKJACK.PY V3

```
from random import randint
        # assign point values to each type of card
card_name=['Ace',2,3,4,5,6,7,8,9,10,'Jack','Queen','King']
card_value=[11, 2, 3, 4, 5, 6, 7, 8, 9, 10, 10, 10, 10]
total = 0                                    # initialize the total points
card = [ ]                                   # this list will hold the random cards
card.append(randint(0,12))
card.append(randint(0,12))
print("You are dealt")
print(card_name[card[0]], "," ,card_name[card[1]])
total += card_value[card[0]] + card_value[card[1]]     # increment total points
hit_me = input("Do you want another card? Y/N ")
if hit_me=="Y":
    card.append(randint(0,12))
    print("You are dealt", card_name[card[2]])
    total += card_value[card[2]]
print("\nYour total is", total)
```

EXERCISE SET 2.4.5

1. CHALLENGE: Modify program **blackjack** v3 so that the player can request more than one extra card. See sample output above the program.

Sometimes you want to make sure not to randomly generate the same number more than once, for example when selecting lottery numbers. The following method demonstrates one way to make sure your random numbers are unique.

PROGRAM 2.4.5

```
from random import *
        # create list of choices [1, 2, 3, ... , 52]
deck = list(range(1,52))
        # shuffle the order of the elements in choices
shuffle(deck)
        # pop off and assign the last element to card1
card1=deck.pop()
        # there are now 51 elements left
        # pop off and assign the last element to card2
card2=deck.pop()
print("First card:", card1, "Second card:", card2)
```

Sample Output

4, 39

A few observations about libraries, functions, and methods:

1. Libraries (such as **math**, **random**) contain a collection of functions and methods. You can either import a single method or function from the module such as:

   from random import randint

   Or you can import all the library items in the module using the wildcard asterisk:

   from random import *

   When writing and running very long programs, it is desirable to only import functions that are needed, rather than the entire library, to conserve memory and computing power.

2. A **function** returns a value or performs an action. For example, **randint**(1,10) returns a random integer between 1 and 10, whereas **shuffle**(deck) does not return a value but performs an action on the list choices. We will learn soon how to define our own functions.

3. A **method** performs an action on an object; for example, decks.**pop()** performs the method **pop** on the object decks. Note the method is appended to the object with a period (.).

EXERCISE SET 2.4.6

1. Modify the program you wrote in Exercise Set 2.4.1 #7 (five lottery numbers between 1 and 49) so that they are unique.
2. Write a program that generates and outputs a batting order for a team of nine baseball players. (You can give the baseball players any names you wish.)

Sample Output

Batting Order:

Axelrod, Barry, Curtis, Damon, Edward, Fahim, Gary, Hiro, Ignacio

Hint: Use a **for** loop.

2.5 The Language of Probability

Consider the case of rolling one six-sided die.

Experiment—An event such as rolling the die one time, flipping a coin, drawing a card, etc.
Outcome—A possible result, for example, rolling a 3.
Sample Space—The set of distinct possible outcomes, for example, {1, 2, 3, 4, 5, 6}.

Each outcome is assigned a probability between 0 and 1. The probabilities of the sample space must add up to 1. In the die roll example, if it's a fair die, each outcome has probability 1/6.

There are a number of ways to partition a deck of cards into sample spaces.

{all 52 cards}—Each card, such as king of diamonds, has probability 1/52.

{H, D, C, S}—Each suit, such as clubs, has probability $13/52 = 1/4$.

{A, 2, 3, ... 10, J, Q, K}—Each value, such as ace, has probability $4/52 = 1/13$.

So far, we have only looked at sample spaces in which each outcome is **equally likely** (i.e., equal probabilities). Is there an experiment in which the outcomes are not all equally likely? Well, take a look at this Big 6 Wheel.

Image 2.18

EXERCISE SET 2.5.1

Regarding the Big 6 Wheel above:

1. What is the experiment?
2. What is the sample space?
3. What is the probability of each outcome in the sample space? (Hint: the probabilities should add up to 1.)

If we were to write a program to simulate spinning the Big 6 Wheel, we would need a way to:

1. Generate only the numbers 1, 2, 5, 10, 20
2. Take into account that each outcome has a different probability.

There are several ways to do this. Here's one of them.
First, note there are 24 slots in the wheel partitioned into:
Eleven 1s
Seven 2s
Three 5s
Two 10s
One 20

We are going to consider the 24 slots as equally likely, so we will load them into a list. When we "spin the wheel," we will randomly generate an outcome between 0 and 23.

PROGRAM 2.5.1 BIGWHEEL.PY V1

```
from random import randint
wheel = [1,1,1,1,1,1,1,1,1,1,1,2,2,2,2,2,2,2,5,5,5,10,10,20]
spin_it = input("Press any key to spin the Big 6 Wheel")
winning_number = randint(0,23)
print("Winning number is", wheel[winning_number], "!")
```

EXERCISE SET 2.5.2

1. In program **bigwheel**, what is the probability of spinning the wheel and getting a 5?
2. A Las Vegas oddsmaker has used team stats to determine the probability of each of the following teams winning the next playoff:

   Honey Badgers: 1/6

   Yetis: 1/3

   Tasmanian Devils: 1/4

   T-Rexes: 1/4

Write a program that generates a winning team based on these probabilities.

Hint: Covert all probabilities to 12ths.

Ready to accept wagers?

All casino games pay out winners based on the odds, or probability. Here's the relationship: If the probability of rolling a "3" on a one-sided die is 1/6, then the odds are 1 to 5.

ODDS

If the probability of an outcome is $\frac{p}{q}$, then the corresponding odds (in favor of the outcome) are:

p to $(q - p)$, alternately $p{:}(q - p)$.

(Some textbooks and websites reverse this.)

If the odds in favor of an outcome are $p{:}q$, then the corresponding probability is $\frac{p}{p+q}$.

More simply, if the odds are win:lose, then the probability of winning is $\frac{win}{total}$ and vice versa.

Example 1

A basketball team has 3:2 odds in favor of winning. What is the probability that they will win?

Solution

The probability is $\frac{3}{3+2} = \frac{3}{5} = 60\%$ or 0.6.

Example 2

A racehorse won 8 out of 11 races. What odds will the bookmaker give the horse?

Solution

The racehorse won 8 and lost 3; therefore the odds in favor of winning are 8:3.

EXERCISE SET 2.5.3

1. The probability of rain tomorrow is 85%. What are the odds in favor of rain?
2. The probability of having three children that are all girls is 1/8. What are the odds?
3. In baseball, the probability of a player getting a hit (out of all his/her turns at bat) is multiplied by 10 to give the batting average. Thus, if a player has a 30% probability of hitting the ball, his/her batting average is .300. What is the batting average of a player who gets 5 out of 20 hits?
4. If a baseball player has a batting average of .650, what are the odds in favor of he/she hitting the ball?
5. The odds of a certain asteroid striking Earth in 2029 are 1:300. What is the probability?
6. The odds of being killed by lightning are 2,000,000 to 1. What is the probability? (Technically, it should be stated as 1 to 2,000,000, but it sounds more dramatic as 2,000,000 to 1.)

Odds in casinos and sports books are usually related to payouts (winnings). For example, if a racehorse has 6:5 odds in favor of winning, then you win \$6 for every \$5 you bet. (Do you see why? 6:5 odds means the probability is $\frac{6}{11}$.)

Program Goal: Modify **bigwheel** to ask the user to input a bet, and if they win, to pay out the winnings based on this table:

20	Pays 20 to 1	**10**	Pays 10 to 1
5	Pays 5 to 1	**2**	Pays 2 to 1
1	Pays 1 to 1		

For example, if they bet \$15 on 5 (which pays 5 to 1) then they win $\$15 \cdot 5 = \$75$.

PROGRAM 2.5.1 BIGWHEEL.PY V2

```
from random import randint
wheel = [1,1,1,1,1,1,1,1,1,1,1,2,2,2,2,2,2,2,5,5,5,10,10,20]
bet = int(input("\nWhat number are you betting on? "))
wager = (input("How much is your wager? ")
winning_number = randint(0,23)
print("Winning number is", wheel[winning_number], "!")
if bet==wheel[winning_number]:
    [Output the amount they won]
else:
    [Calculate and output the amount they lost]
```

EXERCISE SET 2.5.4

1. If winning_number = 10, what is wheel[winning_number]?
2. If a player bets \$5 on 10 and winning_number = 10, did they win or lose and how much?
3. If a player bets \$50 on 20 and winning_number = 23, did they win or lose and how much?
4. Fill in the yellow highlighted code to output the amount they won.
5. Fill in the blue highlighted code to calculate and output the amount they lost.
6. Modify the program you wrote in Exercise Set 2.5.2 #2 so that it asks the user for the amount of their bet, which team they're betting on, and then output how much they won or lost. Assume that if the probability of winning is $1 / N$, the payoff is $N - 1$ to 1.

(Continued)

Sample Output

What is your wager? 5

Which team do you want:

1—Honey Badgers

2—Yetis

3—Tasmanian Devils

4—T Rexes

? 1

Winning Team is Honey Badgers

You won $25

Using Flags and Loops to Control Program Flow

If you have any experience with other programming languages, you may be familiar with the concepts of the **GoTo** statement and labels. For example, here's a snippet of Visual BASIC code:

```
If answer=true Then GoTo Correct
[ code ]
:
Correct:
:
[ other code ]
```

However, the use of **GoTo** statements is discouraged as they are not considered good structured programming. It's sort of like having an organized assembly line of workers doing sequenced tasks and one of the workers decides to run to a different part of the assembly line and push his way in.

Image 2.19

Thus, Python does not include a built-in **GoTo** command. Your installation of Python might offer a **goto** module that includes **goto** and **label** features, but this is not standard so we will not include it in this text. Instead, we'll look at structured programming using the **while** loop.

The snippet of Visual BASIC code could be elegantly rewritten with a while loop as follows.

```
while answer == False:
   [ code ]
[other code]
```

Program Goal: Modify program **bigwheel** again so that (1) if the user enters a number not on the wheel, they are asked to reenter and (2) after the first spin, the user is asked if they want to spin again.

We will accomplish this with the use of the **while** loop.

PROGRAM 2.5.1 BIGWHEEL.PY V3

```
from random import randint
wheel = [1,1,1,1,1,1,1,1,1,1,2,2,2,2,2,2,2,5,5,5,10,10,20]
spin = "Y"                                      # initialize the spin
while spin=="Y":
    bet = 0                                     # initialize the bet amount
    while bet not in [1,2,5,10,20]:
        bet = int(input("\nWhat number are you betting on? "))
        wager = int(input("How much is your wager? "))
        winning_number = randint(0,23)
        print("Winning number is", wheel[winning_number],"!")
        if bet==wheel[winning_number]:
            print("You won $", wager*bet)
        else:
            print("You lost $", wager)
        spin=input("Spin again (Y/N)? ")
```

SAMPLE OUTPUT

What number are you betting on? 3

What is your wager? 100

Winning number is 2!

You lost $100

Spin again (Y/N)? Y

What number are you betting on? 2

What is your wager? 50

Winning number is 2!

You won $100

Spin again? (Y/N)? N

EXERCISE SET 2.5.5

Program **bigwheel** assumes that the player has an endless supply of money. This is not how it works in real life. Modify the program to keep track of the user's money as follows.

A. At the beginning of the program, ask the user to input how much money they have. Assign this value to variable money.

B. After you ask them to make a wager, compare wager to money. If they're trying to bet more money than they have, tell them to wager a lower amount and ask them to make another wager.

C. After generating a winning number, if the user lost, subtract their wager from money.

D. If they won, then add their winnings to money.

E. Output the new value of money.

F. Only ask them to spin again if money > 0.

Sample Output

How much money do you have? 100

What is your wager? 50

What number are you betting on? 2

Winning number is 1!

You lost $50

You now have $50

Spin again? Y

What is your wager? 75

You only have $50.

What is your wager? 50

What number are you betting on? 1

Winning number is 5!

You lost $50

You now have $0

Game over

2.6 Compound Probabilities

Time to turn our attention to the game of **craps,** which has a very simple premise: roll two 6-sided dice. The number of possible outcomes is $6 \cdot 6 = 36$.

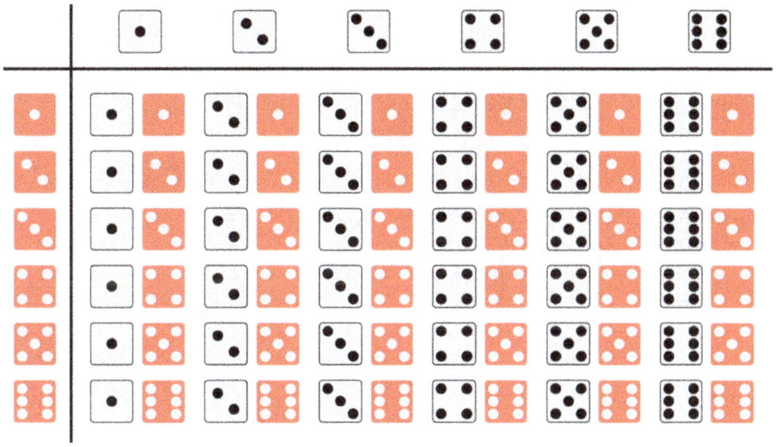

Image 2.20

For simplicity, we will consider "rolling two dice with a sum of 5" as "rolling a 5."

EXERCISE SET 2.6.1

Use the table above to answer the following questions.

1. What is the probability of rolling a 5?
2. What is the probability of rolling two dice and either or both shows a 5?
3. Which number (sum of two dice) is the most likely?
4. What is the probability of rolling an even sum?
5. What is the probability of rolling 7 or 11?
6. What is the probability of rolling 2, 3, or 12?
7. What is the probability of rolling doubles?
8. What is the probability of rolling a 10 or doubles?
9. Write a program that does the following.
   A. Simulates and outputs the rolling of two 6-sided dice
   B. If a 7 or 11 is rolled, output "PASS." If a 2, 3, or 12 is rolled, output "DON'T PASS."

Mutually Exclusive Events

Events with no "overlap" (i.e., $E \cap F = \varnothing$) are called **mutually exclusive** events. They usually refer to a single "experiment" such as rolling two dice or being dealt a single card. Examples of mutually exclusive events:

E = rolling a 7 with two dice, F = rolling an even number with two dice;
E = being dealt a king, F = being dealt a queen

Examples of NOT mutually exclusive events:
E = being dealt a queen, F = being dealt a heart (because you could be dealt the queen of hearts);
E = rolling doubles with two dice, F = rolling a 10 with two dice (because you could roll a pair of 5s);
E = randomly selecting a student who is female, F = randomly selecting a student who is taller than 5 feet (because a female student could be taller than 5 feet)

You might have figured out that if two events are mutually exclusive (such as Exercise #5: rolling 7 or 11), you simply add the probabilities:
The sample space for rolling 7 is { (1,6), (2,5), (3,4), (4,3), (5,2), (6,1) }
The sample space for rolling 11 is { (5,6), (6,5) }

Thus $P(7 \text{ or } 11) = P(7) + P(11) = \frac{6}{36} + \frac{2}{36} = \frac{8}{36} = \frac{2}{9}$

More formally:

PROBABILITY OF A UNION OF MUTUALLY EXCLUSIVE EVENTS

If E and F are events such that $E \cap F = \varnothing$, then

$$P(E \cup F) = P(E) + P(F)$$

Recall $E \cup F$ represents the union of the two sets. In the context of probability events, that would mean event E and/or event F. Also recall $E \cap F$ represents the intersection of two sets.

Let's consider Exercise Set 2.6.1 #8, the probability of rolling a 10 or doubles.

The sample space for rolling doubles is D = { (1,1), (2,2), (3,3), (4,4), (5,5), (6,6) }.

The sample space for rolling a 10 is T = { (4,6), (5,5), (6,4) }.

Note that $D \cap T$ = { (5,5) }. You don't want to count that twice. So the probability of rolling doubles or 10 would be:

$$P(D \cup T) = P(D) + P(T) - P(D \cap T) = \frac{6}{36} + \frac{3}{36} - \frac{1}{36} = \frac{8}{36} = \frac{2}{9}$$

Recall the **Counting Formula** for number of elements in the union of sets:

If A and B are finite sets, then

$$n(A \cup B) = n(A) + n(B) - n(A \cap B)$$

We have a similar rule for probability.

PROBABILITY OF A UNION

If E and F are any two events, then

$$P(E \cup F) = P(E) + P(F) - P(E \cap F)$$

As you can see, this is a more generalized version of the previous formula $P(E \cup F) = P(E) + P(F)$ if $E \cap F = \emptyset$ (i.e., if E and F are mutually exclusive.)

EXERCISE SET 2.6.2

Find the probability of the following.

1. Rolling two 6-sided dice and getting an even number or doubles.
2. Rolling two 6-sided dice and getting 6 or a multiple of 3.
3. Rolling two 6-sided dice and getting an odd number or doubles.
4. Selecting one card from a 52-card deck and getting an ace or a heart.
5. Selecting one card from a 52-card deck and getting a red card or a 7.

Independent Events

Independent events usually refer to more than one "experiment," such as repeatedly rolling a die, tossing a coin, or rolling a roulette wheel. Casinos rely on gamblers' willful ignorance of independent events, because after losing repeatedly, a gambler may feel this increases their probability of winning on the next toss of the dice. However, probabilities are not affected by random events that preceded them. It is often said that "the dice have no memory."

Image 2.21

Let's say we're interested in parents who have three children and all three of them are girls. Since the probability of having a girl without artificial assistance is ½, then do we say the probability of three girls would be $P(G) + P(G) + P(G) = \frac{1}{2} + \frac{1}{2} + \frac{1}{2} = 1\frac{1}{2}$?

First of all, this makes no sense because the probability adds up to more than 1. Secondly, we are not asking if the first is a girl **or** the second is a girl **or** the third is a girl. We want to know the probability of the first is a girl **and** the second is a girl **and** the third is a girl.

INDEPENDENT EVENTS

If E and F are independent events, then

$$P(E \cap F) = P(E) \cdot P(F)$$

This can be extrapolated to any number of independent events:

$$P(E_1 \cap E_2 \cap \ldots E_n) = P(E_1) \cdot P(E_2) \cdots P(E_n)$$

Back to the probability of having three daughters.

$$P(G1 \cap G2 \cap G3) = P(G1) \cdot P(G2) \cdot P(G3) = \frac{1}{2} \cdot \frac{1}{2} \cdot \frac{1}{2} = \frac{1}{8}$$

Example 1

Consider being dealt two cards from a 52-card deck. What is the probability of the first being a king and the second being an ace?

Solution

Image 2.22

$$P(K \cap A) = P(K) \cdot P(A) = \frac{4}{52} \cdot \frac{4}{51} = \frac{\overset{1}{\cancel{4}}}{\underset{13}{\cancel{52}}} \cdot \frac{4}{51} = \frac{4}{663}$$

Notice that $P(A) = \frac{4}{51}$ because after the first card was removed, there are only 51 cards remaining in the deck.

Example 2

A. In the Big Pick lottery in Arizona, you pick 6 numbers between 1 and 44. What is the probability of winning with a single ticket?

B. What is the probability of winning if you buy two tickets?

Solution

A. The number of ways of selecting a subset of 6 out of 44 numbers is ${}_{44}C_6 = 7{,}059{,}052$. Therefore, the probability is $\frac{1}{7{,}059{,}052}$.

B. You've just doubled your chances: $\frac{2}{7{,}059{,}052}$

EXERCISE SET 2.6.3

1. Identify each of the following pairs of events as independent, mutually exclusive, or neither:
   A. Toss a coin twice: the first result is heads, the second is tails.
   B. Choose two balls from a bag of 5 red and 2 green balls: the first is red, the second is green.
   C. The first card dealt from a full deck is a king; the second is also a king.
   D. You are dealt a single card from a full deck: it is an ace or a spade.
   E. You are dealt a single card from a full deck: it is a heart or a diamond.
   F. You choose a sock from your sock drawer, put it back, and choose again.
2. The probability of a flight being delayed is .2. What is the probability of making a round trip with no delays?
3. A coin is tossed four times. What is the probability of getting 4 heads?
4. Determine the probability of having 1 girl and 3 boys in a 4-child family.
5. In the California Powerball lottery, you can pick five lucky numbers from 1 to 69 and one powerball number from 1 to 26.
   A. What is the probability of getting all 5 lucky numbers between 1 and 69?
   B. What is the probability of getting all 5 lucky numbers plus the powerball number?
6. A couple rents 20 DVDs. This table represents a breakdown of each type of movie.

	Movies with Creatures	**Movies with Natural Disasters**
Action-Adventures	7	3
Comedies	3	7

   What is the probability of randomly selecting a DVD . . .
   A. That features a natural disaster?
   B. That is an action-adventure with monsters or a comedy with a natural disaster?
   C. That features creature(s) or is a comedy (or both)?

   What is the probability of randomly selecting two DVDs (order is not important) . . .
   D. In which one features a creature, and the other features a natural disaster?
   E. That are both comedies?
7. Write a program that asks the user to input how many children they would like to have (or already have). Then calculate and output the probability of all girls. *Hint: use the fractions library.*

Complement of an Event

Sometimes it is easier to calculate the probability of an event NOT happening than happening. That event is called the **complement** and is defined more formally as follows.

If E is an event in sample space S for an experiment, then its complement is defined as the set of all outcomes in S that are not in E, and its probability is calculated as:

$$P(\overline{E}) = 1 - P(E)$$

Example 1

A. When flipping a coin, what is the probability of getting tails?

B. When tossing a 6-sided die, what is the probability of not getting 3?

C. When tossing two 6-sided dice, what is the probability of not getting doubles?

Solutions

A. $P(\text{heads}) = ½$, $P(\text{not heads}) = 1 - ½ = ½$.

B. $P(3) = \frac{1}{6}$, $P(\text{not } 3) = 1 - \frac{1}{6} = \frac{5}{6}$

C. $P(\text{doubles}) = \frac{6}{36} = \frac{1}{6}$, $P(\text{not doubles}) = 1 - \frac{1}{6} = \frac{5}{6}$

Example 2

Let's return to the scenario of the family with four children. What is the probability that at least one of the offspring is a boy?

Solution

One way to calculate is:

$P(\text{exactly one boy}) + P(\text{exactly two boys}) + P(\text{exactly 3 boys}) + P(\text{exactly 4 boys})$

However, this is a bit tedious and involved. It is simpler to consider "at least one boy" as "not all girls." We can then take advantage of the complement formula:

$P(\text{at least one boy}) = 1 - P(\text{all girls}) = 1 - \frac{1}{2} \cdot \frac{1}{2} \cdot \frac{1}{2} \cdot \frac{1}{2} = 1 - \frac{1}{16} = \frac{15}{16}$

EXERCISE SET 2.6.4

1. A coin is tossed four times. What is the probability of getting at least one tail? (Hint: Use your answer to Exercise Set 2.6.3 #3.)
2. Refer to the table of movies in Exercise Set 2.6.3 #6. What is the probability that the couple randomly selects one DVD and it is **not** a disaster comedy movie?
3. Modify the program you wrote in Exercise Set 2.6.3 #7 to also calculate and output the probability of having at least one boy.

The Birthday Problem

A well-known math question is this: In a room of 23 (or some number of) people, what is the probability that at least two of them have the same birthday?

Program Goal: Generate 23 random birthdates (between 1 and 365). Store the selected birthdays in a list and keep track of any numbers that are a match.

This will also be our introduction to **Boolean** variables. A Boolean variable has two possible values: "True" or "False." We will use a Boolean variable match to indicate whether a birthday match has been found.

PROGRAM 2.6.1 BIRTHDAY.PY

```
from random import randint
birthdays = [ ]
match = False
for n in range(1,24):
    next_bd = randint(1,365)
    if next_bd in birthdays:
            match = True
    birthdays.append(next_bd)
print(birthdays)
print("match found: ", match)
```

Run the program several times. If you run it at least 5 times, the probability is you will see a birthday match at least once. Why?

Let's figure out the actual probability of at least 2 people in a room of 23 people sharing the same birthday. We will do this by figuring out the probability of NONE of the people sharing a birthday and then take the complement of that.

The probability of none of the people sharing a birthday is

$$\frac{\textit{number of ways to have 23 different birthdays}}{\textit{total number of ways to have 23 birthdays}}$$

First, figure out the number of ways to assign unique birthdays to 23 people. That would be ${}_{365}P_{23}$.
Then, the number of ways to assign birthdays to 23 people, allowing repeats. That would be 365^{23}.

Thus, the probability of no shared birthdays is $\frac{365P_{23}}{365^{23}} \approx 0.497$

The complement of that, which is at least one shared birthday, is $1 - .497 \approx .507$.

Believe it or not, the probability of at least two people sharing a birthday in a room full of 23 people is greater than 50%! This would be a great bet to make; you're guaranteed to win unless you make the bet with a mathematician.

EXERCISE SET 2.6.5

1. Modify program **birthday** so that if there is a match, it outputs the matching day.

Sample Output

[316, 98, 266, 224, 326, 335, 354, 166, 134, 40, 151, 15, 322, 269, 359, 230, 69, 272, 225, 306, 322, 294]

Match found: 322

2. Although it is less likely, there could be more than two matches in the room. Modify program **birthday** so that it outputs all the matches.

Sample Output

[345, 101, 339, 233, 142, 235, 108, 345, 54, 137, 82, 76, 185, 157, 186, 169, 323, 113, 123, 209, 123, 174, 7]

Matches found: [345, 123]

3. What is the probability of at least two people sharing a birthday in a room full of 30 people?

4. A. Write a program that asks the user to input how many people are in a room, and then calculate and output the probability that at least two of them will have the same birthday. Hint: ${}_nP_k = \dfrac{n!}{(n-k)!}$.

   B. What happens if you run the program with too large a number?

2.7 Conditional Probability

Let's expand on the movie breakdown in Exercise Set 2.6.3 #6 by throwing in several suspense films. A couple decides to choose a movie randomly.

	Movies with Creatures	Movies with Natural Disasters
Action-Adventures	7	3
Comedies	3	7
Suspense	1	4

Chris chooses a movie, looks at the title, and tells Pat, "It's a movie with a creature." *Hmm,* Pat thinks, *there's a 3/11 chance that it's a comedy.* Pat is correct; the movie is in fact *Sharknado.*

Chris puts the movie back and now Pat takes a turn. Pat randomly selects a DVD telling Chris only that it's an action-adventure. Aha, thinks Chris, there's a 7/10 chance that it has a monster. Chris is pleased to find out that the movie is *King Kong.*

It's pretty obvious how to deduce these probabilities when they are laid out in a table like this. There is, however, a formula.

For example, the probability that a movie is a comedy given that it has a creature is expressed as:

$$P(\text{comedy}|\text{creature}) = \frac{P(\text{comedy and creature})}{P(\text{creature})} = \frac{3}{11}$$

More generally:

CONDITIONAL PROBABILITY

The probability of event E given event F is:

$$P(E \mid F) = \frac{P(E \cap F)}{P(F)}$$

Example 1

Recall the layout of a standard deck of 52 cards:

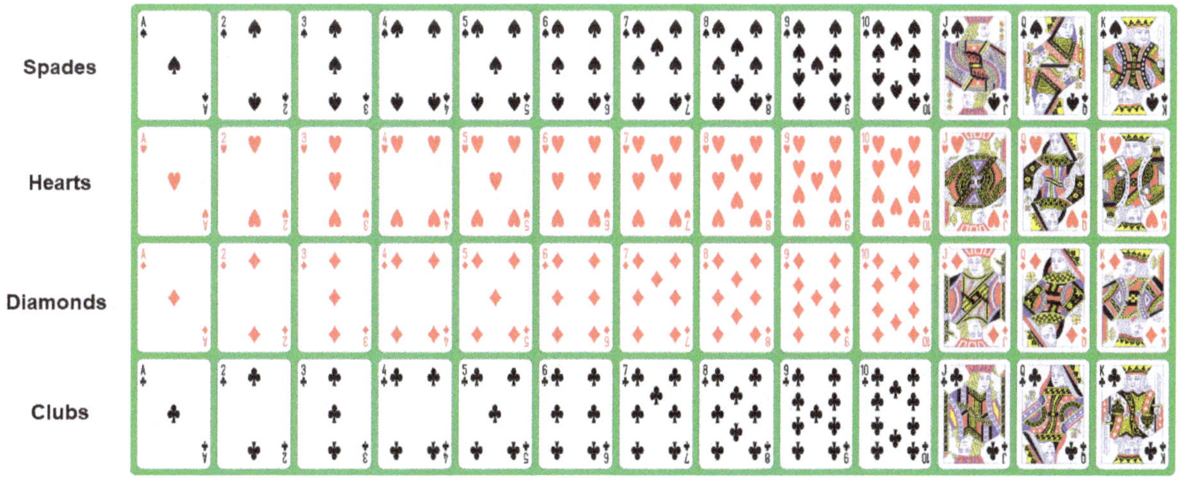

Image 2.1

Answer the following.

A. $P(\text{diamond})$

B. $P(5)$

C. $P(\text{5 of diamonds})$

D. $P(\text{diamond} \mid 5)$

E. $P(5 \mid \text{diamond})$

F. $P(\text{ace} \mid \text{red})$

G. $P(\text{queen} \mid \text{not jack or king})$

H. $P(\text{queen or king} \mid \text{not a heart})$

Solutions

A. $P(\text{diamond}) = 13/52 = 1/4$

B. $P(5) = 4/52 = 1/13$

C. $P(\text{5 of diamonds}) = 1/52$

D. $P(\text{diamond} \mid 5) = \dfrac{P(\text{5 of diamonds})}{P(5)} = \dfrac{1/52}{4/52} = \dfrac{1}{4}$

E. $P(5 \mid \text{diamond}) = \dfrac{P(\text{5 of diamonds})}{P(\text{diamond})} = \dfrac{1/52}{13/52} = \dfrac{1}{13}$

F. $P(\text{ace} \mid \text{red}) = \dfrac{P(\text{red ace})}{P(\text{red})} = \dfrac{2/52}{26/52} = \dfrac{2}{26} = \dfrac{1}{13}$

G. $P(\text{queen} \mid \text{not jack or king}) = \dfrac{P(\text{queen})}{P(\text{not jack or king})} = \dfrac{4/52}{44/52} = \dfrac{4}{44} = \dfrac{1}{11}$

H. $P(\text{queen or king} \mid \text{not a heart}) = \dfrac{P(\text{queen or king of clubs, diamonds, spades})}{P(\text{clubs, diamonds, spades})} = \dfrac{6/52}{39/52} = \dfrac{6}{39} = \dfrac{2}{13}$

Example 2

A family has three children. Assuming that boys and girls are equally likely, we have the following sample space of $2^3 = 8$ possible outcomes.

{ GGG, GGB, GBG, GBB, BGG, BGB, BBG, BBB }

Determine the following probabilities.

A. At least two girls given the first is a girl.

B. Two girls given the first is a boy.

C. Exactly two boys given the middle child is a boy.

Solutions

A. $P(2G \mid \text{1st is girl}) = \dfrac{P(\text{2G and 1st is G})}{P(\text{1st is G})} = \dfrac{3/8}{4/8} = \dfrac{3}{4}$

B. $P(2G \mid \text{1st is boy}) = \dfrac{P(\text{2G and 1st is B})}{P(\text{1st is B})} = \dfrac{1/8}{4/8} = \dfrac{1}{4}$

C. $P(2B \mid \text{2nd is boy}) = \dfrac{P(\text{2B and 2nd is B})}{P(\text{2nd is B})} = \dfrac{2/8}{4/8} = \dfrac{2}{4} = \dfrac{1}{2}$

EXERCISE SET 2.7.1

Find the given conditional probabilities.

For #1–4, refer to the following table of students surveyed at a college.

	Tattooed	Not Tattooed
Male	36	144
Female	288	32

If one of the surveyed students is selected at random, find:

1. $P(\text{female} \mid \text{tattooed})$
2. $P(\text{male} \mid \text{not tattooed})$
3. $P(\text{not tattooed} \mid \text{female})$
4. $P(\text{not tattooed} \mid \text{male})$

For #5–9, assume 1 card is randomly dealt from a 52-card deck. Find:

5. $P(\text{club} \mid \text{black})$
6. $P(\text{jack} \mid \text{red})$
7. $P(5 \mid \text{red})$
8. $P(\text{black} \mid \text{ace})$
9. $P(\text{face card} \mid \text{not } 10)$

   The face cards are jack, queen, and king.

For #10–11, assume a family has four children with boys and girls being equally likely. Find the probability that:

10. Two of the children are boys, given that the first is a boy.
11. The youngest is a girl, given that at least one of the children is a boy.

We can rewrite the conditional probability formula as follows:

PROBABILITY OF AN INTERSECTION

The probability of $E \cap F$, if the conditional probability is known, is:

$$P(E \cap F) = P(F) \cdot P(E \mid F)$$

or

$$P(E \cap F) = P(E) \cdot P(F \mid E)$$

This can be visualized using a tree diagram, as in Example 3.

Example 3

A group of symptomatic people (high fever, cough) were tested for a certain highly contagious disease. Out of that group, it was known that only 10% actually had the disease. However only 89% of that group tested positive. Out of the group who did not have the disease, 2% of those had tests that returned a false positive.

A. Make a tree diagram.

B. Use the diagram to determine the probability that a diseased person tested negative.

C. Use the tree diagram to determine the probability that a symptomatic person is both diseased and tested negative.

D. Use the tree diagram to determine the probability that a symptomatic person didn't have the disease and tested positive.

Solutions

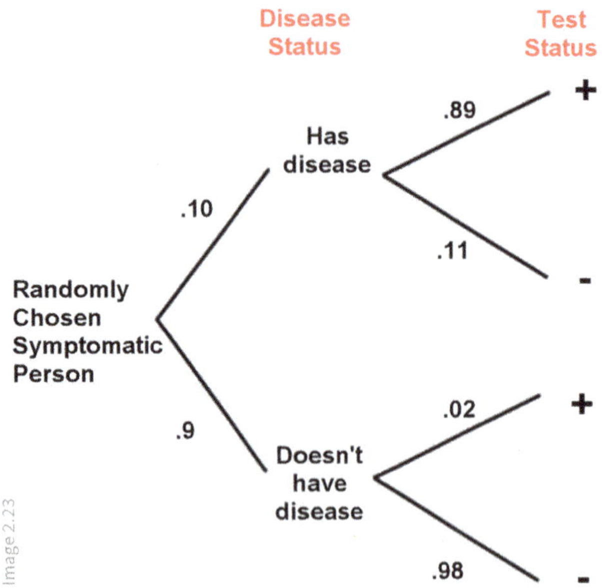

Image 2.23

E. $P(\text{negative}|\text{diseased}) = 0.11$

F. $P(\text{diseased person and tested negative}) = P(\text{diseased}) \bullet P(\text{negative} \mid \text{diseased}) = (.10)(.11) = .011$

G. $P(\text{not diseased and positive}) = P(\text{not diseased}) \bullet P(\text{positive} \mid \text{not diseased}) = .90(.02) = 0.18.$

Do you see the difference between "a person who is both diseased and tested negative" and "a diseased person who tested negative"? The first is $D \cap N$, in which a random person is chosen out of the entire symptomatic population. The second is $N \mid D$, in which a random person is chosen out of the diseased group.

From entire symptomatic population, calculate $P(D \cap N)$

.011

Image 2.24

From symptomatic diseased people, calculate $P(N \mid D)$

.11

Image 2.25

EXERCISE SET 2.7.2

For #1–7, refer to the following:

Out of a group of students sitting down to take a test, 5/8 were prepared. A prepared student had a 1/10th probability of not passing the test. An unprepared student had a 7/8th probability of not passing.

1. Make a tree diagram.
2. Find the probability that a prepared student passed the test.
3. Find the probability that a randomly chosen student was both prepared and passed the test.
4. Find the probability that an unprepared student passed the test.
5. Find the probability that a randomly chosen student passed the test.

   Hint: P(*passed and prepared*) + P(*passed and unprepared*)
6. Find the probability that a randomly chosen student did not pass the test.
7. Do the probabilities in #5 and #6 add up to 1?

(Continued)

For #8–11, refer to the following:

A survey of 10,000 Americans about UFO encounters revealed the statistics in the table on the right. What is the probability that a random respondent:

	UFO Encounter	No UFO Encounter
Live in Rural Area	1,290	605
Listen to Conspiracy Theory Radio	525	390
Live in Rural Area and Listen to Conspiracy Radio	895	115
Neither Factor	445	5,735

8. Lived in a rural area and listened to Conspiracy Theory Radio?
9. Lived in a rural area and listened to Conspiracy Theory Radio, given that they had a UFO encounter?
10. Neither lived in a rural area nor listened to Conspiracy Theory Radio, given that they had no UFO encounter?
11. Had a UFO encounter, given that they lived in a rural area or listened to Conspiracy Theory Radio or both?

Program Goal: Back in section 2.1, we wrote program 2.1.1 that:

1. Asked the user to input $n(A)$, $n(B)$, and $n(A \cap B)$.
2. Calculated and output $n(A \cup B)$, $n(A \setminus B)$ and $n(B \setminus A)$.

Enhance the program so that it calculates the following probabilities:

3. $P(A)$, $P(B)$, $P(A \cap B)$, $P(A \mid B)$, $P(B \mid A)$

Assume the total sample space is $A \cup B$. It might help to think of set A as "Art" and set B as "Baroque Items" and you choose one item randomly.

Express probabilities as simplified fractions, not decimals.

Image 2.26

SAMPLE OUTPUT

How many art pieces? 35

How many baroque items? 15

How many pieces of baroque art? 5

$P(\text{art}) = 7/9$

$P(\text{baroque}) = 1/3$

$P(\text{baroque art}) = 1/9$

$P(\text{art} \mid \text{baroque}) = 1/3$

$P(\text{baroque} \mid \text{art}) = 1/7$

PROGRAM 2.7.1

```
from fractions import Fraction
nA = int(input('How many art pieces? '))
nB = int(input('How many baroque items? '))
nI = int(input('How many pieces of baroque art? '))          # intersection
nU = nA + nB – nI                                             # number of elements in union
pA = Fraction(nA,nU)
print('P(art)=',pA)
pB = Fraction(nB,nU)
print('P(baroque)=',pB)
[ calculate and output P(A ∩ B) ]
[ calculate and output P(A | B ]
```

EXERCISE SET 2.7.3

1. Fill in the yellow highlighted code to calculate and output $P(A \cap B)$.
2. Fill in the pink highlighted code to calculate and output $P(A \mid B)$ and $P(B \mid A)$.

Let's say we want to enhance program 2.7.1 so that rather than inputting the number of elements in A, B, $A \cap B$, the user inputs the actual sets. This first requires a means of inputting a set. Recall, the variable type returned by the **input** statement is always a string, which is why we use **int(input())** to input integers.

TABLE 2.3

To input a list of strings: input_str = **input**("Enter list elements separated by commas: ") input_list = input_str.**split**(",")	**Example:** Enter list elements separated by commas: A,B,C Result: input_list = ['A', 'B', 'C']
To input a list of integers: input_str = **input**("Enter integers separated by commas: ") input_list = input_str.**split**(",") size = **len**(input_list) **for** k **in range**(0,size): input_list[k] = **int**(input_list[k])	**Example:** Enter integers separated by commas: 1,2,3,4,5 Result: Before **for** loop: input_list = ['1', '2', '3', '4', '5'] After **for** loop: input_list = [1, 2, 3, 4, 5]
To input a set of strings: input_str = **input**("Enter list elements separated by commas: ") input_list = input_str.**split**(",") input_set = **set**(input_list)	**Example:** Enter list elements separated by commas: A,B,C Result: input_list = ['A', 'B', 'C'] input_set = {'A', 'B', 'C'} (not necessarily in that order)

EXERCISE SET 2.7.4

In the following exercises, variable names are in blue font.

1. If sentence is a **string**, what does sentence.**split**(",") do?
2. What variable type is returned by *foobar* = **input**()?
3. If alphabet is a **list**, what variable type is returned by scruffy = **set**(alphabet)?
4. If alphabet is a **list** of non-numeric **strings**, what variable type is returned by numbers = **int**(alphabet[0])?
5. If alphabet is a **list** of numeric strings, what variable type is returned by numbers = **int**(alphabet[0])?
6. What sequence of code will input a set of integers?

Keep in mind that our goal is to rewrite program 2.7.1 so that the user inputs two **sets**: setA and setB. The elements can be of any type. The procedure to input a **set** requires several steps, as you saw in Table 2.3. Rather than copy and paste code, we are going to introduce the user-defined **function,** which can be invoked as often as you want.

User-Defined Functions

Our first **function** is going to use the Pythagorean Theorem to find the hypotenuse of a triangle with side lengths *a* and *b*. Here is the flow of code between main program and function:

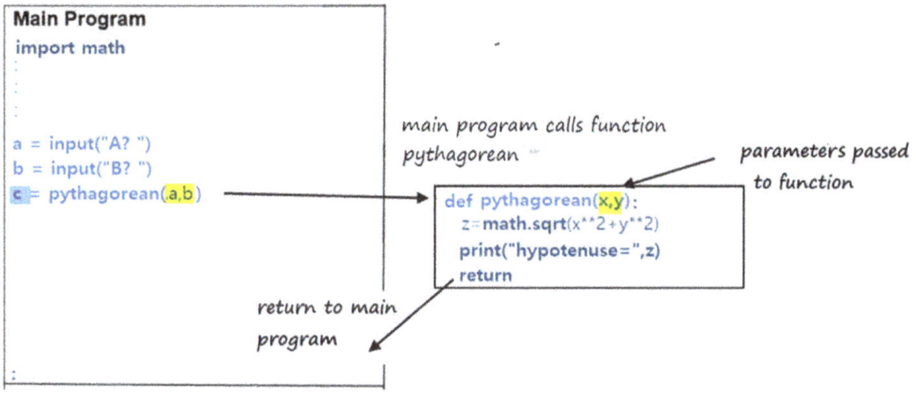

Image 2.27

The advantage of a function is that it enables you to perform the same block of code as many times as you like, for example if you wanted to process 100 triangles.

The variables in parentheses after the function name (in this case *x*, *y*) are called the **parameters** of the function. These are values passed to the function. You can have as many as you like or none at all. The values passed to the parameters can be variables or constants.

Traditionally, user-defined functions are included in the beginning of the program, after the imported Python modules. To avoid confusion, give the variables within the function unique names that do not match variable names in the main program, and should be as generic as possible.

Here's how the function and main program fit together.

PROGRAM 2.7.2

```
import math
# define function
def pythagorean(x,y):                      # end the line with a colon (:)
    z = math.sqrt(x**2+y**2)               # indent the body of the function
    print("hypotenuse=",z)
    return                                 # function definition ends with return
# main program
a = int(input("A? "))
b = int(input("B? "))
pythagorean(a,b)
```

SAMPLE OUTPUT

A? 5

B? 12

hypotenuse = 13

A function can also return a value back to the main program. Here's an example of a function call that returns a random day of the week.

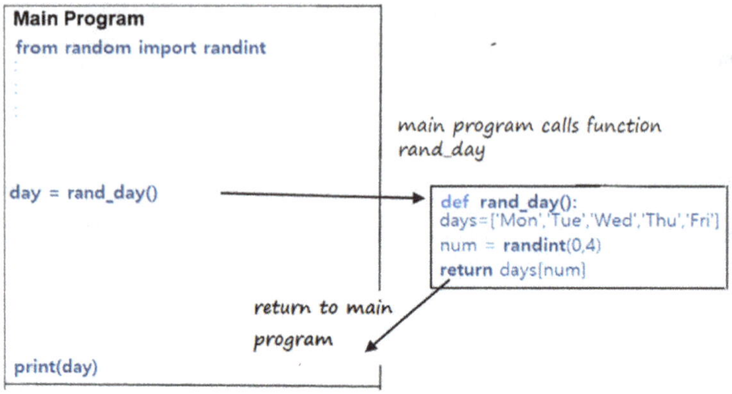

Image 2.28

Note that the function rand_day() takes no parameters from the main program and it returns one parameter to the main program.

EXERCISE SET 2.7.5

1. In the program that selects a random weekday, can the days=['Mon', 'Tue', 'Wed', 'Thu', 'Fri'] statement be moved out of the function into the main program? If so, does it matter whether it comes before or after the function definition?
2. Write the program shown in the diagram above that prints out a random weekday. Modify the program so that it prints out ten weekdays instead of one weekday.
3. Write a program that makes use of a function that squares the integer parameter. Ask the user to input a list of *x*-values, then call the function to calculate and output each ordered pair.

Sample Output

Domain? −2, −1,0,1,2

Ordered Pairs:

(−2,4)

(−1,1)

(0,0)

(1,1)

(2,4)

4. Write a program that makes use of a function that randomly generates a month and day. (Keep in mind January has 31 days, February has 28 days, etc.) Ask the user how many random birthdays they want generated. Call the function to generate and output each birthday.

Sample Output

How many birthdays? 5

Jul 3

Sep 22

Mar 9

Oct 16

Jul 15

Let's return to the task of converting a comma-delimited string (such as "Apple,Banana,Cherry") into a set of strings (such as { "Apple", "Banana", "Cherry" }). Here's an overview of how the function interacts with the main program:

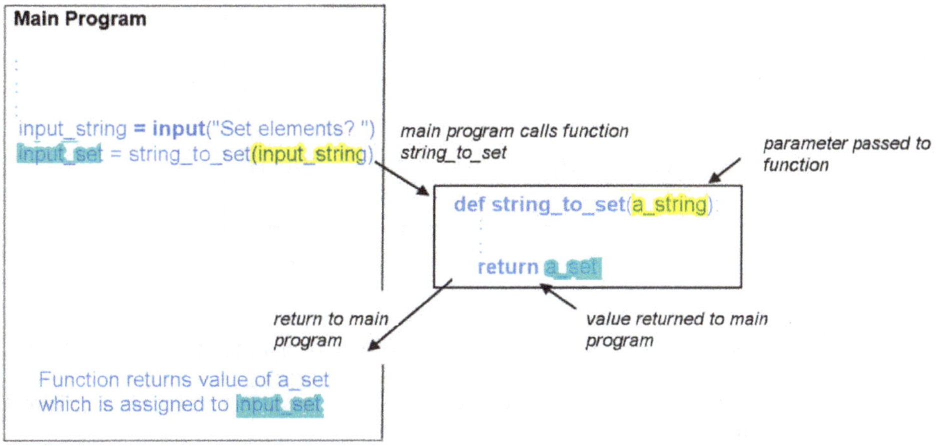

Image 2.29

Thus our function string_to_set() will look like this:

```
def string_to_set(a_string):
    a_list = a_string.split(",")
    a_set = set(a_list)
    return a_set
```

We will incorporate this function into the following.

Program Goal: Enhance program 2.7.1 so that rather than inputting the number of elements in A, B, and $A \cap B$, the user inputs the actual sets. Use the string_to_set() function to convert input strings to sets.

PROGRAM 2.7.3

```
from fractions import Fraction
# define function
def string_to_set(a_string):
    a_list = a_string.split(",")
    a_set = set(a_list)
    return a_set
# main program
input_str = input("Enter list A elements separated by commas: ")
setA = string_to_set(input_str)
input_str=input("Enter list B elements separated by commas: ")
setB = string_to_set(input_str)
A_and_B = setA.intersection(setB)
# find size of each set
nSetA = len(setA)
nSetB = len(setB)
nA_and_B = len(A_and_B)
nA_U_B = nSetA + nSetB – nA_and_B
# find probabilities
[ fill in code to calculate and output P(A), P(B), P(A ∩ B), P(A U B), P(A | B),
P(B | A) ]
```

Sample Output

Enter list *A* elements separated by commas: *A,B,C,D,E,F,G*
Enter list *B* elements separated by commas: *D,E,F,G,H,I*
$P(A) = 7/9$
$P(B) = 2/3$
$P(A \text{ and } B) = 4/9$
$P(A \cup B) = 1$
$P(A \mid B) = 2/3$
$P(B \mid A) = 4/7$

EXERCISE SET 2.7.6

1. Fill in the yellow highlighted code to calculate and output $P(A)$, $P(B)$, $P(A \cap B)$, $P(A \cup B)$, $p(A \mid B)$ and $p(B \mid A)$.
2. What happens if you input the same element twice, for example *A,A,B,C,D,E,F,G*. Does that affect any of the counts or probabilities?
3. Write a function **num_string_to_set()** that converts a comma-delimited string (such as “45,36,18,92,36”) into a set of integers with no repeats.
4. Write a function **num_string_to_frac** that converts a comma-delimited string into a set of fractions.

We will write many more functions in this chapter and future chapters. Functions are a hallmark of structured programming.

Local and Global Variables

It's important to note the difference between **local** and **global** variables.

Local variables are only defined within their **scope** of the program. Variables defined and manipulated within a function essentially do not exist outside of the function, except for variables returned to the main program with the **return** statement. Even then, it is the *value* that is returned to the main program, not the variable itself.

Global Variables defined and manipulated in the main program can be referenced inside of the function; however, if they are changed within the function, that does not affect the value of the variables in the main program. To avoid confusion, it is best to give variables in a function different names than the variables in the main program.

Here's an example.

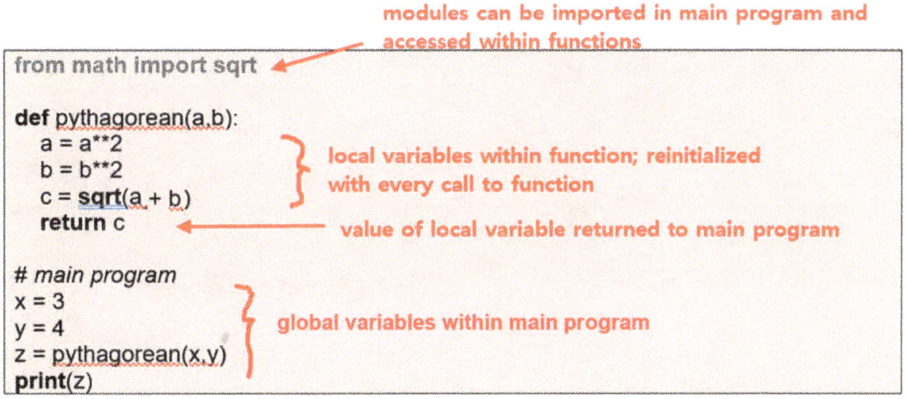

Image 2.30

EXERCISE SET 2.7.7

Answer the following questions regarding this program. Identify whether there is output or an error message. If there is an error message, then "comment it out" (i.e., precede with #) so that you can test the remainder of the program.

```
from math import sqrt
def pythagorean(a,b):
    a = a**2
    b = b**2
    print("INSIDE FUNCTION x=",x, "y=",y)          ← 1. Output or Error?
    c = sqrt(a + b)
    return c
# main program
x = 3
y = 4
z = pythagorean(x,y)
print("MAIN PROGRAM a=",a, "b=",b)                 ← 2. Output or Error?
print("MAIN PROGRAM z=",z)
```

(Continued)

Make the following changes and answer the questions.

```
from math import sqrt
def pythagorean(a,b):
    a = a**2
    b = b**2
    c = sqrt(a + b)
    x = 5
    y = 12
    print("INSIDE FUNCTION x=",x, "y=",y)        ← 3. Output or Error?
    return c
# main program
x = 3
y = 4
z = pythagorean(x,y)
print("MAIN PROGRAM x=",x, "y=",y)               ← 4. Output or Error?
```

2.8 Experiential and Geometric Probability

So far we have only looked at what is called **theoretical probability,** which is based in mathematical theory. Another type of probability is **experiential probability**. This is pretty intuitive, so we will not spend a lot of time with it.

EXERCISE SET 2.8.1

1. If a sports team won 3 out of the last 5 games, then what is the probability that they will win the next game?
2. If 30% of the patients who took an experimental medication suffered from a dangerous side effect, what is the probability that the average person will suffer from a dangerous side effect?
3. If a certain state has voted for a Democratic presidential candidate every election the last 100 years, what is the probability that they will vote Democratic in the next election?
4. Which type of probability do you think is more reliable—theoretical or experiential, and why?

Another type of probability is **geometric probability**, also very intuitive. Here's an example. The standard dartboard divides a circle into 20 sectors (pie-shape slices). The probability of randomly throwing a dart and it landing in the sector labeled "18" is 1/20. (If you're having a hard time imagining how to randomly throw a dart, just imagine the dartboard spinning.)

Image 2.31

EXERCISE SET 2.8.2

1. A dartboard is composed of three concentric circles:

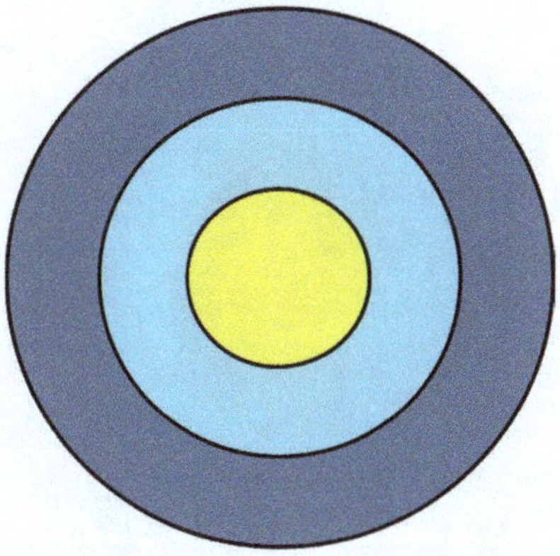

Image 2.32

The inner radius is 1 cm, the middle radius is 2 cm, and the outer radius is 3 cm. What is the probability of randomly throwing a dart and it landing in the middle ring? Assume the dart lands somewhere on the dartboard.

(*Continued*)

2. A family decides to randomly choose one of the 48 contiguous United States for their next summer vacation, so they throw a dart at a map.

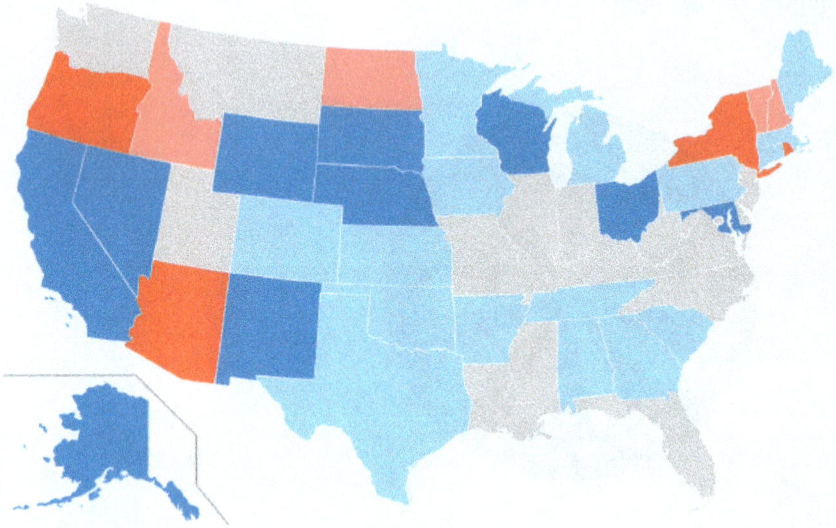

Image 2.33

If the size of Texas is 268,820 sq. miles and the size of the contiguous United States is about 3,119,885 sq. miles, what is the probability the family will be vacationing in Texas next summer? Express your answer as a percent rounded to 1 decimal place.

3. The specifications for the official flag of Japan state the ratio of width to height be 3:2. The ratio of the diameter of the circle to the height is 3/5 to 2.

What is the probability of a fly randomly landing in the red circle?

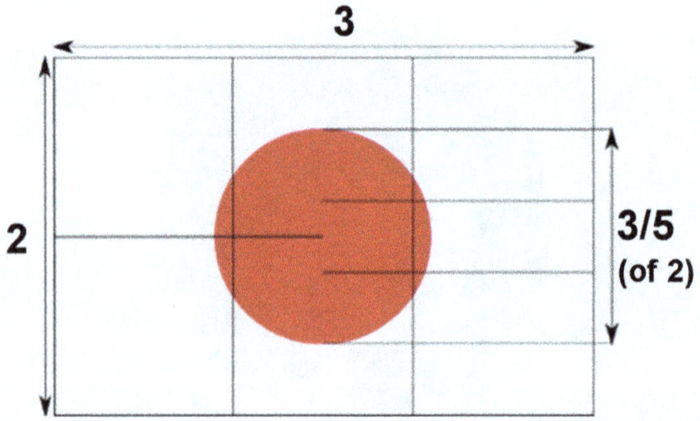

Image 2.34

Now that you understand odds and probability, and know how to generate random numbers, you're ready to program your first game. Check out the Chapter Projects to get started.

Chapter Projects

I. Rock-Paper-Scissors, aka Roshambo

This is a 2-player hand gesture game in which each player simultaneously forms a paper, scissors, or rock. The following diagram illustrates which gesture wins in each pair:

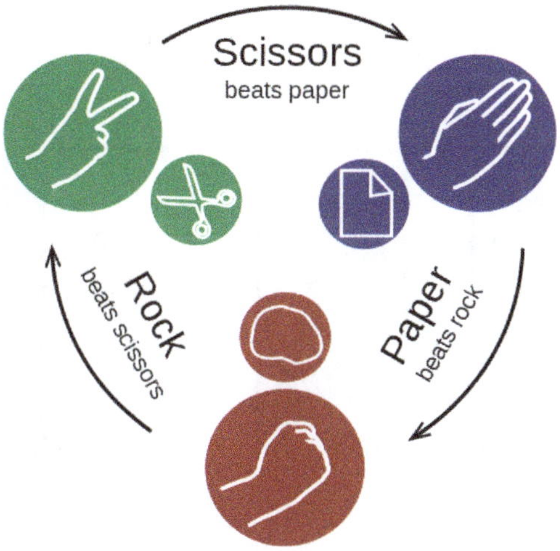

Image 2.35

Write a program that asks the user (Player 1) to input either *P* (paper), *S* (scissors) or *R* (rock). Randomly generate *P*, *S*, or *R* for the hypothetical 2nd player. Determine and output the winner.

SAMPLE OUTPUT

Player 1

P/S/R? R

Player 2

S

Rock crushes scissors.

Player 2 wins.

II. Monty Hall Problem

A well-known (and highly disputed) math problem is the "Monty Hall", aka "Let's Make a Deal" problem. The problem has appeared in several movies and TV shows in which screenwriters attempt to demonstrate that math is fascinating (*21,* starring Kevin Spacey, is one of them.)

The question is based on a TV game show from the 1960s and 1970s (hosted by Monty Hall) in which a contestant is shown 3 closed doors. He/she is informed that behind two of the doors are goats, and one of the doors is a new car.

Image 2.36

The contestant requests a door, so the host opens one of the OTHER doors to show a goat behind it. He then asks the contestant if he wants to stick with his original guess, or change his answer.

What is the probability of winning the car if the contestant sticks?

It seems the most intuitive answer is ½ or 50%, because there are only two doors left. Or is it? The question was first posed in *Parade* magazine in 1990 in a column written by Marilyn vos Savant, who was believed to have the highest recorded IQ. (I beg to differ—the person with the highest IQ would be the person who composed the IQ test questions.) Vos Savant's answer inspired collective mathematical outrage throughout the land and continues to confound.

Write a program to simulate the game show as follows.

1. Ask user to input a number n, which is number of trials (i.e., number of games).
2. Initialize variable **wins** to 0. This will keep track of how many times the contestant wins a car.
3. Repeat the following steps n times.
4. Initialize a list **doors** with {0, 0, 0}. *(0 will represent goat, 1 will represent car.)*
5. Assign to variable r a random number between 1 and 3. Set doors $[r - 1] = 1$ to represent that it hides a car.
6. Select a random number between 1 and 3, which is the contestant's choice of doors.

7. Reveal what is behind one of the other two doors (make sure it is a goat).
8. Assume the contestant wants to stick with the original door choice. Reveal the other of the two doors that has not been opened yet.
9. If it is a car, increment variable **wins** by 1.
10. Go back to Step 4 and repeat.
11. After looping *n* times, output the percent of "wins." This represents the probability of winning a car when the contestant sticks with the original choice.

    Are the results what you expected?

III. Zeckendorf's Theorem

A. Recall the **Fibonacci sequence**: 1, 1, 2, 3, 5, 8, 13, 21, 34, ...

Zeckendorf's theorem asserts that every positive integer can be expressed as the sum of one or more Fibonacci numbers. For example:

$$64 = 55 + 8 + 1$$

$$64 = 34 + 21 + 8 + 1$$

$$64 = 55 + 5 + 3 + 1$$

Write a program in which the user can enter any positive integer between 1 and 1,000. Calculate and output every possible Zeckendorf sum.

B. There is a second part of the theorem: uniqueness. This says that there is only one representation in which the sum does not include any two consecutive Fibonacci numbers. Indicate this sum with an asterisk (use multiplication symbol).

SAMPLE OUTPUT

n? 64

*64 = 55 + 8 + 1

64 = 34 + 21 + 8 + 1

64 = 55 + 5 + 3 + 1

IV. Find the Intersection and Union

Write a program that asks the user to input two sets: A and B. By comparing sets, calculate and output A U B and $A \cap B$.

SAMPLE OUTPUT

Set A? 1, 4, 9, 12

Set B? 2, 3, 4, 6, 8, 9, 12

A U B = {1, 2, 3, 4, 6, 8, 9, 12}

A ^ B = {4, 9, 12}

Bonus:
Ask the user to input a third set U (a universal set that contains both A and B). Calculate and output A', B', $A' \cup B'$ and $(A \cap B)'$.

SAMPLE OUTPUT

Set U? 1, 2, 3, 4, 5, 6, 7, 8, 9, 10, 11, 12

Set A? 1, 4, 9, 12

Set B? 2, 3, 4, 6, 8, 9, 12

A' = {2, 3, 5, 6, 7, 8, 10, 11}

B' = {1, 5, 7, 10, 11}

A' U B' = {1, 2, 3, 5, 6, 7, 8, 10, 11}

(A ^ B)' = {1, 2, 3, 5, 6, 7, 8, 10, 11}

V. Twenty-Four Game

A popular math game (which some believe is a contradiction in terms) involves being given four randomly generated integers between 1 and 9. The objective is to form an expression using add, subtract, multiply, divide, and/or exponents to form exactly 24.

Here are some examples:

Random Numbers:	Possible Solution:
1, 2, 3, 4	(1 + 2 + 3) * 4
1, 5, 6, 9	(6 * (9 – 5)) * 1
6, 7, 8, 9	(6 * 8) / (9 – 7)

Write a program that will randomly generate 4 integers between 1 and 9. Ask the user to input an expression using each of the 4 numbers exactly once. Verify that their expression indeed sums to 24. Also make sure that only those 4 numbers are used: no more, no less.

SAMPLE OUTPUT

Your job is to form 24 using each number in [2, 7, 9, 6] exactly once

Expression that equals 24? 9 + 6 + 7 + 2

Great job!

Your job is to form 24 using each number in [2, 3, 4, 5] exactly once

Expression that equals 24? 2 * 3 + 4 * 5

Sorry, 26 does not equal 24

Your job is to form 24 using each number in [7, 8, 9, 4] exactly once

Expression that equals 24? 7 + 8 + 9

Sorry, you did not use 4

Your job is to form 24 using each number in [2, 6, 8, 4] exactly once

Expression that equals 24? (6/2 + 8/4)**2

Sorry, you cannot use 2 more than once

VI. Sigma Derby

Sigma Derby is a vintage gambling machine in which 5 mechanical horses "gallop" around a horse racing track. At the start of each game, winning odds are randomly chosen for each pair of horses and displayed (in red LED letters!) Each player then studies the odds (in 30 seconds or less) and bets any number of quarters on which horses they think will place first and second (in any order). The miniature horses then race around the track using gears and motors.

The odds range from 2-1 (a sure thing) to 200-1 (a long shot). As you know, the number of ways to choose 2 out of 5 horses is the combination $\binom{5}{2} = 10$, which is why 10 sets of odds are displayed.

Image 2.37

Your task as a programmer is to write a program that simulates this exciting game as follows:

A. Generate 10 random sets of odds for each pair of horses, anything between 2-1 and 200-1. The lower odds (such as 1:3, 2:3, etc.) should be more probable than the longshot odds (such as 100:1, 200:1).

   Output the odds for each pair of horses, and ask the player to place a bet (must be one or more quarters).

B. Randomly choose a winning pair of horses, using the probabilities associated with each set of odds generated in step A. (For example, if the odds are 1:3, the horse combination has a 1 out of 4 chance of winning.)

C. If the player chose the winning combination, award him/her with winnings based on the odds.

D. If the player did not win, tell them they lost, and which horse combination actually won.

As game manufacturers rarely, if ever, reveal their algorithms, it is up to you to devise a reasonable simulation.

CHAPTER SUMMARY

PYTHON CONCEPTS

Set	A set of immutable (cannot be changed) unordered elements. The elements can be any type. Sets are enclosed in braces. *Ex.:* *A = {1, 2, 'three'}*
Set Properties and Methods	A.**union**(B) A.**intersection**(B) A.**issubset**(B) A.**difference**(B) A.**isdisjoint**(B) **len**(A)
List	A set of ordered elements. The elements can be any type. Each element is indexed starting with 0. Lists are enclosed in square brackets. *Ex:* *If L = [1, 2, 'three'] then L[0] = 1, L[1] = 2, L[2] = 'three'*
List Properties and Methods	L.**append**(x) L.**clear**() L.**pop**() **len**(L)
Splitting a String	string.**split**('delimiter') splits a longer string into a list of shorter string(s) separated by the given delimiter. *Ex:* *If name = 'edgar,allan,poe' then* *name.split(',') returns ['edgar', 'allan', 'poe']*
Factorial	**factorial**(n) *(math library)* *Ex: **factorial**(5) = 120*
Random Integer	**randint**(a,b) *(random library)* generates a random integer between *a* and *b*.

(*Continued*)

User-Defined Functions	A function is defined in this form:
	def function_name (variable1, variable2, etc.):
	    statement 1
	    statement 2
	    etc.
	    return [optional variable(s)]
	Ex:
	***def** cost(age):*
	    if age < 18:
	        c = 10
	    else:
	        c = 15
	    ***return** c*
	
	# main program
	***print**(cost(21))*

PRECALCULUS CONCEPTS

Sets	A list of unordered elements. *Ex:* $L = \{1, 2, 4, 3\}$
Operations on Sets: **Intersection, Union**	*Ex:* *If* $A = \{1, 2, 3\}$ *and* $B = \{3, 4, 5\}$ *Then* $A \cap B = \{3\}$ *and* $A \cup B = \{1, 2, 3, 4, 5\}$
Elements and Subsets	*Ex:* *If* $A = \{1, 2, 3\}$ *Then* $1 \in A$ *and* $\{1\} \subset A$
Null (Empty) Set	{ } or Ø
Equal Sets	*If* $A = \{1, 2, 3\}$, $B = \{3, 2, 1\}$ *Then* $A = B$
Disjoint Sets	Sets with no common elements, i.e. $A \cap B = \emptyset$

Difference of Two Sets	A \ B refers to elements in A that are not in B. *Ex:* *If* $A = \{1, 2, 3\}$ *and* $B = \{3\}$ *Then* $A \setminus B = \{1, 2\}$
Number of Subsets	A set with n elements has 2^n subsets. *Ex:* $A = \{1, 2, 3, 4, 5\}$ *has* $2^5 = 32$ *subsets.*
Number of Elements in a Set	$n(A)$
Counting Formula	If A and B are finite sets, then $n(A \cup B) = n(A) + n(B) - n(A \cap B)$ *Ex:* *If* $A = \{1, 2, 3\}$, $B = \{3, 4, 5\}$ *Then* $n(A \cup B) = 3 + 3 - 1 = 5$.
Multiplication Principle of Counting	If a menu consists of n categories, and each category has p_i choices, then the total possible choices for choosing one item from each category is the product $p_1 \cdot p_2 \cdots p_n$. *Ex:* *You have 5 shirts, 3 pants, and 2 pairs of shoes. You have* $5 \cdot 3 \cdot 2$ *different outfits.*
Factorial	$n! = n \cdot (n-1) \cdot (n-2) \cdots 2 \cdot 1$ *Ex:* $5! = 5 \cdot 4 \cdot 3 \cdot 2 \cdot 1 = 120$ $0! = 1$ *(special case)*
Permutation	${}_nP_k$ refers to the number of ordered subsets of k elements chosen from n elements. $${}_nP_k = \frac{n!}{(n-k)!}$$ or $${}_nP_k = n \cdot (n-1) \cdot (n-2) \cdots (n-k+1)!$$ *Ex:* $${}_5P_3 = \frac{5!}{2!} = \frac{5 \cdot 4 \cdot 3 \cdot 2!}{2!} = 60$$ *or:* $${}_5P_3 = 5 \cdot 4 \cdot 3 = 60$$

(Continued)

Combination	${}_nC_k$ refers to the number of unordered subsets of k elements chosen from n elements. $${}_nC_k = \frac{nP_k}{k!}$$ or $${}_nC_k = \frac{n!}{(n-k)!k!}$$ *Ex:* $${}_5C_3 = \frac{5P_3}{3!} = \frac{60}{6} = 10$$ *or:* $${}_5C_3 = \frac{5!}{(5-3)!3!} = \frac{120}{2!3!} = \frac{120}{12} = 10$$
Probability Terms: **Experiment, Outcome, Sample Space**	**Experiment**—an event such as flipping a coin **Outcome**—a possible result of the experiment such as "heads" **Sample Space**—set of all possible outcomes, such as {"heads", "tails"}
Odds	If the probability of an outcome is p / q, then the odds in favor of the outcome are p to $(q - p)$. *Ex:* *If the probability of a horse winning a race is 3/5, the odds in favor of winning are 3:2.*
Probability of a Union of Mutually Exclusive Events	If E and F are mutually exclusive events (i.e., $E \cap F = \emptyset$) then $$P(E \cup F) = P(E) + P(F)$$ *Ex:* *The probability of rolling a 1 or a 6 with a 6-sided die is:* $$P(1) + P(6) = \frac{1}{6} + \frac{1}{6} = \frac{1}{3}$$

Probability of a Union of Two Events	$P(E \cup F) = P(E) + P(F) - P(E \cap F)$ *Ex:* *The probability of selecting a queen or a heart from a standard 52-card deck is:* $P(Q) + P(H) - P(Q \text{ of } H) =$ $\frac{4}{52} + \frac{13}{52} - \frac{1}{52} = \frac{16}{52} = \frac{4}{13}$
Probability of an Intersection of Two Events	$P(E \cap F) = P(E) \cdot P(F)$ *Ex:* *The probability of tossing a coin twice and getting "heads" both times is:* $P(H) \cdot P(H) = \frac{1}{2} \cdot \frac{1}{2} = \frac{1}{4}$
Complement of an Event	If E' is the complement of E, then: $P(E') = 1 - P(E)$ *Ex:* *If a jar with 100 jelly beans contains 7 white jelly beans, then the probability of randomly selecting a jelly bean that is not white is* $1 - \frac{7}{100} = \frac{93}{100}$
Conditional Probability	The probability of event E given event F is: $P(E \mid F) = \frac{P(E \cap F)}{P(F)}$ *Ex:* *You randomly select a card from a standard deck of 52 cards and you know that it is red. The probability that it is a red queen is:* $P(RQ \mid R) = \frac{P(R \cap Q)}{P(R)} = \frac{2/52}{13/52} = \frac{2}{13}$

(Continued)

Experiential Probability	Probability based on previous experience or statistics. *Ex:* *A vaccine was effective on 950 out of 1000 test subjects. Therefore, the effectiveness is 95%.*
Geometric Probability	Probability based on the ratio of the area of a figure to the area of the larger figure that contains it. *Ex: If the area of the red circle in the Japanese flag is 9π / 25, and the area of the flag is 6, then the probability of randomly choosing a point inside the circle is* $\frac{9\pi/25}{6} = \frac{3\pi}{50}$.

Credits

IMG 2.1: Adapted from Dmitry Fomin, https://commons.wikimedia.org/wiki/File:English_pattern_playing_cards_deck.svg, 2017.

IMG 2.2a: Copyright © 2021 Depositphotos/Mhpixshot.

IMG 2.2b: Copyright © 2013 Depositphotos/eldadcarin.

IMG 2.3: Copyright © by Marcelo Reis (CC BY-SA 3.0) at https://commons.wikimedia.org/wiki/File:Set_union.png.

IMG 2.4: Copyright © by Marcelo Reis (CC BY-SA 3.0) at https://commons.wikimedia.org/wiki/File:Set_intersection.png.

IMG 2.5: すじにくシチュー, https://commons.wikimedia.org/wiki/File:Universal_set_and_complement.svg, 2019.

IMG 2.9: Adapted from Ines, https://commons.wikimedia.org/wiki/File:People.svg, 2006.

IMG 2.15: Personeoneste, https://commons.wikimedia.org/wiki/File:Dice.svg, 2010.

IMG 2.16: Film8ker, https://commons.wikimedia.org/wiki/File:American_roulette_wheel_layout.png, 2007.

IMG 2.17: Copyright © 2012 Depositphotos/mannaggia.

IMG 2.18: Copyright © 2015 Depositphotos/destinacigdem.

IMG 2.19: Copyright © 2019 Depositphotos/plahotya.

IMG 2.20: Copyright © 2022 Depositphotos/Piscine.

IMG 2.21: Copyright © 2016 Depositphotos/OtaraevChermen.

IMG 2.24: Copyright © 2021 Depositphotos/macrovector.

IMG 2.25: Copyright © 2016 Depositphotos/Polly_Grimm.

IMG 2.26: Peter Paul Rubens, https://commons.wikimedia.org/wiki/File:Baroque_Rubens_Assumption-of-Virgin-3.jpg, 1626.

IMG 2.31: Copyright © by Mtcv (CC BY-SA 3.0) at https://commons.wikimedia.org/wiki/File:Dart_board.png.

IMG 2.33: Copyright © by Elli (CC BY-SA 4.0) at https://commons.wikimedia.org/wiki/File:1958_United_States_gubernatorial_elections_results_map.svg.

IMG 2.35: Copyright © by Enzoklop (CC BY-SA 3.0) at https://commons.wikimedia.org/wiki/File:Rock-paper-scissors.svg.

IMG 2.36: Adapted from Cepheus, https://commons.wikimedia.org/wiki/File:Monty_open_door.svg, 2006.

IMG 2.39: Copyright © by daryl_mitchell (CC BY-SA 2.0) at https://www.flickr.com/photos/daryl_mitchell/23612364253.

CHAPTER 3

Operations on Polynomials

3.1 Basics: Definition and Evaluation

The Basics of Functions

It is assumed that you learned the basics of functions in a prerequisite course. Here are the key points.

A **function** is a relation (correspondence between two sets X and Y) in which each element of set X is paired with at most one element of set Y.

A function can be represented in the following ways:

Set of Ordered Pairs

e.g., $\{(-2, 6), (-1, 3), (0, 2), (1, 3), (2, 6)\}$

Formula

e.g., $f(x) = x^2 + 2$

Graph

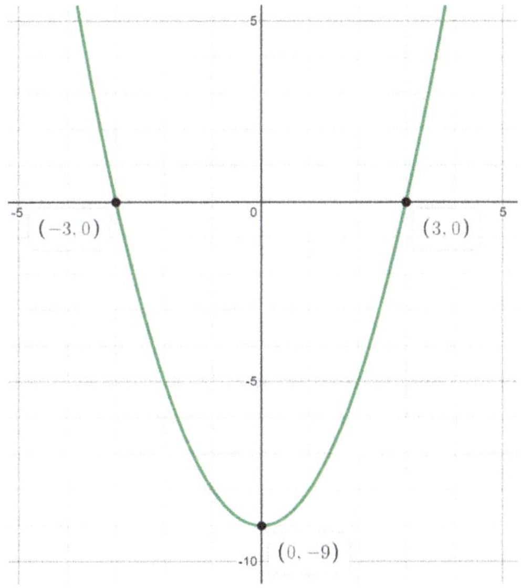

Image 3.1

$f(x)$ is **function notation** read "f of x." It is *not* the product of f and x. $f(x)$ represents a single value, specifically the y-value that corresponds to the x-value.

For example if $f(x) = 3x^2 - 12|x|$, then $f(-2) = 3(-2)^2 - 12|-2| = -12$. This corresponds to the ordered pair $(-2, -12)$.

The **vertical line test** can be used to determine if a graph represents a function. If any vertical line crosses the graph at more than one point, it is *not* a function.

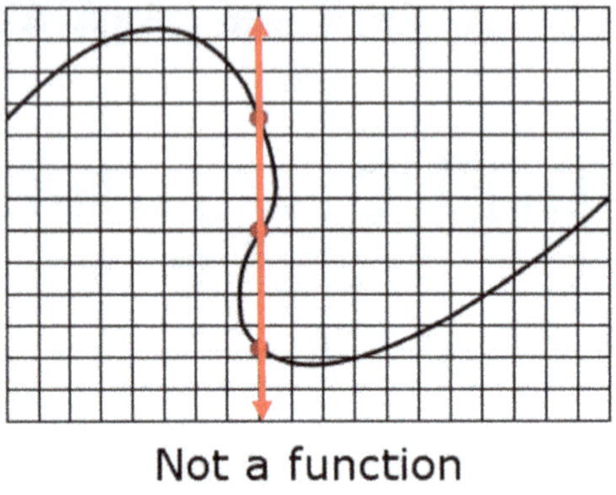

Image 3.2

The **domain** refers to the set of x-values; the **range** refers to the set of y-values. The domain and range are often expressed in interval notation:

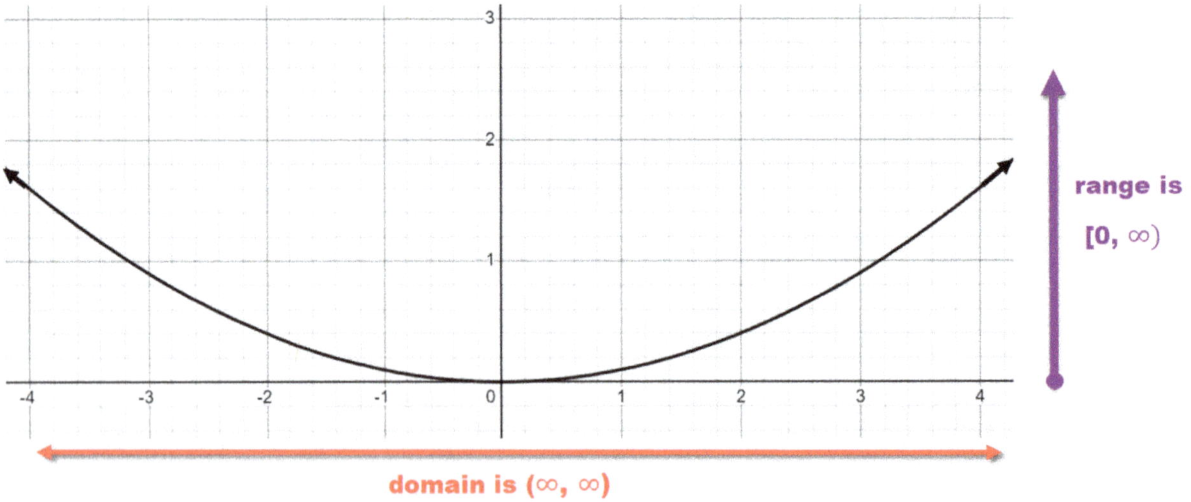

Image 3.3

Functions can be combined in the following ways.

OPERATIONS ON FUNCTIONS

Sum: $(f+g)(x) = f(x) + g(x)$

Difference: $(f-g)(x) = f(x) - g(x)$

Product: $(fg)(x) = f(x) \cdot g(x)$

Quotient: $\left(\frac{f}{g}\right)(x) = \frac{f(x)}{g(x)}$, if $g(x) \neq 0$

Example

Let $f(x) = x^2 + x - 1$ and $g(x) = 2x + 1$.

A. Find $(f + g)(-3)$

B. Find $(f - g)(0)$

C. Find $\left(\frac{f}{g}\right)\left(-\frac{1}{2}\right)$

D. Find $f(-x)$

E. Find $f(x + h)$

Solutions

A. $(f + g)(-3) = f(-3) + g(-3) = [(-3)^2 + (-3) - 1] + [2(-3) + 1] = 5 + -5 = 0$

B. $(f - g)(0) = f(0) - g(0) = [(0)^2 + (0) - 1] - [2(0) + 1] = -1 - 1 = -2$

C. $\left(\frac{f}{g}\right)\left(-\frac{1}{2}\right) = \frac{f(-1/2)}{g(-1/2)} = \frac{\left(-\frac{1}{2}\right)^2 + \left(\frac{1}{2}\right) - 1}{2\left(-\frac{1}{2}\right) + 1} = \frac{-\frac{1}{4}}{0} =$ undefined (can't divide by zero)

D. $f(-x) = (-x)^2 = x^2$

E. $f(x + h) = (x + h)^2 + (x + h) - 1 = x^2 + 2xh + h^2 + x + h - 1$

EXERCISE SET 3.1.1

In #1–2, identify if the graphs represent functions.

1.

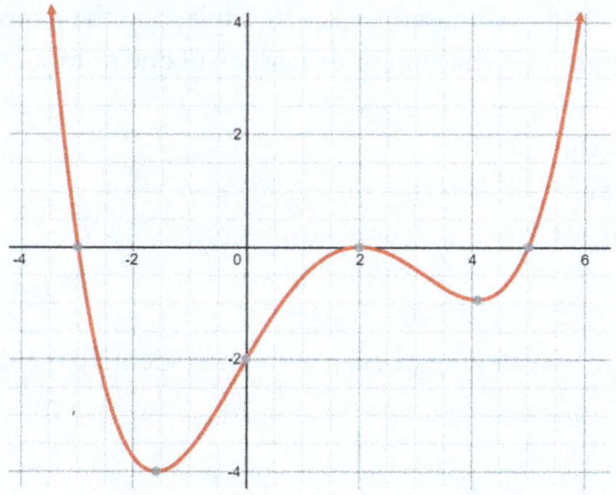

Image 3.4

2.

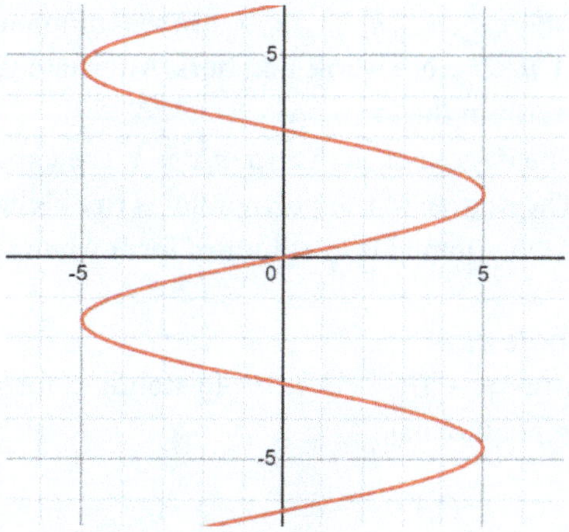

Image 3.5

(*Continued*)

3. Find the domain and range of the graph in Exercise 1 above.

In #4–9, assume $f(x) = 2x^3$ and $g(x) = 1 - 3x$.

4. Find $f(-2)$
5. Find $g(-x)$
6. Find $f(0) \cdot g(0)$
7. Find $g(x + h)$
8. Find $f(-x) - g(-x)$
9. Find $-g(x)$
10. Is $f(-x)$ the same as $-f(x)$?

In this chapter, we focus on a specific type of function—polynomial functions. Examples of polynomials are:

$3x^2 + 9x - 11$

$5x - 99$

$4x^5 + 4x^3 - 9x$

17

A **polynomial** is a mathematical expression in the form

$$a_n x^n + a_{n-1} x^{n-1} + a_{n-2} x^{n-2} + \ldots + a_1 x + a_0$$

where $a_n, a_{n-1}, a_{n-2}, \ldots a_1, a_0$ are real numbers (called the **coefficients**), x is a variable, and the exponents $n, n-1, n-2, \ldots$ are whole numbers. Alternately, a polynomial is the sum or difference of one or more terms, each term in the form $a_k x^k$.

The **degree** of each term $a_k x^k$ is k, the exponent of the variable.

The **degree** of the polynomial is the highest degree of all its terms.

A polynomial is in **standard form** when the terms are written in order from greatest to least degree.

Example

Rewrite $4x^3 - 17 - 9x^5 + 5x^{10}$ in standard form, identify the degree of each term, and identify the degree of the polynomial.

Solution

Standard form: $5x^{10} - 9x^5 + 4x^3 - 17$

Term	Degree of Term
$5x^{10}$	10
$-9x^5$	5
$4x^3$	3
-17	0

Thus, the degree of the polynomial is 10.

Note that any constant, such as -17, has degree zero because $-17 = -17(1) = -17x^0$.

Some polynomials have special names, depending on how many terms they have and their degree. Here are the most common.

# of terms	Name
1	Monomial
2	Binomial
3	Trinomial

Degree	Example	Name
0	11	Constant
1	$5x + 4$	Linear
2	$7x^2 - 3x + 5$	Quadratic
3	$41x^3 + 4x^2 - 9$	Cubic
4	$2x^4 - x$	Quartic

Some math textbooks consider the number 0 to have no degree, rather than 0 degree.

EXERCISE SET 3.1.2

Identify whether each expression is a polynomial. If so, identify the name of the polynomial (based on number of terms and degree) and write it in standard form.

1. $11x^2 + 47x^3 - 31 + 8x^4$
2. $3x^3 - \pi x$
3. $\dfrac{1}{12x^2 - 14x + 2}$
4. $\sqrt{12}x^2 + 3x^3$
5. $4x$

Evaluating a Polynomial at a Given Value

To introduce us to the concept of inputting and manipulating a polynomial, we will write a program that asks the user to input a polynomial and a value of x, and then evaluate the polynomial at the given value of x.

For example, if the polynomial is $5x^4 - 7x^3 + x^2 - 11$, its value at $x = 2$ would be $5(2)^4 - 7(2)^3 + (2)^2 - 11 = 17$.

Python has an extremely powerful library of symbolic mathematics functions and methods called **SymPy.** Here is a simple program that evaluates $5x^4 - 7x^3 + x^2 - 11$ at $x = 2$.

PROGRAM 3.1.1 EVALPOLY.PY

```
from sympy import *
x = Symbol('x')                        # identifies x as an algebraic symbol
f = sympify('5*x**4-7*x**3+x**2-11')   # identifies f as a SymPy type expression
a = 2
value = f.subs({x:a})                  # evaluates f(a)
print(value)
```

The output will of course be 17. Notice that we must identify the **symbol** (x) and the **SymPy** expression (f).

Recall our "dot" notation: **.subs()** is a *method* performed on object f. It is similar to a function.

This program will be infinitely more useful if we allow the user to input a polynomial f and value a. (Changes shown in blue font.)

PROGRAM 3.1.1 EVALPOLY.PY REVISED

```
from sympy import *
x = Symbol('x')
f = sympify(input('Polynomial? '))
a = input('a? ')
value = f.subs({x:a})
print('f(', a, ') = ', value)
```

Sample Output

Polynomial? 17*x**4-3*x**2+4*x-11
a? -5
f(-5) = 10519

EXERCISE SET 3.1.3

Use program **evalpoly** to evaluate the following. You may need to modify the symbol.

1. $3x^3 + 8x^2 + 7x + 1$ at $x = -2$
2. $3n^2 + 7n - 10$ at $n = -3$
3. $2a^3 - 3a^2 - 3a$ at $a = 2$
4. $-n^2 + 9$ at $n = -3$
5. $y^3 - 5y^2 + 9y - 13$ at $y = 3$
6. $f(3)$ given $f(x) = x^2 - 6x + 7$
7. $f(-4)$ given $f(n) = n^4 + 10n^3 + 22n^2 - 15n - 22$

Here is a summary of the most basic **SymPy** functions and methods:

TABLE 3.1 Symbolic Manipulation (SymPy library)

To define x as an algebraic variable (also known as a **symbol**):

```
x = Symbol ('x')
```

To create a **SymPy** expression from a string:

```
poly = sympify('x+2*x+x**2+7+4*x**2')
```

sympify will automatically combine like terms

To "pretty print" the output:

```
>>>pprint(poly)
5·x² + 3·x + 7
```

Note the exponent is superscripted and the terms are padded with spaces in between.

To evaluate an expression for a specific value of the symbol (in this example −1):

```
>>>value = polynomial.subs({x:-1})
9
```

To input an algebraic expression as a **SymPy** type:

```
poly = sympify(input("Polynomial? "))
```

A polynomial such as $152x^7 - 81x^6 + 14x^5 - 35x^4 + 6x^3 - 11x^2 + 45$ would be rather time-consuming to input: 152*x**7-81*x**6+14*x**5-35*x**4+6*x**3-11*x**2+45. One keyboard typo or bug in a program would necessitate you inputting this cumbersome string again (and again ...). It would be much more elegant to simply input the coefficients:

152, −81, 14, −35, 6, −11, 0, 45

Since we will be inputting and manipulating quite a few polynomials in this chapter, it will be worth our while to develop a function that will convert a list of coefficients to a **SymPy** polynomial expression.

First, we will introduce the concept of **string concatenation**, which is accomplished with the plus (+) sign.

STRING CONCATENATION

To join together two string constants:

```
message = "hello" + "world"
```

To join together a string constant and a string variable:

```
name = "Lisa"
message = "hello" + name
```

To join together a string and a number:

```
age = 21
message = "I am " + str(age) + " years old"
```

Next, we need a means to convert a string input '152, −81,14, −35, 6, −11, 0, 45' into a list of integers: [152, −81, 14, −35, 6, −11, 0, 45]. We saw in Chapter 2 that we can use the **.split()** method to split a string into component strings at a given delimiter. For example:

```
co_str = '152,−81,14,−35,6,−11,0,45'
a = co_str.split(',')
```

would result in a = ['152', '−81', '14', '−35', '6', '−11', '0', '45'], which is a list of strings.

We could then use a **for** loop to convert each string in the list to an integer, like so:

```
a = [ ]
for n in co_str:
    n = int(n)
    a.append(n)
```

However, Python has a technique called **List Comprehension** that can combine all four of those lines into a single command:

```
a = [int(n) for n in co_str.split(',')]
```

Thus, streamlining the conversion process as follows:

```
# convert a string of coefficients to a list of integers
co_str = '152,–81,14,–35,6,–11,0,45'
a = [int(n) for n in co_str.split(',')]
```

EXERCISE SET 3.1.4

1. Write a program that asks the user to input a string of integers separated by commas. Convert to a list of integers and output the list.
2. Modify the program so that it handles float numbers in addition to integers.
3. Write a program that asks the user to input a list of dogs, separated by forward slashes. Convert to a list of strings and output the list.

Sample Output

Dogs? Corgi/Beagle/Poodle/Labrador

list of dogs: ['Corgi', 'Beagle', 'Poodle', 'Labrador']

4. Write a program that assigns to list **coeff** the following values: [6, 7, 8, 9, 10]. Use those values to concatenate and output the following polynomial:

   6*x**5 + 7*x**4 + 8*x**3 + 9*x**2 + 10

Now that we can convert a string of integers to a list of integers, we would like to convert that list to a polynomial. Here is an example:

String List	List of Integers	Polynomial
co_str = '5, –3, 2, 0, 1'	coeff = [5, –3, 2, 0, 1]	poly = $5x^4 - 3x^3 + 2x^2 + 0x + 1$

Observe that **coeff** [0] is the coefficient of x^4, **coeff** [1] is the coefficient of x^3, **coeff** [2] is the coefficient of x^2, etc. The polynomial could then be parsed as:

poly = **coeff** [4]*x**4 + **coeff** [3]*x**3 + **coeff** [2]*x**2 + ...

A list of 5 integers thus results in a 4th-degree polynomial, so we begin by setting the first exponent to one less than the degree.

PROGRAM 3.1.2 NUM_STR_TO_POLY.PY

```
from sympy import *
x = Symbol('x')
co_str = input('Coefficients? ')
coeff = [int(n) for n in co_str.split(',')]         # convert the string to a list of integers
degree = len(coeff) - 1                             # degree is one less than the size of coeff
poly_str = ''                                       # initialize the polynomial string
for n in range(len(coeff)):
    if n!=0:                                        # if not the first term, append a '+' sign
        poly_str +='+'
    poly_str += str(coeff[n])+'*x**' + str(degree)
    degree -= 1
poly = sympify(poly_str)
print(poly)                                         # convert to a SymPy type
```

SAMPLE OUTPUT

Coefficients? 7, −9, 11, 0, 1

7*x**4 − 9*x**3 + 11*x**2 + 1

EXERCISE SET 3.1.5

1. How would the following polynomials be represented as a list of digits?

   A. $x^3 - x + 1$
   B. $-x^5 + x^4 + 1$
   C. x^4

2. Change the **print**() function in the program so that the output is in this form:

   $5 \cdot x^4 - 7 \cdot x^3 + x^2 - 11$

3. Run the program with the following lists of coefficients. What is the resulting polynomial?

   A. 3, 11, 0, 19, 4
   B. −6, 0, 0, 5
   C. 1, 0, 0, 0, 0, 0, 0

4. Does **sympify**() always produce a polynomial in standard form? (i.e., exponents in decreasing order)
5. Modify the program so that the user is also asked to input a value **a** to be substituted into the polynomial.

Sample Output

Coefficients? 5,6,7,8

a? −19

5*x**3 + 6*x**2 + 7*x + 8

f(−19) = −32,254

Because we will be writing many programs involving the input of polynomials, let's place this code in a function called **num_list_to_poly**, which will convert a list of numeric coefficients to a **SymPy** polynomial.

```
def num_list_to_poly(coeff):
    # convert the list of coefficients to a polynomial
        degree = len(coeff) – 1
        poly_str = ' '
        for n in range(len(coeff)):
            if n!=0:
                poly_str +='+'
            poly_str += str(coeff[n])+'*x**' + str(degree)
            degree -= 1
        poly = sympify(poly_str)
        return poly
```

The function would be called from a main program like so:

PROGRAM 3.1.3 INPUT_POLYNOMIAL.PY

```
from sympy import *
x = Symbol('x')
def num_list_to_poly(coeff):
        :
        :                               ← body of function included here
# main program
co_str = input('Coefficients? ')
# convert the string to a list of integers
coeff = [int(n) for n in co_str.split(',')]
poly = num_list_to_poly(coeff)
print(poly)
```

EXERCISE SET 3.1.6

1. Modify program **input_polynomial** so that it asks for a second list of coefficients and outputs a second polynomial.
2. Write a program that asks the user to input the coefficients of a polynomial and a constant. Multiply the polynomial by that constant.

Sample Output

Coefficients? 3,–4,6,10

Constant? –2

–6*x**3 + 8*x**2 –12*x –20

3.2 Sum, Difference, Product of Polynomials

It is assumed you know from a prerequisite course how to add, subtract, and multiply polynomials. We will do a quick review so that when you write a program to perform these operations, you will know if it is working correctly or not.

Adding and subtracting polynomials is pretty straightforward. You simply add (or subtract) corresponding terms. For simplicity, from here on, when we refer to addition, it will also include subtraction because subtraction is actually defined in terms of addition:

$$a - b = a + (-b)$$

One can only add **like terms**, which are terms having the same degree, for example $7x^5$ and $-10x^5$.

EXERCISE SET 3.2.1

Add or subtract the following polynomials.

1. $(-5x^2 - 10x + 2) + (3x^2 + 7x - 4)$
2. $(3x^2 + 2x - 7) + (7x^2 - 4x + 8)$
3. $(15x^2 + 12x + 20) - (9x^2 + 10x + 5)$
4. $(14x^3 + 3x^2 - 5x + 14) - (7x^3 + 5x^2 - 8x + 10)$
5. $(4a^3 - 5a + 7) - (8a^3 - 3a - 2)$

Find the products:

6. $3a(a^3 - 4a + 11)$
7. $(14x^5 - 2x^4 + 3x^3 - 9)(-4x^2)$
8. $(7u + v)^2$
9. $(3x - 5)(2x^2 - 4x + 7)$
10. $(x + 1)(x^3 + 2x^2 + 3x - 8)$
11. $(2a^2 + 5a + 3)(a^2 - 3a - 4)$
12. Find the volume of a box with length $(x + 1)$, width $(x + 2)$, and height $(x + 3)$.
13. Find the area of a rectangle with width $(2x^2 - x - 3)$ and height $(x + 4)$.

Maintaining a Function Library

To avoid copying and pasting useful functions into each new program file, we are going to create a function library in a separate file. Let's start with **num_list_to_poly,** which we will place in a file called **poly_fcn.py**.

POLY_FCN.PY

```
# built-in modules
from sympy import sympify
def num_list_to_poly(coeff):
# converts a list of integers to a polynomial using the integers as coefficients
    degree = len(coeff)-1
    poly_str = " "
    for n in range(0,len(coeff)):
        if n!=0:
            poly_str += "+"
        poly_str += str(coeff[n]) + "*x**" + str(degree)
        degree -= 1
    poly = sympify(poly_str)
    return poly
```

Note this is not a program! It is a library of functions that will be included in other programs (files) as demonstrated in the following. When invoking functions in other files, precede the function name with the file name, like this: **file.function**().

Program Goal: Write a program **mathpoly** that will ask the user to input two polynomials, then output their sum.

SAMPLE OUTPUT

Polynomial 1? 3, 7, −11

Polynomial 2? −5, 1

Sum:

$3 \cdot x^2 + 2 \cdot x - 10$

To add two polynomials, we will use:

expand(polynomial1 + polynomial2)

The **expand**() function can be used to simplify and expand many different types of expressions.

PROGRAM 3.2.1 MATHPOLY.PY

```
# built-in modules
from sympy import *
# user-defined functions
import poly_fcn                                    # will include functions in file poly_fcn.py
x = Symbol('x')
coeff_str1 = input("Polynomial 1? ")
coeff_list1 = [int(n) for n in coeff_str1.split(',')]
polynomial1= poly_fcn.num_list_to_poly(coeff_list1)
coeff_str2 = input("Polynomial 2? ")
coeff_list2 = [int(n) for n in coeff_str2.split(',')]
polynomial2 = poly_fcn.num_list_to_poly(coeff_list2)
sum = expand(polynomial1 + polynomial2)
pprint(sum)
```

EXERCISE SET 3.2.2

1. Use the program to add $45x^6 - 17x^5 + 11x$ and $62x^6 + 120x^5 - 36x + 12$.
2. Modify program **mathpoly** so that in addition to the sum, it will also output the difference and product of the polynomials.

Sample Output

Polynomial 1? 1,0,–4

Polynomial 2? 1,–2

Sum:

$x^2 + x - 6$

Difference:

$x^2 - x - 2$

Product:

$x^3 - 2 \cdot x^2 - 4 \cdot x + 8$

*If **pprint** is not working for you (interpreter issue), try using **print** instead.*

3. Write a program that asks the user to enter expressions representing the length, width, and height of a rectangular solid ("box"). Calculate and output expressions for the volume and surface area.

   Hint: The formula for surface area of a rectangular solid is $2LW + 2LH + 2WH$.

Sample Output

Length? 3,2

Width? 5,-4

Height? 2,0,0

Volume:

$30 \cdot x^4 - 4 \cdot x^3 - 16 \cdot x^2$

Surface Area:

$32 \cdot x^3 + 22 \cdot x^2 - 4 \cdot x - 16$

3.3 Quotients of Polynomials

A ratio of polynomials $\frac{p(x)}{q(x)}$ is called a **rational function**. Let's start with the easiest case, which is a monomial divided by another monomial.

Recall that when dividing monomials, you divide the coefficients and subtract the exponents. If the resulting exponent is negative, the term with the negative exponent moves to the denominator.

$$\frac{26x^3}{2x} = 13x^2 \qquad \frac{4x}{8x^3} = \frac{1}{2x^2} \qquad \frac{15x^2}{45x^2} = \frac{1}{3}$$

Dividing a Polynomial by a Monomial

Example

Divide:

$$\frac{12x^4 - 15x^3 + 9x^2 + 6x + 3}{3x}$$

Solution

Simply divide each term in the numerator (called the **dividend**) by the monomial in the denominator (called the **divisor**).

$$\frac{12x^4 - 15x^3 + 9x^2 + 6x + 3}{3x} = \frac{12x^4}{3x} - \frac{15x^3}{3x} + \frac{9x^2}{3x} + \frac{6x}{3x} + \frac{3}{3x}$$

$$= 4x^3 - 5x^2 + 3x + 2 + \frac{1}{x}$$

EXERCISE SET 3.3.1

Divide:

1. $\frac{10x^5}{15x^3}$
2. $\frac{44x}{66x^2}$
3. $\frac{44x^3+8x^2-12x+16}{4x}$
4. $\frac{25x^4-15x^3+10x^2-5x}{10x^2}$
5. $(81x^4 + 54x^3 - 36x + 45) \div (9x)$
6. A rectangle has area $18x^3 - 36x^2 + 3x - 1$ and width $6x$. Find the length.

Dividing by a Polynomial

Example 1

Simplify $\frac{4x^2+13x-66}{4x-11}$

Solution

Factor and cancel common factors:

$$\frac{4x^2+13x-66}{4x-11} = \frac{\cancel{(4x-11)}(x+6)}{\cancel{4x-11}} = x+6$$

Image 3.6

Example 2

Simplify $\frac{3x^3-2x^2-12x+8}{x+2}$

Solution

Factor and cancel common factors:

$$\frac{3x^3-2x^2-12x+8}{x+2} = \frac{(3x-2)\cancel{(x+2)}(x-2)}{\cancel{x+2}} = (3x-2)(x-2)$$

Image 3.7

Not a pro with factoring higher degree polynomials? Try this in your Python command prompt:

```
>>>from sympy import factor
>>>factor("3*x**3-2*x**2-12*x+8")
```

Example 3

Divide $x^3 - 2x^2 - 4$ by $x - 3$.

Solution

The numerator and denominator have no common factors, thus we cannot factor and cancel. Use long division:

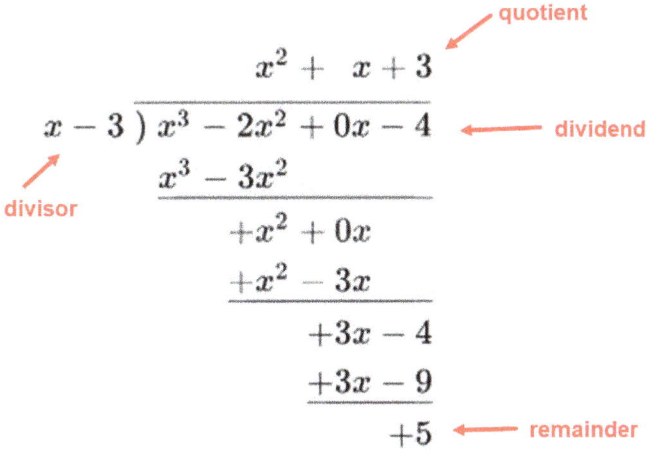

Image 3.8

This technique can be used whenever the degree of the divisor is less than the degree of the dividend. Note the degree of the remainder is less than the degree of the divisor.

The result of the long division can be expressed in either of two forms:

Form 1	$\dfrac{\textit{dividend}}{\textit{divisor}} = \textit{quotient} + \dfrac{\textit{remainder}}{\textit{divisor}}$	$\dfrac{x^3 - 2x^2 - 4}{x - 3} = x^2 + x + 3 + \dfrac{5}{x - 3}$
Form 2	$\textit{dividend} = (\textit{divisor})(\textit{quotient}) + \textit{remainder}$	$x^3 - 2x^2 - 4 = (x - 3)(x^2 + x + 3) + 5$

EXERCISE SET 3.3.2

Divide. Express your results using both forms.

1. $\dfrac{2x^3 - 3x^2 + 4x + 5}{x + 2}$

2. $\dfrac{8x^3 - 10x^2 - x + 3}{x - 1}$

3. $\dfrac{x^3 + 3x^2 - 4x - 12}{x^2 + x - 6}$

4. $\dfrac{4x^3 - 8x^2 + 7x - 11}{2x + 1}$

5. $(3x^3 - 2x^2 + 5) \div (x^2 - 1)$

Program Goal: Write a program **dividepoly** that will ask the user to input two polynomials, then divide the first by the second.

SAMPLE OUTPUT

Dividend: 1, 0, 0, −1

Divisor: 1, 0, −1

Quotient:

$$\frac{x^2 + x + 1}{x + 1}$$

Dividend: 1, 4, 4

Divisor: 1, 2

Quotient:

x + 2

PROGRAM 3.3.1 DIVIDEPOLY.PY

```
# built-in modules
from sympy import *
# user-defined functions
import poly_fcn as pf
x = Symbol('x')
coeff_str1 = input("Dividend? ")
coeff_list1 = [int(n) for n in coeff_str1.split(',')]
dividend = pf.num_list_to_poly(coeff_list1)
coeff_str2 = input("Divisor? ")
coeff_list2 = [int(n) for n in coeff_str2.split(',')]
divisor = pf.num_list_to_poly(coeff_list2)
print("Quotient:")
pprint(factor(dividend/divisor))
```

Here, **pf** is used as an **alias** (sort of a nickname) to **poly_fcn**. The program must reside in the same directory on your computer as **poly_fcn.py**.

EXERCISE SET 3.3.3

1. Use the program to simplify $x^3 + 3x^2 - 4x - 12$ divided by $x^2 + x - 6$.
2. How does the program handle a rational function such as $\frac{x^3 - 2x^2 - 4}{x - 3}$ in which the numerator and denominator have no common factors?

3. Is this program helpful in identifying the quotient and remainder when there is a remainder?
4. Replace the **pprint**(...) line with the following:

   q,r = **div**(dividend,divisor)

   pprint(q)

   print("\nRemainder:")

   pprint(r)

   What is the output for $\frac{x^3 - 2x^2 - 4}{x - 3}$?
5. Modify the program to output the result as *dividend = (divisor)(quotient) + remainder.*

Sample Output

x**3 - 2*x**2 – 4 = (x – 3)(x**2 + x + 3) + 5

The polynomial variables are treated as string variables by the **print()** function. Unfortunately **pprint()** cannot be used to parse together a combination of strings and polynomials.

As you can see, **factor**(dividend/divisor) is only useful for finding the quotient when the remainder is 0. Otherwise you will want to use **div**(dividend,divisor), which returns both the quotient and remainder as a list.

Let's improve on the output of a polynomial (using **print()**) so that something like:

x**3 – 2*x**2 – 4

is output as the slightly more readable, using the more standard carat ^ symbol to represent exponentiation:

x^3 – 2x^2 – 4

This will require two string substitutions: every occurrence of ** will be replaced with ^ and then every occurrence of * will be removed. This is accomplished with the following three function calls:

```
poly = str(poly)                      # convert poly to a string type
poly = poly.replace("**","^")
poly = poly.replace("*","")
```

Place these three statements in a function:

```
def poly_to_string(p):
    p = str(p)
    p = p.replace("**","^")
    p = p.replace("*","")
    return p
```

EXERCISE SET 3.3.4

1. Place the **poly_to_string**() function in the function library **poly_fcn**. Then use function **poly_to_string()** in program **dividepoly** so that the output for $\frac{x^3 - 2x^2 - 4}{x - 3}$ is as follows:

Sample Output

Dividend? 1,–2,0,–4

Divisor? 1,–3

x^3 – 2x^2 – 4 = (x – 3)(x^2 + x + 3) + 5

If the remainder is zero, then don't print it.

Sample Output

Dividend? 1,0,0,–1

Divisor? 1,–1

x^3 – 1 = (x – 1)(x^2 + x + 1)

3.4 Synthetic Division

If a rational function cannot be simplified (i.e., the numerator and denominator have no common factors), then long division is necessary. There is an easier alternative to long division in the special case that the divisor is in the form $x - a$.

Let's revisit:

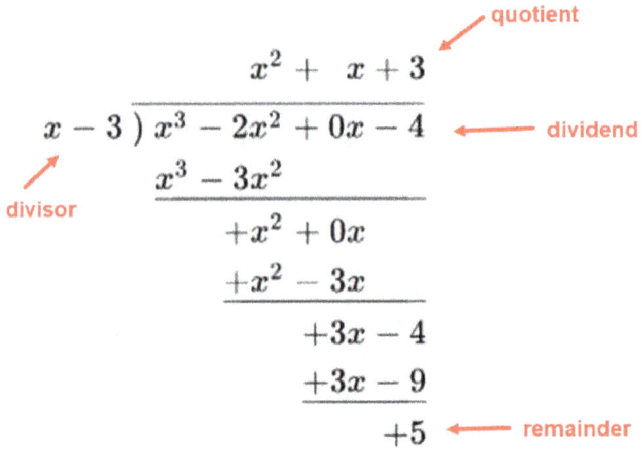

Image 3.8

The easier technique is called **synthetic division** and we'll show you how it works on the example above, which is: $(x^3 - 3x^2 - 4)$ divided by $(x - 3)$.

Set it up as a grid. The number of columns is the degree of the dividend plus 1. The number of rows is always 3.

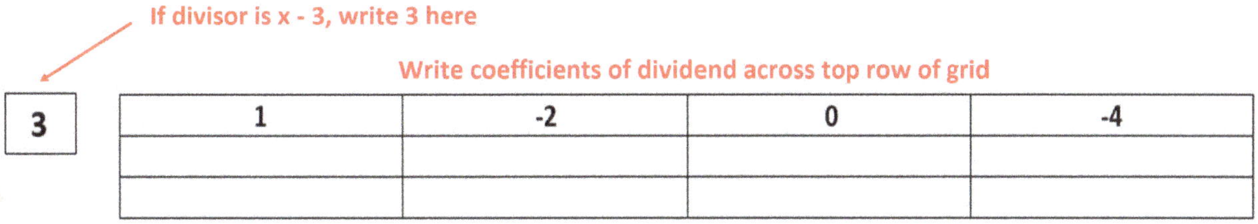

Image 3.9

We will refer to the 3 as the "corner number." Bring down the first coefficient.

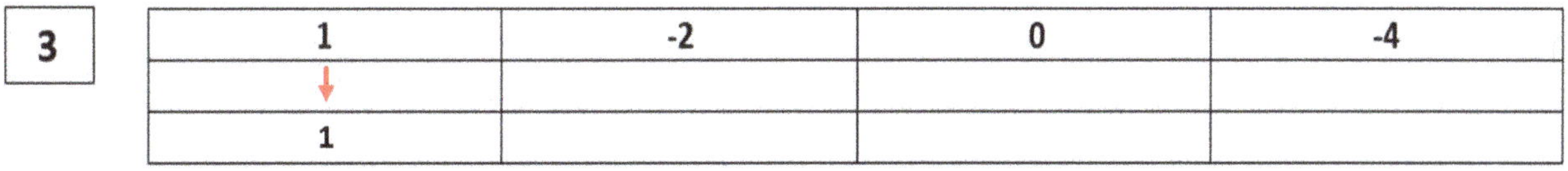

Image 3.10

Multiply the number you just brought down by the corner number, and place it in the next column as so:

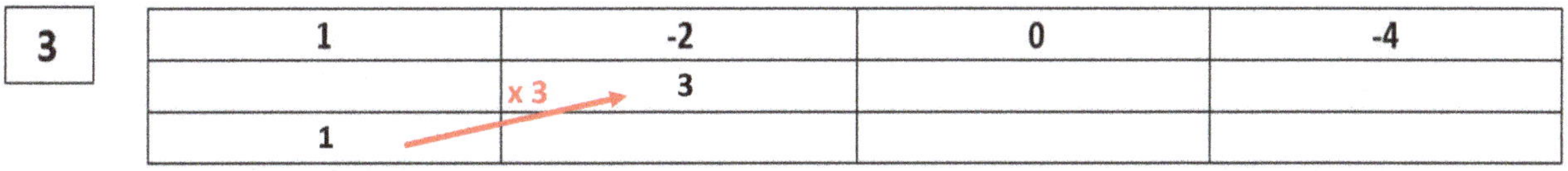

Image 3.11

You will then alternate adding down the column, then multiplying by the corner number.

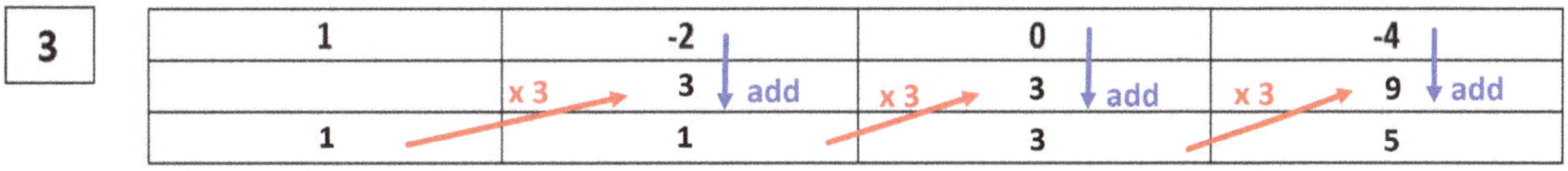

Image 3.12

Remember:

1. You must include a "placeholder" 0 for any missing term, as in the missing x term above.
2. The corner number is the number that comes after "$x -$ " in the divisor, that is, if the divisor is $x - a$, the corner number is a.

Now, how do we interpret the answer?

3	1	-2	0	-4
		3	3	9
	1	1	3	5

The first 3 columns (all except the last) give the coefficients of the quotient:

$x^2 + x + 3$

The degree of the quotient is one less than the degree of the dividend.

The number in the last column gives the remainder. Put the remainder over the divisor:

$\frac{5}{x - 3}$

Image 3.13

Thus the answer to $\frac{x^3 - 2x^2 - 4}{x - 3}$ is $x^2 + x + 3 + \frac{5}{x - 3}$.

Alternately, we can write $x^3 - 2x^2 - 4 = (x - 3)(x^2 + x + 3) + 5$.

EXERCISE SET 3.4.1

Perform synthetic division to find the quotient and remainder, if there is one.

1. $x^3 - 4x^2 + 2x - 3$ divided by $x + 2$
2. Divide $x^2 - 9x - 10$ by $x + 1$
3. $\frac{2x^3 + 5x^2 + 9}{x + 3}$
4. $\frac{3x^3 + 5x - 1}{x + 1}$
5. $x^4 - 1 \div (x + 1)$
6. Is $(x - 3)$ a factor of $x^3 - 5x^2 - 2x + 24$?
7. Is $(x - 1)$ a factor of $x^3 - 3x^2 - 10x + 24$?

Evaluating a Polynomial at a Given Value

Another interesting use for synthetic division is not for division at all. It can be used to evaluate a polynomial function at a given value.

For example, if $f(x) = x^3 - 2x^2 - 4$ and we want to evaluate $f(3)$, simply make 3 the corner number and proceed as usual:

3	1	-2	0	-4
		3	3	9
	1	1	3	5

This is f(3)!

Image 3.14

This might seem like overkill using our algorithm for synthetic division when we can simply type into our calculator: $3^3 - 2(3)^2 - 4$ and get 5. However, when a calculator is not handy and you are dealing with higher degree polynomials, this is often much simpler.

More formally:

THE REMAINDER THEOREM

The value of $f(a)$ is the remainder when $f(x)$ is divided by $(x - a)$.

EXERCISE SET 3.4.2

Use synthetic division to evaluate the following. Remember, all you care about is the remainder.

1. If $f(x) = -3x^4 + 6x^3 + 10x^2 - x$, find $f(-2)$.
2. If $f(x) = 7x^3 + 11$, find $f(5)$.
3. If $f(x) = 12x^3 - 13x^2 + 2x + 1$, find $f(7)$.
4. Write a program that asks the user to input a polynomial and a value a. Use polynomial division (not **.subs()**) to evaluate the polynomial at that value.
5. Write a program that asks the user to input a dividend and divisor. Based on the remainder, identify if the divisor is a factor of the dividend.

3.5 Binomial Expansion

Let's explore a very interesting pattern in polynomial multiplication—the case of binomial expansion.

$(a + b)^0 = 1$

$(a + b)^1 = a + b$

$(a + b)^2 = a^2 + 2ab + b^2$

$(a + b)^3 = a^3 + 3a^2b + 3ab^2 + b^3$

$(a + b)^4 = a^4 + 4a^3b + 6a^2b^2 + 4ab^3 + b^4$

$(a + b)^5 = a^5 + 5a^4b + 10a^3b^2 + 10a^2b^3 + 5ab^4 + b^5$

etc.

Notice what happens when we arrange the coefficients of the products in a pyramid:

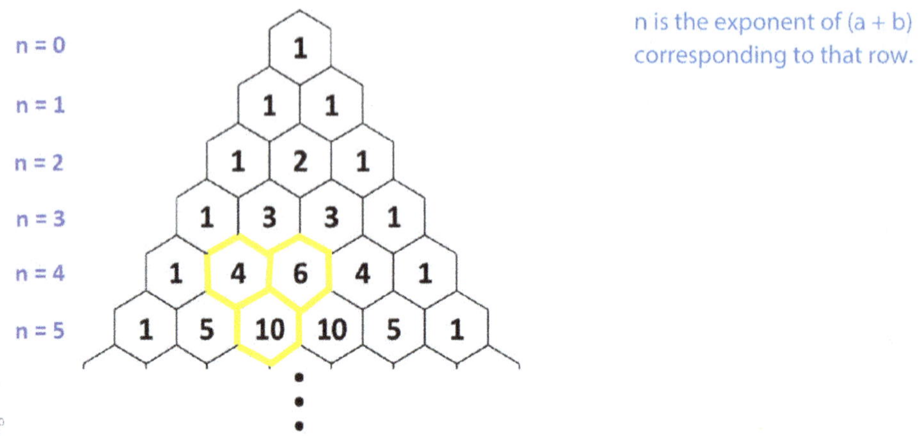

Image 3.15

If I could add another row, I would know the coefficients of the expansion of $(a + b)^6$. So how do we add on another row? Notice the three hexagons outlined in yellow. If you consider those a mouse head, then notice that the ears (4 and 6) add up to the face (10). In this manner we can obtain the next row by adding the two hexagons ("ears") above each next hexagon:

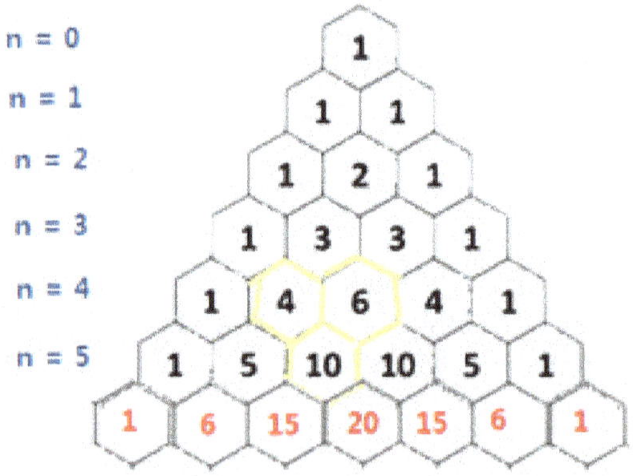

Image 3.16

This is called Pascal's triangle, credited to 17th-century French mathematician Blaise Pascal; however, it was known for centuries before that in various countries around the world. I could write an entire chapter on the various patterns and symmetry found in the triangle, but for now we'll focus on the application to binomial expansion.

EXERCISE SET 3.5.1

1. Extend Pascal's triangle down 2 more rows ($n = 7$ and $n = 8$).
2. Describe at least one other pattern found within Pascal's triangle.

We know how to obtain each new number by adding the two numbers above it, making it kind of recursive. However, there's a much easier method. Remember combinations in Section 2.3:

$$_{n}C_{k} = \frac{n!}{(n-k)!k!}$$

This function was used to calculate the number of ways to take a subset of k objects from a set of n objects, in which the order is not important. An alternate notation for $_{n}C_{k}$ is $\binom{n}{k}$. It just so happens that the combinations fit into Pascal's triangle as so:

$$\begin{array}{ccccccccccc} & & & & & \binom{0}{0} & & & & & \\ & & & & \binom{1}{0} & & \binom{1}{1} & & & & \\ & & & \binom{2}{0} & & \binom{2}{1} & & \binom{2}{2} & & & \\ & & \binom{3}{0} & & \binom{3}{1} & & \binom{3}{2} & & \binom{3}{3} & & \\ & \binom{4}{0} & & \binom{4}{1} & & \binom{4}{2} & & \binom{4}{3} & & \binom{4}{4} & \\ \binom{5}{0} & & \binom{5}{1} & & \binom{5}{2} & & \binom{5}{3} & & \binom{5}{4} & & \binom{5}{5} \end{array}$$

Image 3.17

$\rightarrow$

$$\begin{array}{ccccccccccc} & & & & & 1 & & & & & \\ & & & & 1 & & 1 & & & & \\ & & & 1 & & 2 & & 1 & & & \\ & & 1 & & 3 & & 3 & & 1 & & \\ & 1 & & 4 & & 6 & & 4 & & 1 & \\ 1 & & 5 & & 10 & & 10 & & 5 & & 1 \end{array}$$

Image 3.18

EXERCISE SET 3.5.2

1. Express the 7th row using combinations.
2. What can you deduce about $\binom{n}{0}$ and $\binom{n}{n}$ for any value of n? (Hint: Look at first and last numbers in each row.)
3. What can you deduce about $\binom{n}{1}$ and $\binom{n}{n-1}$ for any value of n? (Hint: Look at the second and second-to-last numbers in each row.)
4. Write an equation in which any number $\binom{n}{k}$ is expressed as the sum of the two "mouse ears" above it for any $\binom{n}{k}$ in which $k \neq 0$ and $k \neq n$ (i.e., not the first or last number in the row).

Remember, our interest in Pascal's triangle is that it provides the coefficients for binomial expansion. For example, the coefficients for $(a + b)^2$ are 1, 2, 1 which is the same as $\binom{2}{0}, \binom{2}{1}, \binom{2}{2}$. More generally, the coefficients for $(a + b)^n$ are $\binom{n}{0}, \binom{n}{1}, \ldots \binom{n}{n}$.

Program Goal: Write a program **binomial_exp** that will calculate the coefficients for any power of $(a + b)$.

Use the formula $\binom{n}{k} = \frac{n!}{(n-k)!\,k!}$.

SAMPLE OUTPUT

n? 4

Coefficients of $(a + b)$^ 4:

1 4 6 4 1

PROGRAM 3.5.1 BINOMIAL_EXP.PY

```
import math
# user-defined function
def n_C_k(n,k):
    return int((math.factorial(n))/((math.factorial(n-k))*(math.factorial(k))))
# main program
n = int(input("n? "))
print("Coefficients of (a+b)^",n,":")
[insert code to calculate and output coefficients]
```

EXERCISE SET 3.5.3

1. Fill in the yellow highlighted code to calculate and output the coefficients of $(a + b)^n$.
2. What is the output for $n = 15$?

Next we're going to figure out the exponents of a and b in each term. Let's look at $(a + b)^4$ for example:

$(a + b)^4 = a^4 + 4a^3b + 6a^2b^2 + 4ab^3 + b^4$, which is the same as $(a + b)^4 = a^4b^0 + 4a^3b^1 + 6a^2b^2 + 4a^1b^3 + a^0b^4$

Notice that the exponents for a start at 4 and descend down to 0, and the exponents for b start at 0 and increase up to 4. Let's generalize.

BINOMIAL EXPANSION

$$(a+b)^n = \binom{n}{0}a^n + \binom{n}{1}a^{n-1}b + \binom{n}{2}a^{n-2}b^2 + \cdots \binom{n}{n-1}ab^{n-1} + \binom{n}{n}b^n$$

$$= \sum_{k=0}^{n} \binom{n}{k} a^{n-k}b^k$$

The kth term of $(a+b)^n$ for $k = 0, 1, \ldots n$ is:

$$\binom{n}{k} a^{n-k}b^k$$

A special case in which $a = x$, $b = 1$ would be:

$$(x+1)^n = \sum_{k=0}^{n} \binom{n}{k} x^{n-k}$$

Example 1

Expand $(3n+2)^3$

Solution

Substitute $n = 3$, $a = 3n$, $b = 2$:

$$\binom{3}{0}(3n)^3 + \binom{3}{1}(3n)^2(2)^1 + \binom{3}{2}(3n)^1(2)^2 + \binom{3}{3}(2)^3$$

$$= 1(27n^3) + 3(9n^2)(2) + 3(3n)(4) + 1(8)$$

$$= 27n^3 + 54n^2 + 36n + 8$$

Example 2

Expand $(2x-1)^4$

Solution

Substitute $a = 2x$, $b = -1$, $n = 4$:

$$\binom{4}{0}(2x)^4 + \binom{4}{1}(2x)^3(-1)^1 + \binom{4}{2}(2x)^2(-1)^2 + \binom{4}{3}(2x)^1(-1)^3 + \binom{4}{4}(-1)^4$$

$$= 1(16x^4) + 4(8x^3)(-1) + 6(4x^2)(1) + 4(2x)(-1) + (1)(1)$$

$$= 16x^4 - 32x^3 + 24x^2 - 8x + 1$$

Example 3

Find the 3rd term of $(7 - y)^4$.

Solution

Substitute $a = 7, b = -y, n = 4, k = 2$ (for the 3rd term):

The 3rd term is $\binom{n}{k}a^{n-k}b^k = \binom{4}{2}(7)^2(-y)^2 = 6(49)(y^2) = 294y^2$

EXERCISE SET 3.5.4

Expand the following binomials using the formula for binomial expansion:

1. $(a + b)^3$
2. $(x + 1)^5$
3. $(u + 2v)^4$
4. $(2a - 1)^4$
5. Find the last term of $(x + 2y)^6$
6. Find the 3rd term of $(6n - 2)^5$
7. Modify program **binomial_exp** so that it outputs the complete binomial expansion of $(x + 1)^n$.

Sample Output

n? 4

(x + 1)^4 =

1x^4 + 4x^3 + 6x^2 + 4x + 1

8. Modify program **binomial_exp** so that it outputs the complete binomial expansion of $(a + b)^n$.

Sample Output

n? 4

(a + b)^4 =

1(a^4) + 4(a^3)(b) + 6(a^2)(b^2) + 4(a)(b^3) + 1(b^4)

9. CHALLENGE: Write a program that asks the user to input values for a, b, n, and k. Calculate and output the kth term of $(a + b)^n$.

Sample Output

a? u

b? -2v

n? 5

k? 2

40(u^3)(v^2)

3.6 Solving Polynomial Equations

Recall, a **polynomial function** is in the form:

$$f(x) = a_n x^n + a_{n-1} x^{n-1} + a_{n-2} x^{n-2} + \ldots + a_1 x + a_0$$

where the coefficients $a_1, a_2, \ldots$ are real numbers, x is a variable, and the exponents $n, n-1, \ldots$ are positive integers. We already wrote a program to evaluate $f(x)$ at a specific value of x: **evalpoly**.

We will continue our study of polynomial functions by exploring features of their graphs. One important feature is the ***x*-intercept**, also called a **zero**. (That's because $f(x) = 0$ at an x-intercept.)

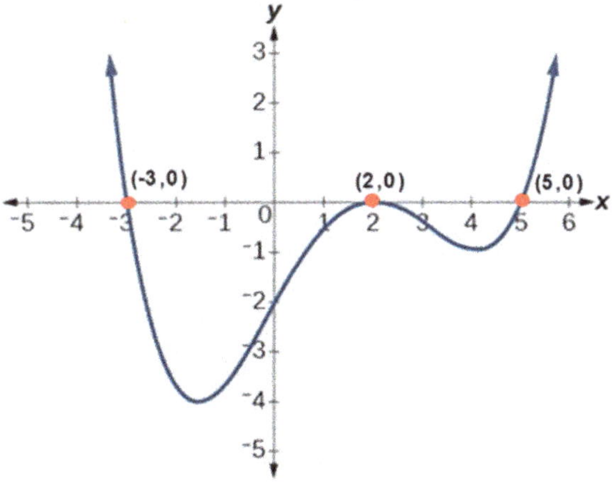

Image 3.19

To find the x-intercept (zero) of a function, simply set it equal to zero and solve for x.

Example 1

Find the zero(s) of $f(x) = 3x - 6$.

Solution

This is a linear function, so it only has one real solution.

Setting it equal to zero:

$3x - 6 = 0$

$3x = 6$

$x = 2$

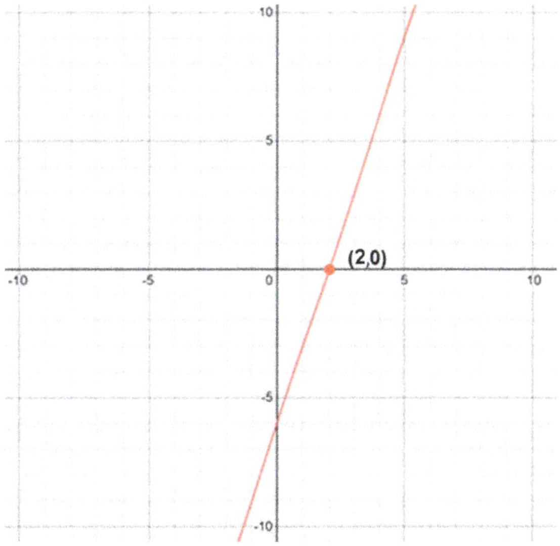

Image 3.20

Linear Functions are easy to solve. In general, the solution of $ax + b = 0$ is $x = -\frac{b}{a}$. We have a number of methods to solve quadratic equations. Here is a summary of the different methods, which you should have covered in a prerequisite course:

Square Root Method

Example 1

Solve $x^2 = 50$

Solution

$x = \pm 5\sqrt{2}$ ← *Take square root of both sides*

Example 2

Solve $(x - 7)^2 + 4 = 29$

Solution

$(x - 7)^2 = 25$

$x - 7 = \pm 5$ ← *Take square root of both sides*

$x = 7 \pm 5$

$x = 12, 2$

Example 3

Solve $x^2 + 9 = 4$

Solution

$x^2 = -5$

$x = \pm\sqrt{-5}$ ← *Take square root of both sides*

$\therefore$ No real solution

Factoring Method

Example 1

Solve $x^2 - 7x + 12 = 0$

Solution

$(x - 4)(x - 3) = 0$

$x = 4, 3$

Example 2

Solve $x^2 - 4x = 0$

Solution

$x(x - 4) = 0$

$x = 0, 4$

Example 3

Solve $x^4 - 16 = 0$

(*Technically not a quadratic, but could be considered quadratic in the form* $(x^2)^2 - (4)^2$.)

Solution

$(x^2 + 4)(x^2 - 4) = 0$

$(x^2 + 4)(x + 2)(x - 2) = 0$

$x = -2, 2$

(The equation also has two nonreal solutions, but we'll get to that later.)

Completing the Square

Example 1

Solve $x^2 + 6x - 8 = 0$

Solution

$x^2 + 6x = 8$ ← *Move constant term to right side of equal sign*

$x^2 + 6x + 9 = 8 + 9$ ← *add* $\left(\frac{b}{2}\right)^2$ *to both sides of equation*

$(x + 3)^2 = 17$ ← *Factor left side of equal sign*

$x + 3 = \pm\sqrt{17}$ ← *Take square root of both sides*

$x = -3 \pm\sqrt{17}$ ← *Subtract 3 from both sides*

Example 2

Solve $2x^2 + 20x + 8 = 0$

Solution

$2x^2 + 20x = -8$ ← *Move constant term to right side*

$2(x^2 + 10x) = -8$ ← *Factor coefficient of* x^2 *out of left side terms*

$2(x^2 + 10x + 25) = -8 + 50$ ← *add 2*25 to both sides*

$2(x + 5)^2 = 42$ ← *Factor left side*

$(x + 5)^2 = 21$ ←*Divide both sides by 2*

$x + 5 = \pm\sqrt{21}$ ← *Take square root of both sides*

$x = -5 \pm\sqrt{21}$ ← *Subtract 5 from both sides*

Example 3

Solve $x^2 + x - 5 = 0$

Solution

$$\begin{aligned}
x^2 + x &= 5 \\
x^2 + x + \frac{1}{4} &= 5 + \frac{1}{4} \\
\left(x + \frac{1}{2}\right)^2 &= \frac{26}{4} \\
x + \frac{1}{2} &= \pm\sqrt{\frac{26}{4}} \\
x + \frac{1}{2} &= \pm\frac{\sqrt{26}}{2} \\
x &= -\frac{1}{2} \pm \frac{\sqrt{26}}{2}
\end{aligned}$$

Quadratic Formula

If $ax^2 + bx + c = 0$, then:

$$x = \frac{-b \pm \sqrt{b^2 - 4ac}}{2a}$$

The solution(s) are real if $b^2 - 4ac \geq 0$

Example 1

Solve: $x^2 - x = 6$

Solution

$x^2 - x - 6 = 0$

$$\begin{aligned}
x &= \frac{-(-1) \pm \sqrt{(-1)^2 - 4(1)(-6)}}{2(1)} \\
&= \frac{1 \pm \sqrt{1 + 24}}{2} \\
&= \frac{1 \pm \sqrt{25}}{2} \\
&= \frac{1 \pm 5}{2} \\
&= 3, -2
\end{aligned}$$

Example 2

Solve $2x^2 + 2x - 7 = 0$

Solution

$$x = \frac{-2 \pm \sqrt{(2)^2 - 4(2)(-7)}}{2(2)}$$

$$= \frac{-2 \pm \sqrt{4 + 56}}{4}$$

$$= \frac{-2 \pm \sqrt{60}}{4}$$

$$= \frac{-2 \pm 2\sqrt{15}}{4}$$

$$= -\frac{2}{4} \pm \frac{2\sqrt{15}}{4}$$

$$= -\frac{1}{2} \pm \frac{\sqrt{15}}{2}$$

EXERCISE SET 3.6.1

Solve.

1. $x^2 - 2x - 8 = 0$
2. $2x^2 + 5x = 12$

Find the zero(s), if they exist. Use whichever method you prefer. Express answers in simplest radical form.

3. $y = -x^2 + 6x - 8$
4. $y = -2x^2 + 12x - 18$
5. $y = x^2 + 5x + 1$
6. $y = x^2 - 49$
7. Write a program that asks the user to input the coefficients *a* and *b* of linear function $f(x) = ax + b$. Calculate and output the zero of the function.
8. Write a program that asks the user to input the coefficients *a*, *b*, and *c* of a quadratic function $f(x) = ax^2 + bx + c$. Calculate and output the solutions using the quadratic formula. Express answers as decimals rounded to 2 decimal places. Check for, and output, whether there is no real solution.
9. CHALLENGE. Write a program that (A) asks the user to input the vertex (h, k) and a point (x, y) on a parabola. Find the equation of the parabola in vertex form.

(Continued)

Sample Output

h,*k*? 1,–5

x,*y*? –1,3

f(*x*) = 2*(*x* – 1)**2 – 5

Using Factor() and Solve()

Now that you know how to find zeros by hand, here's a few Python functions to make it easier.

MORE SYMBOLIC MANIPULATION (SYMPY MODULE)

To factor a polynomial:

factor(poly)

To solve a polynomial equation poly = 0:

solve(poly)

Program Goal: Write a program **factor_and_solve** that will ask user to input the coefficients of a polynomial (as a comma-separated list of coefficients) and then print out its factorization and zeros.

SAMPLE OUTPUT

Coefficients of Polynomial? 2,–5,–11, 20, 12

$2 \cdot x^4 - 5 \cdot x^3 - 11 \cdot x^2 + 20 \cdot x + 12$

Factorization:

$(x - 3) \cdot (x - 2) \cdot (x + 2) \cdot (2 \cdot x + 1)$

Zeros:

[–2, –1/2, 2, 3]

PROGRAM 3.6.1 FACTOR_AND_SOLVE.PY

```
from sympy import *
import poly_fcn as pf
x = Symbol('x')
poly_str = input("Coefficients of Polynomial? ")
poly_list = [int(n) for n in poly_str.split(',')]
poly = pf.num_list_to_poly(poly_list)
zeros = solve(poly)
[fill in code to output original polynomial, its factorization, and zeros]
```

EXERCISE SET 3.6.2

1. Fill in the yellow highlighted code to output the original polynomial, its factorization, and zeros.

For #2–4, find the factorization and zeros of the following polynomial functions using **solve**().

2. $f(x) = 3x^3 + 32x^2 - 303x + 340$
3. $f(x) = x^2 - 2$
4. $f(x) = x^4 - 17x^3 - 52x^2 + 2{,}048x - 8{,}448$
5. Modify program **factor_and_solve** so that it can solve any polynomial equation in the form $p(x) = q(x)$, where p and q are polynomial functions.

Sample Output

Coeff of Left Side? 1,10,9,0,100

Coeff of Right Side? 5,40,35,–68

Equation:

x^4 + 10x^3 + 9x^2 + 100 = 5x^3 + 40x^2 + 35x – 68

Zeros:

$[-8, 3, -\sqrt{7}, \sqrt{7}]$

For #6–8, use the modified program you wrote in #5 above to solve.

6. $x^4 - 3x^3 - 20x^2 + 50x - 19 = -x^4 + 2x^3 - 3x^2 + 9x + 2$
7. $4x^5 + 10x^4 + 7x^3 - 2x^2 + 300x - 100 = x^5 + 20x^4 + 100x^3 - 70x^2 - 24x + 12$
8. $-110x^4 + 1213x^3 - 2100x^2 + 100x - 3000 = -3x^5 - 6x^4 + 117x^3 + 262x^2 + 11903x + 630$

3.7 Linear Regression

A widespread and practical application of polynomial functions in statistics is using data to predict future values. The data is collected and analyzed for the best fit: is it linear, quadratic, exponential, etc.? Once the "shape" is determined, then the data is used to find the coefficients of an approximating function.

The most basic type of relationship between data (x-y pairs) is linear. It is assumed you are familiar with the various forms of linear equations:

Slope-intercept $y = mx + b$

Point-Slope $(y - y_1) = m(x - x_1)$

Standard Form $Ax + By = C$

Only one of these is expressed as a function: $y = mx + b$. Some examples of real-life linear functions:

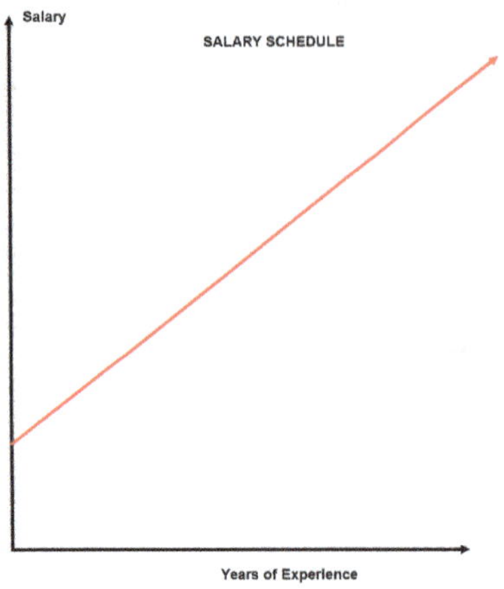

Image 3.21

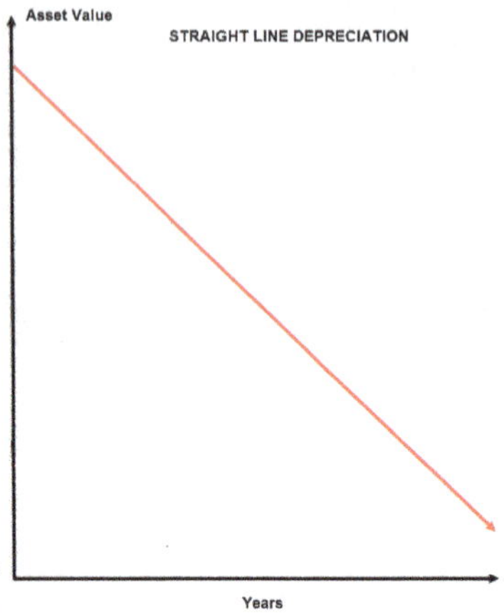

Image 3.22

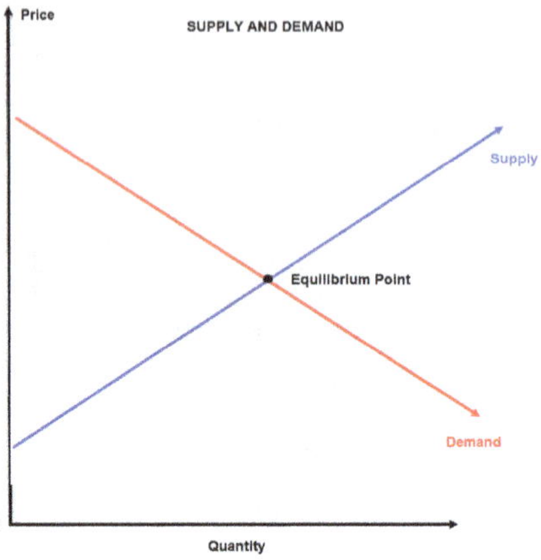

Image 3.23

Sometimes the linear relationship (or whether there is a relationship at all) is not clear cut, and we need to extrapolate the equation from a given set of data.

Example 1

A professor wishes to determine if there is a linear relationship between the number of absences and a student's final grade in the course. She analyzes a sample size of 10 students:

# of absences (*X*)	Grade in class (*Y*)
1	84
1	78
2	88
2	74
2	79
3	75
3	72
3	69
4	71
5	65

Solution

First, she plots the points:

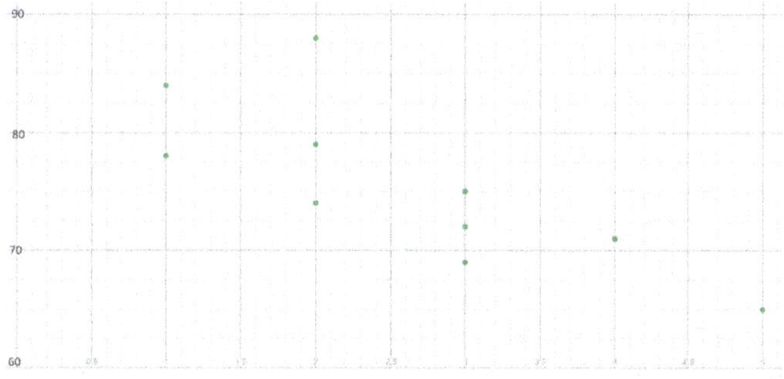

Image 3.24

It appears there may be a linear relationship with the line of best fit about here:

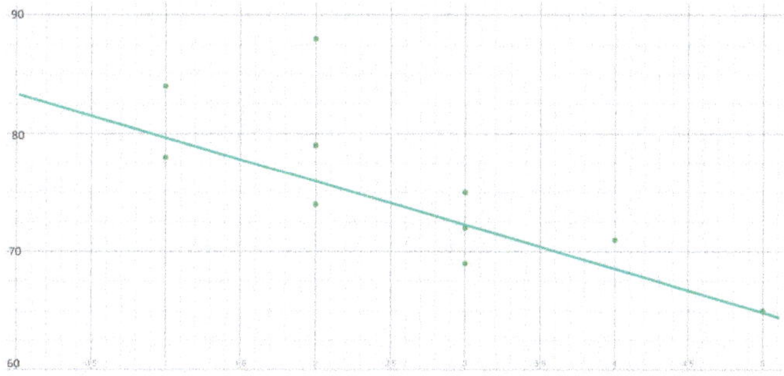

Image 3.25

But what does "about here" mean? From a visual perspective, the line of best fit could be any of these:

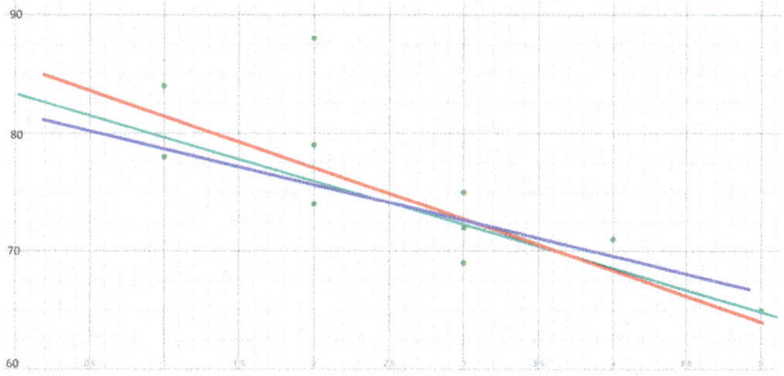

Image 3.26

Of course, there is an algorithm for line of best fit. Actually, there's a number of them, but this method—called **least squares regression**—is the most ubiquitous:

LEAST SQUARES REGRESSION MODEL FOR LINEAR FUNCTION

$$m = \frac{N\sum(xy) - \sum x \sum y}{N\sum(x^2) - \left(\sum x\right)^2},\ b = \frac{\sum y - m\sum x}{N}$$

where N is the number of ordered pairs

Applying these formulas to the data above, we get:

$$N = 10$$

$$\sum xy = 1 \cdot 84 + 1 \cdot 78 + \ldots + 5 \cdot 65 = 1{,}901$$

$$\sum x = 1 + 1 + \ldots + 5 = 26$$

$$\sum y = 84 + 78 + \ldots + 65 = 755$$

$$\sum x^2 = 1^2 + 1^2 + \ldots + 5^2 = 82$$

$$\left(\sum x\right)^2 = 26^2 = 676$$

Thus, our slope and y-intercept are:

$$m = \frac{10 \cdot 1{,}901 - 26 \cdot 755}{10 \cdot 82 - 676} = \frac{19{,}010 - 19630}{820 - 676} = \frac{-620}{144} \approx -4.3$$

$$b = \frac{755 - (-4.3)26}{10} = \frac{866.8}{10} \approx 86.7$$

$$\therefore y = -4.3x + 86.7$$

An important note about statistics: A line of best fit does not necessarily guarantee a correlation or cause-effect relationship. Also keep in mind that the greater your sample size, the more accurate is your linear function. This sample size only had 10 students.

EXERCISE SET 3.7.1

1. Use the resulting linear model in Example 1 to predict the final grade of a student with 10 absences.
2. An **outlier** is a statistic (in this case ordered pair) that is clearly "off the beaten path." What would you say is the outlier(s) in Example 1?
3. A very rudimentary approximation of line of best fit is the line that passes through the first and last points of a data set. Calculate the line of best fit for the above example and compare it to the linear regression result of

$$y = -4.3x + 86.7$$

4. Here is a data set comparing age (X) to systolic blood pressure (Y). Calculate the line of best fit using linear regression. Round m and b to 3 decimal places.

39	144
47	220
45	138
47	145
65	162
46	142
67	170
42	124
67	158
56	154

5. Write a program that calculates $\sum x$, $\sum y$, $\sum xy$, $\sum x^2$ *and* () $\left(\sum x\right)^2$ for the following ordered pairs: (1, 3), (2, 6), (3, 11), (4, 12).

Desired Output

sum of $x = 10$

sum of $y = 32$

sum of $xy = 96$

sum of x^2 $= 30$

(Sum of x)^2 $= 100$

Hint: **sum**(x) will calculate $\sum x$.

The tedious calculations involved in the formulas for m and b should really make one appreciate the power of computer programs. Which is exactly what you're going to write next.

EXERCISE SET 3.7.2

1. Write a program **linreg** to execute the following steps that, when combined, will accomplish the goal of linear regression. Hint: **sum**(x) will calculate $\sum x$.
   A. Assign data in Example 1 to lists x and y.
   B. Calculate m and b using the formulas above.
   C. Output the linear equation in the form $y = mx + b$. Round values of m and b to 2 decimal places.
2. Modify the program you wrote in question 1 by replacing the assignment statements with input statements. Make sure the size of both lists is the same.

Test your program using the sample data in the previous example (absences vs. grades).

One of the benefits of linear regression is making predictions. For example, you could use $y = -4.3x + 86.7$ to make a prediction of a student's grade based on their number of absences. Far more meaningful applications would be to predict stock prices, the housing market, etc.

A useful variable to measure the relative accuracy of a linear model is called the **correlation coefficient**. Chapter Project II in this chapter gives you the opportunity to calculate the correlation coefficient for a linear model.

Not all relationships between data are linear. They could be parabolic (next section) or exponential (Chapter 10).

3.8 Graphing a Quadratic Function

Vertex Form

Prior to computers and graphing calculators, mathematicians needed a quick way to sketch the graph of a polynomial function. A table of values is one method, but it is time-consuming and tedious. Thus, the following set of "shortcuts" were developed.

Example 1

Find the x- and y-intercepts for $y = x^2 - 9$.

Solution

The y-intercept is $y(0) = 0^2 - 9 = -9$. The ordered pair is $(0, -9)$.

The x-intercept is found by solving $y = 0$:

$x^2 - 9 = 0$

$(x + 3)(x - 3) = 0$

$x = -3, x = 3$

$\therefore$ Ordered pairs: $(-3, 0)$ and $(3, 0)$.

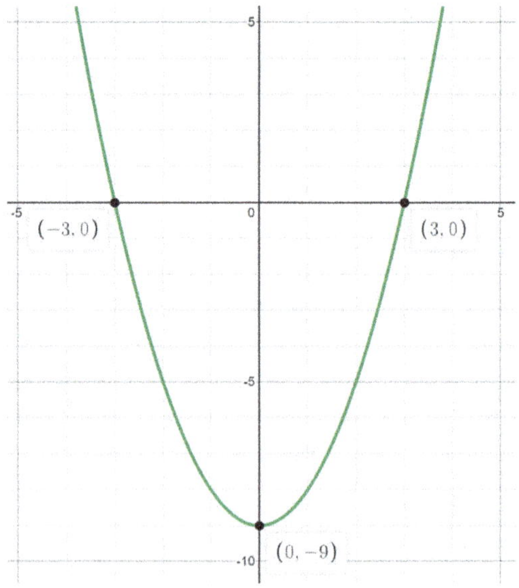

Image 3.1

A quadratic function is in **vertex form** if it is expressed as $\boldsymbol{y = a(x - h)^2 + k}$, $a \neq 0$. It is called vertex form because the vertex is at (h, k). The **axis of symmetry** is the vertical line that runs through the vertex.

Features of Vertex Form

Direction	Face up if $a > 0$ Face down if $a < 0$
Vertex	(h, k)
Axis of Symmetry	$x = h$ (Not just h. Must be equation of vertical line.)
y-intercept	$y(0)$
x-intercept(s)	Solve $y = 0$

Note, some textbooks specify that the intercepts must be expressed as ordered pairs, however we will be liberal in accepting either $x = a$ or $(a, 0)$; likewise $y = b$ or $(0, b)$.

Example 2

Find the vertex, axis of symmetry, intercepts, and direction of $y = 2(x - 3)^2 + 5$. Then sketch graph.

Solution

Direction	$a > 0 \rightarrow$ face up
Vertex	$(h, k) = (3, 5)$
Axis of Symmetry	$x = h \rightarrow x = 3$
y-intercept	$y(0) = 2(0 - 3)^2 + 5 = 23$
x-intercept(s)	Solve $y = 0$: $2(x - 3)^2 + 5 = 0$ $(x - 3)^2 = -\frac{5}{2}$ $x - 3 = \pm\sqrt{-\frac{5}{2}}$ $x = 3 \pm\sqrt{-\frac{5}{2}}$ Not a real number. $\therefore$ No x-intercepts

Since there are no x-intercepts, we only have two points to plot. The graph might look a little better if we plot another point. Let's find the y-coordinate for $x = 6$:

$y(6) = 2(6 - 3)^2 + 5 = 23 \rightarrow$ ordered pair is $(6, 23)$.

It was not random that I chose $x = 6$. Notice that (6,23) is the same horizontal distance from the vertex as (0,23). By symmetry, $x = 0$ and $x = 6$ have the same y-coordinate.

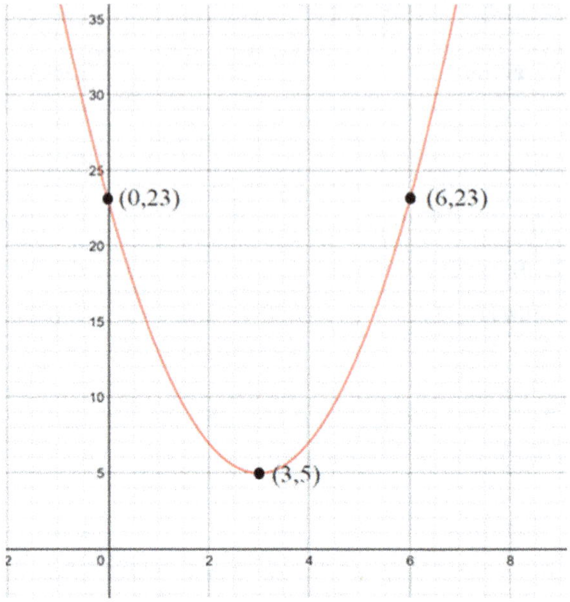

Image 3.27

Example 3

Find the vertex, axis of symmetry, intercepts, and direction of $f(x) = 2x^2 - 12x + 3$. Then sketch graph.

Solution

Note this quadratic function is not expressed in vertex form. Therefore, our first step is to convert to vertex form. This will involve completing the square.

$$y = 2x^2 - 12x + 3$$

$$y = 2(x^2 - 6x) + 3$$ ← *factor 2 out of the* x-*terms, leave space to complete the square*

$$y = 2(x^2 - 6x + 9) + 3 - 18$$ ← *Add 9 to complete the square, subtract 18 to "balance" the right side of the equation*

$$y = 2(x - 3)^2 - 15$$ ← *vertex form*

Direction	$a > 0 \rightarrow$ face up
Vertex	$(h, k) = (3, -15)$
Axis of Symmetry	$x = h \rightarrow x = 3$
***y*-intercept**	$y(0) = 2(0-3)^2 - 15 = 3$
***x*-intercept(s)**	Solve $y = 0$: $$2(x-3)^2 - 15 = 0$$ $$2(x-3)^2 = 15$$ $$(x-3)^2 = \frac{15}{2}$$ $$x - 3 = \pm\sqrt{\frac{15}{2}}$$ $$x = 3 \pm\sqrt{\frac{15}{2}}$$ $$x \approx 5.7, 0.3$$

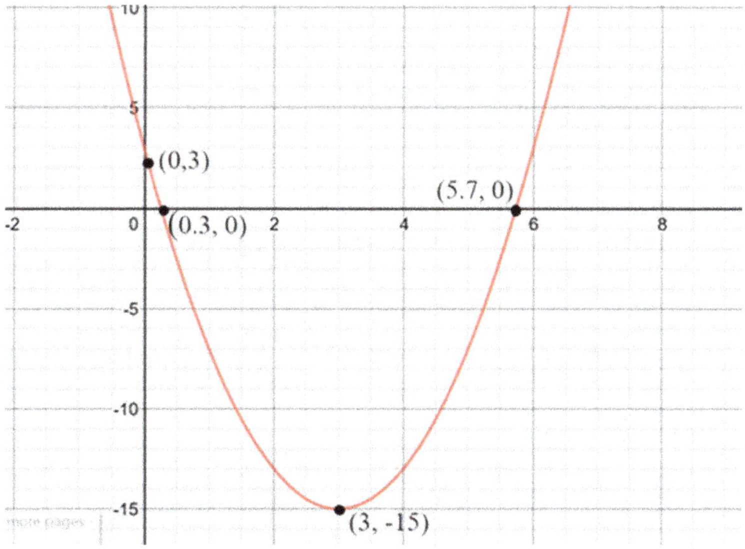

Image 3.28

EXERCISE SET 3.8.1

Find the vertex, axis of symmetry, intercepts, and direction of the parabolas below. Then sketch graph.

1. $y = 2(x + 3)^2 - 8$
2. $y = -(x - 2)^2 - 1$
3. $y = x^2 + 4$
4. $y = x^2 + 4x + 4$
5. $y = -x^2 + 8x$
6. $y = 3x^2 - 6x + 5$
7. $y = -x^2 + 10x - 1$
8. Write a program to calculate the x- and y-intercepts of a quadratic function in the form $y = a(x - h)^2 + k$. Use **SymPy solve**.

Sample Output

a? -1

h? 3

k? 4

x-int: [1, 5]

y-int: −5

Standard Form

We saw in the previous section that the vertex form of a quadratic function is $y = a(x - h)^2 + k$. In this section, we examine the **standard form** (some textbooks call it **general form**):

$$y = ax^2 + bx + c,\ a \neq 0$$

This form shares some features with vertex form but has a few unique features as well.

Features of Standard Form

Direction	Face up if $a > 0$ Face down if $a < 0$
Vertex	$\left(-\frac{b}{2a}, f\left(-\frac{b}{2a}\right)\right)$
Axis of Symmetry	$x = -\frac{b}{2a}$
***y*-intercept**	$(0, c)$
***x*-intercept(s)**	Solve $y = 0$

Example 1

Find the vertex, axis of symmetry, interceptions, and direction of $f(x) = -x^2 + 4x + 21$. Then sketch graph.

Solution

Direction	$a < 0 \rightarrow$ face down
Vertex	$-\frac{b}{2a} = -\frac{4}{2(-1)} = 2$ $f\left(-\frac{b}{2a}\right) = f(2) = -2^2 + 4(2) + 21 = 25$ $\therefore$ (2, 25)
Axis of Symmetry	$x = -\frac{b}{2a} \rightarrow x = 2$
***y*-intercept**	$(0, c) =$ (0, 21)
***x*-intercept(s)**	Solve $y = 0$: $-x^2 + 4x + 21 = 0$ $-(x^2 - 4x - 21) = 0$ $-(x + 3)(x - 7) = 0$ $x = -3, 7$

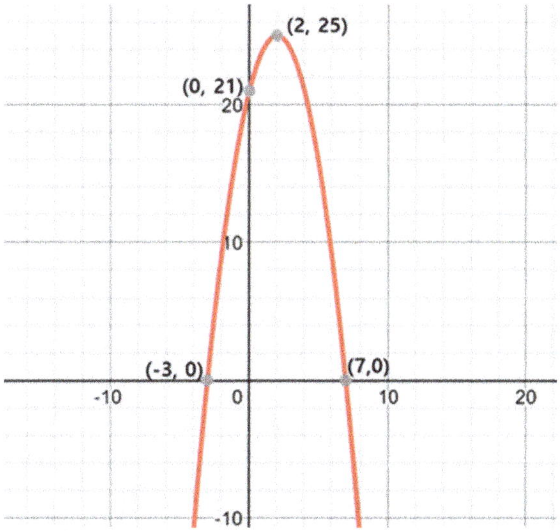

Image 3.29

Example 2

A projectile is fired from a cliff 200 ft. above the water at an inclination of 45° to the horizontal, with a muzzle velocity of 50 ft./sec. The height h of the projectile is modeled by:

$h(x) = \frac{-32}{50^2}x^2 + x + 200$, where x is the horizontal distance from the base of the cliff.

A. At what horizontal distance from the base of the cliff is the height of the projectile a maximum?

B. Find the maximum height of the projectile.

C. At what horizontal distance from the face of the cliff will the projectile strike the water?

Solution

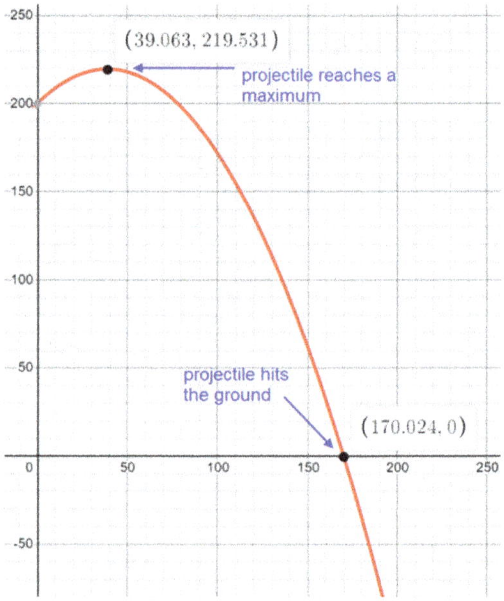

Image 3.30

A. The projectile is a maximum at the vertex. The horizontal distance is given by the x-coordinate of the vertex:

$$x = -\frac{b}{2a} = -\frac{1}{2\left(-\frac{32}{50^2}\right)} = -\frac{1}{-\frac{64}{50^2}} = \frac{50^2}{64} \approx 39\ ft.$$

B. The maximum height of the projectile is given by the y-coordinate of the vertex:

$$y = h(39) == \frac{-32}{50^2}(39)^2 + 39 + 200 \approx 219.5$$

C. The projectile strikes the water when the height is 0, so set $h(x) = 0$ and solve for x:

$$\frac{-32}{50^2}x^2 + x + 200 = 0$$

Use the quadratic formula or a calculator to get $x \approx 170$ ft. from the base of the cliff.

EXERCISE SET 3.8.2

Find the vertex, axis of symmetry, intercepts, and direction of the following quadratic functions in standard form. Then sketch graph.

1. $y = x^2 + 12x + 38$
2. $y = 2x^2 + 6x$

3. $y = -x^2 + 6x - 5$

4. An object is dropped from a height of 256 ft. Its position s (in feet) at time t seconds is given by:

   $s(t) = -16t^2 + 256.$

   A. What is its height after 2 seconds?

   B. When does it hit the ground?

5. A projectile is fired at an inclination of 45° to the horizontal, with a muzzle velocity of 100 ft./sec. The height h of the projectile is given by:

$$h(x) = -\frac{32}{100}x^2 + x$$

   where x is the horizontal distance of the projectile from the firing point.

   A. At what horizontal distance from the firing point is the height of the projectile a maximum?

   B. Find the maximum height of the projectile.

   C. At what horizontal distance from the firing point will the projectile strike the ground?

6. A manufacturer determines that the revenue R based on unit price p dollars is:

$$R(p) = -\frac{1}{2}p^2 + 2{,}800p$$

   A. What unit price should be charged to maximize revenue?

   B. What is the maximum revenue?

7. Write a program that does the following: (1) Asks user to input a, b, and c. (2) Calculates and outputs the vertex, axis of symmetry, y-intercept, and direction of graph.

Sample Output

a? 1

b? 12

c? 38

Vertex: (−6, 2)

Axis of Symmetry: $x = -6$

y-intercept: 38

direction: face up

The **SymPy** module includes tools to help you identify the degree and coefficients of a polynomial.

IDENTIFYING DEGREE AND COEFFICIENTS USING SYMPY

To find the degree of a polynomial expression:

degree(expr)

To get a list of the coefficients of a polynomial expression:

poly(expr).**all_coeffs()**

EXERCISE SET 3.8.3

1. What is the output of the following:

```
>>>from sympy import *
>>>p = sympify('7*x**3 – 15*x**4 + 11*x – 5')
>>>degree(p)
>>>poly(p).all_coeffs()
```

2. Modify the program you wrote in Exercise Set 3.8.2 #7 so that it accepts a polynomial input rather than the coefficients. Check to make sure the degree is 2.

Sample Output

Quadratic? x**2 + 12*x + 38

Vertex: (–6, 2)

Axis of Symmetry: $x = -6$

y-intercept: 38

direction: face up

Quadratic? x**3 – x**2 + 7

That was not a quadratic expression

In Chapter Project IV you will write a program that converts a quadratic equation in standard form to vertex form.

Modeling with Quadratics

If you sell 100 products at \$15 each, your revenue is:

$$R = 100(15) = 1{,}500$$

In general, your **Revenue** is:

$$R = xp$$

in which $x =$ quantity sold and $p =$ price. This represents gross proceeds before costs are deducted.
The **Demand** function expresses the quantity a manufacturer is willing or able to manufacture, and this is often based on several factors including price, market demand, etc.

Image 3.31

Example 1

The business analyst for a manufacturer devises a demand equation relating quantity to price as $x = -6p + 600$.

A. Express revenue as a function of p.

B. Graph $R(p)$ and specify the domain.

C. What price p maximizes revenue?

D. What is the maximum revenue?

E. How many units are sold at this price?

Solutions

A. Substituting $x = -6p + 600$ into $R = xp$:

$$R(p) = (-6p + 600)p = -6p^2 + 600p$$

B. This is a face-down parabola with a y-intercept of zero. The p-intercepts are found by solving $R(p) = 0$:

$$(-6p + 600)p = 0$$

$$p = 100 \text{ or } p = 0$$

The domain is therefore [0, 100].

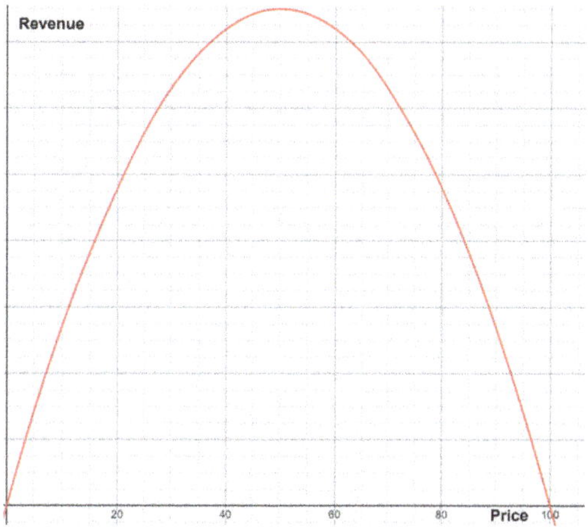

Image 3.32

C. The price that maximizes revenue refers to the x-coordinate (actually p-coordinate) of the vertex:

$$P = -\frac{b}{2a} = -\frac{600}{2(-6)} = 50$$

D. The maximum revenue is:

$$R(50) = -6(50)^2 + 600(50) = \$15{,}000$$

E. To find the quantity x at $p = 50$, substitute into $x = -6p + 600$:

$$x = -6(50) + 600 = 300 \text{ units}$$

Example 2

A street vendor has weekly fixed costs of $540 (cost of permit, employees, etc.) and variable costs of $5.40 for each meal prepared.

Image 3.33

A. Let x represent the quantity of meals prepared and sold each week. Write the weekly cost function, C.

B. The function $R(x) = -0.001x^2 + 8.45x$ models the weekly revenue. Write a function representing the weekly profit.

C. Graph the profit function.

D. Determine the quantity of meals that should be prepared and sold each week to maximize profit.

E. What is the maximum weekly profit?

Solutions

A. The weekly cost is the sum of fixed and variable costs:

$$C(x) = 540 + 5.40x$$

B. The weekly profit is the difference between the revenue and costs:

$$P(x) = R(x) - C(x)$$

$$P(x) = -0.001x^2 + 8.45x - (540 + 5.40x) = -.001x^2 + 3.05x - 540$$

C. This is a face-down parabola with a y-intercept of -540.

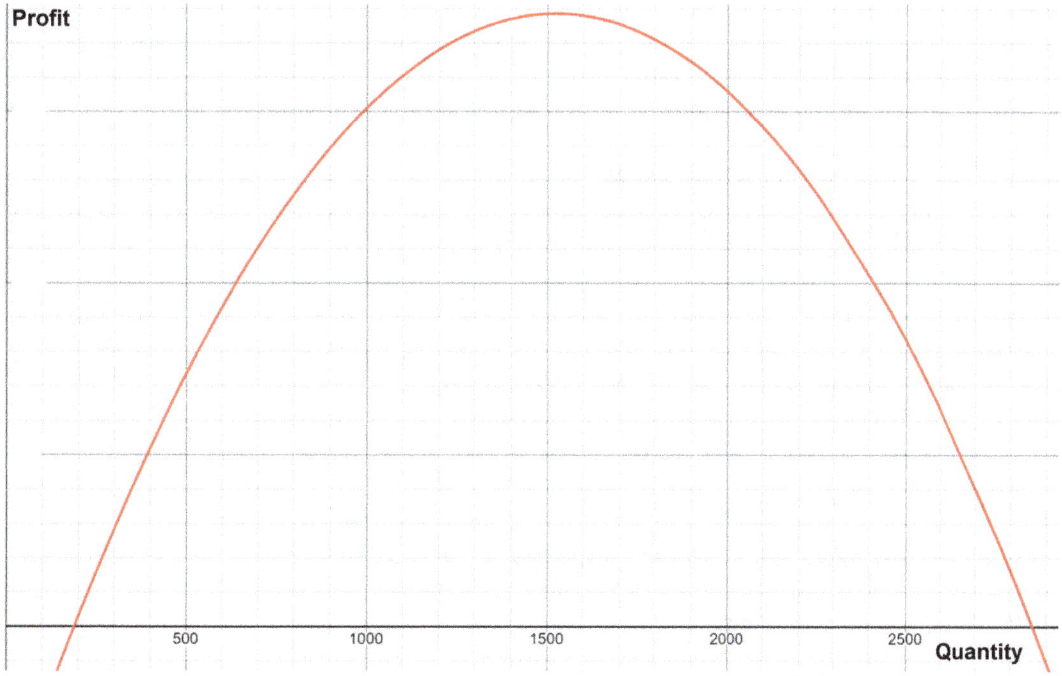

Image 3.34

D. The weekly quantity that maximizes weekly profit is found at the x-coordinate of the vertex:

$$x = -\frac{b}{2a} = -\frac{3.05}{2(-0.001)} = 1,525$$

E. The maximum weekly profit at a quantity of 1,525 is:

$$P(1{,}525) = -0.001(1{,}525)^2 + 3.05(1{,}525) - 540 = \$1{,}785.63 \text{ (rounded to cents)}$$

EXERCISE SET 3.8.4

1. The price p (in dollars) and the quantity x sold of a certain product satisfy the demand equation $x = -6p + 600$.

   A. Find a model expressing the revenue R as a function of p.

   B. Sketch a graph of $R(p)$ including the p-intercepts. Specify the domain.

   C. What price p maximizes the revenue?

   D. What is the maximum revenue?

   E. How many units are sold at this price?

(Continued)

2. The price p (in dollars) and the quantity x sold of a certain product satisfy the demand equation $p = -\frac{1}{3}x + 100$.

   A. Find a model expressing the revenue R as a function of x.

   B. Sketch the graph of $R(x)$ including the x-intercepts. Specify the domain.

   C. What quantity x maximizes the revenue?

   D. What is the maximum revenue?

   E. What price should the company charge to maximize revenue?

3. A company manufacturing widgets has a demand function relating price and quantity of $p = 15 - \frac{x}{1000}$. It has \$10,000 fixed costs and variable costs of \$8 per widget.

   A. Express the revenue and costs as functions of quantity x.

   B. Express the profit as a function of quantity x.

   C. What quantity results in a maximum profit for the company?

   D. What is the maximum profit?

   E. What price should the company charge for each widget, to maximize the profit?

   F. The break-even point is when revenue equals costs, i.e., the profit is zero. What is the break-even point for this company?

4. Write a program that asks the user to enter a demand function relating x to p. Calculate and output the revenue function, the domain, the quantity that yields the maximum revenue, the maximum revenue, and the price that yields the maximum revenue.

Sample Output

Demand function $x(p)$= 3*p + 360

$R(p) = -3*p**2 + 360*p$

domain is [0, 120]

quantity to maximize revenue is 180

maximum revenue is \$10,800

price to maximize revenue is \$60

3.9 Higher Order Polynomial Functions

We have already looked at these two types of polynomial functions:

Linear: $y = ax + b$

Quadratic: $y = ax^2 + bx + c$

We are going to look at some common features of polynomials for their historical significance, i.e., graphing shortcuts that were handy in the centuries prior to computers and calculators.

The graph of a polynomial function is smooth and continuous. We are only able to define these concepts informally at this point in our education. More rigorous definitions involving limits and derivatives await you in calculus.

Smooth means there are no corners v or ^. **Continuous** means you can draw the entire graph without having to lift your pencil from the paper (allowing for an infinite amount of time to draw such a graph, because the domain of all polynomial functions is $(-\infty, \infty)$).

Smooth and Continuous

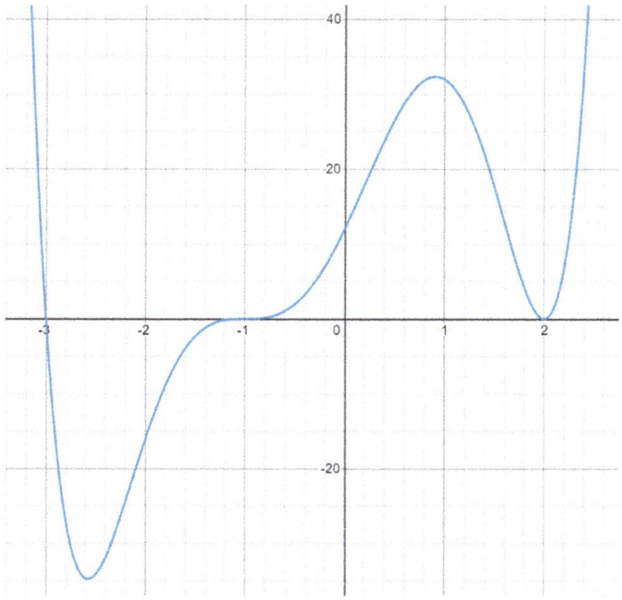

Image 3.35

Neither Smooth nor Continuous

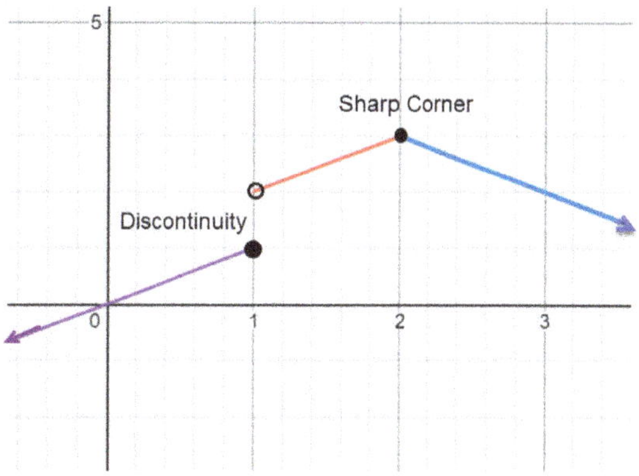

Image 3.36

These are features common to all polynomial functions:

y-intercept	$(0, a_0)$ *(the constant term)* or find $f(0)$
x-intercept(s)	Solve $y = 0$
Max number of turning points (maxes and mins)	$n - 1$ (one less than the degree)

Example 1

Find the x- and y-intercepts and number of turning points of $f(x) = (x - 2)(x + 1)(x - 4)$.

Solution

y-intercept:	$f(0) = (0 - 2)(0 + 1)(0 - 4) = 8$
x-intercept:	Solve $(x - 2)(x + 1)(x - 4) = 0$ $x = 2, -1, 4$
# of turning points:	You can easily verify that if you were to multiply out $(x - 2)(x + 1)(x - 4)$, you would end up with a 3rd degree polynomial. Therefore, $n = 3$ and the maximum number of turning points is $n - 1 = 2$.

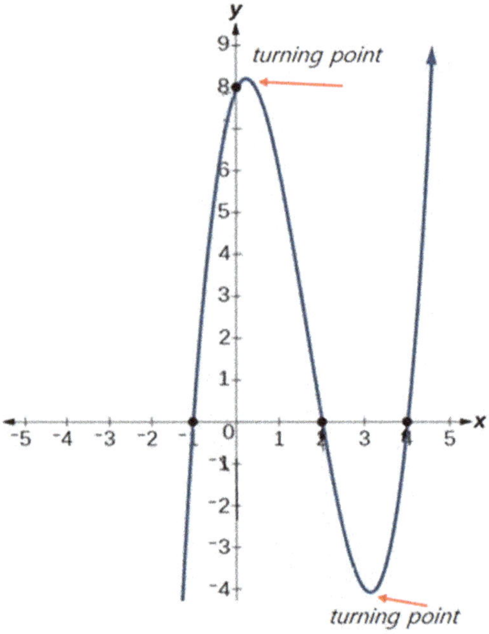

Image 3.37

We already know several ways to find the x-intercepts of quadratic functions (simply solve $f(x) = 0$). The following formulas may come in handy when finding x-intercepts for cubic functions:

Sum of Cubes: $x^3 + a^3 = (x + a)(x^2 - ax + a^2)$

Difference of Cubes: $x^3 - a^3 = (x - a)(x^2 + ax + a^2)$

Example 2

Find the zeros of $f(x) = x^3 - 8$.

Solution

$x^3 - 8 = 0$

$x^3 - (2)^3 = (x - 2)(x^2 + 2x + 2^2) = 0$

$(x - 2)(x^2 + 2x + 4) = 0$

$x - 2 = 0$ or $x^2 + 2x + 4 = 0$

$\therefore\ x = 2$ is the only real solution.

Attempting to solve $x^2 + 2x + 4 = 0$ yields a nonreal answer. Thus, there is only one x-intercept.

All polynomial function graphs have exactly one y-intercept, which we have already learned is at $f(0)$.

Not all cubic functions are in the form of sum of cubes or difference of cubes. In Chapter 8, we will look at other ways to factor and find zeros of higher degree polynomials.

EXERCISE SET 3.9.1

Factor completely if not already factored. Find the degree, y-intercept, x-intercept(s) (if possible), and maximum number of turning points of the following polynomial functions.

1. $y = (x - 2)(x + 3)(x - 4)$
2. $f(x) = x^2 - 81$
3. $f(x) = -3x^2 + 6x + 105$
4. $f(x) = x^5 - 5x^3 + 4x$
5. $y = x^3 - 5x^2 - x + 5$
6. $y = (x - 2)^2(2x + 3)$
7. $y = x^3 + 64$
8. $f(x) = 8x^3 - 125$
9. $f(x) = x^4 + 16$
10. How many x-intercepts can an nth-degree polynomial function have? How many y-intercepts?
11. Is it possible for an odd degree function not to cross the x-axis?
12. Write a program that does the following:
    A. asks user to input a factored polynomial
    B. expands the polynomial
    C. outputs the degree, coefficients, and maximum number of turning points.

    *Use **SymPy** functions.*

(Continued)

Sample Output

factored polynomial? (x − 2)*(x + 3)**2

expanded polynomial: x**3 + 4*x**2 − 3*x − 18

degree: 3

coefficients: [1, 4, −3, 18]

max number of turning points: 2

Absolute and Local Extrema

The turning points of a polynomial function, also referred to as **extrema**, can be classified as either a maximum or minimum as follows.

Think of the graph of a polynomial function as a roller coaster.

A **local** (or relative) **minimum** is the point at which your car hits the bottom of a "valley" and prepares to climb uphill.

In the graph to the right, the local min is 1.75 (the y-value).

A **local** (or relative) **maximum** is the point at which your car hits the peak of the roller coaster and you brace yourself to plummet downward.

In the graph to the right, the local max is 5 (the y-value).

An **absolute** (or global) **minimum** is the lowest point on the graph.

In the graph to the right, there is no absolute min because the graph continues to negative infinity on both sides.

An **absolute** (or global) **maximum** is the highest point on the graph.

In the graph to the right, the absolute max is 10 (the y-value).

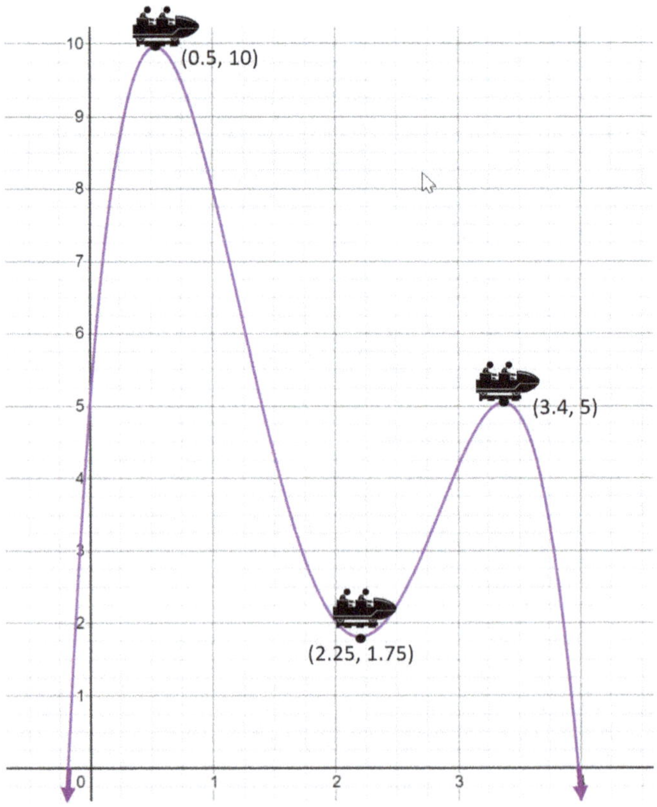

Image 3.38

Formal Definitions

Local minimum—$f(m)$ is a local minimum of polynomial function $f(x)$ if there exists an open interval containing m such that $f(m) \leq f(x)$ for any x in the open interval.

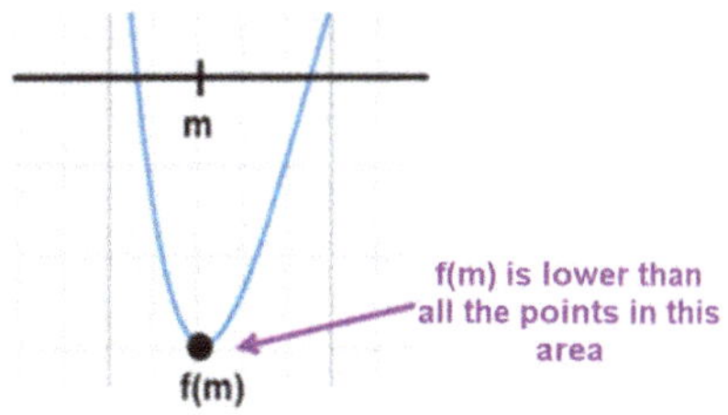

Image 3.39

Local maximum—$f(M)$ is a local maximum of polynomial function $f(x)$ if there exists an open interval containing M such that $f(M) \geq f(x)$ for any x in the open interval.

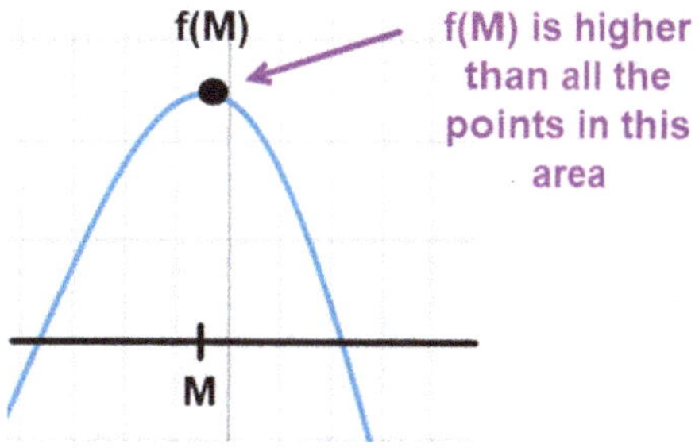

Image 3.40

Absolute minimum—$f(m)$ is an absolute minimum if $f(m) \leq f(x)$ for all x in the domain of the function.

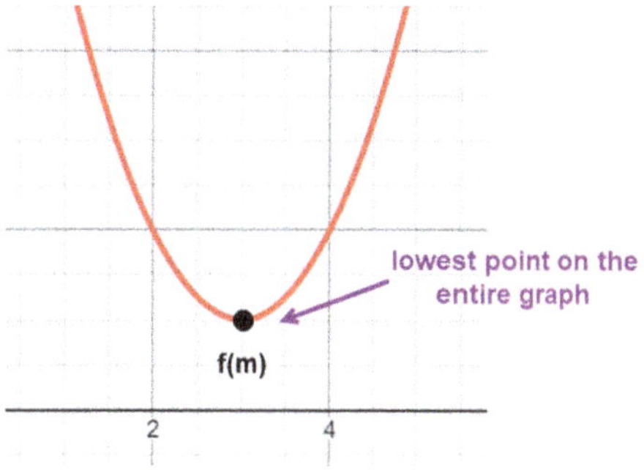

Image 3.41

Absolute maximum—$f(M)$ is an absolute maximum if $f(M) \geq f(x)$ for all x in the domain of the function.

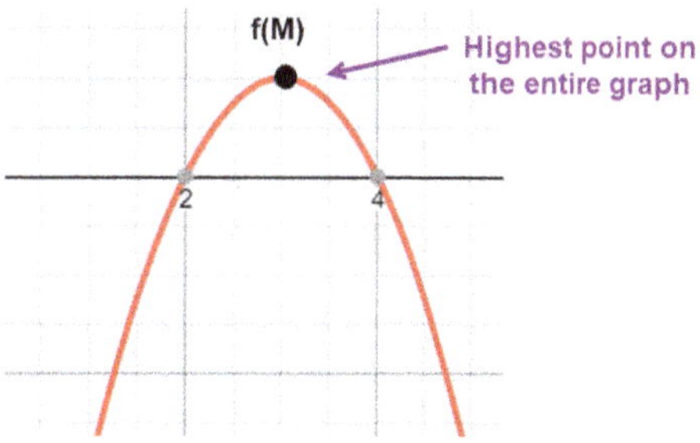

Image 3.42

If a point is in an absolute max[min], then it is **not** also a local max[min]. (Some textbooks will consider an absolute max[min] as also being a local max[min].)

Another arbitrary textbook designation is whether a local max or min can occur at an endpoint. We will adopt the convention that a local max or min **cannot** occur at an endpoint.

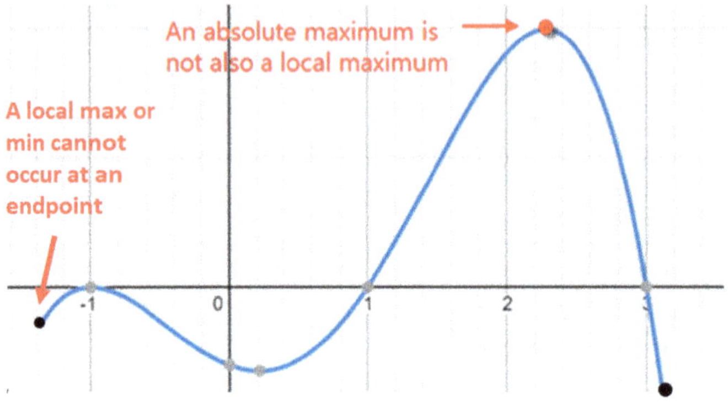

Image 3.43

Recall that the maximum number of turning points for the graph of an nth-degree polynomial is $n - 1$. Likewise, the minimum degree of a polynomial whose graph has m turning points is $m + 1$. In the above example, there are 3 turning points, so the minimum degree is 4.

EXERCISE SET 3.9.2

For the graphs in #1–4, identify any extrema. Classify as local max, local min, absolute max, or absolute min. Do your best to estimate the y-values. Also find the minimum degree of the polynomial based on the number of turning points.

1.

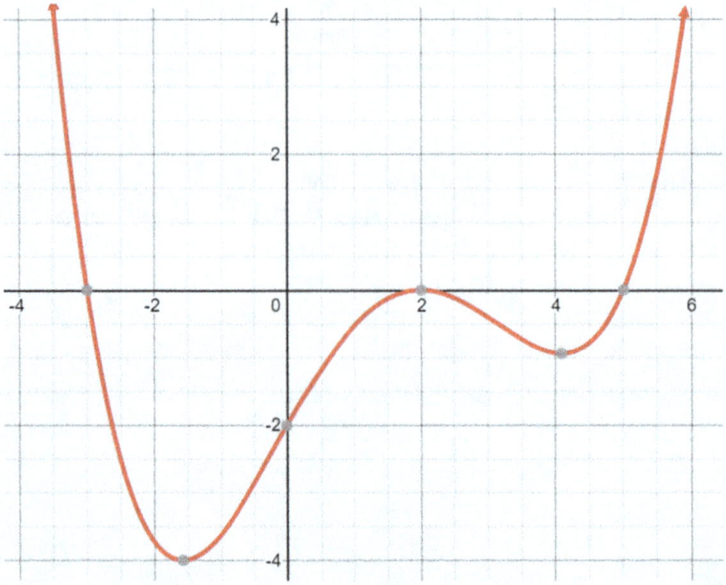

Image 3.4

2.

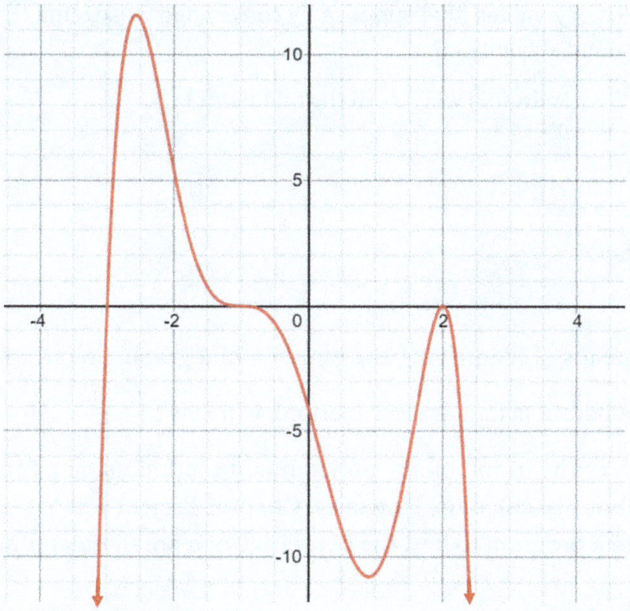

Image 3.44

3.

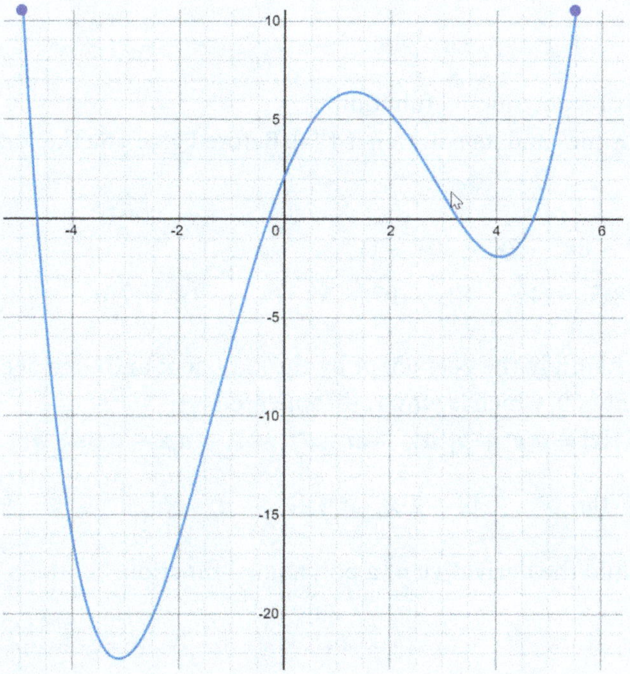

Image 3.45

4. If a polynomial graph has two endpoints (for example #3 above), does it always have at least one absolute maximum and one absolute minimum?
5. Can a polynomial graph have more than one local maximum or minimum?
6. Can a polynomial graph have neither a local maximum nor minimum?

(Continued)

7. Write a program in which the user inputs a list of y-values (that supposedly correspond to extrema points of a graph). Identify the absolute max and absolute min.

   (Hint: **max**(list) will return the maximum of a list; **min**(list) will return the minimum.)

Sample Output

y-values: 12, 7, 8, 11, −3, 4, 0, −9

The absolute min is −9 and the absolute max is 12

Observing a graph is not a very reliable means of identifying the extrema of a graph. So far, we only know a precise formula for the max or min of a parabola (quadratic function), which is $f\left(-\frac{b}{2a.}\right)$. However, there is a precise way to identify the extrema of the graph of *any* polynomial function using a function called the **derivative**, which you will learn about in calculus. However, Chapter Project I at the end of this chapter gives you the opportunity to write a program to find the derivative of a polynomial function.

Chapter Projects

I. Derivative

Write a program that will find the **derivative** of a polynomial function.

The derivative of a function f is called "f prime" and its notation is $f'(x)$. Before I give you the formula, here is an example:

$$f(x) = 7x^4 - 8x^3 + 5x^2 - 10x + 11$$

$$f'(x) = 28x^3 - 24x^2 + 10x - 10.$$

Do you see how it works? For each term, multiply the coefficient by the exponent and then decrease the exponent by 1. The derivative of a constant is 0, which is why 11 disappeared.

The formal definition of the derivative of $f(x) = a_n x^n + a_{n-1}x^{n-1} + a_{n-2}x^{n-2} + ... + a_1 x + a_0$ is:

$$f'(x) = n \cdot a_n x^{n-1} + (n-1)a_{n-1}x^{n-2} + (n-2)a_{n-2}x^{n-3} + ... + a_1$$

Your task is to write a program that will find the derivative of a polynomial function.

SAMPLE OUTPUT

Coefficients of Polynomial Function? 10,−13, 0, 3,−1,0

Derivative:

$50x^4 - 52x^3 + 6x - 1$

II. Correlation Coefficient

The "strength" of a linear relationship is given by what is called the **correlation coefficient** r. r has a value between -1 and 1. A value of zero, as shown in the last case below, indicates absolutely no linear relationship. However the closer $|r|$ is to 1, the stronger the relationship. A positive value indicates a positive slope, and a negative value indicates a negative slope.

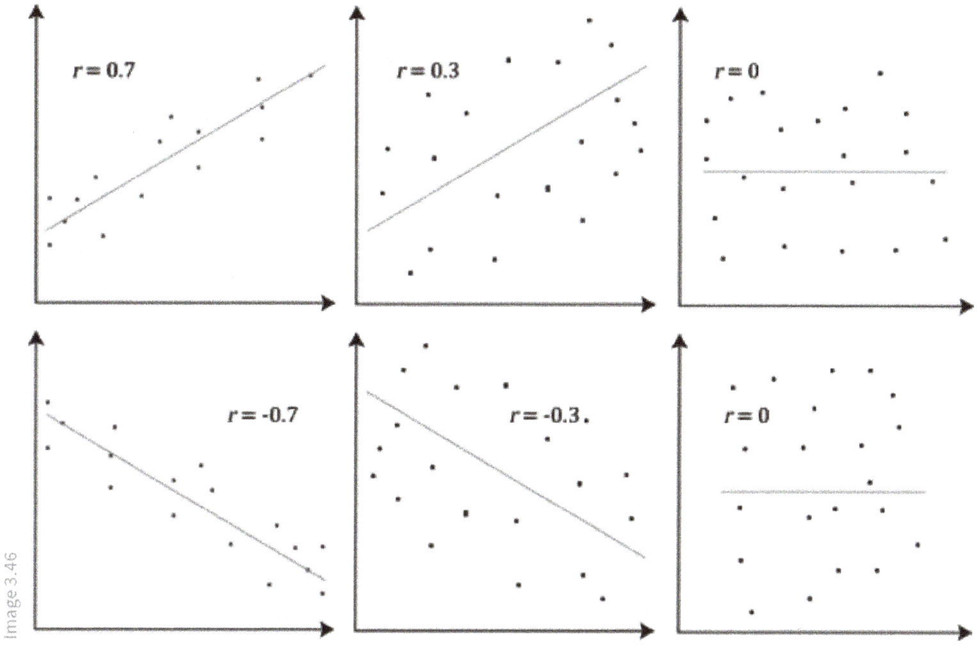

The formula is $r = \dfrac{\sum(x-\bar{x})(y-\bar{y})}{\sqrt{\sum(x-\bar{x})^2}\sqrt{(y-\bar{y})^2}}$ where $\bar{x}$ and $\bar{y}$ represent the means of x and y respectively.

Modify program **linreg** (Exercise Set 3.7.2) so that it also outputs the correlation coefficient and a brief description ("strong positive correlation," etc.).

III. Binomial Expansion

Modify program **binomial_exp** (Section 3.5) so that rather than expanding $(x+1)^n$, it expands $(a+b)^n$. The user will be asked to input values for a, b, and n. a and b can be any monomial expression.

SAMPLE OUTPUT

a? x

b? 2

n? 4

(x + 2)^4 =

x^4 + 8x^3 + 24x^2 + 32x + 16

Alternately: Ask the user to also input an integer value for k. Calculate and output the kth term of $(a + b)^n$.

IV. Converting to Vertex Form

Write a program that asks the user to input a quadratic function in standard form $y = ax^2 + bx + c$ and then calculates and outputs the function in standard form $y = a(x - h)^2 + k$. This involves an algebraic technique called **completing the square**.

SAMPLE OUTPUT

Coefficients of Quadratic Function? 2, −12, 3

Vertex Form:

$2 \cdot (x - 3)^2 - 15$

V. Equilibrium Price and Quantity

You may have heard of supply and demand with respect to retail sales of a product. Business analysts study the effect of the product's price and availability to determine the optimal price.

Business analysts will formulate a **demand function** $D(p)$, which provides a forecast of the quantity demanded by consumers based on price p. They will also formulate a **supply function** $S(p)$, which forecasts the quantity supplied by the manufacturer based on price p.

Image 3.47

The **equilibrium price** is that price in which demand equals eupply, thus satisfying both the manufacturer and the consumers. It is quite easy to determine the equilibrium price: simply set $D(p) = S(p)$ and solve for p. The **equilibrium quantity** is found by substituting the equilibrium price $\boldsymbol{p}$ into either the supply or demand function.

Write a program that asks the user to input both the demand and supply functions. Determine the equilibrium price and quantity (which together form the **equilibrium point**).

SAMPLE OUTPUT

Demand function? $1200 - 25p$

Supply function? $-600 + 50p$

The equilibrium price is \$24

The equilibrium quantity is 600

VI. Zeros of a Quadratic Function

A quadratic function can have zero, one, or two zeros (x-intercepts).

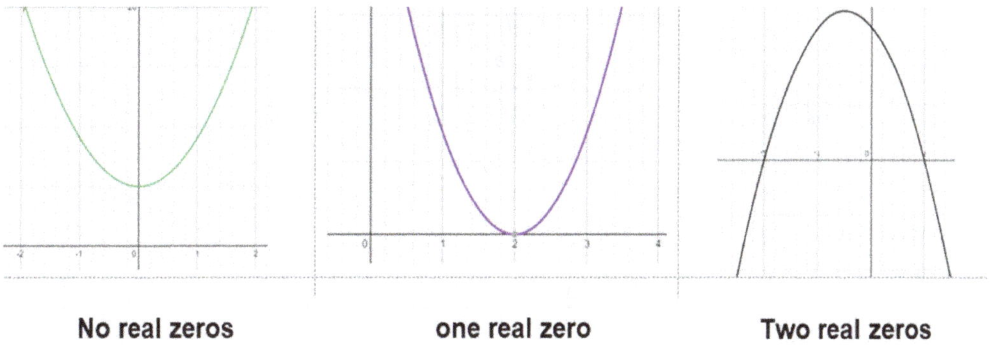

Image 3.48

There is actually a straightforward way to identify the number and type of zeros without graphing. Simply use the **discriminant** $b^2 - 4ac$. (This is the radicand of the quadratic formula.)

If the discriminant is ...	Number of real zeros
negative	None
zero	1 rational zero
positive perfect square	2 rational zeros
positive, not a perfect square	2 irrational zeros

Write a program that asks the user to input a quadratic function in the form $ax^2 + bx + c$. Calculate and output the number of real zeros.

SAMPLE OUTPUT

quadratic? 2*x**2 + 4

There are no real zeros

quadratic? 4*x**2–4* –-24

There are 2 rational zeros:

[–2.0, 3.0]

In Chapter 8 you will learn that a quadratic function can have complex (imaginary) zeros and what that means graphically.

VII. Difference Quotient

The difference quotient for a function is defined as:

$$\frac{f(x+h) - f(x)}{h}$$

In calculus, it sets the stage for the definition of the **derivative**.

Example

The difference quotient for $f(x) = x^2 + 7x - 3$ is:

$$\frac{[(x+h)^2 + 7(x+h) - 3] - [x^2 + 7x - 3]}{h}$$

$$= \frac{[x^2 + 2xh + h^2 + 7x + 7h - 3] - [x^2 + 7x - 3]}{h}$$

$$= \frac{x^2 + 2xh + h^2 + 7x + 7h - 3 - x^2 - 7x + 3}{h}$$

$$= \frac{2xh + h^2 + 7h}{h}$$

$$= \frac{h(2x + h + 7)}{h}$$

$$= 2x + h + 7$$

Write a program that asks user to input the coefficients of a polynomial function. Calculate and output the difference quotient for that function.

VIII. Midpoint Method of Approximating the Zero of a Polynomial Function

Recall that to find the x-intercepts (zeros) of any function, you set $f(x) = 0$.

If a polynomial can be factored, for example, $x^3 - 2x^2 - 11x + 12 = (x - 4)(x + 3)(x - 1)$, then one can easily find the zeros:

$$(x - 4)(x + 3)(x - 1) = 0$$
$$x = 4, -3, 1$$

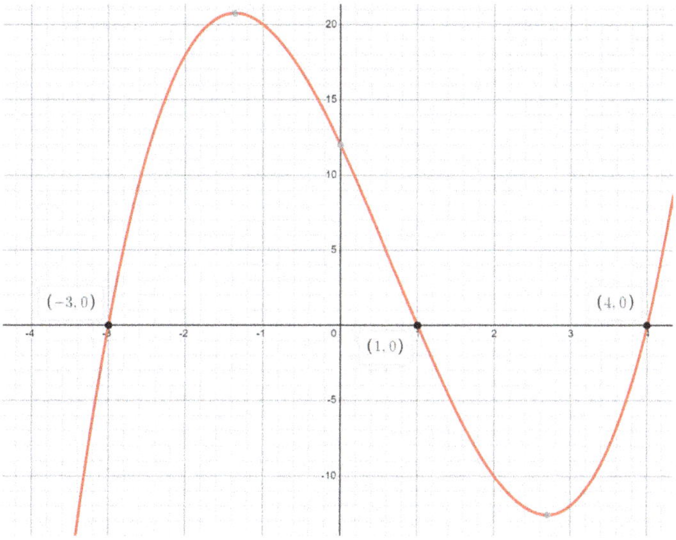

Image 3.49

However, the solution to $f(x) = 0$ is not always an integer and/or the polynomial cannot always be factored. Often in real life problems the solution to $f(x) = 0$ cannot be found algebraically so we resort to approximation techniques. One such technique is the Midpoint Method.

Let's say we want to find the zero of $f(x) = x^5 + x^3 - 3x - 2$. We know the graph looks something like this:

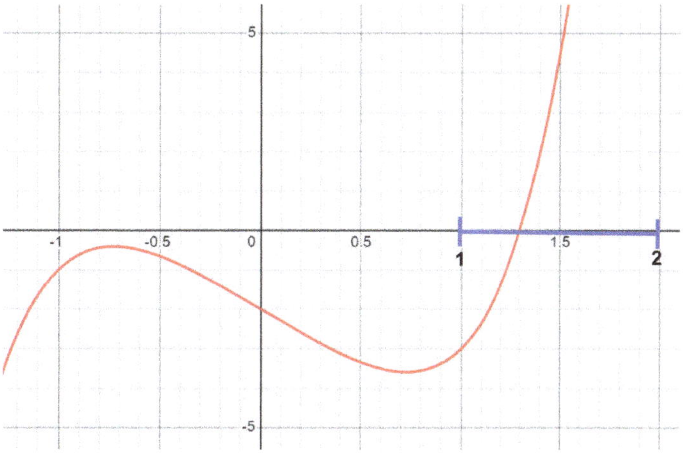

Image 3.50

As you can see, the zero lies somewhere between 1 and 2.

If we are unable to determine the zero algebraically, there are several methods to approximate the zero using an iterative method well-suited to computer programming.

One such method is the **midpoint method**. Research the midpoint method and write a program that will approximate a zero if the user inputs a polynomial and two values that "bracket" the estimated zero.

You will need to decide how many decimal places of accuracy you wish.

SAMPLE OUTPUT

Polynomial function? x**5+x**3−3*x−2

Left bound? 1

Right bound? 2

The approximate zero between 1 and 2 is:

1.298828125

IX. Quadratic Inequalities

Solving an equation such as $x^2 > 9$ is not as simple as taking the square root of both sides. You would end up with $x \geq 3$, which is only half the solution, because $x < 3$ is also a solution. This is more obvious if you convert $x^2 > 9$ to:

$$x^2 - 9 > 0$$

The solution would be the portion of the graph of $f(x) = x^2 - 9$ that lies above the x-axis, which is $(-\infty, -3) \cup (3, \infty)$.

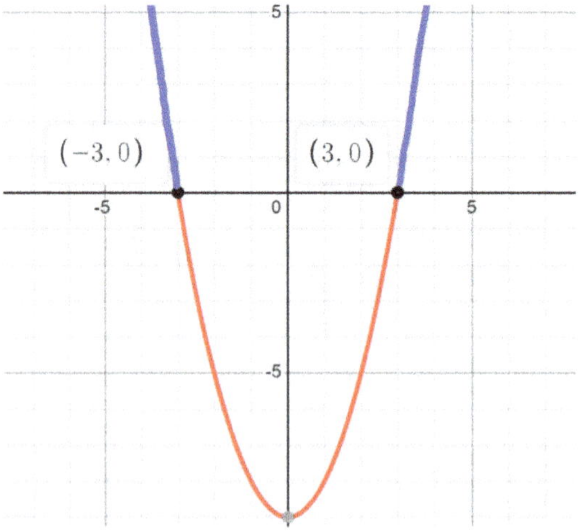

Image 3.51

Similarly, the solution to a quadratic function in the form $ax^2 + bx + c < 0$ would be the portion of the graph that lies below the x-axis. In the example above, the solution to $x^2 - 9 < 0$ would be $(-3, 3)$.

Write a program that asks the user to input a quadratic inequality in the form $ax^2 + bx + c < 0$ or $ax^2 + bx + c > 0$. Use the x-intercepts to formulate the solution.

Keep in mind that if $a < 0$, the parabola is reflected through the x-axis.

SAMPLE OUTPUT

Quadratic Inequality? x**2–7*x+10 < 0

The solution is (2, 5)

Quadratic Inequality? –x**2 + 2*x + 8 < 0

The solution is (–infinity, –2) U (4, infinity)

BONUS: Allow the user to input the inequality in factored form (such as $(x - 7)*(x + 4) > 0$) or in the form $ax^2 + bx < c$ or $ax^2 + bx > c$.

CHAPTER SUMMARY

PYTHON CONCEPTS

SymPy Library	A library of symbolic mathematics, including polynomial manipulation. See Table 3.2. *Ex:* *from sympy import **
To declare a variable as a symbol in SymPy	**Symbol**('string') *Ex:* x = ***Symbol***('x')
To convert a string to a SymPy expression	**sympify**(string representing numeric expression) *Ex:* *poly* = ***sympify***('x+2*x+x**2+7+4*x**2')
To "pretty print" a SymPy expression	**pprint**(Sympy type expression) *Ex:* ***pprint***(*poly*)
To expand (combine like terms or multiply out) a SymPy polynomial expression	**expand**(Sympy type expression) *Ex:* *poly* = ***expand***(*poly1 + poly2*) *or* *poly* = ***expand***(*poly1*poly2*)
To factor a polynomial in SymPy	**factor**(Sympy type polynomial) *Ex:* ***factor***(*polynomial*)

(Continued)

To find the quotient and remainder of polynomial division	**div**(dividend,divisor). Returns a list in the form [quotient, remainder]. *Ex:* *div('x**3–2*x**2–4','x–3')* *will return:* *(x**2 + x + 3, 5)*
To solve an algebraic equation in SymPy	To solve *f* = 0, use **solve**(*f*). Returns a list of complex (both real and imaginary) solutions. *Ex:* *zeros=solve(x**3+1)* *will return:* *[-1, 1/2 - sqrt(3)*I/2, 1/2 + sqrt(3)*I/2]*
To evaluate a SymPy expression for a given value of the variable	To find *f*(*a*), in which *f* is a SymPy type expression: f.**subs**({x: a}) *Ex:* *value = polynomial.subs({x:x_value})*
To factor a polynomial in SymPy	**factor**(polynomial)
Identifying degree and coefficients using SymPy	To find the degree of a SymPy type expression: **degree**(expr) To get a list of the coefficients of a SymPy type expression: **poly**(expr).**all_coeffs()**
String concatenation	Join two or more strings (and/or string variables) with a plus (+) sign. *Ex:* *greeting = "hello"+"world"*
Finding the sum of a list	**sum**(mylist) *Ex:* *>>>mylist = [1, 2, 3, 4]* *>>>sum(mylist)* *10*
Finding the max and min of a list	**max**(mylist) **min**(mylist) *Ex:* *>>>mylist = [1, 2, 3, 4]* *>>>max(mylist)* *4* *>>>min(mylist)* *1*

PRECALCULUS CONCEPTS

Polynomial	A mathematical expression in the form $a_nx^n + a_{n-1}x^{n-1} + a_{n-2}x^{n-2} + \ldots + a_1x + a_0$ where $a_n, a_{n-1}, a_{n-2}, \ldots a_1, a_0$ are real numbers (called the **coefficients**), x is a variable, and the exponents $n, n-1, n-2, \ldots$ are whole numbers. *Ex:* $3x^5 - 7x^4 + 5x^2 - x + 101$
Degree of a term	The degree of term a_kx^k is k, the exponent of the variable. *Ex:* *The degree of* $86x^{47}$ *is 47.* *The degree of 75 is 0.*
Degree of a polynomial	The highest degree of all its terms. *Ex:* *The degree of* $3x^5 - 7x^4 + 5x^2 - x + 101$ *is 5.*
Remainder theorem	The value of $f(a)$ is the remainder when $f(x)$ is divided by $(x - a)$. *Ex:* *If* $f(x) = x^3 - 2x^2 - 4$, *then* $f(-3)$ *is the remainder when* $x^3 - 2x^2 - 4$ *is divided by* $x + 3$.
Linear regression (aka least squares regression)	A method of approximating a line of best fit given a list of data. $m = \dfrac{N\sum(xy) - \sum x \sum y}{N\sum(x^2) - (\sum x)^2}$ $b = \dfrac{\sum y - m\sum x}{N}$
Vertex form of a quadratic equation	$y = a(x - h)^2 + k$ where (h, k) is the vertex.
Standard form of a quadratic equation	$y = ax^2 + bx + c, a \neq 0$ The x-coordinate of the vertex is $x = -\dfrac{b}{2a}$ The y-intercept is $(0, c)$.
Factoring sum and difference of cubes	$x^3 + a^3 = (x + a)(x^2 - ax + a^2)$ $x^3 - a^3 = (x - a)(x^2 + ax + a^2)$

(Continued)

Extrema of a graph	The **absolute maximum** is the y-value of the highest point on the graph. The **absolute minimum** is the y-value of the lowest point on the graph. A **local maximum** is the highest point in a local area of the graph. A **local minimum** is the lowest point in a local area of the graph. 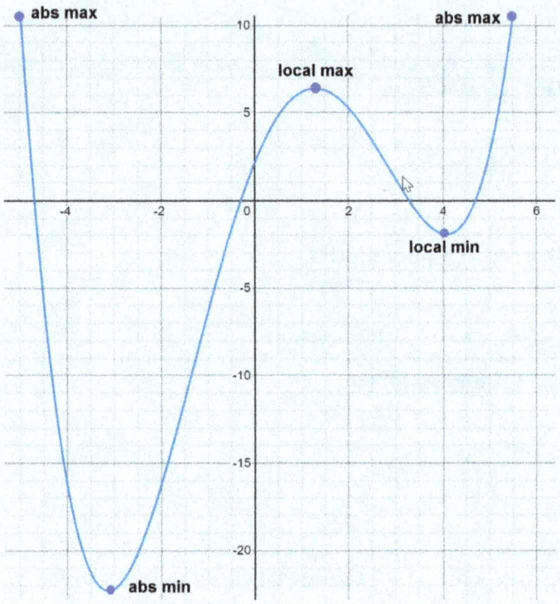 Image 3.52

Credits

IMG 3.8: Adapted from Copyright © by Wikimedia Rest API (CC BY-SA 3.0) at https://wikimedia.org/api/rest_v1/media/math/render/svg/a72e0ccc1a771f9725311b6346ecad57f9c2d02e.

IMG 3.17: Copyright © by Gjacquenot (CC BY-SA 4.0) at https://commons.wikimedia.org/wiki/File:PascalsTriangleCoefficient.svg.

IMG 3.18: Copyright © by Drini (CC BY-SA 3.0) at https://commons.wikimedia.org/wiki/File:Pascal%27s_triangle_5.svg.

IMG 3.31: Copyright © 2014 Depositphotos/Olivier26.

IMG 3.33: Copyright © 2014 Depositphotos/glowonconcept.

IMG 3.38a: Copyright © by Gan Khoon Lay (CC BY 3.0) at https://commons.wikimedia.org/wiki/File:Icon-roller-coaster-881334.svg.

IMG 3.46: Copyright © by Laerd Statistics (CC BY-SA 4.0) at https://commons.wikimedia.org/wiki/File:Pearson_Correlation_Coefficient_and_associated_scatterplots.png.

IMG 3.47: Copyright © 2012 Depositphotos/xtrekx.

CHAPTER 4

Systems of Equations and Matrices

4.1 Matrix Fundamentals

Matrices have many applications in mathematics, amongst them solving linear systems, optimization, and computer animation. Spreadsheets are essentially matrices.

So far, we have used lists for storing sequences of numbers such as [2, 4, 8, 16, 32], coefficients of polynomials, etc.

A **matrix** can be thought of as a 2-dimensional list of numbers:

n columns

m rows

$$\begin{bmatrix} a_{1,1} & a_{1,2} & \cdots & a_{1,n} \\ a_{2,1} & a_{2,2} & \cdots & a_{2,n} \\ \vdots & \vdots & & \vdots \\ a_{m,1} & a_{m,2} & \cdots & a_{m,n} \end{bmatrix}$$

Matrix element $a_{i,j}$ is located at row i, column j

Image 4.1

We typically refer to a matrix with m rows and n columns as an $m \times n$ matrix.

A $1 \times n$ matrix is called a **row matrix**.

$$[a_{1,1} \; a_{1,2} \; \cdots \; \cdots \; a_{1,n}]$$

An $m \times 1$ matrix is called a **column matrix**.

$$\begin{bmatrix} a_{1,1} \\ a_{2,1} \\ \vdots \\ \vdots \\ a_{m,1} \end{bmatrix}$$

A matrix with the same number of rows and columns is called a **square matrix**.

$$\begin{bmatrix} 9 & 13 & 5 \\ 1 & 11 & 7 \\ 3 & 7 & 4 \end{bmatrix}$$

A matrix is usually named with an uppercase letter (such as A), and its **elements** (aka **entries**) would be described with subscripted lowercase letters (such as $a_{i,j}$).

Two matrices are considered **equal** if they have the same dimensions and corresponding elements are equal.

$$\underbrace{\begin{bmatrix} 1 & 2 & 3 \\ 4 & 5 & 6 \\ 7 & 8 & 9 \end{bmatrix}}_{\text{Matrix A}} = \underbrace{\begin{bmatrix} 1 & 2 & 3 \\ 4 & 5 & 6 \\ 7 & 8 & 9 \end{bmatrix}}_{\text{Matrix B}}$$

Image 4.2

Matrix addition and Subtraction is intuitive. The matrices must have the same dimensions. Simply add or subtract corresponding elements.

$$\begin{bmatrix} 1 & -4 & 5 \\ 2 & 0 & -8 \end{bmatrix} + \begin{bmatrix} -4 & -2 & 1 \\ 0 & 1 & 5 \end{bmatrix} = \begin{bmatrix} -3 & -6 & 6 \\ 2 & 1 & -3 \end{bmatrix}$$

EXERCISE SET 4.1.1

1. Given $A = \begin{bmatrix} 7 & 6 & 3 \\ 4 & -2 & 0 \\ 1 & 11 & -5 \end{bmatrix}$

   A. What is $a_{1,1}$?

   B. What is $a_{2,3}$?

2. Identify the dimensions (row × column) of the following matrices.

   A. $\begin{bmatrix} 1 & 2 & 3 \\ 4 & 5 & 6 \end{bmatrix}$

   B. $\begin{bmatrix} 1 & 0 \\ 0 & 1 \end{bmatrix}$

   C. $\begin{bmatrix} 5 \\ 7 \\ -3 \end{bmatrix}$

   D. $[6 - 2 - 5]$

   E. $\begin{bmatrix} -1 & 0 & 2 \\ 0 & 4 & 7 \\ 3 & -4 & -2 \end{bmatrix}$

   F. $[44]$

3. Perform the following matrix operations.

   A. $\begin{bmatrix} 1 & 2 & 3 \\ 4 & 8 & 5 \end{bmatrix} + \begin{bmatrix} 0 & 1 & 8 \\ 1 & 7 & 0 \end{bmatrix}$

   B. $\begin{bmatrix} 2 & -18 \\ 20 & -5 \end{bmatrix} - \begin{bmatrix} -4 & 2 \\ -5 & 1 \end{bmatrix}$

In Python, a matrix is considered a 2-dimensional list, or a list of lists. For example, the 2×3 matrix $A = \begin{bmatrix} 1 & 2 & 3 \\ 4 & 5 & 6 \end{bmatrix}$ would be represented as:

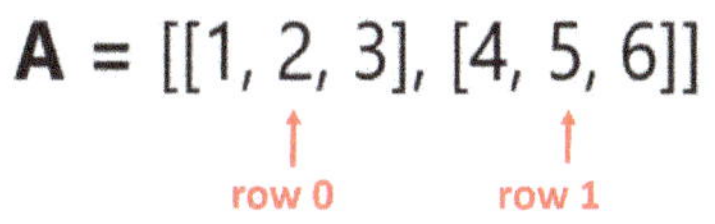

Image 4.3

Just like lists, the first index in each row and column is 0. Thus, in the above example:

$A[0][0] = 1$ $\quad$ $A[0][1] = 2$ $\quad$ $A[0][2] = 3$

$A[1][0] = 4$ $\quad$ $A[1][1] = 5$ $\quad$ $A[1][2] = 6$

Do not confuse this indexing schema with the mathematical convention of starting the first row and column at 1—i.e., $A[0][0] = a_{1,1}$ and so on.

TABLE 4.1 Matrix Notation and Functions in Python

To initialize an empty matrix:	A = [ ]
Each element in A is $A[\ i\ ]\ [\ j\ ]$ *(extra spaces added here for readability)* Assigning a value to a single matrix element:	$A[\ i\ ]\ [\ j\ ] = x$
Number of rows in matrix A:	**len**(A)
Number of columns in matrix A:	**len**(A[0])
The ith row of matrix A (which is a list):	$A[i]$
To append a new row to matrix A (where row is a list of the appropriate length):	A.**append**(row)

Here is a simple program that (*A*) assigns $\begin{bmatrix} 1 & 2 & 3 \\ 4 & 5 & 6 \end{bmatrix}$ to matrix *A*, (*B*) asks the user to input a row and column, (*C*) performs an error check, and (*D*) outputs the desired entry in that row, column.

PROGRAM 4.1.1

```
A = [[ 1, 2, 3], [4, 5, 6]]
r = int(input("row? "))
c = int(input("column? "))
if r in [0,1] and c in [0,1,2]:
    print(A[r][c])
else:
    print("out of range")
```

Because we will be working with programs that manipulate matrices of various sizes, it will be handy to have a program (or function) that (*A*) asks the user to input the dimensions of a matrix, (*B*) initializes each element to 0, then (*C*) outputs the contents of the matrix.

PROGRAM 4.1.2

```
r = int(input("Rows? "))
c = int(input("Columns? "))
A = [ ]                           # initialize the matrix
for i in range(r):                # for each row
    row = [ ]                     # initialize row i
    for j in range(c):            # for each column
        row.append(0)             # append 0 to row i
    A.append(row)                 # append row to matrix A
print(A)
```

Notes:

1. The matrix must be initialized with $A = [\,]$ before any elements are assigned to it.
2. $A[i][j] = \text{value}$ cannot be used to assign a value to a matrix element that doesn't already exist. It can only be used to reassign.

SAMPLE OUTPUT

Rows? 2

Columns? 3

[[0, 0, 0], [0, 0, 0]]

The **SymPy** library contains a number of matrix attributes and methods that would render most of this chapter useless. Likewise **NumPy** is a popular matrix-handling library. A thorough understanding of matrices as mathematical structures is best gained by writing your own functions. Also, the **SymPy** library is probably not available for other programming languages, and you will want to be a versatile programmer.

EXERCISE SET 4.1.2

1. Modify program 4.1.1 so that $A = \begin{bmatrix} 1 & 2 \\ 3 & 4 \\ 5 & 6 \end{bmatrix}$.
2. Modify program 4.1.2 so that instead of assigning 0s to each matrix element, it assigns a random number between 1 and 10 to each element.
3. Write a program that calculates and outputs the sum of the following two matrices:

$$A = \begin{bmatrix} 1 & 0 \\ -2 & 5 \end{bmatrix} \quad B = \begin{bmatrix} -3 & 10 \\ 4 & -7 \end{bmatrix}$$

4. If we were to have the program assign $A = \begin{bmatrix} 1 & 0 \\ -2 & 5 \end{bmatrix}$ *and* $B = \begin{bmatrix} -3 & 10 \\ 4 & -7 \end{bmatrix}$, what would be the result of **print**$(A + B)$?

List Comprehension

Notice program 4.1.2 requires a pair of nested loops (**for** *j* and **for** *i*) to traverse every element in the matrix. As we saw in Chapter 3, Python offers a wonderful alternative technique that streamlines this process.

In general, the syntax is:

list_name = [*f*(*x*) **for** *x* **in** {domain}]

where *f*(*x*) is an expression (such as a constant or mathematical operation) and {domain} is another list or a range. In the case of a matrix, the list_name can be a row of a matrix.

Here are some examples:

```
>>>integer_list = [int(x) for x in [1.618, 2.712, 3.1415]]
>>>integer_list
[1, 2, 3]
>>>c_list = [a**2+b**2 for a in range(3) for b in range(3)]
>>>c_list
[0, 1, 4, 1, 2, 5, 4, 5, 8]
>>>x_list = [0, 1, 2, 3]
>>>y_list = [4, 5, 6, 7]
>>>points = [[x, y] for x in x_list for y in y_list]
>>>points
[[0, 4],[0,5], [0,6], [0,7], [1,4], [1,5], [1,6], [1,7], [2,4], [2,5], [2,6], [2,7], [3,4], [3,5], [3,6], [3,7]]
```

EXERCISE SET 4.1.3

1. Write a list comprehension statement that assigns to list *A* the first ten multiples of 5.
2. Write a list comprehension statement that assigns to list *B* the square of each number in a list called list *C*.
3. Write a list comprehension statement that assigns to list *C* the product of each element in list *D* with each element in list *E*.

   Sample output with list *D* = [6,7,8] and list *E* = [3,2,1]:

   [18, 12, 6, 21, 14, 7, 24, 16, 8]

Going back to program 4.1.2, we can streamline it as follows:

```
r = int(input("Row? "))
c = int(input("Columns? "))
A = [ ]                                    # initialize the matrix
for i in range(r):                         # for each row
    row = [0 for j in range(c)]
    A.append(row)                          # append row to matrix A
print(A)
```

But wait! We can streamline even further with an elegant "nested" list comprehension:

```
r = int(input("Rows? "))
c = int(input("Columns? "))
A = [ [ 0 for j in range(c) ] for i in range(r) ]
print(A)
```

Note that it is no longer necessary to initialize matrix *A*; the nested list comprehension both creates and assigns values.

EXERCISE SET 4.1.4

1. Write a brief function called **init_matrix**(r,c) that returns a matrix M with dimensions $r \times c$ in which all the entries are 0.

Formatting Output of a Matrix

The default output for a matrix in Python is not very readable. For example, if $A = \begin{bmatrix} 2 & 7 \\ -8 & 9 \\ 0 & -12 \end{bmatrix}$ then **print**(A) would yield:

[[2, 7], [−8, 9], [0, −12]]

Let's take advantage of Python's formatting features so that each number would be centered in a field width of 6. Rows would actually be printed out in rows. For the above example, *A* would be output thusly:

Col [1 - 6]	Col [7 - 12]
2	7
−8	9
0	−12

Formatting the output of matrix *A* would require the following four lines:

```
fstring = '{:^6n} {:^6n}'
print(fstring.format(2,7))
print(fstring.format(−8,9))
print(fstring.format(0,−12))
```

An entire chapter could be written about string formatting. Suffice to say that **fstring** defines a format for two numbers to be output in fields of width 6 characters.

We incorporate these concepts into the following function.

```
def print_matrix(M):
    r = len(M)                                  # number of rows
    c = len(M[0])                               # number of columns
# create format string: one {:^6n} for each column
    fstring = ''
    for j in range(c):
        fstring += '{:^6n}'
    # print the matrix
    for i in range(r):
        print(fstring.format(*M[i]))            # print row i using format
    return
```

Create a new file **matrix_fcn.py** and put in it functions **print_matrix**() (above) and **init_matrix**() (that you wrote in Exercise Set 4.1.4). We can then include **matrix_fcn.py** in future matrix programs.

Scalar Multiplication

A **scalar** is simply a number. Scalar multiplication is sort of a distributive property for matrices:

$$2\begin{bmatrix} 1 & 2 \\ 3 & 4 \end{bmatrix} = \begin{bmatrix} 2(1) & 2(2) \\ 2(3) & 2(4) \end{bmatrix} = \begin{bmatrix} 2 & 4 \\ 6 & 8 \end{bmatrix}$$

Program Goal: Write a program **scalar_mult** to perform scalar multiplication using the example above.

PROGRAM 4.1.3 SCALAR_MULT.PY

```
# include user-defined functions
import matrix_fcn as mf
A = [[1,2],[3,4]]
r = len(A)                                        # of rows
c = len(A[0])                                     # of columns
# create and initialize matrix B same size as A
B = mf.init_matrix(r,c)
scalar = 2
for i in range(r):                                # for each row in A
    B[i] = [scalar*A[i][j] for j in range(c)]     # generate next row of B
mf.print_matrix(B)
```

Desired Output

```
4    8
12   16
```

Notice we are generating one row at a time of matrix *B*.

EXERCISE SET 4.1.5

1. Perform the following matrix operations.

   A. $2\begin{bmatrix} 1 & 5 \\ 0 & -6 \end{bmatrix}$

   B. $\frac{1}{3}\begin{bmatrix} 6 & 4 \\ -9 & -1 \end{bmatrix}$

   C. $3\begin{bmatrix} 1 & 5 \\ -1 & -5 \end{bmatrix} + 4\begin{bmatrix} -4 & -3 \\ -.2 & -1 \end{bmatrix}$

   D. $\frac{3}{4}\begin{bmatrix} 8 & 12 \\ -16 & 20 \end{bmatrix} + \frac{2}{3}\begin{bmatrix} 27 & -9 \\ 54 & -18 \end{bmatrix}$

2. In program **scalar_mult**, couldn't we just shorten the program as follows:

```
import matrix_fcn as mf
A = [[2,4],[6,8]]
scalar = 2
```

(Continued)

```
B = scalar*A
mf.print_matrix(B)
```

Why or why not?

3. Write a program that squares all the entries of $\begin{bmatrix} 11 & 12 & 13 \\ 14 & 15 & 16 \end{bmatrix}$.

   Use list comprehension if possible.

4. Modify program **scalar_mult**, or write a new one, that performs the matrix operations in exercise 1C above. Use list comprehension if possible.

5. Write a program that generates a multiplication table for integers up to 9.

Desired Output:

0	1	2	3	4	5	6	7	8	9
1	1	2	3	4	5	6	7	8	9
2	2	4	6	8	10	12	14	16	18
3	3	6	9	12	15	18	21	24	27
4	:	:	:	:	:	:	:	:	:
5									
6									
7									
8									
9	9	18	27	36	45	54	63	72	81

Our matrix programs would be far more useful if we allowed the user to input the matrices, rather than hard-coding the matrix values. This function will ask the user to input the rows one at a time:

```
def input_matrix():
    r = int(input("# of rows? "))
    c = int(input("# of columns? "))
    M = [ ]
    for i in range(r):
        row_str = input("Row "+str(i+1)+"separated by commas:")
        row_list = [int(k) for k in row_str.split(",")]
        M.append(row_list)
    return M
```

For the sake of simplicity, we will work only with matrices with integer values, not float.

EXERCISE SET 4.1.6

1. Modify the **input_matrix**() function so that it accepts two parameters: *r* and *c*. If either parameter is 0, then prompt for the dimensions. Otherwise, use those values for the rows and columns.

2. Modify program **scalar_mult** so that the user is prompted to enter a scalar, the dimensions of a matrix, and the matrix. Perform scalar multiplication.

   Sample Output

   Scalar? –3

   # of rows? 2

   # of columns? 2

   Row 1 separated by commas: 3,0

   Row 2 separated by commas: 2, –5

–9	0
–6	15

3. Write a program that asks the user for a single number *N*. Create a square matrix with *N* rows and *N* columns. Each element of the matrix will be 0 EXCEPT for the elements along the diagonal, which will be 1. For example, a 3×3 matrix would look like this:

$$\begin{bmatrix} 1 & 0 & 0 \\ 0 & 1 & 0 \\ 0 & 0 & 1 \end{bmatrix}$$

This is in fact a special kind of square matrix called the **identity matrix**.

Program Goal: Write a lottery "lucky number" generating program. In the lottery ticket shown, there are 6 rows (representing 6 wagers), and 6 columns. The first 5 numbers in each row are randomly chosen between 1 and 47. The 6th number, called the "MEGA" number, is a randomly chosen number between 30 and 50. Output as a matrix.

Image 4.4

PROGRAM 4.1.4 LOTTERY.PY

```
from random import randint
# user-defined functions
import matrix_fcn as mf
lotto = mf.init_matrix(6,6)
[ assign numbers to row i of lotto matrix ]
mf.print_matrix(lotto)
```

EXERCISE SET 4.1.7

1. Fill in the yellow highlighted code to generate each row of the lotto matrix.

The limitation of program **lottery** is that it is possible to generate duplicates within the rows, which will get you barred from the lottery programmer's union. We need a function that will generate a list of unique numbers within a specified range.

```
def randint_norepeat(m,k,n):
    # returns a list of m unique random numbers in the range [k,n]
    from random import shuffle
    numbers = list(range(k,n+1))
    shuffle(numbers)
    random_list = [ ]
    for i in range(m):
        random_list.append(numbers.pop())
    return random_list
```

EXERCISE SET 4.1.8

1. Incorporate the use of **randint_norepeat**() into program **lottery**.
2. Modify the program so that the first 5 numbers in each row are sorted least to greatest. Use the list.**sort**() method, which sorts a list.

4.2 Matrix Multiplication

I'd like to tell you that matrix multiplication is completely intuitive, as is matrix addition: simply multiply corresponding entries.

But I'd be lying.

The best I can do is tell you that matrix multiplication is an acquired taste. Let's start with multiplying a row matrix by a column matrix:

$$\begin{bmatrix} 1 & 2 & 3 \end{bmatrix}\begin{bmatrix} 4 \\ 5 \\ 6 \end{bmatrix} = [1*4 + 2*5 + 3*6] = [32]$$

Some important notes:

1. The answer is not 32. It is a 1×1 matrix consisting of a single element, which is 32.
2. The number of entries in the row matrix must be the same as the number of entries in the column matrix.
3. The row matrix always precedes the column matrix.

In general, if $A = \begin{bmatrix} a_{1,1} & a_{1,2} & \cdots & a_{1,n} \end{bmatrix}$ and $B = \begin{bmatrix} b_{1,1} \\ b_{2,1} \\ \vdots \\ b_{n,1} \end{bmatrix}$ then their product is:

$$AB = \begin{bmatrix} a_{1,1} & a_{1,2} & \cdots & a_{1,n} \end{bmatrix}\begin{bmatrix} b_{1,1} \\ b_{2,1} \\ \vdots \\ b_{n,1} \end{bmatrix} = \begin{bmatrix} a_{1,1} \cdot b_{1,1} + a_{1,2} \cdot b_{2,1} + \cdots + a_{1,n} \cdot b_{n,1} \end{bmatrix}$$

This is also called the **dot product** and has applications to vectors.

EXERCISE SET 4.2.1

Multiply the following, if possible.

1. $\begin{bmatrix} -1 & 5 \end{bmatrix}\begin{bmatrix} 3 \\ 2 \end{bmatrix}$
2. $\begin{bmatrix} 3 \\ 2 \end{bmatrix}\begin{bmatrix} -1 & 5 \end{bmatrix}$
3. $\begin{bmatrix} 7 & 2 & -43 \end{bmatrix}\begin{bmatrix} 5 \\ -2 \\ 0 \end{bmatrix}$
4. $\begin{bmatrix} 6 & -11 & 5 \end{bmatrix}\begin{bmatrix} 2 \\ 5 \\ -2 \end{bmatrix}$
5. $\begin{bmatrix} -2 & -7 & 8 \end{bmatrix}\begin{bmatrix} 4 \\ -9 \end{bmatrix}$
6. The first row of A times the second column of B.

$$A = \begin{bmatrix} 1 & 2 & 3 \\ 4 & 5 & 6 \end{bmatrix} \quad B = \begin{bmatrix} 10 & 11 \\ 20 & 21 \\ 30 & 31 \end{bmatrix}$$

Program Goal: Write a program to multiply a row matrix by a column matrix. Use the example above:

$$\begin{bmatrix} 1 & 2 & 3 \end{bmatrix} \begin{bmatrix} 4 \\ 5 \\ 6 \end{bmatrix}.$$

PROGRAM 4.2.1

```
A = [1,2,3]
B = [4,5,6]
result = 0
for k in range(3):
    result += A[k]*B[k]
print("Product is",result)
```

Notice Python does not distinguish between a "row" and "column" matrix. It considers both as 1-dimensional lists.

In the following Exercise Set, you will expand this program to accept input of any size row and column matrices.

EXERCISE SET 4.2.2

1. Modify program 4.2.1 as follows: (1) Rather than assigning values for A and B, request user to input values of A and B. (2) If A and B are not the same size, output an error message. (3) Otherwise, proceed to multiply and output the result.

Sample Output

A? 8,9,–3

B? 0,1,6

Product is –9

A? 7,32,–11

B? 2,–3

Sorry, *A* and *B* must be the same size.

Now that we've mastered multiplying a row matrix by a column matrix (also known as the **dot product**), let's expand to larger matrices. The process is essentially repeating the above row-column multiplication for each row of the first matrix.

This is best illustrated with an example.

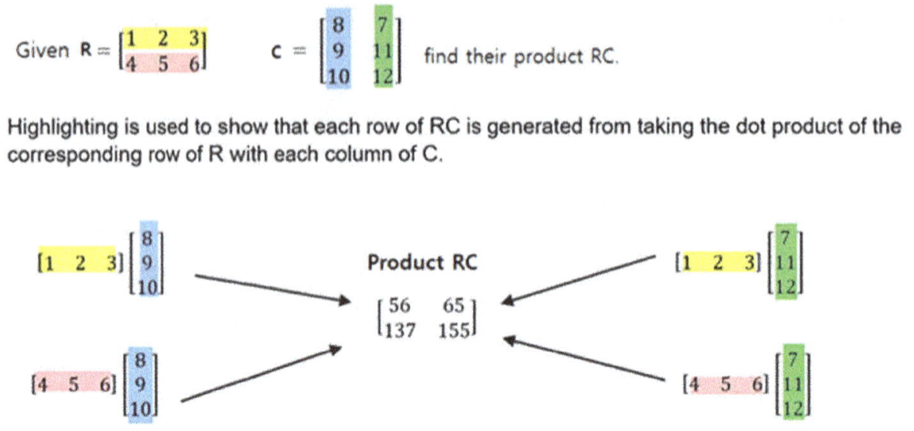

Image 4.5

You cannot multiply any two matrices. You can only multiply an $r \times k$ matrix with a $k \times c$ matrix. The result is an $r \times c$ matrix.

Example

Find the product:

$$\begin{bmatrix} 1 & 4 \\ 7 & 9 \end{bmatrix}\begin{bmatrix} 5 & 2 \\ -1 & 0 \end{bmatrix}$$

Solution

We are multiplying a 2 × 2 matrix by a 2 × 2 matrix, so the result is a 2 × 2 matrix.

$$\begin{bmatrix} 1 & 4 \\ 7 & 9 \end{bmatrix}\begin{bmatrix} 5 & 2 \\ -1 & 0 \end{bmatrix} = \begin{bmatrix} 1\cdot 5 + 4\cdot -1 & 1\cdot 2 + 4\cdot 0 \\ 7\cdot 5 + 9\cdot -1 & 7\cdot 2 + 9\cdot 0 \end{bmatrix} = \begin{bmatrix} 1 & 2 \\ 26 & 14 \end{bmatrix}$$

EXERCISE SET 4.2.3

In #1–5, if possible, multiply the following matrices.

1. $\begin{bmatrix} 8 & 5 \\ -2 & -1 \end{bmatrix}\begin{bmatrix} 6 & 10 \\ 2 & -4 \end{bmatrix}$

2. $\begin{bmatrix} 6 & 10 \\ 2 & -4 \end{bmatrix}\begin{bmatrix} 8 & 5 \\ -2 & -1 \end{bmatrix}$

3. $\begin{bmatrix} 6 & 7 & 0 \\ 4 & 2 & 15 \end{bmatrix}\begin{bmatrix} 1 & 2 \\ 9 & 8 \\ 3 & -5 \end{bmatrix}$

4. $\begin{bmatrix} 1 & 2 \\ 9 & 8 \\ 3 & -5 \end{bmatrix}\begin{bmatrix} 6 & 7 & 0 \\ 4 & 2 & 15 \end{bmatrix}$

5. $\begin{bmatrix} 1 & 11 & -8 \\ 7 & 7 & 2 \\ -1 & 10 & -4 \end{bmatrix}\begin{bmatrix} 1 & 0 & 0 \\ 0 & 1 & 0 \\ 0 & 0 & 1 \end{bmatrix}$

(Continued)

6. Is matrix multiplication commutative?
7. Referring to your answer to #5 above, what is the result of multiplying by the identity matrix?
8. A college professor uses a weighted average to award a final course grade as follows:

   Homework 15%

   Tests 30%

   Final Exam 40%

   Semester Project 15%

   As this is a graduate-level course, there are only 5 students in the class. Their final scores in each category are as follows:

	Homework	Tests	Final Exam	Semester Project
Amelia	84	72	75	91
Bradley	82	68	74	85
Kimiko	99	85	86	95
Wesley	60	74	68	75
Antwone	89	91	96	0

   Set this up as a matrix multiplication problem to calculate each student's final grade in the course. (Hint: You will be multiplying a 5×4 matrix by a 4×1 matrix.)

Program Goal: Write a program **matrix_product** to multiply two matrices with compatible dimensions (i.e., the number of columns of the first matrix is the same as the number of rows in the 2nd matrix).

Let's start with a specific example:

$$\begin{bmatrix} 1 & 2 & 3 \\ 4 & 5 & 6 \end{bmatrix} \begin{bmatrix} 7 & 8 \\ 9 & 10 \\ 11 & 12 \end{bmatrix} = \begin{bmatrix} 58 & 64 \\ 139 & 154 \end{bmatrix}$$

To derive a formula for each element in the product (which we'll call P), take a look at the 154 in the example above:

$$154 = 4{*}8 + 5{*}10 + 6{*}12$$

Using variables and subscripts:

$$p_{2,2} = a_{2,1} \cdot b_{2,2} + a_{2,2} \cdot b_{2,2} + a_{2,3} \cdot b_{3,2}$$

Using variables in the subscripts:

$$p_{i,j} = a_{i,1} \cdot b_{1,j} + a_{i,2} \cdot b_{2,j} + a_{i,3} \cdot b_{3,j}$$

Generalizing to n terms:

$$p_{i,j} = a_{i,1} \cdot b_{1,j} + a_{i,2} \cdot b_{2,j} + \ldots + a_{i,n} \cdot b_{n,j}$$

Using summation notation, the element in the ith row, jth column of product P is:

$$p_{i,j} = \sum_{k=1}^{n} a_{i,k} \cdot b_{k,j} \text{ where } n \text{ is the \# of columns in } [A] \text{ (or number of rows in } [B])$$

PROGRAM 4.2.2 MATRIX_PRODUCT.PY

```
import matrix_fcn as mf
A = [[1,2,3],[4,5,6]]
B = [[7,8],[9,10],[11,12]]
A_num_rows = len(A)
A_num_cols = len(A[0])
B_num_rows = len(B)
B_num_cols = len(B[0])
if A_num_cols != B_num_rows:
    print("Sorry, # of A columns must equal # of B rows")
else:
    P = mf.init_matrix(A_num_rows,B_num_cols)
    for i in range(A_num_rows):                    # for each row in A
    for j in range(B_num_cols):                    # for each column in B
    [ Fill in code to calculate value of p_i,j ]
    mf.print_matrix(P)
```

There are various Python libraries such as **NumPy** that perform matrix multiplication with a single function call; however, we gain more programming acumen by writing our own algorithm.

EXERCISE SET 4.2.4

1. Fill in yellow highlighted code to calculate value of $p_{i,j}$.
2. Modify the program so that user can input any two compatible matrices *A* and *B*.
3. Convert the program to a function **matrix_product**(*A*,*B*) that accepts two matrices *A* and *B* as parameters then returns their product *P* to the main program. If the matrices do not have compatible dimensions, then return the zero matrix [[0]]. Add the function to file **matrix_fcn.py**.

4.3 Determinant and Inverse of a 2 × 2 Matrix

Let's say we would like to solve a matrix equation that looks something like this:

$$\begin{bmatrix} 2 & -3 \\ 5 & 3 \end{bmatrix} \cdot \begin{bmatrix} x \\ y \end{bmatrix} = \begin{bmatrix} 15 \\ 27 \end{bmatrix}$$

One of the concepts needed to solve a matrix equation is that of the inverse. Recall when solving the algebraic equation $ax = b$ using real numbers:

$$x = \frac{b}{a} = \frac{1}{a} \cdot b = a^{-1} \cdot b$$

where a^{-1} is the multiplicative inverse of a.

We will likewise define the **multiplicative inverse** of a 2 × 2 matrix as follows.

The **multiplicative inverse** of $A = \begin{bmatrix} a & b \\ c & d \end{bmatrix}$ is:

$$A^{-1} = \frac{1}{ad - bc}\begin{bmatrix} d & -b \\ -c & a \end{bmatrix}$$

provided $ad - -bc \neq 0$

If $ad - bc = 0$, then we say that A is **noninvertible**.

$ad - bc$ is called the **determinant** and its notation is $|A|$ or $\begin{vmatrix} a & b \\ c & d \end{vmatrix}$.

Example

Find A^{-1} if possible.

A. $\begin{bmatrix} 2 & -3 \\ 0 & 11 \end{bmatrix}$

B. $\begin{bmatrix} 4 & 2 \\ 8 & 4 \end{bmatrix}$

Solutions

A. $ad - bc = 2(11) - (-3)(0) = 22$

$$A^{-1} = \frac{1}{22}\begin{bmatrix} 11 & 3 \\ 0 & 2 \end{bmatrix} = \begin{bmatrix} 1/2 & 3/22 \\ 0 & 1/11 \end{bmatrix}$$

B. $ad - bc = 4(4) - (2)(8) = 0$ so the matrix is noninvertible. No inverse exists.

EXERCISE SET 4.3.1

If possible, find the inverse of each of the following matrices. Express your answers using rational, not decimal, numbers.

1. $\begin{bmatrix} 5 & 2 \\ -7 & 3 \end{bmatrix}$

2. $\begin{bmatrix} -3 & -2 \\ 7 & 5 \end{bmatrix}$

3. $\begin{bmatrix} -3 & 1 \\ 5 & -2 \end{bmatrix}$

4. $\begin{bmatrix} 4 & 2 \\ 2 & 1 \end{bmatrix}$

5. $\begin{bmatrix} 1 & 2 \\ 3 & 4 \end{bmatrix}$

6. Write a program **matrix_inverse** that asks the user to input a 2×2 matrix A and then calculates and outputs the inverse. Include a check to see if the matrix is invertible. Hint: **round**(n, 2) will round to 2 decimal places.

Sample Output

Row 1 separated by commas: 2,–3

Row 2 separated by commas: 0,11

[[0.5, 0.14], [−0.0, 0.09]]

Program Goal: Convert program **matrix_inverse** (Exercise Set 4.3.1 #6) so that the main program calls a function **matrix_inverse**(A) that accepts 2×2 matrix A as a parameter and returns the inverted matrix to the main program. Output the inverse.

SAMPLE OUTPUT

Row 1 separated by commas? 12,1

Row 2 separated by commas? 11,2

Inverse:

[[0.15, −0.08], [−0.92, 0.92]]

PROGRAM 4.3.1 MATRIX_INVERSE.PY

```
import matrix_fcn as mf
def matrix_inverse(A):
    I = [[0,0],[0,0]]
    d = A[0][0] *A[1][1]-A[0][1]*A[1][0]
    I[0][0] = round((1/d)*A[1][1],2)
    I[0][1] = round( (1/d)*-1*A[0][1],2)
    I[1][0] = round( (1/d)*-1*A[1][0],2)
    I[1][1] = round( (1/d)*A[0][0],2)
    return I

# main program
A = mf.input_matrix(2,2)
I = mf.matrix_inverse(A)
print("\nInverse:")
print(I)
```

An important skill for programmers is to test each program with a robust selection of test data so that they can handle errors with helpful messages. We explore this in the Exercise Set below.

EXERCISE SET 4.3.2

1. What happens if you replace **print**(I) with mf.**print_matrix**(I)?
2. What happens if you omit the **round**() functions and attempt to find the inverse of

$$\begin{bmatrix} 5 & 2 \\ -7 & 3 \end{bmatrix}?$$

3. What happens when you try to find the inverse of $\begin{bmatrix} 4 & 2 \\ 2 & 1 \end{bmatrix}$? Modify the **matrix_inverse** function so that it returns [[0]] if the matrix is noninvertible. Then modify the main program so that if the matrix in noninvertible, it simply prints "noninvertible."
4. What happens if the user inputs a matrix that is not a 2 × 2? Modify the **matrix_inverse** function to output an error message instead.

Now that we have perfected the **matrix_inverse** function, add it to your function library in **matrix_function.py**.

Back to:

$$\begin{bmatrix} 2 & -3 \\ 5 & 3 \end{bmatrix} \cdot \begin{bmatrix} x \\ y \end{bmatrix} = \begin{bmatrix} 15 \\ 27 \end{bmatrix}$$

If $A = \begin{bmatrix} 2 & -3 \\ 5 & 3 \end{bmatrix}$, $X = \begin{bmatrix} x \\ y \end{bmatrix}$ and $B = \begin{bmatrix} 15 \\ 27 \end{bmatrix}$, then we can think of this matrix equation as:

$$A \cdot X = B$$

Recall that the solution to real number equation $ax = b$ is $x = a^{-1}b$. The matrix counterpart is:

$$X = A^{-1} \cdot B$$

Now that we know how to find A^{-1}, we can solve this equation.

$$A^{-1} = \frac{1}{21}\begin{bmatrix} 3 & 3 \\ -5 & 2 \end{bmatrix} = \begin{bmatrix} \frac{1}{7} & \frac{1}{7} \\ \frac{-5}{21} & \frac{2}{21} \end{bmatrix}$$

$$X = A^{-1} \cdot B = \begin{bmatrix} \frac{1}{7} & \frac{1}{7} \\ \frac{-5}{21} & \frac{2}{21} \end{bmatrix} \cdot \begin{bmatrix} 15 \\ 27 \end{bmatrix} = \begin{bmatrix} 6 \\ -1 \end{bmatrix}$$

Thus $X = \begin{bmatrix} x \\ y \end{bmatrix} = \begin{bmatrix} 6 \\ -1 \end{bmatrix}$

EXERCISE SET 4.3.3

In #1–4, solve the matrix equation, if possible.

1. $\begin{bmatrix} 2 & 1 \\ -4 & 6 \end{bmatrix}\begin{bmatrix} x \\ y \end{bmatrix} = \begin{bmatrix} -1 \\ 42 \end{bmatrix}$

2. $\begin{bmatrix} 2 & 3 \\ -1 & 1 \end{bmatrix}\begin{bmatrix} x \\ y \end{bmatrix} = \begin{bmatrix} 1 \\ -3 \end{bmatrix}$

3. $\begin{bmatrix} 3 & -5 \\ 15 & 5 \end{bmatrix}\begin{bmatrix} x \\ y \end{bmatrix} = \begin{bmatrix} 3 \\ 21 \end{bmatrix}$

4. $\begin{bmatrix} 2 & -1 \\ 1 & \frac{1}{2} \end{bmatrix}\begin{bmatrix} x \\ y \end{bmatrix} = \begin{bmatrix} -1 \\ \frac{3}{2} \end{bmatrix}$

Program Goal: Write a program **matrix_solve** that will solve matrix equation $A \cdot X = B$ using $X = A^{-1} \cdot B$.

SAMPLE OUTPUT

Matrix *A*:

Row 1 separated by commas: 1,1

Row 2 separated by commas: 2,3

Matrix *B*:

Row 1 separated by commas: 5

Row 2 separated by commas: 8

[[7.0], [−2.0]]

PROGRAM 4.3.2 MATRIX_SOLVE.PY

```
import matrix_fcn as mf
print("matrix A:")
A = mf.input_matrix(2,2)
print("matrix B:")
B = mf.input_matrix(2,1)
I = mf.matrix_inverse(A)
X = mf.matrix_product(I,B)
print(X)
```

EXERCISE SET 4.3.4

1. What is the result when you run the program to solve:

$$\begin{bmatrix} 2 & -3 \\ 5 & 3 \end{bmatrix}\begin{bmatrix} x \\ y \end{bmatrix} = \begin{bmatrix} 15 \\ 27 \end{bmatrix}$$

   What is the actual answer when you solve by hand? How can you fix this?

2. What is the result when you run the program to solve:

$$\begin{bmatrix} 1 & 2 \\ 2 & 4 \end{bmatrix}\begin{bmatrix} x \\ y \end{bmatrix} = \begin{bmatrix} 8 \\ 14 \end{bmatrix}$$

   Modify the program so that the output is simply:

   noninvertible
   no solution

3. What is the result when you run the program to solve:

$$\begin{bmatrix} 1 & 2 \\ 1 & 1/2 \end{bmatrix}\begin{bmatrix} x \\ y \end{bmatrix} = \begin{bmatrix} 4 \\ 4 \end{bmatrix}$$

Now that we know how to solve $AX = B$, where A is a 2×2 matrix, could we devise a similar formula for larger matrices?

The answer is yes, but it requires a bit more finesse. First, let's look at the relationship between matrix equations and linear systems.

4.4 Linear Systems

A linear system is one or more linear equations, for example:

$$2x - 3y = 15$$
$$5x + 3y = 27$$

This can be represented as a single matrix equation:

$$\begin{bmatrix} 2 & -3 \\ 5 & 3 \end{bmatrix} \begin{bmatrix} x \\ y \end{bmatrix} = \begin{bmatrix} 15 \\ 27 \end{bmatrix}$$

Note the two equations must be written in the form:

$$a_{1,1}x_1 + a_{1,2}x_2 = b_1$$
$$a_{2,1}x_1 + a_{2,2}x_2 = b_2$$

so that they can be converted to:

$$\begin{bmatrix} a_{1,1} & a_{1,2} \\ a_{2,1} & a_{2,2} \end{bmatrix} \begin{bmatrix} x_1 \\ x_2 \end{bmatrix} = \begin{bmatrix} b_1 \\ b_2 \end{bmatrix}$$

EXERCISE SET 4.4.1

Rewrite the following linear systems as matrix equations and use program **matrix_solve,** if possible, to solve the following linear systems.

1. $3x - 4y = 4$
   $\frac{1}{2}x - 3y = -\frac{1}{2}$
2. $2y + 3x = 2$
   $x - 7y = -30$
3. $4x + 5y = -3$
   $-2y = -8$
4. $4x + 8y = 11$
   $17 - 4y = 2x$
5. $x - y = 4$
   $y = 8 - x$
6. $x + 2y = 4$
   $x - 4 = -\frac{1}{2}y$

Handling Fractions

As we saw in Exercise Set 4.4.1 #6, if we attempt to use **matrix_solve** to solve:

$$\begin{bmatrix} 1 & 2 \\ 1 & \frac{1}{2} \end{bmatrix}\begin{bmatrix} x \\ y \end{bmatrix} = \begin{bmatrix} 4 \\ 4 \end{bmatrix}$$

... we get an error. We can fix this by modifying the **input_matrix** function to handle float numbers input rather than restricting it to integers. However, there is an easier fix on the user side. Keep in mind that multiplying both sides of a linear equation by a constant does **not** change the equation. So $x + \frac{1}{2}y = 4$ is the same as $2x + y = 8$. Thus, you can get the same solution to the linear system using:

$$\begin{bmatrix} 1 & 2 \\ 2 & 1 \end{bmatrix}\begin{bmatrix} x \\ y \end{bmatrix} = \begin{bmatrix} 4 \\ 8 \end{bmatrix}$$

We will briefly look at a few other methods to solve linear systems that do not involve matrices.

Let's focus on a linear system of two equations in two variables. First, recall from a prerequisite course that a linear equation represents the graph of a line. The solution to a linear system is the point of intersection (if there is one) of the two lines.

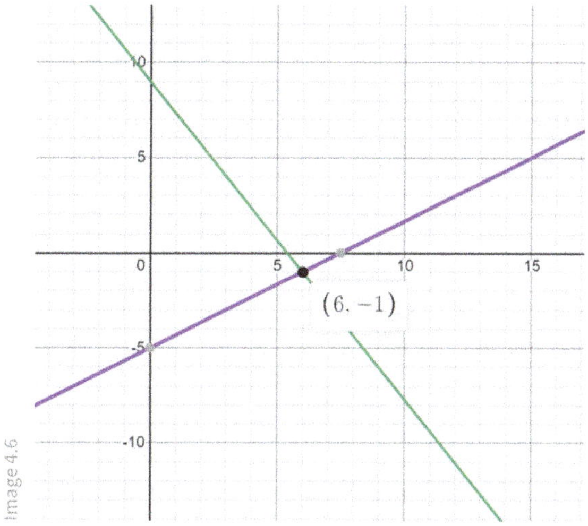

Important: The solution is an ordered pair (x, y).

Graphing is useful only in estimating the solution, therefore we turn to two other nonmatrix methods. You have probably already learned these in the math prerequisite to this course.

Substitution Method of Solving a Linear System

This method is useful if one of the equations is defined explicitly in terms of the other variable, e.g., $y = 3x + 2$ or $x = 5 - 4y$.

Example

Solve the linear system using the substitution method.

$$y = -2x - 1$$
$$-2x + 3y = 21$$

Solution

Notice the first equation has y defined in terms of the other variable, so we will substitute the equivalent expression for y into y of the second equation:

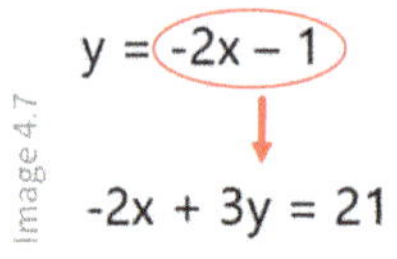

$$-2x + 3(-2x - 1) = 21$$
$$-2x - 6x - 3 = 21$$
$$-8x = 24$$
$$x = -3$$

However, this is only half the solution! Remember, the full solution is an ordered pair. Thus, we need to find y. Just substitute $x = -3$ into either of the two original equations:

$$y = -2(-3) - 1$$
$$y = 6 - 1$$
$$y = 5$$

The complete solution is (–3, 5).

Special Cases

If, during simplification, you end up with a false statement such as this one:

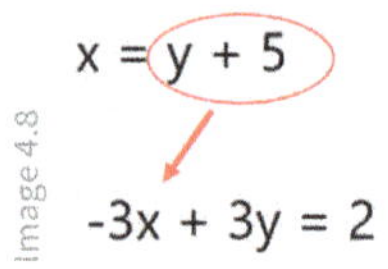

$$-3(y + 5) + 3y = 2$$
$$-3y - 15 + 3y = 2$$
$$-15 = 2 \text{ FALSE}$$

This means there is no solution, i.e., the lines are parallel. This is called an **inconsistent system**.

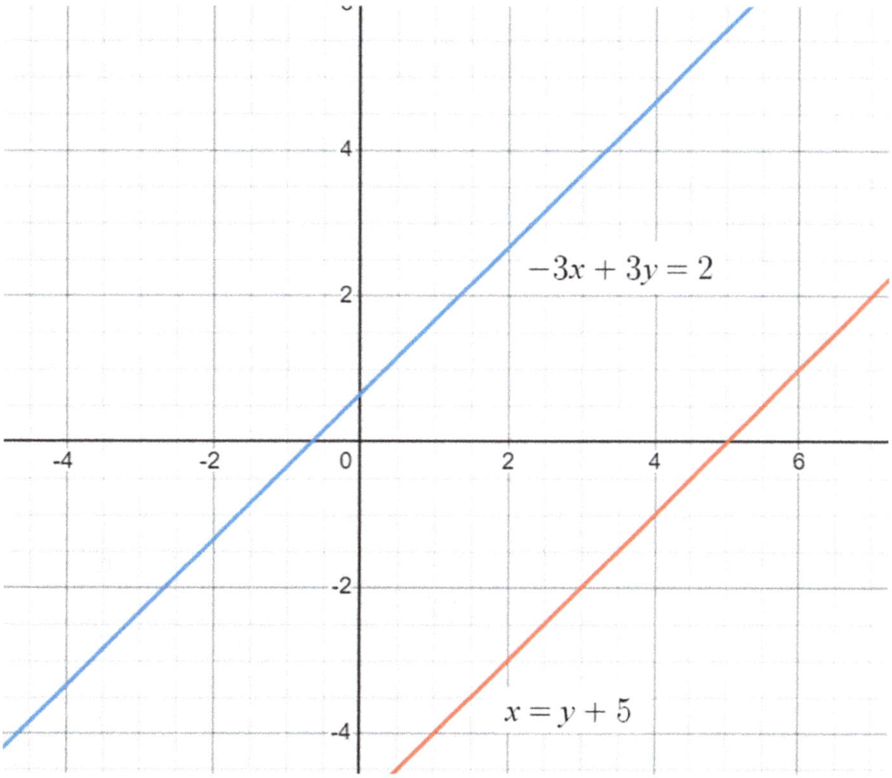

Image 4.9

Another special case is when you end up with a true statement like this one:

$$y = -2x + 4$$

$$x = -\frac{1}{2}y + 2$$

Substituting y into the second equation, we get:

$$x = -\frac{1}{2}(-2x + 4) + 2$$

$$x = x - 2 + 2$$

$$0 = 0 \text{ TRUE}$$

This indicates these are two equations describing the same line (also called **coincident lines**). Since the line contains infinitely many points, there are infinitely many solutions. This is called a **consistent-dependent system**. One way to describe the solution set is:

$\{ (x, y) \mid y = -2x + 4, x \text{ is any real number.}\}$

This is read as "the set of all ordered pairs (x, y) such that y equals $-2x + 4$ and x is any real number."

Linear systems that have one unique solution are called **consistent-independent systems**.

Summary:

Inconsistent	Consistent	
	Independent	**Dependent**
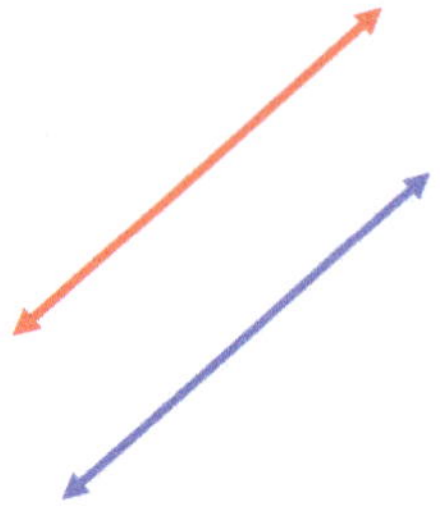 Image 4.10 Parallel Lines No solution	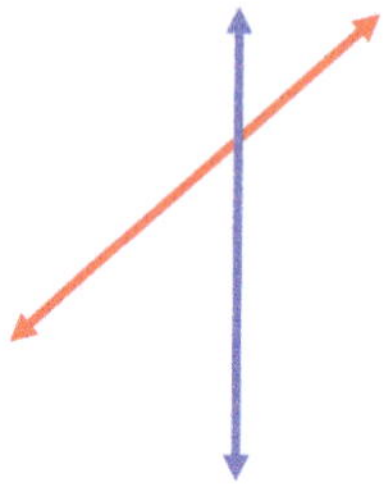 Image 4.11 Lines that intersect in exactly one point One solution	 Image 4.12 Coincident lines Infinite solutions

EXERCISE SET 4.4.2

Use the substitution method to solve the following linear systems, if there is a solution. If there is no solution, identify whether it is inconsistent or consistent-dependent.

1. $y = x + 3$
   $2x + 3y = 1$
2. $2x - y = -1$
   $2x + y = 3$
3. $3x - y = 5$
   $-3x + 3y = 2$
4. $x - 2y = 8$
   $x = 5 - y$
5. $x + 2y = 3$
   $2x + 4y = 6$

Elimination Method of Solving a Linear System

Sometimes it's easier to simply add the columns (like terms) of the linear system.

Example 1

Solve the linear system using the Elimination Method.

$$\begin{aligned} 2x + 3y &= 1 \\ -2x + 2y &= -6 \end{aligned}$$

Solution

$$\begin{aligned} 2x + 3y &= 1 \\ \underline{-2x + 2y} &\underline{= -6} \\ 5y &= -5 \\ y &= -1 \end{aligned}$$

Now that you know the value of one of the variables, substitute into either equation to find the value of the other variable:

$$\begin{aligned} 2x + 3(-1) &= 1 \\ 2x - 3 &= 1 \\ 2x &= 4 \\ x &= 2 \end{aligned}$$

∴ Complete Solution: (2, –1)

Example 2

Solve the linear system using the Elimination Method.

$$\begin{aligned} 5x - y &= 21 \\ 2x + 3y &= -12 \end{aligned}$$

Solution

In this linear system, the *x*s or *y*s do not immediately cancel.

The technique is to find the LCM (least common multiple) of either the *x*'s or *y*'s, and then multiply one or both equations by a constant that will cause one of the variables to "eliminate." In the above example, if we want the *x*'s to eliminate, we multiply as follows:

$$\begin{aligned} 2(5x - y = 21) &\rightarrow 10x - 2y = 42 \\ -5(2x + 3y = -12) &\rightarrow \underline{-10x - 15y = 60} \\ &\qquad -17y = 102 \\ &\qquad y = -6 \end{aligned}$$

Then add the columns of like terms

Image 4.13

Substitute $y = -6$ into either of the two original equations:

$$\begin{aligned} 5x - (-6) &= 21 \\ 5x + 6 &= 21 \\ 5x &= 15 \\ x &= 3 \end{aligned}$$

$\therefore$ Complete solution: $(3, -6)$

As with substitution, if your calculations result in a true statement (e.g., $5 = 5$), then the two equations represent the same line and you have infinitely many solutions. If you end up with a false statement (e.g., $7 = 11$), then the two equations represent parallel lines and you have no solution.

It is not worthwhile to write a program to perform either the substitution or elimination methods when we already have a program for the matrix method.

Example 3

Represent the following mixture problem as a linear system and solve.

A store sells peanuts for \$1.50 per pound and pistachios for \$5.00 per pound. The manager decides to form a 105lb. mixture to sell for \$3.00 per pound. How many pounds of pistachios and how many pounds of peanuts should be mixed so that the mixture will produce the same revenue as selling the nuts separately?

Solution

Let x = pounds of peanuts, y = pounds of pistachios

Quantity Equation: [lbs. of peanuts] + [lbs. of pistachios] = [105 lbs.]

$$x + y = 105$$

Cost Equation: [Cost of Peanuts] + [Cost of Pistachios] = [Cost of Mixture]

$$1.50x + 5.00y = 3.00[105]$$

Thus, the two equations are:

$$\begin{aligned} x + y &= 105 \\ 1.5x + 5y &= 315 \end{aligned}$$

Multiply the first equation by -1.5 to eliminate the x's:

$$\begin{array}{r} -1.5x - 1.5y = -157.5 \\ \underline{1.5x + \quad 5y = \quad 315} \\ 3.5y = 157.5 \end{array}$$

$y = 45$ lbs. of pistachios

Substitute $y = 45$ into $x + y = 105$:

$$\begin{aligned} x + 45 &= 105 \\ x &= 60 \text{ lbs. of peanuts} \end{aligned}$$

EXERCISE SET 4.4.3

Solve each of the linear systems, if possible, using either the substitution or elimination methods.

1. $3x - 5y = 3$
   $15x + 5y = 21$
2. $2x + 4y = \frac{2}{3}$
   $3x - 5y = -10$
3. $2x + \frac{1}{2}y = 0$
   $3x - 4y = -\frac{19}{2}$
4. $x - y = 5$
   $-3x + 3y = 2$
5. $3x + 2y = 2$
   $x = 7y - 30$
6. A movie theater charges $9.00 for adults and $7.00 for military. On a day when 325 people paid for admission, the total receipts were $2,495. How many were adults? How many were members of the military?
7. A chocolatier decides to create a new blend of chocolate using dark chocolate (which sells for $5.00 per pound) and ruby chocolate (which sells for $10.00 per pound). She wants to sell the blended chocolate for $7.00 per pound, which would result in the same income if she were to sell the chocolate separately. If she wants to mix together a 100-pound batch of the new blend, how many pounds each of dark and ruby chocolate should she combine?
8. A recently retired couple needs $12,000 per year to supplement their Social Security. They have $150,000 to invest to obtain this income. They have decided to invest in AA bonds yielding 10% per year and a bank CD yielding 5%. How much should be allocated to each to realize exactly $12,000?

4.5 Cramer's Rule

System of Two Equations in Two Unknowns

We have yet a fourth method to solve a linear system. This one involves matrices, and the beauty is that it can be used with any number of linear equations, not just two as in the $[A][X] = [B]$ method.

Recall that the determinant of matrix $A = \begin{bmatrix} a & b \\ c & d \end{bmatrix}$ is:

$$|A| = \begin{vmatrix} a & b \\ c & d \end{vmatrix} = ad - bc.$$

Let's start with two equations so that we can see how this works. Given linear system:

$$\begin{aligned} ax + by &= s \\ cx + dy &= t \end{aligned}$$

We can solve for x and y as follows:

CRAMER'S RULE FOR 2 VARIABLES

Given $A = \begin{bmatrix} a & b \\ c & d \end{bmatrix}$ *and* $B = \begin{bmatrix} s \\ t \end{bmatrix}$, the solution to $AX = B$ is:

$$x = \frac{|A_x|}{|A|} \text{ and } y = \frac{|A_y|}{|A|}$$

where $A_x = \begin{bmatrix} s & b \\ t & d \end{bmatrix}$, $A_y = \begin{bmatrix} a & s \\ c & t \end{bmatrix}$, *provided that* $|A| \neq 0$.

Notice that A_x is formed by replacing the 1st column of A with $\begin{bmatrix} s \\ t \end{bmatrix}$, and A_y is formed by replacing the 2nd column of A with $\begin{bmatrix} s \\ t \end{bmatrix}$:

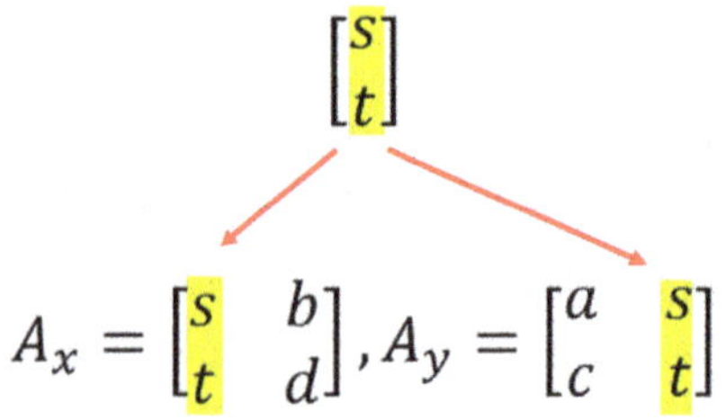

Image 4.14

Example

Solve the linear system using Cramer's Rule.

$$3x - 2y = 4$$
$$6x + y = 13$$

Solution

$$A = \begin{bmatrix} a & b \\ c & d \end{bmatrix} = \begin{bmatrix} 3 & -2 \\ 6 & 1 \end{bmatrix} \text{ and } \begin{bmatrix} s \\ t \end{bmatrix} = \begin{bmatrix} 4 \\ 13 \end{bmatrix}.$$

Thus $|A| = 3 - (-12) = 15$, which means the solution exists.

$$|A_x| = \begin{vmatrix} s & b \\ t & d \end{vmatrix} = \begin{vmatrix} 4 & -2 \\ 13 & 1 \end{vmatrix} = 4 - (-26) = 30$$

$$|A_y| = \begin{vmatrix} a & s \\ c & t \end{vmatrix} = \begin{vmatrix} 3 & 4 \\ 6 & 13 \end{vmatrix} = 39 - 24 = 15$$

$$x = \frac{|A_x|}{|A|} = \frac{30}{15} == 2,\ y = \frac{|A_y|}{|A|} = \frac{15}{15} = 1$$

$\therefore$ The solution is (2, 1).

EXERCISE SET 4.5.1

Re-do Exercise Set 4.4.3 #1–4 using Cramer's Rule. The questions are reprinted here:

1. $3x - 5y = 3$
   $15x + 5y = 21$
2. $2x + 4y = \frac{2}{3}$
   $3x - 5y = -10$
3. $2x + \frac{1}{2}y = 0$
   $3x - 4y = -\frac{19}{2}$
4. $x - y = 5$
   $-3x + 3y = 2$

Whenever the solution is in the form of a formula, we can write a program.

Program Goal: Write a program **cramer** that will solve a linear system using Cramer's Rule. Matrix A refers to $\begin{bmatrix} a & b \\ c & d \end{bmatrix}$ as above, and we'll call $\begin{bmatrix} s \\ t \end{bmatrix}$ matrix ST.

PROGRAM 4.5.1 CRAMER.PY

```
from fractions import Fraction
A = [[3,-2], [-1,1]]
B = [[4], [5]]
a = A[0][0]
b = A[0][1]
c = A[1][0]
d = A[1][1]
s = B[0][0]
t = B[1][0]
detA = a*d-b*c
Ax = s*d-b*t
Ay = a*t-s*c
print("X=",Fraction(Ax/detA))
print("Y=",Fraction(Ay/detA))
```

Output

$X = 14$

$Y = 19$

Note that we used A as a matrix and a as an integer; same with B and b. This should be avoided if possible, to avoid confusion on the part of the human and the computer. However, in this case I wanted to be as consistent as possible with the variables used in the formula.

EXERCISE SET 4.5.2

1. Modify program **cramer** as follows.

   A. Replace the two assignment statements for [*A*] and [*B*] with user input statements.

   B. Delete the 6 assignment statements for *a,b,c,d,s,t* with the following more succinct lines:

   a,b,c,d = *A*[0][0], *A*[0][1], *A*[1][0], *A*[1][1]

   s,t = *B*[0][0], *B*[1][0]

Sample Output

Matrix *A*:

Row 1 separated by commas: 3,–5

Row 2 separated by commas: 15,5

Matrix *B*:

Row 1 separated by commas: 3

Row 2 separated by commas: 21

X = 4/3

Y = 1/5

2. Check if $|A| = 0$ and if so, output "no solution" and end the program.

Sample Output

Matrix *A*:

Row 1 separated by commas: 1,–1

Row 2 separated by commas: –3,3

Matrix *B*:

Row 1 separated by commas: 5

Row 2 separated by commas: 2

no solution

3. Bypass use of variables ***a, b, c, d, s, t*** entirely. Have assignment statements for detA, *Ax* and *Ay* use entries of *A* and *B*. Example:

   detA = *A*[0][0]**A*[1][1] – *A*[0][1]**A*[1][0]

You may be thinking that any of the three previous methods (**matrix_solve**, substitution, elimination) would be easier than Cramer's Rule and you'd probably be right. However, Cramer's Rule can be applied to larger systems (3 equations with 3 unknowns, 4 equations with 4 unknowns, etc.)

System of Three Equations in Three Unknowns

The first thing we'll need to know is how to find the determinant of a 3×3 matrix. This is just a matter of breaking it into smaller (2×2) determinants, and there are several ways to do it.

The first way, called the **cofactor expansion** along the first row, is as follows:

Image 4.15

$$|A| = \begin{vmatrix} a_{1,1} & a_{1,2} & a_{1,3} \\ a_{2,1} & a_{2,2} & a_{2,3} \\ a_{3,1} & a_{3,2} & a_{3,3} \end{vmatrix} = a_{1,1}\begin{vmatrix} a_{2,2} & a_{2,3} \\ a_{3,2} & a_{3,3} \end{vmatrix} - a_{1,2}\begin{vmatrix} a_{2,1} & a_{2,3} \\ a_{3,1} & a_{3,3} \end{vmatrix} + a_{1,3}\begin{vmatrix} a_{2,1} & a_{2,3} \\ a_{3,1} & a_{3,3} \end{vmatrix}$$

The 2×2 matrices are called the cofactors. Note that to find the cofactor of any element in the matrix, you mentally "shade out" the row and column containing that element. The 4 remaining elements are the cofactor. For example, to find the cofactor of $a_{1,2}$, shade out the row and column containing $a_{1,2}$:

Image 4.16

$$\begin{vmatrix} a_{1,1} & a_{1,2} & a_{1,3} \\ a_{2,1} & a_{2,2} & a_{2,3} \\ a_{3,1} & a_{3,2} & a_{3,3} \end{vmatrix}$$

The cofactor is the remaining numbers:

$$\begin{vmatrix} a_{2,1} & a_{2,3} \\ a_{3,1} & a_{3,3} \end{vmatrix}$$

The second way to find the determinant of a 3×3 matrix, called cofactor expansion along the second row, is as follows:

$$|A| = \begin{vmatrix} a_{1,1} & a_{1,2} & a_{1,3} \\ a_{2,1} & a_{2,2} & a_{2,3} \\ a_{3,1} & a_{3,2} & a_{3,3} \end{vmatrix} = -a_{2,1}\begin{vmatrix} a_{1,2} & a_{1,3} \\ a_{3,2} & a_{3,3} \end{vmatrix} + a_{2,2}\begin{vmatrix} a_{1,1} & a_{1,3} \\ a_{3,1} & a_{3,3} \end{vmatrix} - a_{2,3}\begin{vmatrix} a_{1,1} & a_{1,2} \\ a_{3,1} & a_{3,2} \end{vmatrix}$$

Notice the signs of the coefficients have changed. In the first row, the signs were $+ - +$. In the second row the signs are $- + -$.

EXERCISE SET 4.5.3

1. What do you think is the cofactor expansion along the third row of a 3×3 matrix?
2. Find the determinant of

$$\begin{bmatrix} 3 & -5 & 3 \\ 2 & 1 & -1 \\ 1 & 0 & 4 \end{bmatrix}$$

3. Find the determinant of

$$\begin{bmatrix} -1 & -1 & 5 \\ 1 & -1 & -2 \\ 0 & -2 & 3 \end{bmatrix}$$

4. Write a function called **determ(M)** that will calculate the determinant of a 3×3 matrix M. Return the determinant. Test with a brief main program that calls the function and outputs the result. Add the function to file **matrix_fcn.py**.

Sample Output

Determinant of

−5	0	−1
1	2	−1
−3	4	1

is −40

Now we're ready for Cramer's Rule for a system of three linear equations in three unknowns:

$$a_{1,1}x + a_{1,2}y + a_{1,3}z = b_1$$

$$a_{2,1}x + a_{2,2}y + a_{2,3}z = b_2$$

$$a_{3,1}x + a_{3,2}y + a_{3,3}z = b_3$$

which can be converted to a matrix equation $[A][X] = [B]$ as follows:

$$\begin{bmatrix} a_{1,1} & a_{1,2} & a_{1,3} \\ a_{2,1} & a_{2,2} & a_{2,3} \\ a_{3,1} & a_{3,2} & a_{3,3} \end{bmatrix} \begin{bmatrix} x \\ y \\ z \end{bmatrix} = \begin{bmatrix} b_1 \\ b_2 \\ b_3 \end{bmatrix}$$

CRAMER'S RULE FOR 3 VARIABLES

The formulas for x, y, and z are as follows:

$$x = \frac{D_x}{D}, \qquad y = \frac{D_y}{D}, \qquad z = \frac{D_z}{D}$$

where:

$$D_x = \begin{vmatrix} b_1 & a_{1,2} & a_{1,3} \\ b_2 & a_{2,2} & a_{2,3} \\ b_3 & a_{3,2} & a_{3,3} \end{vmatrix} \qquad D_y = \begin{vmatrix} a_{1,1} & b_1 & a_{1,3} \\ a_{2,1} & b_2 & a_{2,3} \\ a_{3,1} & b_3 & a_{3,3} \end{vmatrix} \qquad D_z = \begin{vmatrix} a_{1,1} & a_{1,2} & b_1 \\ a_{2,1} & a_{2,2} & b_2 \\ a_{3,1} & a_{3,2} & b_3 \end{vmatrix}$$

Image 4.17

and $D \neq 0$

Notice how D_x was formed by replacing $\begin{bmatrix} b_1 \\ b_2 \\ b_3 \end{bmatrix}$ into the 1st column of D, D_y was formed by replacing $\begin{bmatrix} b_1 \\ b_2 \\ b_3 \end{bmatrix}$ into the 2nd column of D, and D_z was formed by replacing $\begin{bmatrix} b_1 \\ b_2 \\ b_3 \end{bmatrix}$ into the 3rd column of D.

Example

Solve the linear system using Cramer's Rule.

$$\begin{aligned} x - 4y + z &= 1 \\ 2x + 3y - z &= 4 \\ x + y - z &= 0 \end{aligned}$$

Solution

$$D = \begin{vmatrix} 1 & -4 & 1 \\ 2 & 3 & -1 \\ 1 & 1 & -1 \end{vmatrix}$$

Expanding along 1st row:

$+1(-3-(-1)) - (-4)(-2-(-1)) + 1(2-3) = -2 - 4 - 1 = -7$

$$D_x = \begin{vmatrix} 1 & -4 & 1 \\ 4 & 3 & -1 \\ 0 & 1 & -1 \end{vmatrix}$$

Expanding along 3rd row:

$+0 - 1(-1-4) + (-1)(3-(-16)) = 0 + 5 - 19 = -14$

$$D_y = \begin{vmatrix} 1 & 1 & 1 \\ 2 & 4 & -1 \\ 1 & 0 & -1 \end{vmatrix}$$

Expanding along 3rd row:

$+1(-1-4) - 0 + (-1)(4-2) = -5 - 2 = -7$

$$D_z = \begin{vmatrix} 1 & -4 & 1 \\ 2 & 3 & 4 \\ 1 & 1 & 0 \end{vmatrix}$$

Expanding along 3rd *row:*

$+1(-16-3) - 1(4-2) + 0 = -19 - 2 = -21$

(The choice of row along which you'd like to expand the determinant is up to you, however it's easiest if you choose a row with a 0 in it.)

We can now solve for x, y, and z:

$$x = \frac{D_x}{D} = \frac{-14}{-7} = 2 \qquad y = \frac{D_y}{D} = \frac{-7}{-7} = 1 \qquad z = \frac{D_z}{D} = \frac{-21}{-7} = 3$$

$\therefore$ Solution (**ordered triple**) is (2, 1, 3).

Recall that a system of two linear equations represented two lines in the xy-plane. A system of three linear equations represents three planes in xyz-space:

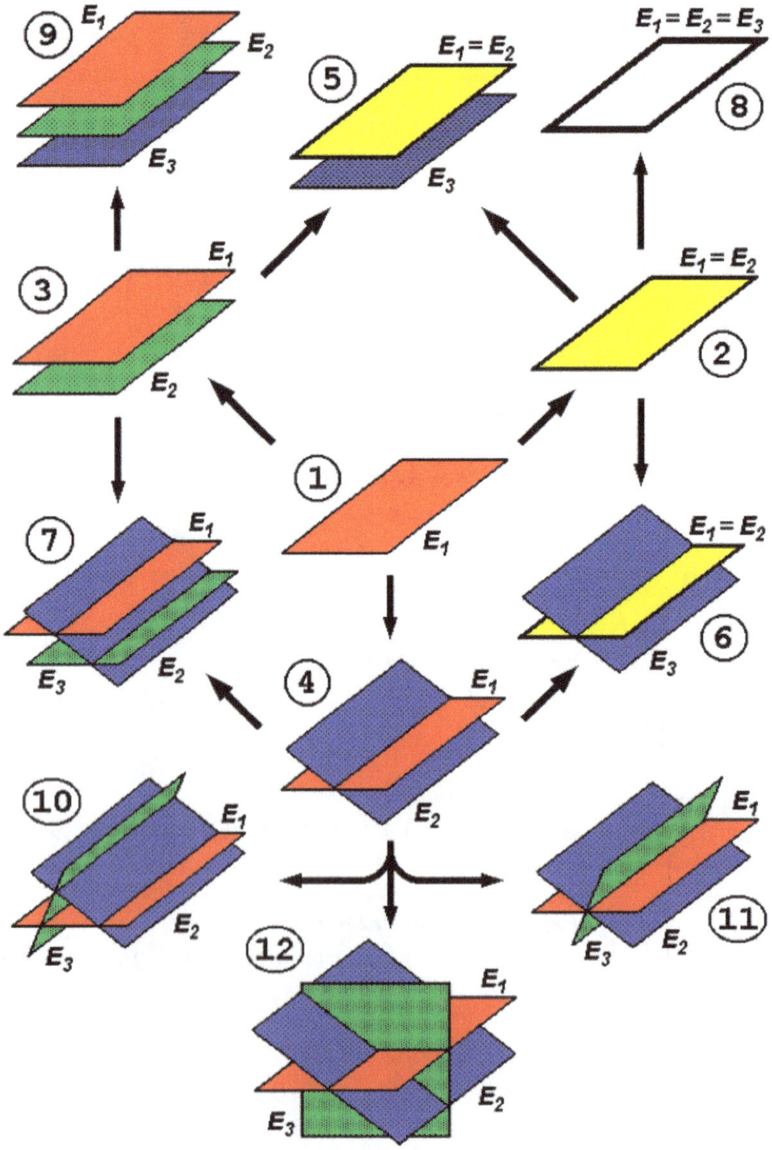

Image 4.18

We are only interested in the last case, which is when all three planes intersect at a single point.

EXERCISE SET 4.5.4

If possible, solve the following linear systems.

1. $2x - y + 3z = 6$
   $3x - 5y + 4 = 7$
   $2x + y + z = -2$
2. $2x + y - z = 3$
   $-x + 2y + 4z = -3$
   $x - 2y - 3z = 4$
3. $x + y - z = 6$
   $3x - 2y + z = -5$
   $x + 3y - 2z = 14$

Computers (and calculators!) are made for these types of calculations—repetitive, tedious, and requiring absolute accuracy.

Program Goal: Write a program **cramer3d** to solve a system of three equations in three unknowns using Cramer's Rule. It will make use of function **determ** that you wrote in Exercise Set 4.5.3 #4, which finds the determinant of matrix A and assigns it to variable det. (If you didn't write the program, or it's not working, use the version in the **matrix_fcn** library in Appendix C.)

This program makes use of manipulating individual columns of matrix A using lists.

PROGRAM 4.5.2 CRAMER3D.PY

```
from fractions import Fraction
from copy import deepcopy
import matrix_fcn as mf
A = [ [ 1, –4, 1], [ 2, 3, –1], [ 1, 1, –1 ] ]
At = deepcopy(A)                              # At is a temporary clone of A
B = [ 1 , 4 , 0 ]                             # B is a list, not a matrix
D = mf.determ(A)
   # Replace A col 1 with B
At[0][0], At[1][0], At[2][0] = B
Dx = mf.determ(At)
   # Replace A col 2 with B
At = deepcopy(A)
At[0][1], At[1][1],At[2][1] = B
Dy = mf.determ(At)
   # Replace A col 3 with B
At = deepcopy(A)
At[0][2], At[1][2], At[2][2] = B
Dz = mf.determ(At)
[ insert code to calculate and output final answers for x,y,z ]
```

Desired Output

(2, 1, 3)

Note that in order to assign the values of matrix At to another matrix A, we could not simply type:

$At = A$

This "binds" the two names together so that they refer to the same matrix. We instead use:

$At =$ **deepcopy**(A)

This creates a new copy of matrix A, and changes to one matrix do not affect the other matrix. Note you must import the **deepcopy** function from the **copy** module.

EXERCISE SET 4.5.5

1. Fill in the yellow highlighted code to calculate and output x,y,z.
2. Replace assignment statements for A and B with input statements.
3. Modify **cramer3d** to check for the following special cases:

   A. If $D = 0$ and at least one of the determinants D_x, D_y, or $D_z \neq 0$, then output "INCONSISTENT" and end the program. Use the following inconsistent system for test data:

   $x - 3y + z = 4$

   $-x + 2y - 5z = 3$

   $5x - 13y + 13z = 8$

   B. If $D = 0$ and all the determinants D_x, D_y, and $D_z = 0$, then output "CONSISTENT DEPENDENT" and end the program. Use the following consistent dependent system for test data:

   $x - 2y + 3z = 1$

   $3x + y - 2z = 0$

There are other various ways to solve matrix equations such as Gauss-Jordan elimination. These can be explored in a higher-level Linear Algebra course.

4.6 Quadratic Curve Fitting

One application of matrices is that of approximating a quadratic function that passes through three given points. The goal is similar to linear regression except we are fitting a parabola rather than a line.

Given $f(x) = ax^2 + bx + c$, our goal is to find the values of a, b, and c. To do so, we need three noncollinear points that lie on the parabola.

Example

Given the following three points that lie on the same parabola: (1, −5), (3, 5), and (4, 16), find a quadratic function whose graph contains them.

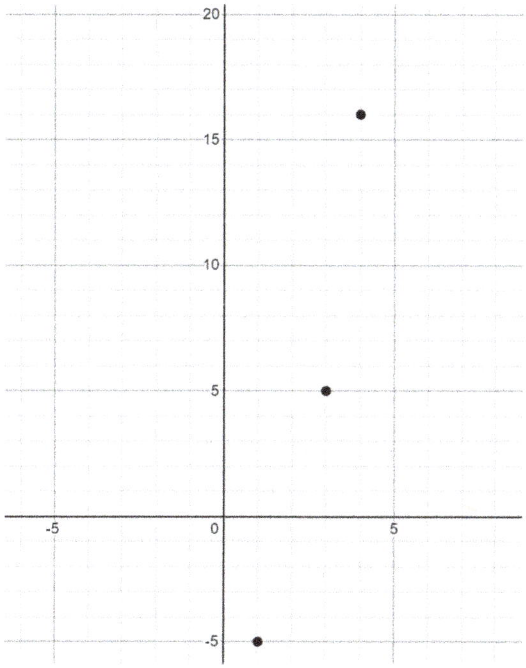

Image 4.19

Solution

We simply substitute each ordered pair into $y = ax^2 + bx + c$:

$$-5 = a(1)^2 + b(1) + c \rightarrow -5 = a + b + c$$

$$5 = a(3)^2 + b(3) + c \rightarrow 5 = 9a + 3b + c$$

$$16 = a(4)^2 + b(4) + c \rightarrow 16 = 16a + 4b + c$$

This gives us a system of three equations in 3 unknowns, which you learned how to solve in Section 4.5 using Cramer's Rule for matrices.

$$\begin{bmatrix} 1 & 1 & 1 \\ 9 & 3 & 1 \\ 16 & 4 & 1 \end{bmatrix} \begin{bmatrix} a \\ b \\ c \end{bmatrix} = \begin{bmatrix} -5 \\ 5 \\ 16 \end{bmatrix}$$

After using **Cramer's Rule**, you will arrive at the solution $a = 2$, $b = -3$, $c = -4$, so that:
$\therefore\ y = 2x^2 - 3x - 4$

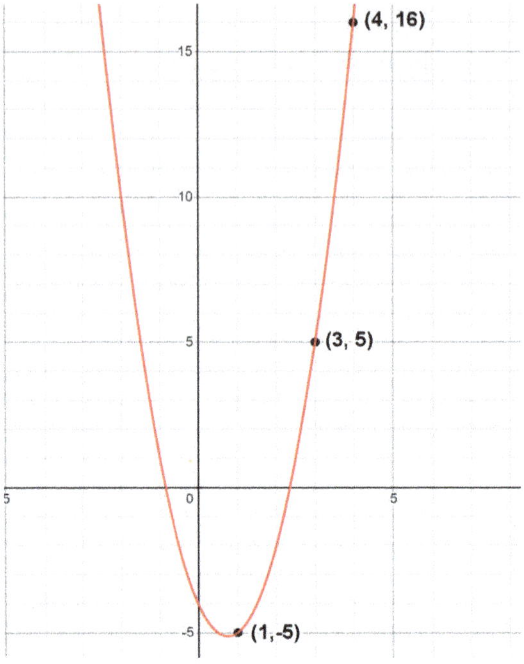

Image 4.20

Among other things, matrices provide the foundations of computer graphics and animation. You are on the frontier of a whole new world.

EXERCISE SET 4.6.1

Find the quadratic function whose graph passes through:

1. Points $(0, -3)$, $(1, 0)$ and $(2, 1)$.
2. Points $(2, 0)$, $(3, -2)$, and $(5, -12)$
3. The table below shows the prices of an ice cream cake, depending on its size. Find a quadratic model for the cost of the cake, given the diameter. Then use the model to predict the cost of an ice cream cake with a diameter of 18 in.

Diameter (in.)	Cost
6	7.50
10	12.50
15	18.50

4. Write a program that asks the user to input three ordered pairs. Convert the ordered pairs to a matrix equation.

Sample Output

1st ordered pair? (1, −5)

2nd ordered pair? (3, 5)

3rd ordered pair? (4, 16)

Matrix *A*:

1	1	1
9	3	1
16	4	1

Matrix *B*:

−5

5

16

5. Extend the program you wrote in exercise #4 above so that it solves the matrix equation using **cramer3d**.

Chapter Projects

I. Area of a Triangle

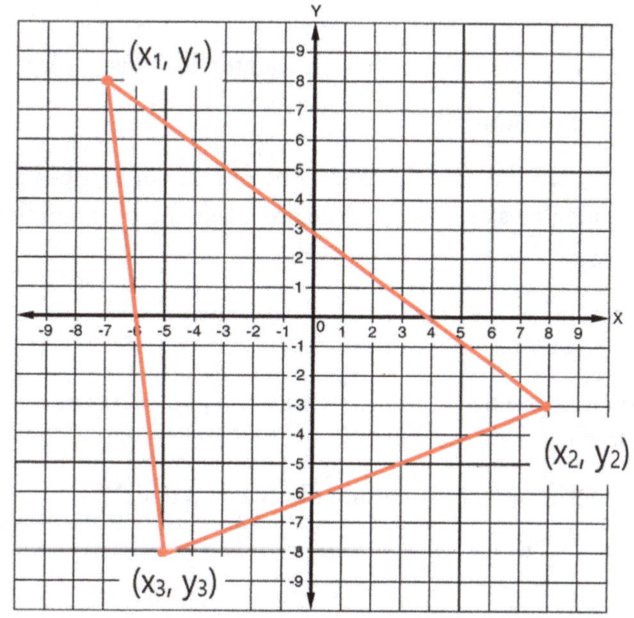

Image 4.21

If a triangle has vertices (x_1, y_1), (x_2, y_2), (x_3, y_3), then the area of the triangle is the absolute value of:

$$\frac{1}{2}\begin{vmatrix} x_1 & x_2 & x_3 \\ y_1 & y_2 & y_3 \\ 1 & 1 & 1 \end{vmatrix}$$

A. Find the area of the triangle with vertices (2,3), (5,2), and (6,5).

B. Write a program that asks the user to input the three vertices of a triangle. Calculate and output the area. Use the program to verify your answer to part A.

II. Area of a Polygon

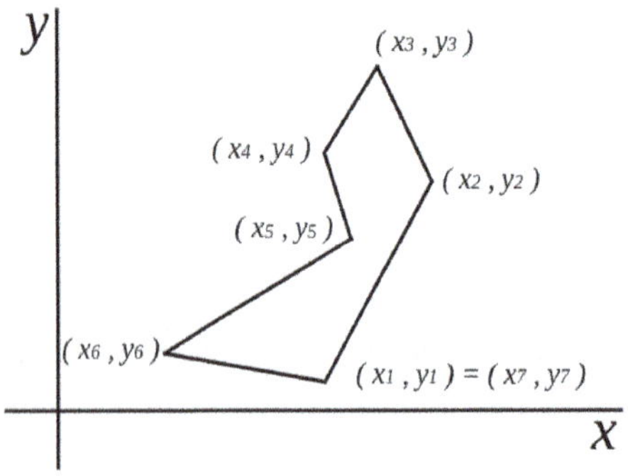

Image 4.22

The following formula will find the area of a polygon using determinants:

$$A = \frac{1}{2}\left(\begin{vmatrix} x_1 & y_1 \\ x_2 & y_2 \end{vmatrix} + \begin{vmatrix} x_2 & y_2 \\ x_3 & y_3 \end{vmatrix} + \begin{vmatrix} x_3 & y_3 \\ x_4 & y_4 \end{vmatrix} \ldots + \begin{vmatrix} x_n & y_n \\ x_1 & y_1 \end{vmatrix}\right)$$

Where (x_1, y_1), (x_2, y_2), ... (x_n, y_n) are the vertices in counterclockwise order.

A. Find the area of the polygon pictured.

B. Write a program that (1) asks the user to input n, the number of vertices, (2) asks the user to input the n vertices, and (3) calculates and outputs the area. Use the program to verify your answer to part A.

III. Cryptography

One method of encryption is to use a square matrix to encrypt a secret message and then use the corresponding inverse matrix to decode the message.

Step 1—Assign each letter of the alphabet a number between 0 and 25, plus 26 for the "space":

A	B	C	D	E	F	G	H	I	J	K	L	M
1	2	3	4	5	6	7	8	9	10	11	12	13

N	O	P	Q	R	S	T	U	V	W	X	Y	Z	SPACE
14	15	16	17	18	19	20	21	22	23	24	25	26	0

For example, if the message is I LOVE MATH, then the corresponding numbers would be 9-0-12-15-22-5-0-13-1-20-8.

Step 2—Select a square matrix "key." It must be invertible, and its determinant cannot be divisible by any factor of 27. (It's safest to use a matrix with a prime number determinant.) For this exercise, we will use

$$K = \begin{bmatrix} 1 & 2 & 3 \\ 0 & 2 & 1 \\ 1 & 1 & 1 \end{bmatrix}$$

Step 3—Arrange the coded message into a matrix M with 3 rows (because K is a 3×3 matrix) padding with spaces as necessary.

$$M = \begin{bmatrix} 9 & 0 & 12 & 15 \\ 22 & 5 & 0 & 13 \\ 1 & 20 & 8 & 0 \end{bmatrix}$$

Step 4—Multiply K by M.

$$KM = \begin{bmatrix} 1 & 2 & 3 \\ 0 & 2 & 1 \\ 1 & 1 & 1 \end{bmatrix} \begin{bmatrix} 9 & 0 & 12 & 15 \\ 22 & 5 & 0 & 13 \\ 1 & 20 & 8 & 0 \end{bmatrix} = \begin{bmatrix} 56 & 70 & 36 & 41 \\ 45 & 30 & 8 & 26 \\ 32 & 25 & 20 & 28 \end{bmatrix}$$

Thus, we send the following encoded message: 56-70-36-41-45-30-8-26-32-25-20-28.

Step 5—The recipient must already know the inverse of the key matrix. In this example:

$$K^{-1} = \begin{bmatrix} -1/3 & -1/3 & 4/3 \\ -1/3 & 2/3 & 1/3 \\ 2/3 & -1/3 & -2/3 \end{bmatrix}$$

The recipient also knows that if it is a 3×3 matrix, then the encoded message must be arranged in a matrix with 3 rows, which we'll call E:

$$E = \begin{bmatrix} 56 & 70 & 36 & 41 \\ 45 & 30 & 8 & 26 \\ 32 & 25 & 20 & 28 \end{bmatrix}$$

Step 6—The recipient decodes the message by multiplying K^{-1} by E:

$$K^{-1}E = \begin{bmatrix} -1/3 & -1/3 & 4/3 \\ -1/3 & 2/3 & 1/3 \\ 2/3 & -1/3 & -2/3 \end{bmatrix}\begin{bmatrix} 56 & 70 & 36 & 41 \\ 45 & 30 & 8 & 26 \\ 32 & 25 & 20 & 28 \end{bmatrix} = \begin{bmatrix} 9 & 0 & 12 & 15 \\ 22 & 5 & 0 & 13 \\ 1 & 20 & 8 & 0 \end{bmatrix}$$

$$= \begin{bmatrix} I & & L & O \\ V & E & & M \\ A & T & H & \end{bmatrix}$$

Your task is to do the following:

A. Given the key matrix $K = \begin{bmatrix} 2 & 1 & 1 \\ 1 & 1 & 0 \\ 1 & 1 & 1 \end{bmatrix}$, find its inverse K^{-1}.

B. Use K^{-1} to decode the encrypted matrix $E = \begin{bmatrix} 47 & 34 & 33 \\ 44 & 36 & 27 \\ 47 & 41 & 20 \end{bmatrix}$.

C. Each entry in your result for part B represents the position of a letter in the English alphabet (A = 1, B = 2 and so on). What is the original message?

D. Write two programs. The first, **encrypt**, will ask the user to input a message, then encrypt the message using $K = \begin{bmatrix} 2 & 1 & 1 \\ 1 & 1 & 0 \\ 1 & 1 & 1 \end{bmatrix}$ (from question A above). The output is a list of encoded letters.

E. The second program **decrypt** will accept as input the list of encoded letters from program **encrypt**, then decodes them using K^{-1}.

IV. Transpose of a Matrix

The **transpose** of square matrix A, having notation A^T, results from reflecting the matrix about the main diagonal, or equivalently, swapping the (i,j)th element and the (j,i)th. (Note that when you transpose the terms of the matrix, the main diagonal [from upper left to lower right] is unchanged.) Another way to think of transposing is that you rewrite the first row as the first column, the middle row becomes the middle column, and the third row becomes the third column.

Example: The transpose of $A = \begin{bmatrix} 1 & 2 & 3 \\ 4 & 5 & 6 \\ 7 & 8 & 9 \end{bmatrix}$ *is* $\begin{bmatrix} 1 & 4 & 7 \\ 2 & 5 & 8 \\ 3 & 6 & 9 \end{bmatrix}$

Write a program that finds the transpose of a square matrix of any size.

V. Fibonacci Sequence Revisited

The nth and $(n + 1)$th Fibonacci numbers can be generated using:

$$\begin{bmatrix} f_{n+1} \\ f_n \end{bmatrix} = \begin{bmatrix} 1 & 1 \\ 1 & 0 \end{bmatrix}^{n-1} \begin{bmatrix} 1 \\ 0 \end{bmatrix}$$

Write a program that asks the user to input n. Generate and output the nth and $(n + 1)^{\text{th}}$ Fibonacci numbers using the matrix equation above.

VI. Bingo Card

Write a bingo card generating program. A sample bingo card looks like this:

Image 4.23

As you can see, the bingo card is essentially a 5×5 matrix (ignoring the letters at the top of each column.) The columns contain the following range of numbers:

	B	I	N	G	O
Column	1	2	3	4	5
Range of numbers	1–15	16–30	31–45	46–60	61–75

VII. Least Squares Quadratic Regression

This is a technique using a set of test data to approximate a quadratic function $y = ax^2 + bx + c$. (This is not the same as the Cramer's Rule method in Section 4.5.)

If the test data is the set of n ordered pairs $\{(x_1, y_1), (x_2, y_2), \dots (x_n, y_n)\}$ and we want to find a, b, and c such that our quadratic model is $y = ax^2 + bx + c$, solve the following matrix equation:

$$\begin{bmatrix} \sum x^4 & \sum x^3 & \sum x^2 \\ \sum x^3 & \sum x^2 & \sum x \\ \sum x^2 & \sum x & n \end{bmatrix} \begin{bmatrix} a \\ b \\ c \end{bmatrix} = \begin{bmatrix} \sum x^2 y \\ \sum xy \\ \sum y \end{bmatrix}$$

Write a function **quadratic_reg**(x_list, y_list) that calculates and returns the approximated quadratic equation using the formulas above. Draw the points corresponding to the test data, and draw the parabola.

Which method do you think is more accurate—the Least Squares Method or the Cramer's Rule Method?

SAMPLE OUTPUT

Ordered pairs? (2, 3), (4, 6), (6, 4)

Quadratic is

−0.625*x**2 + 5.25*x − 5

CHAPTER SUMMARY

PYTHON CONCEPTS

Matrix	Technically a list of lists. Each element in matrix *A* has two subscripts (row,column): *A*[*i*][*j*] Just like lists, the first index in each row and column is 0. *Ex:* *A = [[1,2,3],[4,5,6]]* ***print***(*A[1][2]*) *6*
Initializing and assigning values to a matrix	To assign a value to a specific element: *A*[*i*][*j*] = value To assign a row to a matrix (that already exists): *A*[row_num] = list To initialize an empty matrix: *A* = [ ] *Ex:* A*[0][0] = –5 replaces a single element* A*[0] = [11, 12, 13] replaces a row*

Dimensions of a matrix	Number of rows in matrix A: **len**(A) Number of columns in matrix A: **len**(A[0])
To append a new row to matrix *A*	If row is a list of length n, it can be appended to a matrix with n columns: A.**append**(row)
List comprehension	*A Python statement that efficiently processes all elements in a list:* list_name = [$f(x)$ **for** x **in** {domain}] *Ex:* *A = [[1,2,3],[4,5,6]]* *B = [[0,0,0],[0,0,0]]* *B[i] = [10*A[i][j] for j in **range**(3)]* ***print**(B)* *[[10, 20, 30], [0, 0, 0]]*
Shuffle elements of a list	Use the **shuffle**() function in the **random** library.
Assign one matrix to another	Use the **deepcopy**() function in the **copy** library. *Ex:* *A = **deepcopy**(B)*

PRECALCULUS CONCEPTS

Matrix	An array of numbers. A matrix with m rows and n columns has dimensions $m \times n$. $\begin{bmatrix} a_{1,1} & a_{1,2} & \cdots & a_{1,n} \\ a_{2,1} & a_{2,2} & \cdots & a_{2,n} \\ \vdots & \vdots & \vdots & \vdots \\ a_{m,1} & a_{m,2} & \cdots & a_{m,n} \end{bmatrix}$
Scalar multiplication	A number multiplied by a matrix. $n\begin{bmatrix} a & b \\ c & d \end{bmatrix} = \begin{bmatrix} na & nb \\ nc & nd \end{bmatrix}$

(Continued)

Dot product (multiplying a row matrix by a column matrix)	If $A = \begin{bmatrix} a_{1,1} & a_{1,2} & \cdots & a_{1,n} \end{bmatrix}$ and $B = \begin{bmatrix} b_{1,1} \\ b_{2,1} \\ \vdots \\ b_{n,1} \end{bmatrix}$ then their product is the 1×1 matrix: $AB = [\, a_{1,1} \cdot b_{1,1} + a_{1,2} \cdot b_{2,1} + \ldots + a_{1,n} \cdot b_{n,1} \,]$
Matrix multiplication	An $r \times k$ matrix can be multiplied by a $k \times c$ matrix. The result is an $r \times c$ matrix. Each entry $p_{i,j}$ of their product P can be found by: $p_{i,j} = \sum_{k=1}^{n} a_{i,k} \cdot b_{k,j}$ where n is the # of columns in $[A]$
Determinant of a 2 × 2 matrix	The determinant of $\begin{bmatrix} a & b \\ c & d \end{bmatrix}$ is the number $ad–bc$. *Ex:* *The determinant of* $\begin{bmatrix} 2 & -3 \\ 0 & 11 \end{bmatrix}$ *is* $(2)(11) - (0)(-3) = 22$.
Multiplicative inverse of a 2 × 2 matrix	The multiplicative inverse of 2×2 matrix $A = \begin{bmatrix} a & b \\ c & d \end{bmatrix}$ *is* $A^{-1} = \frac{1}{ad - bc} \cdot \begin{bmatrix} d & -b \\ -c & a \end{bmatrix}$ provided $ad - bc \neq 0$. A matrix with a determinant of 0 is **noninvertible.** *Ex:* *The multiplicative inverse of* $\begin{bmatrix} 2 & -3 \\ 0 & 11 \end{bmatrix}$ *is* $A^{-1} = \frac{1}{22}\begin{bmatrix} 11 & 3 \\ 0 & 2 \end{bmatrix} = \begin{bmatrix} 1/2 & 3/22 \\ 0 & 1/11 \end{bmatrix}$
Solving matrix equation $AX = B$	The solution to matrix equation $AX = B$ is $X = A^{-1}B$ provided the number of columns of A is the same as the number of rows of B. *Ex:* *Given equation* $\begin{bmatrix} 2 & -3 \\ 5 & 3 \end{bmatrix} \cdot \begin{bmatrix} x \\ y \end{bmatrix} = \begin{bmatrix} 15 \\ 27 \end{bmatrix}$, *first* *calculate* $A^{-1} = \begin{bmatrix} 1/7 & 1/7 \\ -5/21 & 2/21 \end{bmatrix}$, *then calculate* $X = A^{-1}B = \begin{bmatrix} 1/7 & 1/7 \\ -5/21 & 2/21 \end{bmatrix} \cdot \begin{bmatrix} 15 \\ 27 \end{bmatrix} = \begin{bmatrix} 6 \\ -1 \end{bmatrix}$

Inconsistent system	A linear system in which there is no solution. In the case of 2 variables, this represents parallel lines. 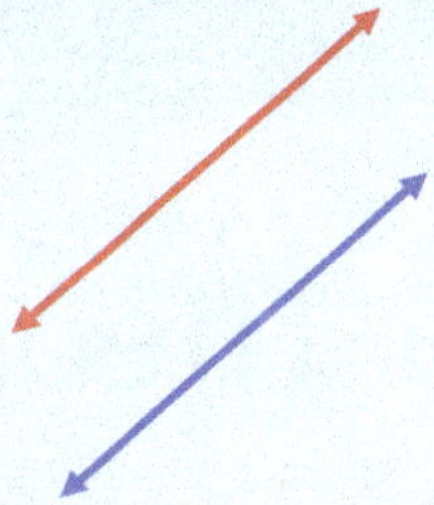 Image 4.10
Consistent independent system	A linear system with one unique solution. In the case of 2 variables, this represents a pair of lines that intersect in exactly one point. 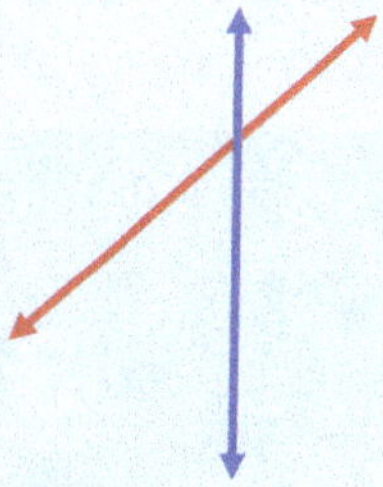 Image 4.11
Consistent dependent system	A linear system with infinitely many solutions. In the case of 2 variables, this represents "coincident" lines. Image 4.12 
Cramer's Rule for 2 variables	Given $A = \begin{bmatrix} a & b \\ c & d \end{bmatrix}$ *and* $B = \begin{bmatrix} s \\ t \end{bmatrix}$, the solution to $AX = B$ is: $x = \dfrac{\begin{vmatrix} s & b \\ t & d \end{vmatrix}}{\lvert A \rvert}$ *and* $y = \dfrac{\begin{vmatrix} a & s \\ c & t \end{vmatrix}}{\lvert A \rvert}$ provided that $\lvert A \rvert \neq 0$.

Credits

CHAPTER 5

Number Theory and Logic

5.1 Real Numbers and Order of Operations

Order of Operations

You may have seen an internet meme such as this one posted on social media:

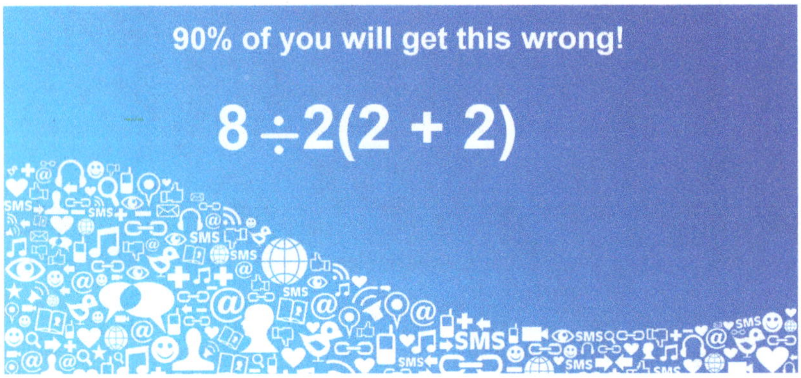

Image 5.1

Surprisingly, social media posts such as this generate thousands of replies and bitter disagreements. What do *you* think the answer is?

Keep in mind the following order of operations, most commonly taught to schoolchildren with the catchy acronym "Please Excuse My Dear Aunt Sally."

ORDER OF OPERATIONS

Parentheses (and any grouping symbol, evaluated from innermost to outermost)

Exponents

Multiplication and **D**ivision (evaluated from left to right)

Addition and **S**ubtraction (evaluated from left to right)

It is imperative that programmers understand how numeric expressions are evaluated to ensure that the operations are carried out in the order intended. For example, 100 + 20 / 5 is a completely different number than (100 + 20) / 5.

Back to our internet meme: $8 \div 2(2 + 2)$

If you answered 1, you are not alone. But you are wrong. It may become more clear if the multiplication is more conspicuous: $8 \div 2 \cdot (2 + 2)$. Thus, we have:

Parentheses: $8 \div 2 \cdot (2 + 2) = 8 \div 2 \cdot (4)$

Division: $8 \div 2 \cdot (4) = 4 \cdot (4)$

Multiplication: $4 \cdot 4 = 16$

Ready to try your hand?

EXERCISE SET 5.1.1

Simplify the following:

1. -1^0
2. $[9 - (3^2 - 3)] \cdot 10$
3. $[(-9) - \{(-15) \div (-5)\}^2] \cdot (-10)$
4. $[4 - (8 \div 4)^3] \cdot 11$
5. $(-2)^2 - [7 \cdot \{(-2) - (-2)\}]$

Using * for multiplication, ** for exponentiation, and / for division, express the following math expressions as Python expressions. Keep in mind that Python doesn't allow the use of [ ] or { } for grouping symbols, only ( ), which can be nested.

Example: $[\{5^2 + 8\} \cdot 5] - 11$ would be represented as ((5**2 + 8)*5) – 11.

6. $[(-9) - \{(-4)^3 - (-4)\}] \cdot (-5)$
7. $[\{90 \div 5\}^2 - 10] \cdot 11$
8. $\dfrac{4 \cdot 10 + 3^2}{2^3 - 1}$

The Python interpreter can essentially act as a calculator using the **eval**() function as follows:

```
>>>eval('((–11) – ((–18) / (–9))**2)* –10')
```

Real Numbers

These are the different types of numbers referenced in elementary mathematics:

Natural Numbers *N* aka Counting Numbers: {1, 2, 3, 4, ... }

Whole Numbers: {0, 1, 2, 3, 4, ... }

Integers *Z*: {... −3, −2, −1, 0, 1, 2, 3, ...}, or the set of positive and negative whole numbers and zero. Zero is considered unsigned (neither positive nor negative).

Rational Numbers *Q*: The set of all numbers that can be expressed as a ratio (fraction) of two integers, for example, 2/3, −1/4, 7, .3333..., $\sqrt{4}$, etc. Any repeating decimal can be expressed as a rational number by first expressing it as a geometric series and then using $S = \frac{a_1}{1-r}$ as we did in Section 1.5.

Example: $.3333... = .3 + .03 + .003 + ... = 3\left(\frac{1}{10}\right) + 3\left(\frac{1}{10}\right)^2 + 3\left(\frac{1}{10}\right)^3 + ...$ has $a_1 = .3$ and $r = \frac{1}{10}$ thus

$$S = \frac{\frac{3}{10}}{1-\frac{1}{10}} = \frac{\frac{3}{10}}{\frac{9}{10}} = \frac{3}{9} = \frac{1}{3}.$$

A handy memory device is that $.1111... = \frac{1}{9}$, $.2222... = \frac{2}{9}$, $.3333... = \frac{3}{9}$, $.8888... = \frac{8}{9}$, $.9999... = \frac{9}{9} = 1$.

The square root of any perfect square, such as $\sqrt{\frac{16}{25}}$, is also a rational number. More generally:

$\sqrt[n]{a^n}$ is a rational number.

Irrational Numbers: The set of all numbers that are not rational—for example, π, *e*, $\sqrt{2}$ ...; essentially all nonterminating, nonrepeating decimals. The "discovery" of irrational numbers proved fatal for one of the followers of Pythagorus (who felt the universe could be exclusively described by whole numbers). Hippasus (approximately 500 BC) is believed to have been drowned by the Pythagorean Society because of his revelation that irrational numbers were a necessary part of mathematics.

Real Numbers *R*: The set of all rational and irrational numbers.

This diagram summarizes the types of numbers and their relationships:

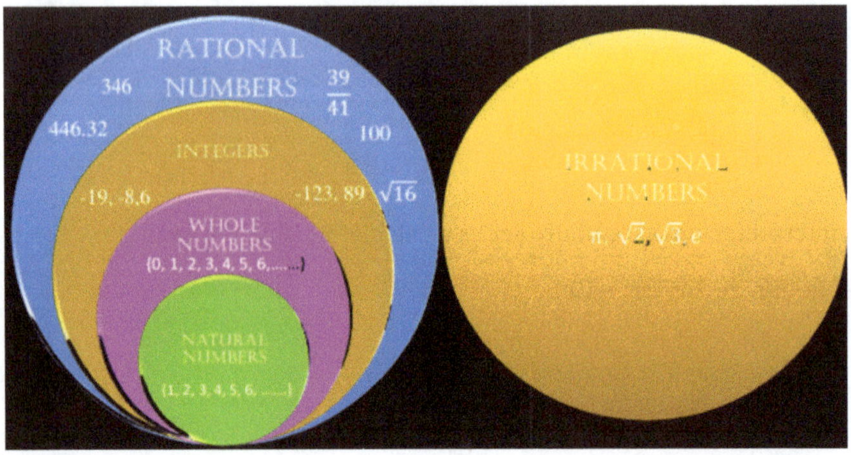

Image 5.2

Do these sets encompass all types of numbers? No, there are also imaginary numbers, which we will discuss in Chapter 8.

EXERCISE SET 5.1.2

1. Classify each of the following, using all labels that apply.
   A. $\sqrt{4}$
   B. $\sqrt{-4}$
   C. $\sqrt[3]{-64}$
   D. 0.123123123...
   E. $-\frac{1}{2}$
   F. 0
   G. $\sqrt{3}$
   H. $\sqrt{.0625}$
2. A. Find a rational number between $\frac{1}{4}$ and $\frac{1}{3}$.
   B. There is a theorem that says that between any two rational numbers is an irrational number, and vice versa. Find an irrational number between $\frac{1}{4}$ and $\frac{1}{3}$.
2. Convert to a fraction: .121212...
3. If a single repeating digit $.nnnn\ldots = \frac{n}{9}$, then what do you think is a handy formula for two repeating digits $.mnmnmn\ldots$?

Number Types in Python

Let's do a recap of the numeric variable types as represented in Python: **integer** and **float**. You do not have to declare the type of the integer in the program; in fact, the type of a variable can even change throughout the program. (There are also **complex** type numbers, which we'll cover in Chapter 8.)

A float type number is any number with a decimal point. For example: 3 is an integer; 3.0 and 3.5 are floats. You can convert from float to integer by using the **int**() function, and integer to float using the **float**() function.

Keeping in mind that the **input**() function always returns a string, you must use **int(input**()) to input an integer and **float(input**()) to input a float.

As we have seen in previous chapters, it is often useful to manipulate and output fractions (rational numbers). For that capability you need to import the **fractions** module. Using this module, a fraction is considered an **instance** rather than a type, but for our purposes, we'll consider it a type that is represented by a numerator and denominator: **Fraction**(n,d), which is output as n/d.

If you'd ever like to know the type of a Python variable, you can use the **type()** function. For example:

```
>>>type((1/3)*7)
<class 'float'>
```

EXERCISE SET 5.1.3

1. What is the resulting type when you add a float and integer?
2. What is the result when you convert the following float numbers to integers:
   A. 3.2
   B. 3.8
   C. −3.2
   D. −3.8
3. Does **int**() round, truncate (i.e., drop the digits after the decimal), or perform the greatest integer function?
4. What is the result when you output the following?
   A. **Fraction**(16, −12)
   B. **Fraction**(42)
   C. **Fraction**('10/20')
   D. **Fraction**(3.141592654)
   E. **Fraction**(.125)
   F. **Fraction**(3.0,5.0)
   G. **Fraction**(1/5, 2/5)
5. What is the result when you output the following?
   A. **int**(**Fraction**(17,8)
   B. **float**(**Fraction**(17,8)
   C. **Fraction**(3.9)
   D. **float(Fraction**(3.9))
6. Can you add a float and fraction? If so, what is the result?
7. Can you add an integer and fraction? If so, what is the result?

As you (hopefully) observed, **Fraction**() does not always give the most succinct representation of a float number. For example, **Fraction**(1.1) yields 2476979795053773/2251799813685248, but the simplest form is clearly 11/10. To ensure simplest form, we need to import yet another module:

```
>>>from decimal import Decimal
>>>f = Fraction(Decimal('1.1')                ← Note the decimal number is input as a string
>>>print(f)
11/10
```

Another workaround is:

```
>>>f = Fraction(1.1)
>>>f.limit_denominator(100)                   ← limits denominator to 100
>>>print(f)
11/10
```

Summary:

TABLE 5.1 Selected Functions in fractions Module

Fraction() converts the parameter to a fraction in the form **Fraction**(*n*,*d*). It is output as *n*/*d*. The parameter can be a pair of integers, e.g., (3,4); a float, e.g., (.75); a string representation of a fraction, e.g., ('3/4'); a string representation of a decimal, e.g., ('.75'); or a single integer, e.g., (7).
***Fraction**(float) is not always simplest form. Use **Fraction(Decimal**('float')) instead.*
f.**numerator** returns the numerator of fraction *f*
f.**denominator** returns the denominator of fraction *f*
f.**limit_denominator**(*n*) limits the denominator to *n* and often returns the most accurate representation of *n*.

Program Goal: Write a program that asks the user to input two fractions *a*/*b* and *c*/*d*. Multiply the fractions. Do not use the shortcut Fraction(*a*,*b*)*Fraction(*c*,*d*); parse the numerators and denominators, then use (*a**b*) / (*c**d*).

PROGRAM 5.1.1 MULTIPLY_FRACTIONS.PY

```
from fractions import *
F1 = Fraction(input('Fraction 1? '))
F2 = Fraction(input('Fraction 2? '))
a, b = F1.numerator, F1.denominator
c, d = F2.numerator, F2.denominator
e = a*c
f = b*d
F3 = Fraction(e,f)
print(F1, "*", F2, "=", F3)
```

Sample Output

```
Fraction 1? 1/2
Fraction 2? 3/4
1/2 * 3/4 = 3/8
```

EXERCISE SET 5.1.4

1. When using f.**numerator** and f.**denominator**, does Python use the original fraction or the simplest form of the fraction? That is, if *f* = Fraction(90,100), does f.**numerator** return 90 or 9?
2. Does program **multiply_fractions** handle input of decimal numbers?
3. What is the output when the inputs are 0.3 and 0.5?
4. Modify program **multiply_fractions** so that it also divides the two fractions (using (*a**d*) / (*b**c*)).
5. Write a program that will add two fractions.
6. Write a function that will convert a string fraction (ex. "2/3") to a float variable rounded to 3 decimal places (e.g., 0.667).

5.2 A Few Math Functions

A Python **math function** is essentially an operation(s) performed on one or more inputs that returns a single output. The **domain** of a function refers to allowable inputs, analogous to mathematical functions. Likewise, the **range** of a function refers to possible outputs.

We'll start with the Python built-in math functions. Unless otherwise stated, assume *x* represents any real number (float, integer, fraction) and *n* represents an integer.

a**b

You probably already know this is the exponentiation operator.

Examples:

4**–1 = 0.25
27**(1/3) = 3.0
–2**3 = –8
–64**(1/2) = –8
(–64)**(1/3) = (2 + 3.464101615137754*j*)

Uh oh, that last answer did not come out as expected. (−64)**(1/3) should be $\sqrt[3]{-64} = -4$, a real number; however, the ** operator will return an (undesirable) complex number whenever you raise a negative number to a fractional exponent.

abs(x)

This is the absolute value function $|x|$. Technically it is defined as a piecewise-defined function:

$$x = \begin{cases} x \text{ if } x \geq 0 \\ -x \text{ if } x < 0 \end{cases}$$

The type of the result is the same as the type of the parameter.

Examples:

abs(5) = 5
abs(−7.2) = 7.2
abs(7/4) = 1.75 ← When you take the **abs** of a fraction, the result is float

round(x,n)

The **round()** function will round x to n decimal places. If n is not specified, the default is 0. The type of the result can be integer or float.

Examples:

round(3.5) = 4
round(10.4523,3) = 10.452
round(−4.1) = −4
round(10/4) = 2

pow(x,y)

This is the same as x**y, which is the same as x^y. The type of the result can be integer or float.

Examples:

pow(2,3) = 8
pow(4,1/2) = 2.0
pow(100,−1/2) = 0.1
pow(−125,1/3) = (2.5000000000000004 + 4.330127018922193*j*)

Again, this last example is not a desirable answer (the real answer is $\sqrt[3]{-125} = -5$). Just like the ** operator, **pow** results in a complex number when you try to raise a negative number to a fractional exponent.

int(x)

This function returns the integer portion of x.

Examples:

int(1.789) = 1
int(10.37) = 10
int(−4.32) = −4

divmod(x,y)

This function returns the quotient and remainder, in the form of a 2-element tuple, when x is divided by y.

Examples:

divmod(11,4) = (2,3) because 11 = 2*4 + 3.
divmod(50,7) = (7,1) because 50 = 7*7 + 1.
divmod(10,−4) = (−3, −2) because 10 = −4*−3 + −2
divmod(3,8) = (0,3) because 3 = 8*0 + 3
divmod(−26,5) = (−6,4) because −26 = 5*−6 + 4

We will restrict our discussion of **divmod** to integer values. The value returned is a tuple of two integers.

x % y

Similar to **divmod**(x,y) except it only returns the remainder (also known as the modulus).

Examples:

45%11 = 1

EXERCISE SET 5.2.1

1. Find the value and type of each of the following expressions.

   A. −4**2

   B. −4**2

   C. 4.0**2

   D. −64**(1/2)

   E. (−64)**(1/2)

   F. **Fraction**(1,4)**2

   G. (1/4)**2

   H. math.**sqrt**(−16)

   I. pow(1/4,2)

2. Find the value and type of each of the following expressions.

   A. **divmod**(17,2)

   B. **divmod**(–17,2)

   C. **divmod**(–17,–2)

   D. **divmod**(17,–2)

   E. 17%2

   F. 17%0

3. Is the remainder returned by the **divmod()** function always a positive number?

4. Find n%2 when *n* is any even number.

   A. Find n%2 when *n* is any odd number.

5. If *a*, *b*, *m*, and *n* are positive integers:

   A. Does $(a + b)$ % $m = a$ % $m + b$ % n?

   B. Does a % $m + a$ % $n = a$ % $(m + n)$?

6. Because a negative number raised to a fractional exponent results in a complex number, what snippet of code can give the true value of a^b?

7. Ignoring the **divmod()** function for a moment, how can you use the **%** function to find the quotient (integer portion) of *m* and *n*? For example, the quotient of 13 and 5 is 2. The remainder is 3.

8. The notation for "*m* divides *n*" is $m|n$ (i.e., *m* is a factor of *n*). How can you use the **%** function to determine if $m|n$ for $m > 0$ and $n > 0$?

9. The 24-hour format time represents midnight to noon as 0–12 and 1 pm to 11 pm as 13–23. Write a program that converts 24-hour time (1–24) to 12-hour time.

Sample Output

hour? 3

minutes? 45

am or pm? pm

15 : 45

10. Write a program that performs the same function as **divmod**, but without using **divmod**. Check to make sure that the inputs are whole numbers.

The selection of built-in Python math functions is rather limited; however, there is a math library that greatly expands the calculating power of the language.

Math Module Functions

Unless otherwise stated, assume x represents any real number (float, integer, fraction) and n represents an integer.

math.sqrt(x)

This is the square root function $\sqrt{x}$. You can only take the square root of a positive real number, and a float number is returned.

Examples:

math.**sqrt**(144) = 12.0
math.**sqrt**(10) = 3.1622776601683795
math.**sqrt**(144/100) = 1.2
math.**sqrt**(−7) = ValueError

Unless x is a perfect square, the result will be an irrational number, which is technically an infinite nonrepeating decimal. Thus, the float number that is returned is either truncated or rounded.

math.floor(x)

Depending on the textbook or calculator you pick up, the **greatest integer function** has notation int(x) or $\lfloor x \rfloor$. It is defined as the greatest integer less than or equal to x. More informally, it refers to the greatest integer to the left of x on the number line.

Examples:

math.**floor**(10.84) = 10
math.**floor**(−5.72) = −6

math.ceil(x)

Returns the ceiling of x, which is the smallest integer greater than or equal to x (i.e., to the right on the number line). Alternate notation: $\lceil x \rceil$.

math.**ceil**() returns an integer.

Examples:

math.**ceil**(8.92) = 9
math.**ceil**(2.1) = 3
math.**ceil**(−4.8) = −4

math.factorial(n)

Returns n!, which is defined as $n \cdot (n-1) \cdot (n-2) \ldots 2 \cdot 1$. The parameter must be a positive integer, and the result is likewise an integer.

Examples:

math.**factorial**(10) = 3628800
math.**factorial**(0) = 1

math.trunc(x)

Truncation function; returns the integer part of a number.

Examples:

math.**trunc**(4.25) = 4
math.**trunc**(12/5) = 2
math.**trunc**(7/8) = 0

math.exp(x)

Natural exponential function e^x (aka Euler's Number) where e is an irrational constant $\approx$ 2.718281828. This constant (which will be explored further in Chapter 10) is an integral component of many formulas in finance, exponential growth and decay, etc.

math.gcd(m,n)

Returns the greatest common divisor (an integer) of integers m and n. The result is always a positive integer, even if one or both of m,n are negative.

Examples:

math.**gcd**(24,30) = 6
math.**gcd**(−12,200) = 4
math.**gcd**(−90, −57) = 3

Recall the convention for importing modules:

Method 1—Import a Single Module

```
>>>from math import gcd
>>>print(gcd(10,12))
```

Method 2—Import the Entire Module

```
>>>import math
>>>print(math.gcd(10,12))                    # precede function with math.
```

Method 3—Import the Entire Module

```
>>>from math import *
>>>print(gcd(10,12))
```

The **import** statement can appear in the main program, even if it only accessed within a user-defined function.

EXERCISE SET 5.2.2

1. Find the value and type of the following:
   A. **abs**(−3.8)
   B. **round**(−3.8)
   C. **int**(−3.8)
   D. math.**floor**(−3.8)
   E. math.**floor**(3.8)
   F. math.**ceil**(−3.8)
   G. math.**ceil**(3.8)
   H. math.**trunc**(−3.8)
   I. math.**trunc**(3.8)
   J. math.**gcd**(12,126)
   K. math.**gcd**(−12,−126)
2. What is the difference between math.**floor**() and math.**trunc**()?
3. Define **round()** as a piecewise-defined function on a positive number using **math.floor()** and **math.ceil()**.
4. Write a brief function that finds the decimal portion of a float number. For example, fractional_part(pi) = 0.14159265358979312.
5. Is there a difference between **int**(x) and math.**trunc**(x)? If so, what is it?
6. Write a program that determines if a positive number is a perfect square.
7. Two positive integers are **relatively prime** if they share no common factors other than 1. How can you use the **gcd** function to determine if two positive integers M and N are relatively prime?
8. Does $\lfloor x \rfloor + \lfloor y \rfloor = \lfloor x + y \rfloor$? Be sure to test with both positive and negative numbers.
9. Write a program that expresses a positive rational number $\frac{m}{n}$ as $m = q \cdot n + r$, where $m > n$ and $0 \le r < n$. For example, $\frac{54}{4} = 13 \cdot 4 + 2$.
10. The number of multiples of positive integer k between integers m and n (inclusive), with $m \le n$, is

$$\left\lfloor \frac{n}{k} \right\rfloor - \left\lfloor \frac{m-1}{k} \right\rfloor$$

(That's the **floor** function, not the **abs** function.) Write a program that asks the user to input values for k, m, and n. Calculate and output the number of multiples of k between m and n.

Sample Output

k? 3

m? 1

n? 20

The number of multiples of 3 between 1 and 20 is 6.

Program Goal: Write a program that (A) asks the user to input the weight of a package in pounds and (B) convert to pounds and ounces.

SAMPLE OUTPUT

Weight in pounds? 2.5

The package weight is 2 lb., 8.0 oz.

PROGRAM 5.2.1

```
weight = float(input("weight in lb? "))
lb = int(weight)
fractional_part = weight - lb
oz = fractional_part*16
print("The package weight is", lb, "lb,", oz, "oz")
```

EXERCISE SET 5.2.3

1. When you run program 5.2.1 with an input of 2.8, do you get exactly 2 lbs., 12.8 oz.? If not, how can you modify the program?
2. A shipping service charges 0.68 per oz. Modify the program so that it converts the total weight to ounces, then calculates and outputs the fee.

Sample Output

weight in lbs.? 2.8

The package weight is 2 lbs., 12.8 oz.

The shipping charge is $30.46

3. The shipping service has a new policy that weights will be rounded up to the nearest oz.; in other words, even a fractional part of an oz. will be charged a full oz.

Sample Output

weight in lb.? 2.8

The package weight is 2 lbs., 13 oz.

The shipping charge is $30.6

(Continued)

4. After excessive negative feedback from customers, the shipping service has decided that it will charge \$3.00 for the first pound and 0.62 for each additional pound. Modify the program accordingly.

Sample Output

weight in lbs.? 2.8

The package weight is 2 lbs., 13 oz.

The shipping charge is \$20.98

5. A professor has grown tired of students submitting late tests online, so she has decided to deduct 5 points for each half hour late. Fractional parts of half hours are rounded up to the next half hour; for example, 39 minutes late would result in a 10-point deduction (5 points for the first 30 minutes and 5 points for the next 9 minutes). Write a program for the professor so that she inputs the number of minutes late and it will output the total points to be deducted.

Sample Output

Minutes late? 75

Points to deduct: 15

Simplifying a Square Root

If you were to ask Python to evaluate a square root, you would get a decimal approximation:

```
>>>math.sqrt(24)
4.898979485566356
```

Compare to the **SymPy sqrt**() function:

```
>>>from sympy import sqrt,pprint
>>>pprint(sqrt(24))
2·√6
```

EXERCISE SET 5.2.4

1. Write a program that asks the user to input any positive integer; calculate and output the simplified square root.

Sample Output

Positive integer? 96
4·√6

2. Modify the program so that it verifies that the input is a positive integer.

3. Use the program to simplify the following:

   A. $\sqrt{492}$

   B. $\sqrt{3456}$

   C. $\sqrt{2020}$

   D. $\sqrt{4753}$

4. Write a program to find the cube root of a number n. Hint: solve $x^3 - n = 0$. Output only the first solution in the list.

Sample Output

Integer? −729

−9

Integer? −81

3_

−3 · √ 3

The "pretty print" doesn't look exactly pretty in this example; it represents $-3\sqrt[3]{3}$.

5.3 Least Common Multiple

As of this writing, Python does not have a built-in function to find the least common multiple of a pair of numbers. However, you probably learned in elementary school one or more methods to find the LCM of a pair of positive integers. One painstaking method is to make a list of multiples for each number until you reach a number in both lists. For example, to find lcm(8,18):

Image 5.3

8: 8, 16, 24, 32, 40, 56, 64, 72, ...

18: 18, 36, 54, 72, ...

This method is straightforward and will give us an opportunity to review some elementary Python techniques.

Program Goal: Write a function lcm that finds the least common multiple of a pair of positive integer inputs m and n. Call the function from a main program.

SAMPLE OUTPUT

First number? 8

Second number? 18

lcm(8,18) = 72

PROGRAM 5.3.1

```
def lcm(m,n):
# make sure m is the larger number
    if m < n:
        temp = m
        m = n
        n = temp
        # iterate through multiples of m until you find a multiple that is divisible by both m and n
    k = m
    result = 0
    while result==0:
        if k%m == 0 and k%n == 0:
            result = k
        else:
            k += m
    return result

# MAIN PROGRAM
a = int(input("First number? "))
b = int(input("Second number? "))
print("\nlcm(", a, ",", b, ") =", lcm(a,b))
```

Python has an awesome way to have two variables "swap" values (long overdue in the programming multiverse). Simply replace the following three lines:

```
temp = m
m = n
n = temp
```

with the following:

```
m, n = n, m
```

Reminder about **local** and **global** variables:

To avoid confusion, give the variables in the body of the function different names from the variables in the main program. The main program does not have access to the values of the variables used within the function. If the same variable name is used in both the main program and the function, then any changes made to the variable in the function will not apply to the same variable name in the main program.

EXERCISE SET 5.3.1

1. Use the program to find the lcm of the following pairs of numbers:
   A. 16, 102
   B. 9, 55
   C. 11, 124
2. What happens if the user inputs one or both negative numbers?
3. What happens if the user inputs zero as one of the numbers?
4. A more direct algorithm for finding the lcm is $lcm(m, n) = \frac{m * n}{gcf(m, n)}$.

   Rewrite the lcm function in program 5.3.1 to take advantage of this.
5. Does the function return an **int** or **float** type variable? If not already **int,** modify the function to return an **int.**
6. Modify the main program in program 5.3.1 to find the lcm of three numbers.

Sample Output

First number: 61
Second number: 80
Third number: 49

lcm(61, 80, 49) = 239120

7. CHALLENGE: Write a program that will add fractions $\frac{a}{b} + \frac{c}{d}$ without using the **Fraction**() function.

   This will involve using the **lcm**() function to find the common denominator. Make sure your result is in lowest terms.

Sample Output

Fraction 1? 3/4
Fraction 2? 5/12
7 / 6

5.4 Euclid's Method for Finding the GCD

One of the objectives of this book is to show you the underlying logic behind the built-in functions. Let's take a look at the greatest common divisor (**gcd**). One of the historic and elegant algorithms for doing so is Euclid's algorithm.

Let's say we wanted to find gcd(125, 50).

gcd(125, 50)
$= \gcd(125 - 50, 50) = \gcd(75, 50)$
$= \gcd(75 - 50, 50) = \gcd(25, 50)$
$= \gcd(25, 50 - 25) = \gcd(25, 25)$
$= 25.$

More generally:

(1) if $m > n$, then $\gcd(m, n) = \gcd(m - n, n)$. Repeat until $m = n$.
(2) if $m < n$, then $\gcd(m, n) = \gcd(m, n - m)$. Repeat until $m = n$.

Program Goal: Find the greatest common divisor of a pair of integers m and n using Euclid's Algorithm.

PROGRAM 5.4.1

```
m = int(input("First number? "))
n = int(input("Second number? "))
print("gcd(",m,",",n,")=")
while m != n:
    if m > n:
            m = m - n
                if n > m:
                n = n – m
print(m)
```

EXERCISE SET 5.4.1

1. Use the program to find the gcd of the following pairs of numbers:
   A. 56, 284
   B. 77, 91
   C. 3725, 4335
   D. 4225, 1625

2. Modify the program so that if the gcd = 1, it outputs “relatively prime.”
3. What happens if one of the inputs is 0? Modify the program so that it correctly handles 0 as an input ($\gcd(M,0) = M$ for any integer M).
4. Does this algorithm effectively handle negative numbers? If not, fix.
5. Simplify: $\gcd(m, 2m + n)$.

5.5 Prime Numbers

Most of us have a notion of prime numbers from back in middle school. However, their significance is vast—in fact, prime number theory forms the foundation of cybersecurity. First some definitions:

Prime Number—a natural number whose only divisors are 1 and itself.

Composite Number—a natural number greater than 1 that is not prime.

Formatting Output in Columns

Before we start generating large lists of prime numbers, we need a tidy way to output a list on the screen.

Program Goal: Write a program to print out the numbers 1 through 100 on the screen—10 numbers per row and 10 numbers per column.

Columns→	1–4	5–9	10–14	15–19	20–24	25–29	30–34	35–39	40–44	45–49
	1	2	3	4	5	6	7	8	9	10
	11	12	13	14	15	16	17	18	19	20
	:	:	:	:	:	:	:	:	:	:
	91	92	93	94	95	96	97	98	99	100

Recall from Chapter 4 that to print a row of 10 integers centered in columns that are 5 characters wide, you would need a formatted **print()** function that looks like this:

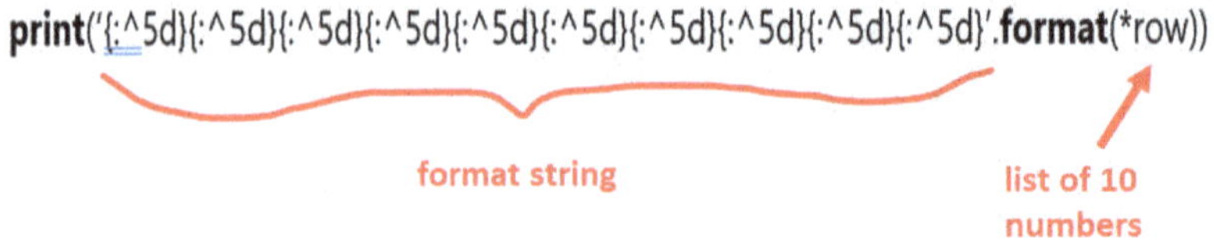

Image 5.4

For reasons that will become more clear later, we will store the numbers 1–100 in a list and call a function **grid()** to print out the numbers in the list.

PROGRAM 5.5.1

```
def grid(num_list):
    fstring=''
    for k in range(10):
        fstring += '{:^5d}'
    for row in range(10):
        # assign 1st 10 entries in num_list to row_to_print
        row_to_print = num_list[:10]
        # remove 1st 10 entries from num_list
        num_list = num_list[10:]
        # print next row
        print(fstring.format(*row_to_print))
    return
# MAIN PROGRAM
# print a list of numbers 1..100
grid(list(range(1,101)))
```

EXERCISE SET 5.5.1

1. Ten columns might be too wide for your display. Modify the function so that only 7 numbers are printed per row. Keep in mind that 7 is not a divisor of 100, so the last row will have fewer than 7 numbers.
2. Make the function more versatile. Pass in another parameter *n* that specifies how many numbers will be printed per row.

You now have a handy function that will output lists of any size. We're ready to generate and output prime numbers.

One simple way of generating a finite table of prime numbers is called the **Sieve of Eratosthenes**. You might remember this from middle school. Let's say we want a table of primes between 1 and 100. We start by writing down all the integers between 1 and 100:

1	2	3	4	5	6	7	8	9	10
11	12	13	14	15	16	17	18	19	20
21	22	23	24	25	26	27	28	29	30
31	32	33	34	35	36	37	38	39	40
41	42	43	44	45	46	47	48	49	50
51	52	53	54	55	56	57	58	59	60
61	62	63	64	65	66	67	68	69	70
71	72	73	74	75	76	77	78	79	80
81	82	83	84	85	86	87	88	89	90
91	92	93	94	95	96	97	98	99	100

Image 5.5

We then start with the first prime, which is 2. "Cross out" all multiples of 2 less than 100 (but not including 2). Then the next prime is 3: cross out all multiples of 3 less than 100. Continue in this manner, crossing out multiples of the remaining odd numbers less than 100. (Why just odd?)

~~1~~	2	3	~~4~~	5	~~6~~	7	~~8~~	~~9~~	~~10~~
11	~~12~~	13	~~14~~	~~15~~	~~16~~	17	~~18~~	19	~~20~~
~~21~~	~~22~~	23	~~24~~	~~25~~	~~26~~	~~27~~	~~28~~	29	~~30~~
31	~~32~~	~~33~~	~~34~~	~~35~~	~~36~~	37	~~38~~	~~39~~	~~40~~
41	~~42~~	43	~~44~~	~~45~~	~~46~~	47	~~48~~	~~49~~	~~50~~
~~51~~	~~52~~	53	~~54~~	~~55~~	~~56~~	~~57~~	~~58~~	59	~~60~~
61	~~62~~	~~63~~	~~64~~	~~65~~	~~66~~	67	~~68~~	~~69~~	~~70~~
71	~~72~~	73	~~74~~	~~75~~	~~76~~	~~77~~	~~78~~	79	~~80~~
~~81~~	~~82~~	83	~~84~~	~~85~~	~~86~~	~~87~~	~~88~~	89	~~90~~
~~91~~	~~92~~	~~93~~	~~94~~	~~95~~	~~96~~	97	~~98~~	~~99~~	~~100~~

Image 5.6

The remaining numbers (circled) are the prime numbers.

Program Goal: Write a program **sieve** to generate a grid of prime numbers. Use list **potential** to store numbers between 2 and *N*. Initially all values will be set to 1 (meaning prime). To "cross out" a composite number, we set the corresponding list entry to 0 (meaning not prime). Here's an overview.

1. Cross out all the even numbers (multiples of 2).
2. Starting with 3, cross off multiples of odd numbers less than *N*.
3. When done with sieve, transfer the numbers not crossed off to a new list **prime**.
4. Output list **prime** using the **grid** function in program 5.5.1.

PROGRAM 5.5.2 SIEVE.PY

```
[ include the definition of function grid here or import from another file ]
# main program
n = int(input("Upper bound? ")
# initialize all the list entries to 1
potential = [1 for i in range(n+1)]
# 1 is not a prime
potential[1] = 0
# set all even numbers to 0
for i in range(0,n+1,2):
    potential[i] = 0
# iterate through all odd numbers ("seeds") in [3, max]
j = 3
while j <= n:
    # iterate through multiples of j
    k = 2
    while k*j <= n:
        # "cross off" that multiple of j
        potential[k*j] = 0
        k += 1
    # get next odd seed
    j += 2
# transfer only numbers that still have a value of 1 to list prime
prime = [ ]
for i in range(2,n+1):
    if potential[i]==1:
prime.append(i)
[ fill in code to output list prime in a grid ]
```

You might notice that the larger value you input for *n*, the longer the program takes to run. We will try to optimize in Exercise Set 5.5.2 below.

EXERCISE SET 5.5.2

1. Fill in the yellow highlighted code to output the list of primes.
2. The list is missing one very unique prime number. Modify the program to include this number in the list.
3. Add an output statement at the end of the program that prints out how many primes are in the list.

Desired Output

Upper bound? 100

2	3	5	7	11	13
17	19	23	29	31	37
41	43	47	53	59	61
67	71	73	79	83	89
97					

Total: 25 primes <= 100

4. A handy estimation for the number of primes less than a given number *n* is $\frac{n}{ln(n)}$. Output this estimation at the end of the program. Round to the nearest whole number using the **round()** function. Use **math.log**(*n*) to find ln(*n*). What is the approximate number of primes less than 100?
5. How can this program be made more efficient? Here's a suggestion. Do we need to consider multiples of any numbers greater than *n* / 2 ? For example, if *n* = 100, are there any multiples of numbers greater than 50 that are less than 100? A bit of research on the Sieve of Eratosthenes will turn up other optimization techniques as well.
6. Modify the program as follows (or write a new one): (1) Instead of asking the user to input *n*, ask them to input a number *m* (to test for primality). (2) Determine if the requested number *m* is prime or not.

   Hint: **if** *m* **in** prime will determine if *m* is in the prime list.

Desired Output

m? 97
97 is prime
m? 98
98 is not prime

Check for Primality Using Factors

A more direct and elegant method of determining whether a number is prime is to check if it has any factors other than 1 and itself.

PROGRAM 5.5.3 PRIME_CHECK.PY

```
n = int(input("Enter a positive integer: "))
comp = False                          # assume that the number
if n > 1:                             # is prime (not composite)
    # check for factors
    for i in range(2, n):
        if (n % i) == 0:
            comp = True
if comp:
    print(n, "is composite")
else:
    print(n, "is prime")
```

Sample Output

Enter a positive number: 7900001
7900001 is a prime number

EXERCISE SET 5.5.3

1. Rewrite program **prime_check** as a main program that asks the user to input positive integer *n*. Call function **prime_check**(*n*) to return true or false.
2. Modify the program as follows: (1) Instead of asking the user to input *n*, ask them to input a number *p* (how many primes they want). (2) Generate and output the first *p* primes.

Sample Output

How many primes do you want? 5

The first 5 primes are:

[2, 3, 5, 7, 11]

3. Modify the program as follows: Ask the user to input *m*, an upper bound. Generate and output all primes less than or equal to *m*. Call this function **prime_generator**.

Sample Output

Upper bound for primes? 17

The primes less than 17 are:

[2, 3, 5, 7, 11, 13]

4. Approximately what percentage of whole numbers is prime? Since there are infinitely many whole numbers AND primes, $\frac{\infty}{\infty}$ does not yield a very meaningful percentage. Instead, write a program that asks the user to input a whole number *n*. Generate *n* random numbers between 1 and 10,000, count the number of primes *p*, and form a percentage *p* / *n*.

Sample Output

How many random whole numbers? 50

The random numbers are:

[7849, 2172, 1443, 2467, 9923, 9931, 7341, 6530, 2139, 52, 819, 6977, 8656, 2406, 6689, 2668, 6521, 8735, 4713, 9270, 235, 3046, 800, 4452, 8771, 2284, 2100, 3969, 2274, 9349, 343, 330, 6026, 5387, 953, 6462, 677, 9441, 6927, 2886, 1906, 9146, 3721, 2006, 6988, 9724, 1540, 7420, 3280, 4573]

The primes in that list are:

[2467, 9923, 9931, 6977, 6689, 6521, 9349, 5387, 953, 677]

The percentage of primes is 0.2

Run the program several times with varying input values of *n*. Does the percentage *p* / *n* vary wildly or remain pretty consistent (within a few percentage points)?

PrimeGrid is a volunteer distributed computing project that searches for very large (up to near-world-record size) prime numbers while also aiming to solve long-standing mathematical conjectures.

As of this writing, another organization—**the Electronic Frontier Foundation**—offers $150,000 and $250,000 for primes with at least 100 million digits and 1 billion digits, respectively.

5.6 Prime Factorization

The **Fundamental Theorem of Arithmetic** states that every integer greater than 1 can be expressed as the product of one or more prime factors.

Prime factorization is the process of factoring a natural number into a product of powers of primes. For example, $120 = 2^3 \cdot 3 \cdot 5$. When you were in elementary school, you may have learned to do this with a "factor tree":

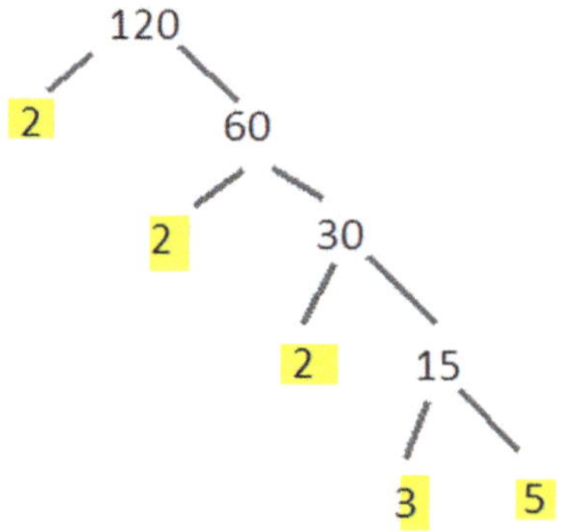

Image 5.7

Program Goal: Write a program that does the prime factorization of input n.

We will maintain a finite list of primes in prime (the size of prime will limit the range of numbers we can factor, but we'll discuss that later). As we "chisel away" at n, we will assign its prime factors to list factors.

PROGRAM 5.6.1 PRIME_FACTORS.PY

```
primes= [2,3,5,7,11,13,17,19,23,29,31,37,41,43,47,53,59,61,67,71,73,79,83,89,97]
n = int(input("n? "))
factors = [ ]
while n > 1:
    for p in primes:
        if n%p==0:
            factors.append(p)
            n = int(n/p)
print(factors)
```

Sample Output

n? 100
[2, 2, 5, 5]

Program **prime_factor** can actually handle quite a few large values of n with the first 25 primes loaded into list prime.

The field of cryptography relies on the fact that no known algorithm exists to efficiently and effectively factor "very large" numbers into primes.

EXERCISE SET 5.6.1

1. Modify the program so that the list factors are sorted (in ascending order) before they are printed.
2. What is an example of a whole number that **cannot** be factored using the program?
3. Modify the program so that it calculates ALL the divisors of a number, not just the prime factors. For example, the prime factorization of 12 is [2, 2, 3]. The divisors of 12 are [1, 2, 3, 4, 6, 12].

Let's improve the output so that the prime factorization is output using exponents. For example, program **prime_factor** will return [2, 2, 5, 5], which means $100 = 2^2 \cdot 5^2$.

First off, after running program **prime_factor**, type the following into the console prompt:

```
>>>factors
[2, 2, 5, 5]
>>>unique_factors = set(factors)
>>>unique_factors
{2, 5}
```

Notice that we've created a set version of the list **factors**. Recall from Chapter 2 that a set is a special type of list in which the elements are unique—i.e., no repeats.

Here's one way to print the factorization, using a list function called factors.**count()** to count the occurrences of each prime factor in the list **factors**. We'll save each unique prime in a list called **each_prime** and the associated count (i.e., exponent) in list **exponent**. For example, to factor $n = 100$, we'd form the following lists:

each_prime	2	5
exponent	2	2
	↓	↓
	2^2	5^2

PROGRAM 5.6.1 PRIME_FACTORS.PY REVISED

```
primes = [2,3,5,7,11,13,17,19,23,29,31,37,41,43,47,53,59,61,67,71,73,79,83,89,97]
n = int(input("n? "))
factors = [ ]
while n > 1:
    for p in primes:
        if n%p==0:
            factors.append(p)
            n = int(n/p)
print(factors)
factors.sort()
unique_factors = set(factors)
each_prime = [ ]
exponent = [ ]
for f in unique_factors:
    each_prime.append(f)
    exponent.append(factors.count(f))
    [fill in code to output final answer ]
```

Sample Output

n? 8403500
(2^2)(5^3)(7^5)

EXERCISE SET 5.6.2

1. Fill in the yellow highlighted code to output the prime factorization.
2. Modify the program so that factors raised to the first power will print without the exponent.

Sample Output

n? 124

(2^2)(31)

3. Use the program to factor the following:
   - A. 1,176,120
   - B. 13,456,625
   - C. 5,498,361
   - D. 532,564,273

More Fun with Primes

Goldbach's Conjecture asserts that every even integer greater than 2 can be written as a sum of two primes. As of 2014, this conjecture has been verified for all numbers up to $4 \cdot 10^{18}$.

Program Goal: Write a program that accepts as input an even integer greater than 2, and outputs the appropriate sum(s) of two primes.

Hint: Use the same list **primes** that we did in program **prime_factors,** which will limit us to positive even numbers far less than $4 \cdot 10^{18}$. This will require nested loops to go through every combination of two numbers in **primes**.

SAMPLE OUTPUT

Positive even integer? 42

$42 = 5 + 37$

$42 = 11 + 31$

$42 = 13 + 29$

$42 = 19 + 23$

Done

PROGRAM 5.6.2 GOLDBACH.PY

```
primes = [2,3,5,7,11,13,17,19,23,29,31,37,41,43,47,53,59,61,67,71,73,79,83,89,97]
n = int(input("Positive Even integer? "))
for p1 in primes:
    for p2 in primes:
        if p1 + p2==n:
            print(n,"=",p1,"+",p2)
            # Exit inner loop when answer found
            break
print("done")
```

Notice the use of the **break** statement that breaks out of the inner loop. If it weren't there, you'd still get the same output but you'd have to wait through 25^2 iterations of the inner loop.

Try running the program. Are you getting each prime pair repeated twice? For example, both $42 = 5 + 37$ and $42 = 37 + 5$? Your job is to fix this in the Exercise Set below.

EXERCISE SET 5.6.3

1. Why doesn't 198 produce a result?
2. Modify the program so that each prime pair is only output once.
3. What is the maximum *n* that can be successfully handled by program **goldbach** with only 25 primes in list primes?
4. Add the following error messages:
   A. If the answer was not found.
   B. If the input *n* is not positive or even and greater than 2.
5. **Vinogradov's theorem** says that every sufficiently large positive odd integer can be written as a sum of three primes. Modify program **goldbach** to implement Vinogradov's theorem including checks that the input is a "sufficiently large" positive odd integer. Eliminate permutations (repeats).

Sample Output

positive odd integer? 17
[2, 2, 13]
[3, 3, 11]
[3, 7, 7]
[5, 5, 7]

(Continued)

6. A **Mersenne prime** is a prime that is one less than a power of two; i.e., it can be written in the form $2^n - 1$ for some positive integer n. The first three Mersenne primes are 3, 7, 31. Write a program to find all Mersenne primes less than 10,000.

   Hint: Rather than using the prime list (which only goes up to 97), use function ***prime_generator*** *in Exercise Set 5.5.3 #3 to generate a larger list of primes.*

7. **Twin primes** are a pair of prime numbers p and $p + 2$. The first three twin primes are (3, 5), (5, 7), (11, 13). Write a program to find all twin primes less than 500.

8. **Bertrand's postulate** says that for every positive integer $n \geq 2$, there exists a positive prime p such that $n < p \leq 2n$. Write a program that asks the user to input n. Find p.

Sample Output

positive integer *n*? 48
53 is between 48 and 96

5.7 Symbolic Logic

In order to compose effective **if** and **while** structures, a programmer must have a firm grasp of how to evaluate logical expressions involving "and," "or," and "not." Never fear, these boil down to simple truth tables and DeMorgan's laws.

Let's say that P and Q represent statements having a true or false value, such as "N is prime" or "$N \neq 0$". In order for the compound expression ***P*** **and** ***Q*** to be true, they both need to be true. Here is the truth table for "**and**" based on the values of P and Q:

P	*Q*	*P* and *Q*
T	T	T
T	F	F
F	T	F
F	F	F

In order for ***P*** **or** ***Q*** to be true, then one or the other or both needs to be true.

P	*Q*	*P* and *Q*
T	T	T
T	F	T
F	T	T
F	F	F

EXERCISE SET 5.7.1

For #1–5, let $N = 72$ and $M = 20$. Determine whether the following statements evaluate to true or false.

1. remainder(N,12) = 0 and $M > 0$
2. remainder(N,M) =12 and gcd(72,20) = 4
3. int(N/M) = N / M or $N > M$
4. $N - M < 0$ or $N > 3*M$
5. ($N > 0$ or $N - M = 50$) and abs(M)= $-M$

Another logical operator is **not *P*,** which basically negates whatever truth value *P* has.

P	Not *P*
T	F
F	T

DeMorgan's laws tell us that the following are equivalent: (we will use the symbol ↔ to mean *is equivalent to*)

not (*P* and *Q*) ↔ not *P* or not *Q*.

Example: "Not (hungry and tired)" is the same as "not hungry or not tired."

not (*P* or *Q*) ↔ not *P* and not *Q*.

Example: "Not (old enough or rich enough)" is the same as "not old enough and not rich enough."

This truth table demonstrates the first equivalence:

P	*Q*	*P* and *Q*	not (*P* and *Q*)	not *P*	not *Q*	not *P* or not *Q*
T	T	T	F	F	F	F
T	F	F	T	F	T	T
F	T	F	T	T	F	T
F	F	F	T	T	T	T

The symbol for **and** is ^. The symbol for **or** is v. The symbol for **not** is usually ~ or ¬.

EXERCISE SET 5.7.2

For #1–5, let $m = 17$. Determine the truth value of each statement.

1. $\neg$ (m is prime)
2. $\neg$ (m is odd $\vee$ $m > 20$)
3. $\neg$ (m is even) $\wedge$ $\neg$ ($m < 0$)
4. $\neg$ ($m < 10$) $\vee$ $\neg$ ($m > 20$)
5. (m is odd) $\wedge$ $\neg$ (m is prime $\vee$ $m < 0$)

For #6–7, construct truth tables to assess the validity of the following statements.

6. $(p \wedge q) \leftrightarrow \neg(\neg p \vee \neg q)$
7. $[p \wedge (q \vee r)] \leftrightarrow [(p \wedge q) \vee (p \wedge r)]$

For #8–9, let $m = 5$ and $n = 6$. Evaluate whether the following compound logical statements are true or false.

8. (m and n are relatively prime) and not (m and n are both odd).
9. (m and n are relatively prime) and (m and n are both odd).

For #10–11, let $m = 5$ and $n = 7$. Evaluate whether the following compound logical statements are true or false.

10. (m and n are relatively prime) and not (m and n are both odd).
11. (m and n are relatively prime) and (m and n are both odd).

For #12–13, write compound logical statements to represent the following.

12. $a = 0$ or $b = 0$ but not both.
13. The absolute value of f minus L is less than e; and e is a positive number less than 1.

Boolean Variables and Values in Python

Variables in Python can have a **Boolean type** (bool), which has a **True** or **False** value (capitalized). However, any variable (or expression) with a value of 1 is evaluated as True and any variable (or expression) with a value of 0 is evaluated as False. Examples:

```
>>>m = 100-99
>>>n = True
>>>m and n
```

True

```
>>>p = True
>>>q = False
>>>p and q
```

False

```
>>>p or q
```

True

EXERCISE SET 5.7.3

1. If *p* has a value of True and *n* has a value of 0, what is the result of **p and n** (both the value and variable type)?
2. If *p* has a value of True and *n* has a value of 10, what is the result of **p and n==1** (both the value and variable type)?
3. If *p* has a value of True and *n* has a value of 0, what is the result of **p or n** (both the value and variable type)?
4. If *p* has a value of True and *n* has a value of 0, what is the result of **p or n==1** (both the value and variable type)?

Program Goal: Write a program to generate and output the truth table for the first version of DeMorgan's law, i.e., **not (*P* and *Q*) ↔ not *P* or not *Q***.

DESIRED OUTPUT

P	Q	P^Q	~(P^Q)	~P	~Q	~P V ~Q
T	T	T	F	F	F	F
T	F	F	T	F	T	T
F	T	F	T	T	F	T
F	F	F	T	T	T	T

For testing purposes, it is recommended that you perform "top-down" programming, which means essentially to start with the "framing" of the program, as though you were putting in the foundation of a building. That usually means your "outermost" **for**, **while**, and **if** structures. In this program, we'll start with two nested **for** loops to cycle through the four combinations of T/F values, then calculate and output the third column (*P* and *Q*).

PROGRAM 5.7.1 DEMORGAN.PY

```
# print column headings
print('{:^5s}{:^5s}{:^7s}{:^7s}{:^5s}{:^5s}{:^7s}'.format('P','Q','P^Q','~(P^Q)','~P','~Q','~PV~Q'))
value = ['F','T']
# print first two columns
for p in [True,False]:
    for q in [True,False]:
        print('{:^5s}{:^5s}{:^7s}'.format(value[p],value[q],value[p and q]))
```

EXERCISE SET 5.7.4

1. What is value[0] and value[1]?
2. What is the type of the variables *p* and *q*?
3. If p = False and q = True, what is *p* or *q*? What is value[*p* or *q*]?
4. Can a bool type variable be used as the index of a list?
5. Replace the final **print()** function so that it prints all seven columns of the truth table.
6. Modify the program so that it outputs the second version of DeMorgan's law:

   not (*P* **or** *Q*) ↔ **not** *P* **and not** *Q*
7. A. Write a program that generates the truth table for *p* **xor** *q* (**xor** is the exclusive "or," i.e., "*one or the other but not both*"). One of its symbols is ⊕.

   B. What is the logical equivalence of **xor**?

5.8 Pythagorean Theorem and Pythagorean Triples

Wikipedia says it best: *The Pythagorean theorem has attracted interest outside mathematics as a symbol of mathematical abstruseness, mystique, or intellectual power; popular references in literature, plays, musicals, songs, stamps and cartoons abound.*

You probably learned in middle school the details of the theorem: The sum of the squares of the two legs of a right triangle equals the square of the hypotenuse.

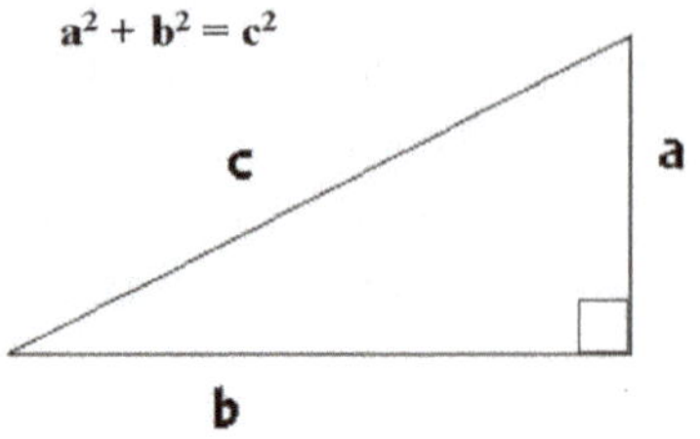

Image 5.8

There are literally hundreds of proofs of this theorem. Although it is attributed to Greek philosopher and mathematician Pythagorus (b. 570 BC), it may have been the work of one of his followers. Pythagorus founded a secret society called the **Pythagorean Brotherhood**.

The society, worthy of a book of its own, followed an interesting mashup of beliefs that tied together religion, philosophy, mathematics, music, numerology, cosmology, metaphysics, and symbolism. Its motto was "All is number."

The Pythagoreans were progressive in that they considered men and women as equals, practiced vegetarianism, and lived communally. Some of their beliefs were rather bizarre, such as abstention from eating beans and not to pick up what has fallen.

We will see in Chapter 6 how the Pythagorean theorem gives rise to the distance formula and the equation of circles.

At his point, keeping our focus on number theory, we are going to examine **Pythagorean triples**—a set of three positive integers that satisfy $a^2 + b^2 = c^2$, making them ideal problems for math students who have not yet learned to deal with square roots.

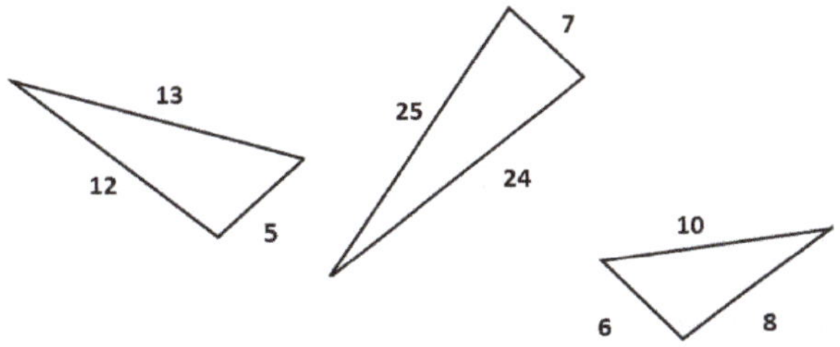

Image 5.9

Note that each Pythagorean triple gives rise to infinitely many "multiples" by multiplying each side of the corresponding triangle by a constant. These represent **similar triangles**. We will call the lengths of the sides of the "smallest" or "originating" triangle a **primitive Pythagorean triple**. In the below set of similar triangles, the primitive Pythagorean triple is (3, 4, 5).

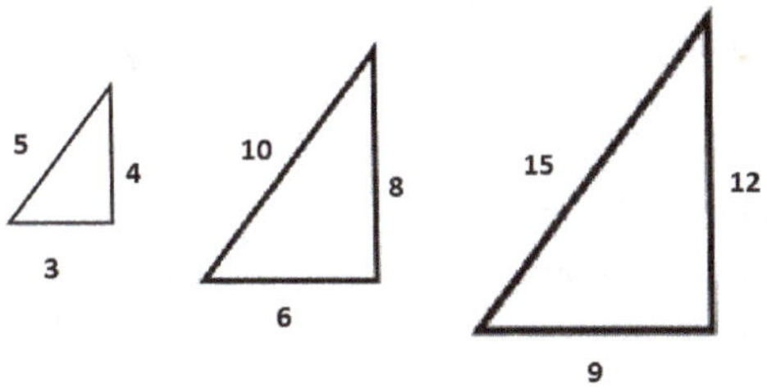

Image 5.10

The three numbers in a primitive Pythagorean triple are relatively prime, which means they share no common factor other than 1.

EXERCISE SET 5.8.1

In #1–5, which of the following sets of three numbers are Pythagorean triples? Which are primitive?

1. 28, 45, 53
2. 28, 96, 100
3. 10, 40, 41
4. 65, 72, 97
5. 24, 70, 74

In #6–9, (a,b,c) refers to a primitive Pythagorean triple. Is each of the following statements true or false? If true, give an example. If false, give a counterexample.

6. At most one of a, b, c is a square.
7. Exactly one of a, b is divisible by 3.
8. All prime factors of c are primes of the form $4n + 1$.
9. The area of the triangle is divisible by 6.
10. Write a program in which the user inputs three positive integers. Determine whether they are a Pythagorean triple and if so, whether it is a primitive Pythagorean triple. Check your answers to #1–5 above.

One could write a program to generate Pythagorean triples using a "brute force" method—simply cycling through every combination of three whole numbers (selected from a finite set) and testing whether they satisfy $a^2 + b^2 = c^2$. This is time-consuming and a waste of computing power.

Euclid's formula is an elegant formula for generating Pythagorean triples given an arbitrary pair of integers m and n with $m > n > 0$. The formula states that the integers

$$a = m^2 - n^2,\ b = 2mn,\ c = m^2 + n^2$$

form a Pythagorean triple. The triple generated by Euclid's formula is primitive if and only if m and n are relatively prime and not both odd. When both m and n are odd, then a, b, and c will be even, and the triple will not be primitive; however, dividing a, b, and c by 2 will yield a primitive triple when m and n are relatively prime and both odd.

The description does not explicitly state if it is possible to generate a primitive Pythagorean triple from m and n if m and n are not relatively prime, and if so, how.

Example

If possible, generate Pythagorean triples based on the following values of m and n.

A. $m = 13, n = 2$

B. $m = 7, n = 5$

Solutions

A. $a = m^2 - n^2 = 13^2 - 2^2 = 165$

$b = 2mn = 2(13)(2) = 52$

$c = m^2 + n^2 = 13^2 + 2^2 = 173$

Check: $165^2 + 52^2 = 173^2$?

Yes, 29,929 = 29,929

B. $m = 17$ and $n = 5$

$a = m^2 - n^2 = 17^2 - 5^2 = 264$

$b = 2mn = 2(17)(5) = 170$

$c = m^2 + n^2 = 17^2 + 5^2 = 314$

However 264, 170, and 314 do not comprise a primitive triple because they are not relatively prime.

Simply divide by 2:

$a = 264 / 2 = 132$, $b = 170 / 2 = 85$, $c = 314 / 2 = 157$

Check: $132^2 + 85^2 = 157^2$?

Yes, 24,649 = 24,649

EXERCISE SET 5.8.2

1. If possible, generate Pythagorean triples based on the following values of *m* and *n*.
   A. $m = 11, n = 6$
   B. $m = 22, n = 17$
   C. $m = 33, n = 25$
2. Write a program that asks the user to input two positive integers in which $m > n$. Generate a Pythagorean triple based on these numbers. Make sure your program checks that inputs *m* and *n* are both positive integers and $m > n$.

   Hint: the gcd() function must be imported from the math library.

Sample Output

```
m,n? 15,17
32, 255, 257
```

3. Is it possible to generate a primitive Pythagorean triple if *m* and *n* are not relatively prime? If so, how? Modify the program accordingly.

Congratulations, you've mastered the most important skills in programming: data types, basic mathematical functions and operations, the order of operations and symbolic logic.

Chapter Projects

I. Perfect Numbers

A **perfect number** is a positive integer that equals the sum of its divisors (excluding the number itself). For example, $6 = 1 + 2 + 3$ and $28 = 1 + 2 + 4 + 7 + 14$. Write a program that generates at least 10 perfect numbers. Exercise Set 5.6.1 #3 (a program that finds all divisors of a number) may be helpful.

II. Square Roots "By Hand"

Prior to handheld calculators (pre-1970s), students were taught to find square roots using a variation of long division. (In fact, the first handheld calculators did not have square root keys.) Sadly, this lesson is no longer found in modern arithmetic textbooks. The good news is that you can find this technique on the internet.

Write a program that will calculate the square root of a positive integer using this long division method. The example shown is illustrative rather than explanatory.

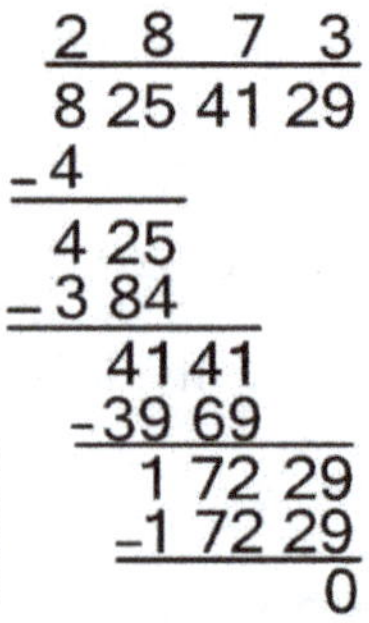

Image 5.11

III. Simplifying Square Roots

Write a program that will simplify a square root. For example, $\sqrt{124} = 2\sqrt{31}$. Hint: This is similar to the prime factorization problem. Do not use **SymPy**.

IV. Is It a Square Root?

Write a program that checks if a decimal number is possibly the square root of a positive integer, and if so, outputs the number in that form. For example, 1.414214 would be output as $\sqrt{2}$. Technically, the decimal number might not be the exact square root, because square roots of integers that are not perfect squares (e.g., $\sqrt{3}$) are irrational and thus would be nonrepeating, nonterminating decimals. Thus, you will want to decide on "how close" the number is to the actual square root.

V. LCM and GCD

Write a program that calculates the gcd and lcm of two integers M and N using prime factorizations.

Here's how it works. If M and N have the following prime factorizations:

$$M = p_1^{e_1} \cdot p_2^{2} \cdots p_m^{e_m}$$

$$N = p_1^{f_1} \cdot p_2^{f_2} \cdots p_m^{f_m}$$

then:

$$gcd(m,n) = p_1^{min(e_1, f_1)} \cdot p_2^{min(e_2, f_2)} \cdots p_m^{min(e_m, f_m)}$$

$$lcm(m,n) = p_1^{max(e_1, f_1)} \cdot p_2^{max(e_2, f_2)} \cdots p_m^{max(e_m, f_m)}$$

SAMPLE OUTPUT

m? 54

n? 24

54 = (2^1)(3^3)

24 = (2^3)(3^1)

gcd(54,24) = (2^1)(3^1) = 6

lcm(54,24) = (2^3)(3^3) = 216

VI. Finding Cube Roots "Mentally"

Finding the cube root of a large positive integer (more than 3 digits) mentally is an impressive party trick (if the party consists of math aficionados or a captive audience).

Example: $\sqrt[3]{103823}$

Step 1: Memorize the first 10 perfect cubes:

$1^3 = 1$	$2^3 = 8$	$3^3 = 27$	$4^3 = 64$	$5^3 = 125$
$6^3 = 216$	$7^3 = 343$	$8^3 = 512$	$9^3 = 729$	$10^3 = 1000$

Note that the last digit of the perfect cube is the same as its cube root, with the exception of 2 and 8, which are swapped, and 3 and 7, which are swapped:

$1^3 = 1$	$2^3 = 8$	$3^3 = 27$	$4^3 = 64$	$5^3 = 125$
$6^3 = 216$	$7^3 = 343$	$8^3 = 512$	$9^3 = 729$	$10^3 = 1000$

Step 2: Use the last digit of the target number to identify the 2nd digit of the cube root, based on the last digit of the perfect cubes, as color coded above:

Given $\sqrt[3]{103823}$: the last digit is 3, so the last digit of its cube root is 7:

$$\sqrt[3]{103823} = \_\ 7$$

Step 3: To find the first digit of the cube root, consider only the first 3 digits of the perfect cube:

$$4^3 < 103 < 5^3$$

Thus, the first digit of the cube root is 4:

$$\sqrt[3]{103823} = \underline{4}\ \underline{7}$$

Write a program that finds the cube root using this method. Do not use any built-in Python functions or raise the number to the 1/3 power. Make sure your program checks that the input is a positive integer of greater than 3 digits. Verify that your resulting cube root, when raised to the third power, indeed equals the input number.

SAMPLE OUTPUT

Perfect cube? 614125

Cube root is 85

Perfect cube? 9800

9,800 is not a perfect cube

Perfect cube? 1061208

1061208 is not the cube of a 2-digit number

VII. Repeating Decimals

Write a program that converts a repeating decimal to a rational number.

SAMPLE OUTPUT

Repeating Decimal? .11111111111

1/9

Repeating Decimal? .343434343434

34/99

VIII. Modulo Arithmetic

Modulo arithmetic can be thought of as the arithmetic of remainders, and has applications in cryptography. The notation $m \equiv n \ mod \ p$ means that m has remainder n when divided by p, $0 \leq n < p$.

For example:

$$13 \equiv 3 \bmod 5 \text{ because } 13 = 2 \cdot 5 + 3$$

$$119 \equiv 19 \bmod 20 \text{ because } 119 = 5 \cdot 20 + 19$$

$$23 \equiv 11 \bmod 12 \text{ because } 23 = 1 \cdot 12 + 11 \text{ (This is why 2300 hours is the same as 11:00 pm)}$$

$$M \equiv N \bmod p \text{ could also be expressed as } M \bmod p = N \text{ (e.g., } 26 \bmod 10 = 6)$$

Write a program that will ask the user to input a positive integer $n \leq 10$. Output an addition table for mod n.

SAMPLE OUTPUT

n? 7

+	0	1	2	3	4	5	6
0	0	1	2	3	4	5	6
1	1	2	3	4	5	6	0
2	2	3	4	5	6	0	1
3	3	4	5	6	0	1	2
4	4	5	6	0	1	2	3
5	5	6	0	1	2	3	4
6	6	0	1	2	3	4	5

IX. Fermat's Little Theorem

Fermat's little theorem states that $n^{p-1} \equiv 1(\text{mod } p)$ if p is prime and n is not a multiple of p. Write a program that asks the user to input a positive integer n. Loop through at least 10 prime integers p that are relatively prime to n and verify that $n^{p-1} \equiv 1(\text{mod } p)$.

SAMPLE OUTPUT

integer? 3

p = 2: 3^(2–1) = 3

is 3 congruent to 1 mod 2? yes

p = 5: 3^(5–1) = 81

is 81 congruent to 1 mod 5? Yes

p = 7: 3^(7–1) = 729

Is 729 congruent to 1 mod 7? Yes

etc.

X. Fermat's Last Theorem

Fermat's last theorem asserted (around 1637) that no three positive integers a, b, and c satisfy the equation $a^n + b^n = c^n$ for any integer value of n greater than 2. This conjecture captivated the math community for centuries. Books were written about this theorem alone (although yet to be proven), and the mathematicians/screenwriters who worked on the long-running animated TV show The Simpsons managed to insert into the background of various episodes "false" instances of $a^n + b^n = c^n$.

Imagine that you have been hired to find three positive integers a, b, and c that satisfy the equation $an + bn = cn$ for an integer value of n greater than 2. Write a program to perform this search. How would you arrange your algorithm? One possibility: start with $n = 3$ and cycle through various a, b, c triples until you reach a solution. Or cycle through various a, b pairs and check if $\sqrt{a^2 + b^2}$ is also an integer.

Spoiler alert: This theorem was not proven until 1994! To avoid an infinite loop, make sure to limit your search to numbers less than a reasonable maximum.

XI. Divisibility Tests

The easiest way to check for divisibility, if you're a programmer, is to use modulo arithmetic.

However, elementary school students are (hopefully) taught how to use mental math to quickly identify the factors of large positive integers. Here are a few of them:

Divisible by:	Condition	Example
2	even number	2,048
3	sum of digits is a multiple of 3	2,046
4	last 2 digits are 00 or divisible by 4	2,044
5	last digit is 0 or 5	2,045
6	even number divisible by 3	2,046
7	Remove the last digit of the number and double that number. Subtract from the remaining number. If the number is 0 or an easily recognizable 2-digit multiple of 7, the original number is divisible by 7. Repeat until you have a 1- or 2-digit number.	2,037
8	last 3 digits divisible by 8	2,168
9	sum of digits divisible by 9	2,043
10	last digit is 0	2,160

Example: check 2037 for divisibility by 7.

Remove 7 from 2037 and double it.

Subtract 14 from 203 to get 189. Not a 2-digit number.

Remove 9 from 189 and double it.

Subtract 18 from 18 to get 0.

When the remainder is 0, the original number 2,037 is divisible by 7.

Write a program that performs the divisibility tests for 1 to 10.

BONUS: Research the algorithm to check for divisibility by 11 and add that to your program.

XII. Palindromes

Numbers can be palindromes as well as words:

161, 2442, 73537, etc.

are all palindromes, i.e., the number is the same if the digits are reversed.

Write a program that generates all the palindromic numbers between 10 and 1000.

XIII. Coin Change Problem

You have decided to open a food stand with only a cash drawer at your disposal. You do not have access to an electronic cash register.

Image 5.12

If a customer purchases something for $7.59 and hands you a ten-dollar bill, it is your job to calculate the exact change using the least number of coins possible.

Write a program that finds the least number of coins necessary to give the customer their change. Assume the cash drawer has an infinite supply of pennies, nickels, dimes, quarters, and dollar bills.

SAMPLE OUTPUT

Sale? 7.59

Customer paid? 10.00

The change is:

1 penny

1 nickel

1 dime

1 quarter

2 dollars

CHAPTER SUMMARY

PYTHON CONCEPTS

Evaluating a numerical expression	To evaluate (calculate) a numeric expression, use: **eval**(string) *Ex:* *>>>eval('3*5**(3–1)*11')* *825*
Fraction type	To convert a numeric variable or value to a fraction type, use: **Fraction**(number) in the **fractions** module. *Ex:* *>>>A =* ***Fraction****(2.5)* *>>>****print****(A)* *5/2* *>>>B =* ***Fraction****(10,12)* *>>>****print****(B)* *5/7*
Ensuring that a Fraction type is output in simplest form	To limit the denominator to a given number *n*, use: frac. **limit_denominator(*n*)** *Ex:* *>>>F =* ***Fraction****(1.2)* *>>>****print****(F)* *5404319552844595/4503599627370496* *>>>F = F.****limit_denominator****(100)* *>>>****print****(F)* *6/5*
Accessing the numerator and denominator of a fraction	To return the numerator and denominator of a fraction in simplest form: f.**numerator** f.**denominator** *Ex:* *>>>F =****Fraction****(90,100)* *>>>****print****(F.****numerator****)* *9* *>>>****print****(F.****denominator****)* *10*

To input a fraction	Use: *F* = **Fraction(input**("Enter fraction: "))
Built-in math functions (not requiring imported modules)	Exponentiation: *a****b* or **pow**(*a*,*b*) Absolute value: **abs**(*x*) Round to *n* decimal places: **round**(number,*n*) Return the quotient and remainder in the form of a 2-element tuple when *x* is divided by *y*: **divmod**(*x*,*y*) Return the remainder (aka modulus) when *x* is divided by *y*: *x* **%** *y* Truncation function: **int**(*x*) *Ex:* *See Section 5.2*
Math functions in the math module	Square root function: math.**sqrt**(*x*) Greatest integer function: math.**floor**(*x*) Least integer >= *x*: math.**ceil**(*x*) Factorial function: math.**factorial**(*n*) Truncation (integer part): math.**trunc**(*x*) Exponential function e^x: math.**exp**(*x*) Greatest common divisor of integers *m* and *n*: math.**gcd**(*m*,*n*) *Ex:* *See Section 5.2*

(*Continued*)

Math constants in the math module	π **math.pi** e **math.e**
Boolean type variable	A Boolean type variable has a value of **True** or **False**. These are not strings; do not enclose them in quotes. Capitalize the first letter.

PRECALCULUS CONCEPTS

Euclid's method for finding the GCD	An iterative process (loop) to find the GCD of a pair of numbers m and n. In each iteration: 1. if $m > n$, then replace gcd(m, n) with gcd($m-n$, n). Repeat until $m = n$. 2. if $m < n$, then replace gcd(m, n) with gcd(m, $n-m$). Repeat until $m = n$.
Sieve of Eratosthenes	An iterative process to filter out composite numbers from a list. The remaining numbers are prime numbers.
Fundamental Theorem of Arithmetic	Every integer greater than 1 can be expressed as the product of one or more prime factors. *Ex:* $45 = 3^2 \cdot 5$
Goldbach's Conjecture	Every even integer greater than 2 can be written as a sum of two primes. *Ex:* $198 = 5 + 193$
Pythagorean Theorem	If a, b, and c are the lengths of the 3 sides of a right triangle, with c being the hypotenuse, then $a^2 + b^2 = c^2$.
Pythagorean Triple	A set of 3 integers that satisfy the Pythagorean Theorem. *Ex:* *3,4,5* *5,12,13*

Credits

IMG 5.1: Copyright © 2012 Depositphotos/cienpies.

IMG 5.2: Adapted from SG0039, https://commons.wikimedia.org/wiki/File:Real_Number_Classification.png, 2022.

IMG 5.8: Softssa, https://commons.wikimedia.org/wiki/File:Pythsats.jpg, 2004.

IMG 5.11: Sciencing Editorial Team, "Square Root Long Division," https://sciencing.com/calculate-square-root-hand-5081134.html. Copyright © 2017 by Sciencing.

IMG 5.12: Copyright © 2015 Depositphotos/elena.lielnors.gmail.com.

CHAPTER 6

Coordinate Geometry

6.1 Basic Graphing

There are several available graphing/plotting packages for Python such as **Processing** (which does translations, rotations, shapes, etc.), but **matplotlib** is the most popular and robust package. Think of a **package** as a collection of modules (which contain objects and their associated functions and methods).

We'll start by graphing a set of ordered pairs. Your introduction to graphing in middle school probably involved a table of values:

x	–2	–1	0	1	2
y	4	1	0	1	4

If we want to plot this set of ordered pairs, we'll assign the *x*-values to a list and the corresponding *y*-values to another list. To do this, we'll import the **pyplot** module from **matplotlib** and give it an **alias** (nickname) of **plt**. We'll then use **plot** and **show** (methods contained in **pyplot**.) Remember, statements preceded with >>> can either be entered directly into the Python console or saved as a program.

```
>>>from matplotlib import pyplot as plt
>>>x_values = [–2, –1, 0, 1, 2]
>>>y_values = [4, 1, 0, 1, 4]
>>>plt.plot(x_values, y_values)
```

Notice that the domain (*x*-axis) was automatically set to [–2, 2] (the least and greatest numbers in the **x_values** list), and the range was automatically set to [0, 4] (the least and greatest numbers in the **y_values** list).

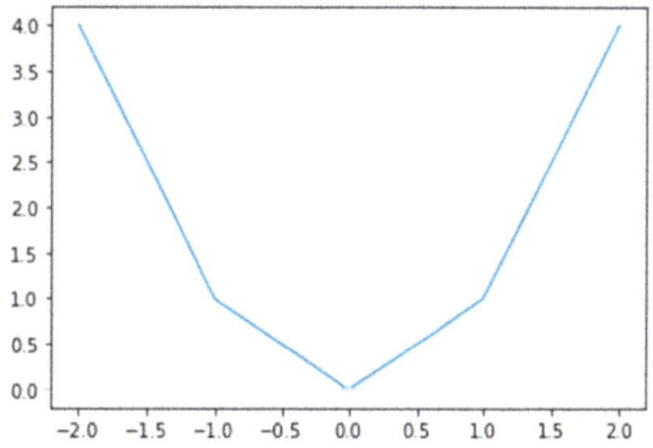

Image 6.1

Also observe that by default, the points have been connected with blue line segments. If you just want the points plotted, try this:

```
>>>plt.plot(x_values,y_values,'o')
```

This has the effect of suppressing the line segments and indicating each ordered pair with a **marker**, in this case a closed blue circle.

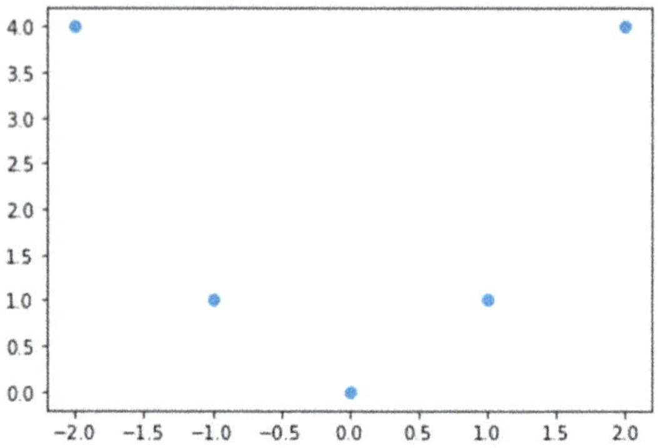

Image 6.2

To set the scaling and labeling of the axes, append these statements:

```
>>>plt.axis(xmin=–10,xmax=10,ymin=-10,ymax=10)
>>>plt.xlabel('x axis')
>>>plt.ylabel('y axis')
```

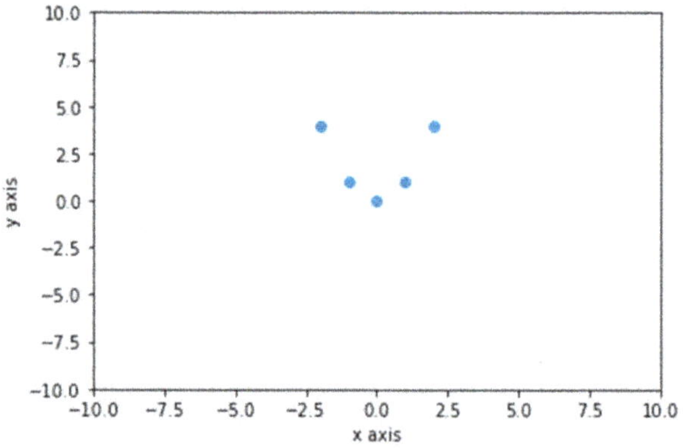

Image 6.3

If you omit the list of x-values in the plot() function, it will assume the x-values are:
[0, 1, 2, ... **len**[y-values].
For example:

```
>>>f_x = [10,9,8,7,6,5,4,3,2,1,0]
>>>plt.plot(f_x, 'o')
```

This plots the ordered pairs: (0, 10), (1, 9), (2, 8), ... (10, 0)

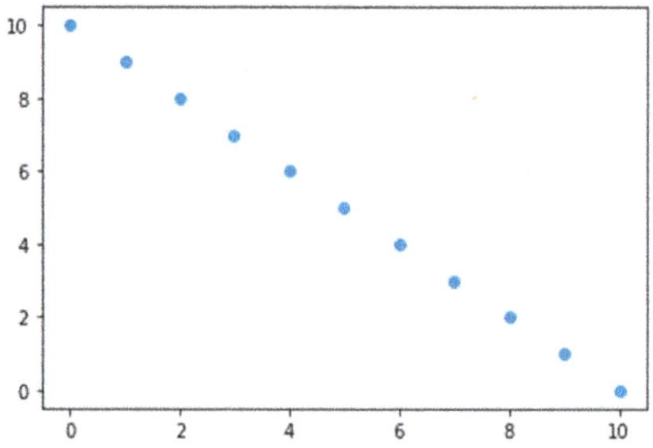

Image 6.4

You will find that in Python, as in real life, there are often several ways to express the same idea. For example, in place of **from** matplotlib **import** pyplot **as** plt, you can use:

```
import matplotlib.pyplot as plt
```

We'll stick with the former.

TABLE 6.1

matplotlib methods

Assume ***pyplot*** *has been imported as* ***plt****.*

To plot a set of ordered pairs, connected with line segments (default color blue):

```
plt.plot(x_values,y_values)
plt.plot([1,2,3],[4,5,6])
plt.plot(y_values)
```

To plot a set of ordered pairs, connected with black line segments:

```
plt.plot(x_values,y_values,color='k')
```

To plot a set of discrete ordered pairs (not connected):

```
plt.plot(x,values,y_values,marker)
```

where marker can be 'o', '', 'x', '.' or '+'.*

Optional: precede the marker with a color (r, b, g) for example 'r+'

To plot a set of ordered pairs using a marker and connected by line segments:

```
plt.plot(x_values,y_values,marker='o')
```

To set the scaling of the axes:

```
plt.axis(xmin=value1,xmax=value2,ymin=value1,ymax=value2)
```

or

```
plt.axis([xmin,xmax,ymin,ymax])
```

To label the axes:

```
plt.xlabel('label')
plt.ylabel('label')
```

To give the graph a title:

```
plt.title('This is the title')
```

To show the graph:

```
plt.show()
```

EXERCISE SET 6.1.1

1. What happens when you try to plot a point that is outside the range you set with the **axis**() function?

Example

```
>>>plt.axis([-10,10,-10,10])
>>>plt.plot([20],[40])
```

2. What command(s) will plot the following set of ordered pairs, not connected, in color green?

   {(−5, 5), (−4, 4), (−3, 3), (−2, 2), (−1, 1), (0, 0), (1, 1), (2, 2), (3, 3), (4, 4), (5, 5)}

3. An instructor wishes to plot a set of ordered pairs representing the number of days a student is absent (x) and score on final exam (y).

   A. If the following ordered pairs are plotted without specifically scaling the axes, what does Python choose for xmin, xmax, xscale, ymin, ymax, yscale?

   {(0, 96), (0, 88), (1, 72), (1, 99), (1, 87), (2, 86), (2, 54), (2, 77), (2, 71), (2, 74), (3, 72), (3, 68), (3, 59), (4, 62)}

   B. Does this set of ordered pairs represent a function?

   C. What **pyplot** command would provide a more informative scaling of the axes?

4. Scale the axes at [−20, 20] x [−20, 20] and draw the xy axes in black in the middle of the screen.

5. Write a program that will plot the following ordered pairs, connected with blue lines:

   (1, 1000), (2, 750), (3, 1200), (4, 1450), (5, 1600), (6, 2000)

Label the x-axis as 'Month', the y-axis as 'Revenue', and the title of the graph 'Six Month Projection'.

6.2 Graphing Lines and Functions

First, a quick review of slopes and linear equations.

Slopes

The slope of a line determines its steepness and direction. For example:

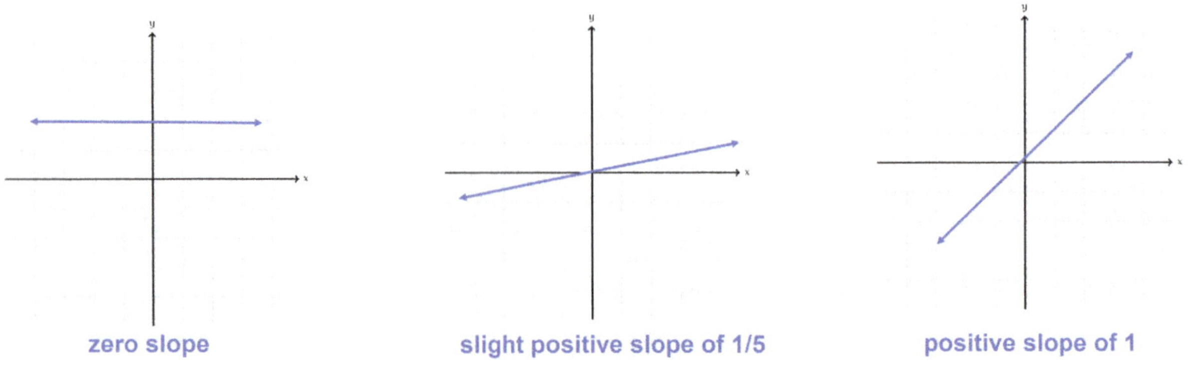

Image 6.5

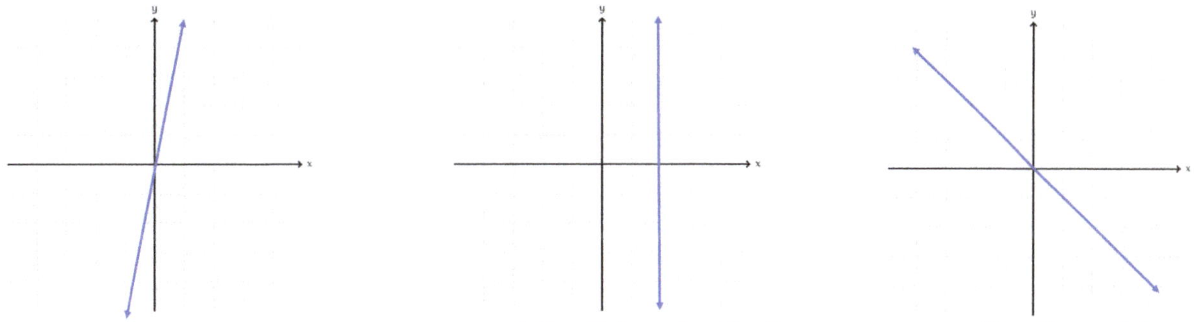

steep positive slope of 5 | undefined slope (vertical) | negative slope of -1

Image 6.6

The slope of a line is described by $\frac{rise}{run}$ or $\frac{\Delta y}{\Delta x}$. The **rise** is the vertical change, and the **run** is the horizontal change.

SLOPE OF A LINE

The slope of the line through points (x_1, y_1) and (x_2, y_2) is given by:

$$m = \frac{y_2 - y_1}{x_2 - x_1}$$

It is unknown why the letter m is used to represent slope. It just is.

Example 1

Find the slope of the line through the given pair of points, and describe the direction.

A. (3, 6) and (–7, 1)

B. (4, 3) and (4, –5)

C. (2, 12) and (0, 12)

D. (–5, 9) and (–3, 2)

Solutions

A. $m = \frac{1-6}{-7-3} = \frac{-5}{-10} = \frac{1}{2}$. Positive slope.

B. $m = \frac{-5-3}{4-4} = \frac{-8}{0} =$ undefined. Vertical line.

C. $m = \frac{12-12}{0-2} = \frac{0}{-2} = 0$. Horizontal line.

D. $M = \frac{2-9}{-3-(-5)} = \frac{-7}{2} = -\frac{7}{2}$. Negative slope.

The slopes of lines can help us determine if a pair of lines are parallel, perpendicular, or neither.

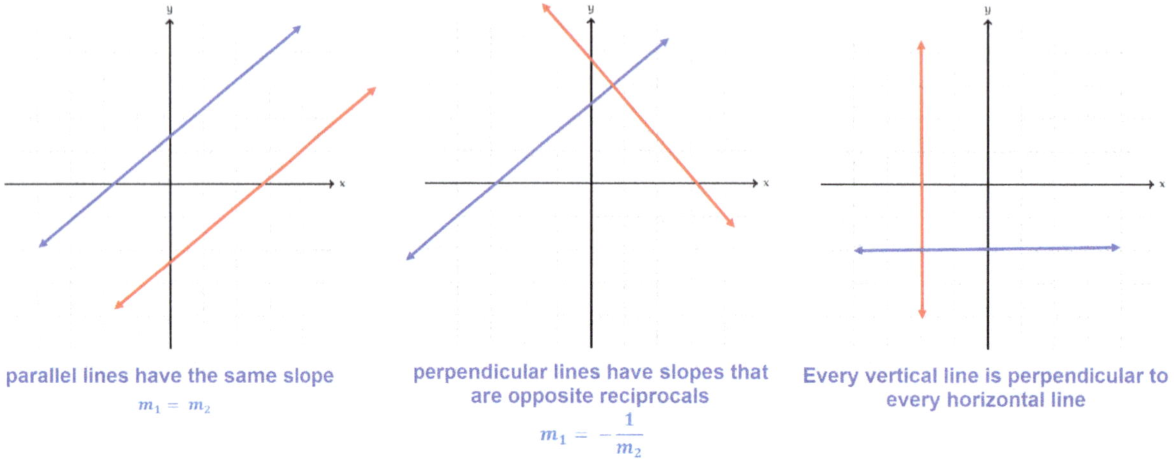

Image 6.7

Example 2

Given the following slopes, identify whether the lines are parallel, perpendicular, or neither.

A. $m_1 = 6$ and $m_2 = \frac{1}{6}$

B. $m_1 = 1$ and $m_2 = -1$

C. $m_1 = \frac{2}{3}$ and $m_2 = \frac{4}{6}$

Solutions

A. Neither. Although the slopes are reciprocals, they are not opposites.

B. Perpendicular. 1 and –1 are opposite reciprocals.

C. Parallel. The two fractions are equivalent.

EXERCISE SET 6.2.1

In #1–4, find the slope of the line between the given pair of points, and describe the direction.

1. $(4, 7)$ and $(-2, 3)$
2. $\left(\frac{1}{3}, \frac{2}{5}\right)$ and $\left(\frac{5}{6}, \frac{1}{5}\right)$
3. $(7, -8)$ and $(8, -8)$
4. $(2, 4)$ and $(3, -4)$

In #5–8, compare the slopes of the given lines and identify whether they are parallel, perpendicular, or neither.

5. $m_1 = 2$ and $m_2 = -2$
6. $m_1 = \frac{3}{9}$ and $m_2 = -3$
7. $m_1 = 0$ and $m_2 =$ undefined

As a programming warm-up, let's write a brief program that asks the user to input two ordered pairs (x_1, y_1) and (x_2, y_2), then calculates and outputs the slope.

INPUT TIP!

You will be asked in this chapter to input ordered pairs, etc. Here is one efficient way to input the ordered pair in the form "x,y" and have it converted to a pair of integers:

```
point = input("x,y ? ")
x, y = [int(i) for i in point.split(',') ]
```

PROGRAM 6.2.1

```
p1 = input('x1,y1? ')
x1,y1 = [int(i) for i in p1.split(',')]
p2 = input('x2,y2? ')
x2,y2 = [int(i) for i in p2.split(',')]
m = (y2-y1)/(x2-x1)
print('m=',m)
```

Sample Output

```
x1,y1? 3,5
x2,y2? 5,3
m= -1.0
```

EXERCISE SET 6.2.2

1. Does the program work with fractions? If not, modify. *(Hint: Use the **fractions** module and **Fraction()** function.)*
2. Does the program work with inputs (2, 6) and (2, −4)? If not, modify so that a helpful message is output.
3. Write a program that asks the user to input the slopes of two lines, *m*1 and *m*2. Compare the slopes. Output whether they are parallel, perpendicular, or neither. *(Hint: a pair of slopes m1 and m2 are opposite reciprocals if their product is −1.)*
4. Write a program (or modify program 6.2.1) that asks the user to input two ordered pairs (x_1, y_1) and (x_2, y_2). Calculate the slope of the line through the two points, and also calculate and output the parallel and perpendicular slopes.

(Continued)

Sample Output

x1,y1? 3,5

x2,y2? 7,2

$m = -0.75$

Parallel slope is $m = -0.75$

perpendicular slope is $m = 1.3333333333333333$

(Even better if your program handles fractions)

Lines

There are three forms of nonvertical lines, with special cases for horizontal and vertical lines:

Slope-Intercept Form

$y = mx + b$

m = slope, b = y-intercept

Point-Slope Form

$y - y_1 = m(x - x_1)$

m = slope, (x_1, y_1) is any point on the line

Standard Form (aka General Form)

$Ax + By = C$

Where A, B, and C are relatively prime constants; $A > 0$

Horizontal Line

$y = b$

Vertical Line

$x = a$

We will focus on slope-intercept form because it lends itself most readily to programming.

Example 1

Find the equation of the line in slope-intercept form that passes through the two points: (−7, 5) and (4, 2).

Solution

The slope is $\frac{y_2 - y_1}{x_2 - x_1} = \frac{2-5}{4-(-7)} = -\frac{3}{11}$.

To find b, substitute the slope and either point into $y = mx + b$:

$$2 = -\frac{3}{11} \cdot 4 + b$$

$$b = 2 + \frac{3}{11} \cdot 4 = \frac{23}{11}$$

$$\therefore y = -\frac{3}{11}x + \frac{23}{11}$$

Example 2

Find the equation of the line that passes through the two points: (6, 17) and (6, 7).

Solution

The slope is $\frac{7-17}{6-6} = \frac{-10}{0} =$ undefined.

∴ The equation of the vertical line through (6, 17) is $x = 6$.

Example 3

Find the equation of the line perpendicular to $y = 3x + 7$, through point (2, 5).

Solution

The slope of the given line is $m = 3$. The slope of the perpendicular line is the opposite reciprocal, which is $m = -\frac{1}{3}$.

To find the equation of the perpendicular line through the given point, substitute $m = -\frac{1}{3}$ and (2, 5) into $y = mx + b$:

$$5 = -\frac{1}{3}(2) + b$$

$$5 = -\frac{2}{3} + b$$

$$\frac{17}{3} = b$$

$$\therefore y = -\frac{1}{3}x + \frac{17}{3}$$

EXERCISE SET 6.2.3

In #1–4, find the equation of the line through the following pairs of points, in slope-intercept form if possible.

1. $(-8, 8)$ and $(-6, 3)$
2. $\left(\frac{3}{5}, \frac{7}{8}\right)$ and $\left(\frac{2}{5}, \frac{3}{4}\right)$
3. $(7, 8)$ and $(-2, 8)$
4. $\left(\frac{5}{4}, 3\right)$ and $\left(\frac{5}{4}, -2\right)$
5. Convert from point-slope form to slope-intercept form: $y - 7 = \frac{2}{3}\left(x + \frac{1}{2}\right)$.
6. Convert from standard form to slope-intercept form: $3x - 7y = 21$.
7. Find the equation of the line parallel to $y = 17$ through $(2, 5)$.
8. Find the equation of the line perpendicular to $y = \frac{1}{7}x - \frac{2}{21}$ through $(5, -11)$.
9. Find the equation of the line perpendicular to the x-axis through $(-7, 8)$.
10. Find the equation of the line parallel to $x = 5$ through the origin.
11. Solve the equation for b: $y = mx + b$.
12. According to the Centers for Disease Control and Prevention (CDC), to calculate one's target heart rate for moderate physical activity, subtract your age from 220 and then take 70% of that number. Rewrite this formula as a linear equation in slope-intercept form using the variable x to represent their age.
13. Write a program that asks the user to input their age. Calculate and output their target heart rate for moderate physical activity.

Now that we know how to calculate equations of lines by hand, let's write a program to do the heavy lifting for us. This program will also be a good review of working with randomly generated numbers and fractions.

Program Goal: Write a program **random_line** that will randomly generate two endpoints of a line segment and then calculate and output the slope-intercept form of the line.

PROGRAM 6.2.2 RANDOM_LINE.PY

```
from matplotlib import pyplot as plt
from random import randint
from fractions import Fraction
x1 = randint(-10,10)
y1 = randint(-10,10)
x2 = randint(-10,10)
y2 = randint(-10,10)
plt.plot([x1,x2],[y1,y2])
m = Fraction(y2-y1,x2-x1)
b = y1 - m*x1
heading = 'y = ('+str(m)+')x + '+str(b)
plt.title(heading)
plt.show()
```

Sample Output

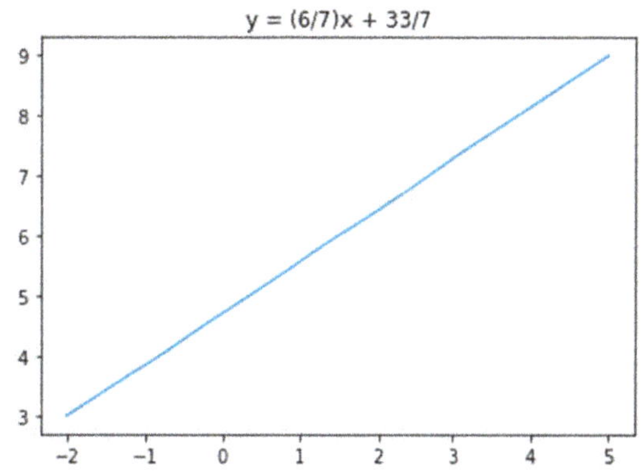

Image 6.8

EXERCISE SET 6.2.4

1. If program **random_line** randomly generates $(x1, y1) = (7, -5)$ and $(x2, y2) = (2, -9)$, what values are assigned to variables m and b?
2. Modify the program so that it also outputs the values of $x1, y1, x2$, and $y2$. (Using **print**() will output the values to the console rather than the plot.)
3. If you run the program enough times, you might get an error message if it randomly generates $x1 = x2$. Why is that? Modify the program so that it outputs a helpful message rather than an error message.
4. What kind of line is generated if the program randomly generates $y1 = y2$?
5. Modify the program so that instead of assigning random values to $x1, y1, x2, y2$, it allows the user to enter values.
6. Modify the program so that it generates a random line color. *(Hint: use a list ['b', 'g', 'r', 'c', 'm', 'y', 'k', 'w'].)*

Program Goal: Write a program **function_plot** that asks the user to input the linear function in the form $mx + b$. Plot the points on this line in domain [–10, 10].

PROGRAM 6.2.2 RANDOM_LINE.PY

```
from matplotlib import pyplot as plt
from sympy import *
x = Symbol('x')
x_list = range(-10,11)
y_list = [ ]                                    # initialize y_list
f = sympify(input('function? '))
# For each a in x_list, find f(a) and assign to y_list
y_list = [ f.subs({x:a}) for a in x_list ]
plt.plot(x_list,y_list,'o')
plt.show()
```

Sample Output

function? 3*x+1

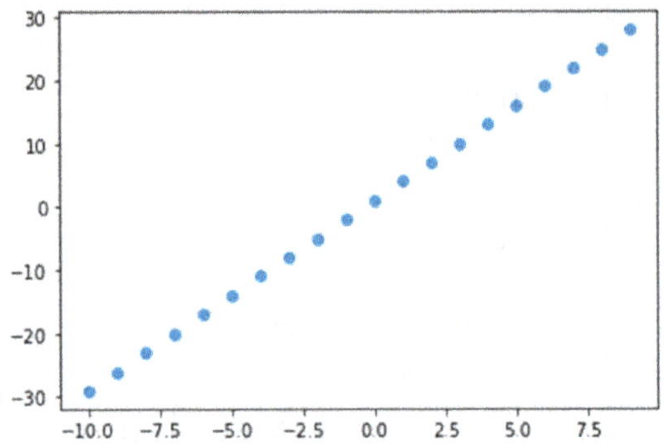

Image 6.9

EXERCISE SET 6.2.5

1. One way to generate more points (and hence a more continuously appearing line) is to expand the domain. Modify the program so that the user is prompted to enter xmin, xmax values.
2. Modify the program so that the title of the graph is "$f(x) = mx + b$ over [xmin,xmax]," in which the linear function, xmin and xmax, reflect the actual values that were input.

Sample Output

function? 4*x+5

domain? –25, 25

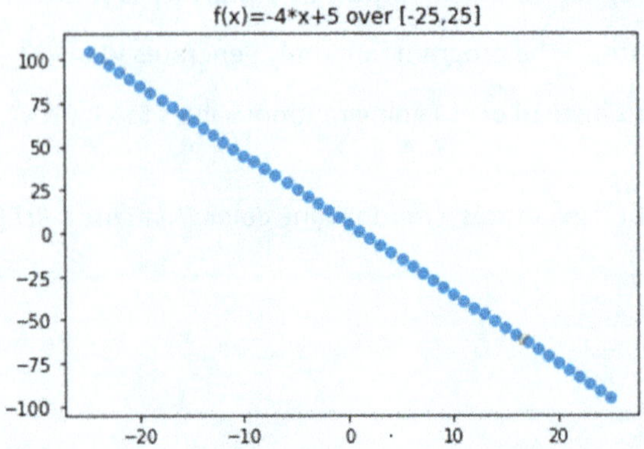

Image 6.10

3. If 3*x – 17 is input, and a domain of [–25, 25], what are the coordinates of the left and right endpoints?
4. Modify the program so that a line is drawn instead of discrete dots. (Hint: You only need to connect the two endpoints.)

Sample Output

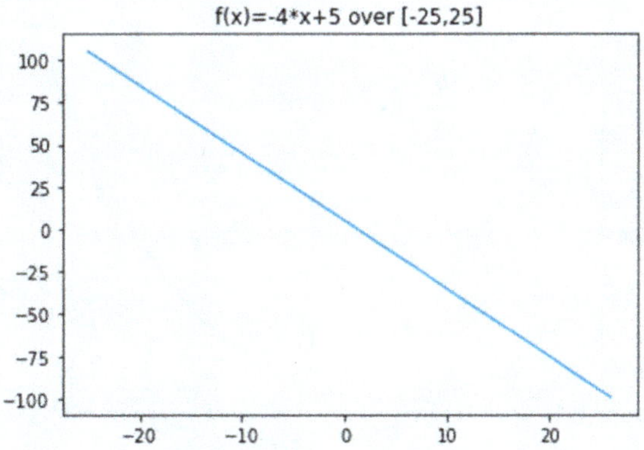

Image 6.11

5. CHALLENGE: Modify the program so that the user is also prompted to input a point. Draw a line through that point, parallel to the given line. Title the graph with the equation of the parallel line.

Sample Output

function? 5*x–11

x, y? –4, –5

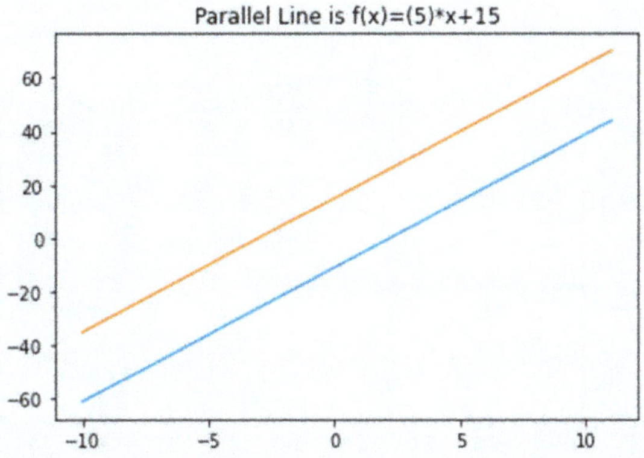

Image 6.12

Program **function_plot** will actually draw the graph of any polynomial function, not just linear functions!

Sample Output

function? 3*x**2+4*x−5

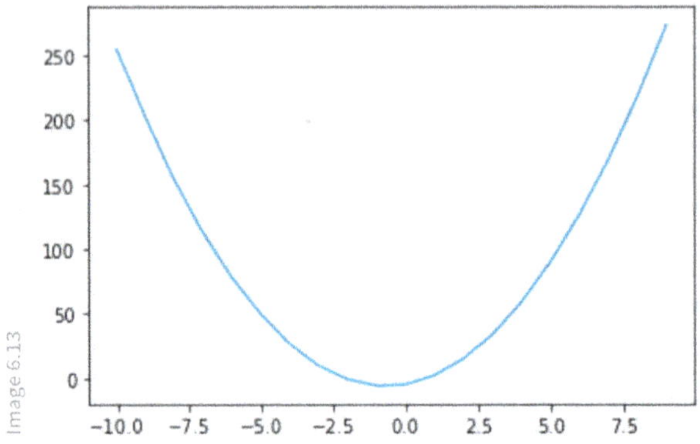

Enhancing the Graph with Axes and Annotations

A graph is more informative if it includes the x- and y-axes. For example, here is a Python-generated plot of $f(x) = \frac{1}{10}(x^3 - 6x^2 - 27x + 140)$.

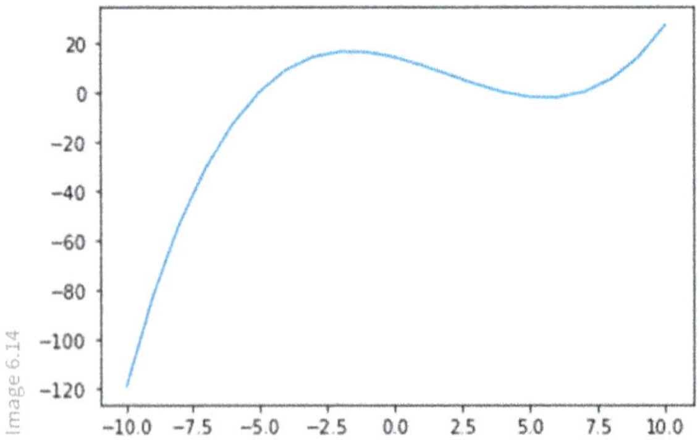

On a subconscious level, we perceive this graph as lying entirely in the first quadrant. However, on close inspection of the labeling of the axes, this graph lies in all four quadrants.

It's easy to add the x-axis in black by connecting (xmin, 0) to (xmax, 0) using:

```
plt.plot([xmin,xmax],[0,0],c='k')
```

resulting in:

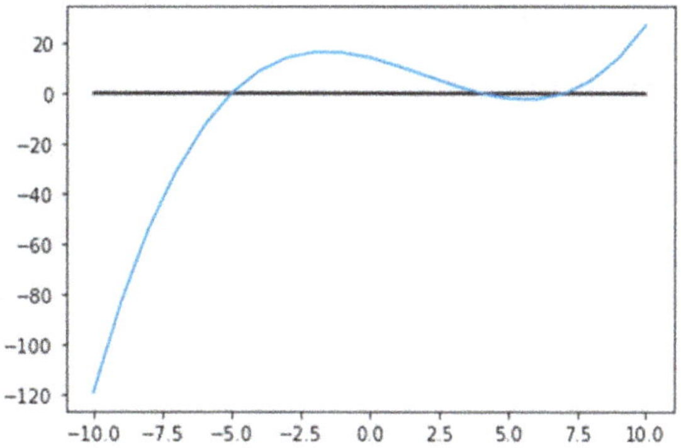

Image 6.15

However, to draw the y-axis, we need to know the default ymin and ymax employed by Python. Python simply selects the minimum and maximum y-values in **y_list** using the min() and max() list functions:

```
y_min = min(y_list)
y_max = max(y_list)
```

Thus we will use the same y_min and y_max to draw the y-axis in black by connecting (0,ymin) to (0,ymax):

```
plt.plot([0,0],[ymin,ymax],c='k')
```

resulting in a much more informative graph:

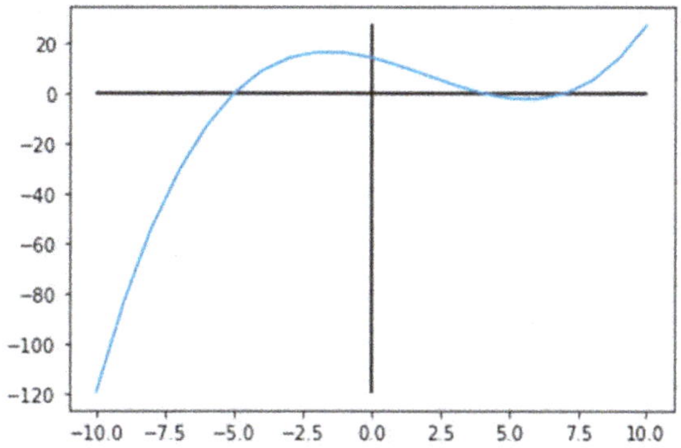

Image 6.16

One can plot a single point and print the ordered pair next to it as follows:

```
plt.plot([x], [y], 'o')
plt.text(x,y,text)
```

for example:

```
plt.plot([0],[0], 'o')
plt.text(0,0,'origin')
```

EXERCISE SET 6.2.6

1. Given $f(x) = \frac{1}{10}(x^3 - 6x^2 - 27x + 140)$, what are the ymin and ymax?
2. What statement will print your first name at position (10, –10)?
3. Modify program **function_plot** so that if the polynomial is a quadratic, the vertex is labeled with the coordinates. Round to 1 decimal place.

Sample Output

function? x**2 – 4*x + 2

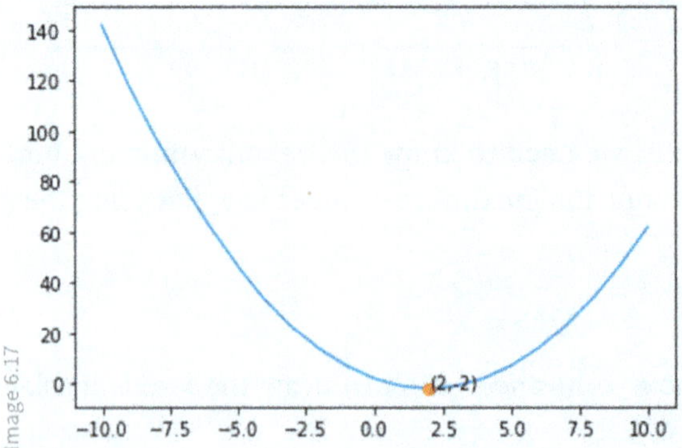

Image 6.17

4. Modify program **function_plot** to draw the *x*- and *y*-axes in black, using the Python commands discussed above.
5. Write a program (or modify **function_plot**) that asks the user to input a function and a point. Plot both the graph of the function and the point, labeled with the coordinates. Determine if the point lies on the graph of the function.

Sample Output

function? –x**3

point? 2, 3

(2, 3) is not on the graph

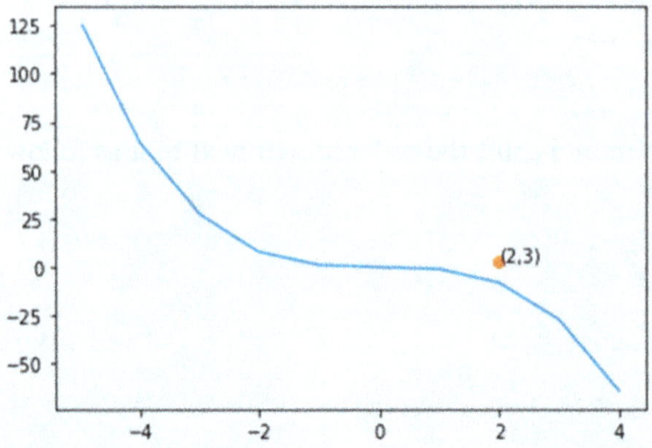

Image 6.18

Colors and Line Styles

As you can see, the default color for plotting is blue. (This may change over time or by interpreter.) You can change the color by adding an additional parameter to the **plot**() function call.

The color can be specified in any of the following formats, to name a few:

TABLE 6.2

Specifying Color
*Assume **pyplot** was imported as alias **plt.*** Example to specify a base color ('b','g','r','c','m','y','k','w') (see chart below): plt.**plot**(x_list,y_list,**c**='m')
Example to specify a hex RGB or RGBA string (approximately 140 of them): plt.**plot**(x_list,y_list, ' #808000')
Example to specify a Tableau Color from the T10 palette (see chart below): plt.**plot**(x_list,y_list,**color**='pink')
Example to specify a CSS4 color name (see Appendix B; too many to list here): plt.**plot**(x_list,y_list,**color**='olivedrab')

Image 6.19

The line style can also be customized by adding a **linewidth** and/or **linestyle** parameter to the **plot()** function call.

Linewidth values (or lw) (in pixels) can be set to 1 for a thin line, 2 for a medium line, 4 for a thick line, or more if you want a really thick line.

Linestyle (or ls) values can be set to the following:

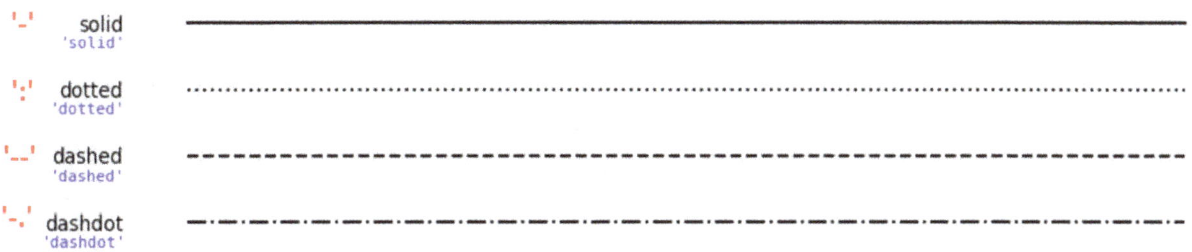

Image 6.20

For example:

```
plt.plot(x_list,y_list,color='magenta',linewidth=1,linestyle='-.')
```

will result in the following:

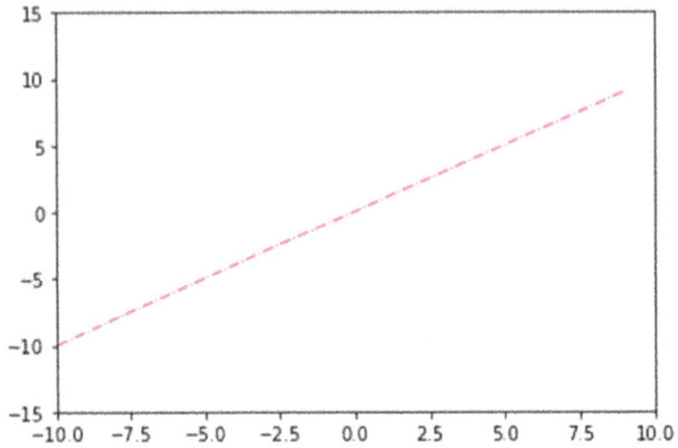

Image 6.21

EXERCISE SET 6.2.7

1. Write a program that overlays the following 3 graphs:

   $y = x^2$ in a solid narrow width line

   $y = 2x^2$ in a dashed ('--') medium width line

   $y = 3x^2$ in a dash-dot heavy weight line

   Use any 3 colors of your choosing.

This is by no means an exhaustive lesson on the features of **matplotlib**; the User's Manual for version 0.99.1.1 has over 800 pages!

6.3 Distance Formula

The **distance formula** is a handy derivation of the Pythagorean theorem. Let's say we have a line segment with endpoints (x_1, y_1) and (x_2, y_2):

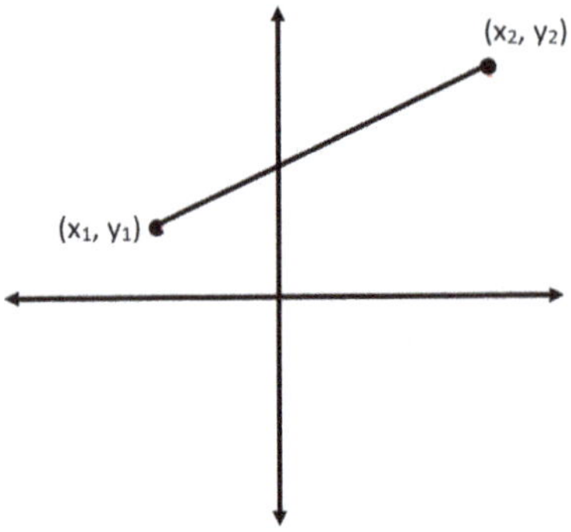

Image 6.22

To find its length, simple draw a right triangle above or below it:

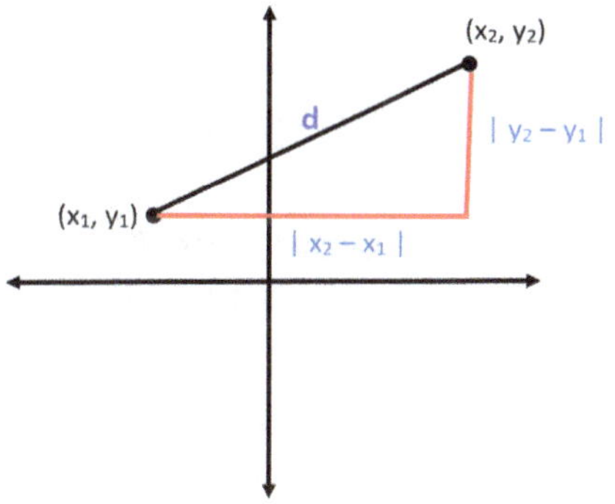

Image 6.23

Note the length of the horizontal side is $|x_2 - x_1|$, the length of the vertical side is $|y_2 - y_1|$, and the length of the hypotenuse we'll call d. Substituting into the Pythagorean theorem $a^2 + b^2 = c^2$:

$$|x_2 - x_1|^2 + |y_2 - y_1|^2 = d^2$$

The absolute value is no longer needed because squaring will ensure that the lengths are positive. Replace the absolute value symbols with parentheses and take the square root of both sides:

$$\sqrt{(x_2 - x_1)^2 + (y_2 - y_1)^2} = \sqrt{d^2}$$

Which leads to our distance formula:

The **distance** from (x_1, y_1) to (x_2, y_2) is:

$$d = \sqrt{(x_2 - x_1)^2 + (y_2 - y_1)^2}$$

Example

Find the distance from (8, –1) to (3, –9).

Solution

$$d = \sqrt{(3 - 8)^2 + (-9 + 1)^2} = \sqrt{(-5)^2 + (-8)^2} = \sqrt{89}$$

EXERCISE SET 6.3.1

1. Find the distance between (−6, 14) and (−3, 5). Express your answer in simplest radical form, not decimal form.

2. Find the length of the line segment with endpoints (3, 20) and (16, 11). Express your answer in simplest radical form, not decimal form.

3. Find the lengths of the 3 sides of the triangle with vertices (0, 0), (60, 40), and (−60, 90). Express your answers as decimals rounded to 1 decimal place. Is it a right triangle? (Hint: two lines are perpendicular if the product of their slopes = −1).

4. Write a function **distance** that calculates and returns the distance between two points (as a float). Call the function from a main program. Output the answer as both a decimal and a simplified radical expression. Use integer inputs only.

Sample Output

Point 1? 11,8

Point 2? 8,2

6.708203932

$3\sqrt{5}$

*Hint: Use the **SymPy sqrt**() function to return a simplified radical expressions. Use the **math sqrt**() function to return a float value.*

5. Modify the main program you wrote in #4 so that it also graphs the line segment between the two points. Output the distance as a title, rounded to 1 decimal point.

Sample Output

Point 1? 1,3

Point 2? 4,5

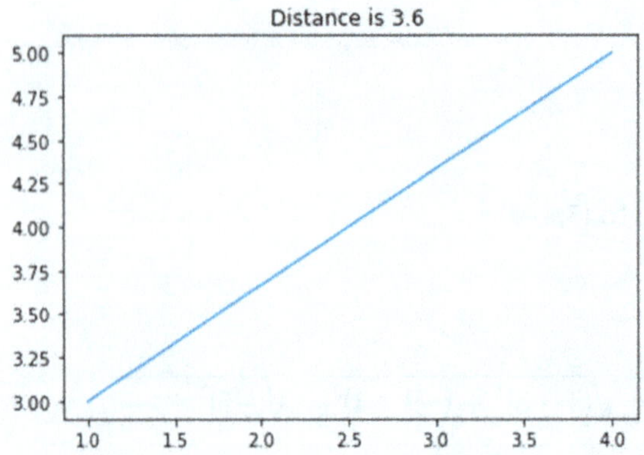

Image 6.24

6. Use the **distance**() function you wrote in #4 above to find the perimeter of a pentagon with vertices: (0, 8), (6, 5), (6, −1), (−2, −4), (−5, 3). Express your answer as a float.

3D Distance Formula

The foundation of animation, computer games, and movie special effects is of course 3-dimensional graphing. Basically, we add a 3rd axis (the z-axis). In most renderings, we consider the x and z axes to be lying “flat” on the paper (or screen), with the y-axis projecting out of the paper perpendicular to x and z. Points are now described with ordered triples (x, y, z) rather than ordered pairs (x, y).

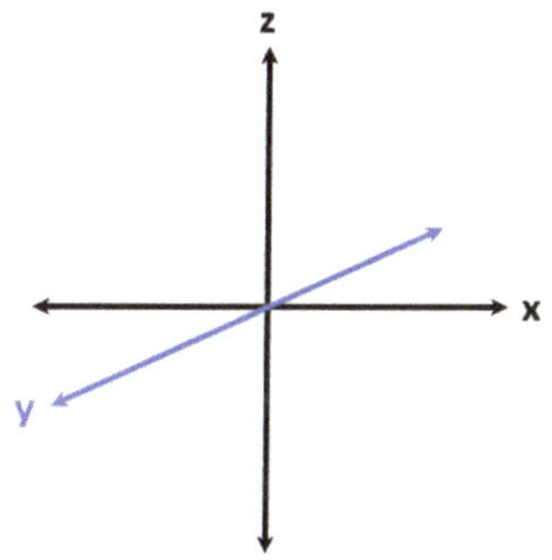

Image 6.25

Can Python plot in 3 dimensions? Indeed it can.

Program Goal Write a program that draws a line segment from (0, 0, 0) to (10, 10, 10) on a 3D graph.

PROGRAM 6.3.1 3DPLOT.PY

```
from matplotlib import pyplot as plt
ax = plt.axes(projection="3d")          # declare ax as an alias for plt.axes
ax.plot3D([0,10],[0,10],[0,10])
ax.set_xlabel('x axis')
ax.set_ylabel('y axis')
ax.set_zlabel('z axis')
```

Desired Output

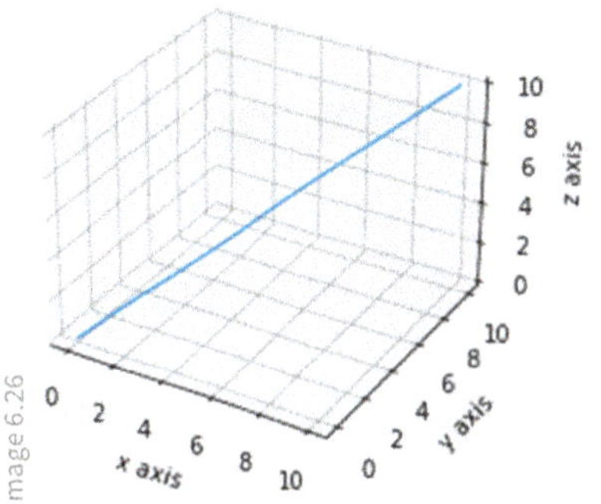

Image 6.26

Note the xyz axes may not be consistent with many advanced mathematics textbooks, which traditionally use the following orientation:

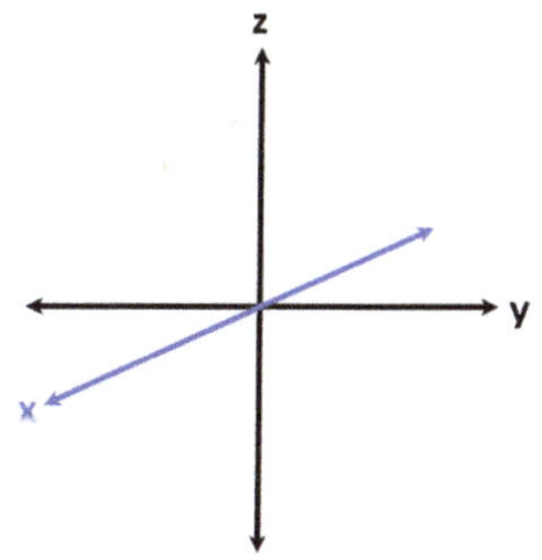

Image 6.27

but ... if you are writing your programs using a Python interpreter that allows for the creation of a pylab console, then you can interact with the figure using your mouse to rotate, etc.

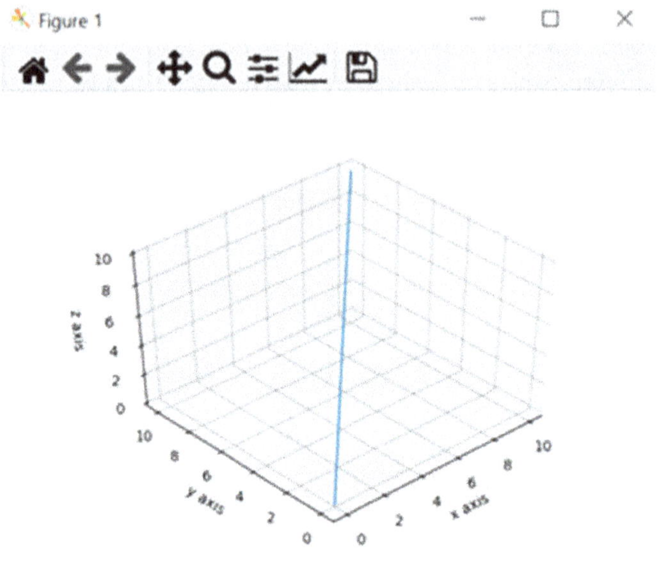

Image 6.28

Back to precalculus. The distance formula in 3D is a simple extension of the 2D distance formula:

The **distance** from (x_1, y_1, z_1) to (x_2, y_2, z_2) is:

$$d = \sqrt{(x_2 - x_1)^2 + (y_2 - y_1)^2 + (z_2 - z_1)^2}$$

If you are familiar with vectors, this formula can also be used to find what is called the **magnitude** of the vector.

Example

Find the distance from (16, −4, 5) to (18, 10, 0).

Solution

$$d = \sqrt{(18 - 16)^2 + (10 + 4)^2 + (0 - 5)^2} = \sqrt{4 + 196 + 25} = \sqrt{225} = 15$$

EXERCISE SET 6.3.2

1. Use the 3D distance formula to find the distance between the following ordered triples:

   A. (2, 1, −2) to (7, −1, 11)

   B. (8, −9, 4) to (0, 16, 5)

2. Modify program **3dplot** so that it also prints the distance between the two points. Simplify the radical using the **SymPy sqrt**() function.

3. Modify program **3dplot** so that it allows the user to input any two ordered triples (x_1, y_1, z_1) and (x_2, y_2, z_2).

4. CHALLENGE! Write a program that asks the user to input a positive number *x*. Draw a cube with each edge length *x*. Also draw a diagonal inside the cube and calculate the length of the diagonal.

   Hint: It's easiest if you position one of the vertices at the origin.

Sample Output

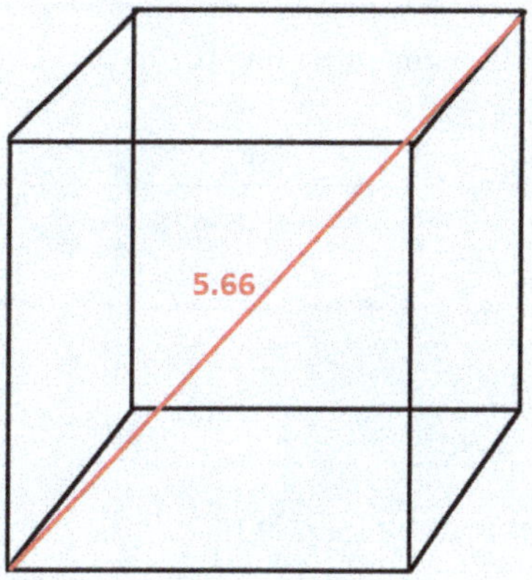

Image 6.29

6.4 Midpoint

On the real number line, the **midpoint** between coordinates a and b is $\frac{a+b}{2}$, or simply the "average" of the endpoints.

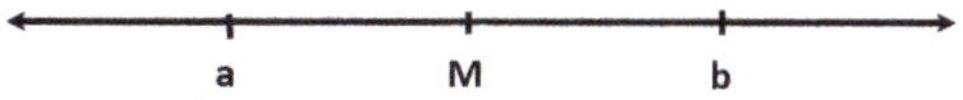

Image 6.30

On the Cartesian plane, we would like to find the coordinates of the midpoint of the line segment joining (x_1, y_1) and (x_2, y_2). We will derive the midpoint formula similar to how we derived the formula for distance—by drawing a right triangle above or below the line segment and then "fusing together" the midpoints of the two legs of the triangle:

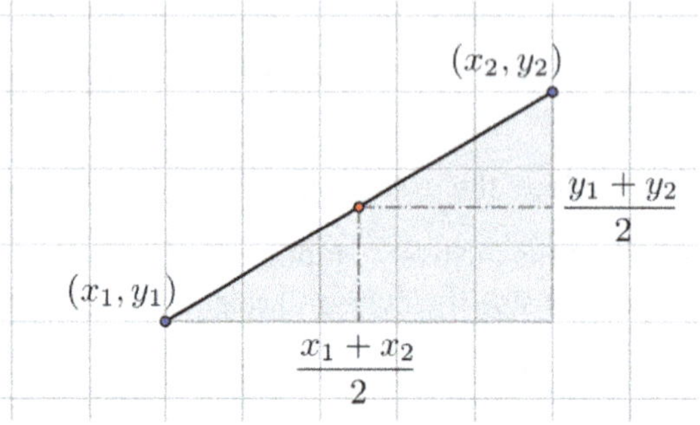

Image 6.31

The coordinates of the **midpoint** of the line segment joining (x_1, y_1) to (x_2, y_2) are:

$$\left(\frac{x_1 + x_2}{2}, \frac{y_1 + y_2}{2}\right)$$

Example 1

Find the midpoint of the line segment joining (–9, 5) to (–7, 1).

Solution

$$\left(\frac{-9 + -7}{2}, \frac{5 + 1}{2}\right) = (-8, 3)$$

Example 2

The midpoint of a line segment is (–7, –4) and one of the endpoints is (–4, –3). Find the other endpoint.

Solution

$$\left(\frac{-4 + x_2}{2}, \frac{-3 + y_2}{2}\right) = (-7, -4)$$

Solve $\frac{-4 + x_2}{2} = -7$:

$$-4 + x_2 = -14$$

$$x_2 = -10$$

Solve $\frac{-3 + y_2}{2} = -4$:

$$-3 + y_2 = -8$$

$$y_2 = -5$$

∴ The other endpoint is (–10, –5).

EXERCISE SET 6.4.1

Express your answers as simplified rational numbers, not decimal form.

1. Find the midpoint:

   A. between (−6, 14) and (−3, 5)

   B. between (3, −120) and (46, 111)

2. Find the coordinates of point R if $M(8, -3)$ is the midpoint of RS and S has coordinates (−1, 5).

3. What is the measure of the line segment from A to C if B is the midpoint?

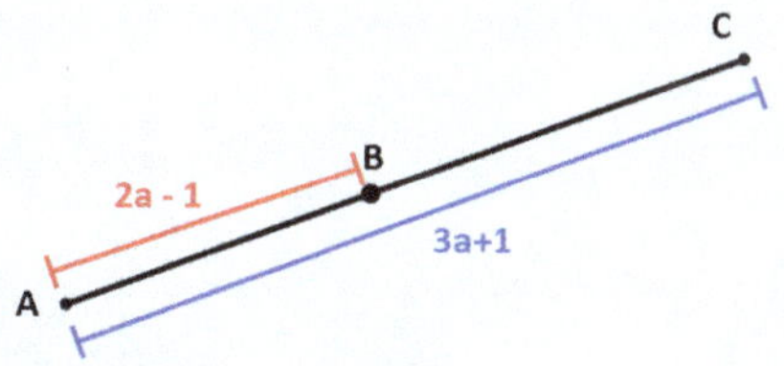

Image 6.32

4. Write a program that asks the user to input two ordered pairs and calculates the midpoint between them. Output the coordinates as integers or simplified fractions.

5. Write a program (or modify the program in the previous exercise) that calculates the midpoint of a line segment. Draw the line segment and label the **distance** at the coordinates of the midpoint.

   *(Hint: Use plt.**text**(x,y,text) to output text at coordinates x,y)*

Sample Output

Point 1? 2,8

Point 2? 7,4

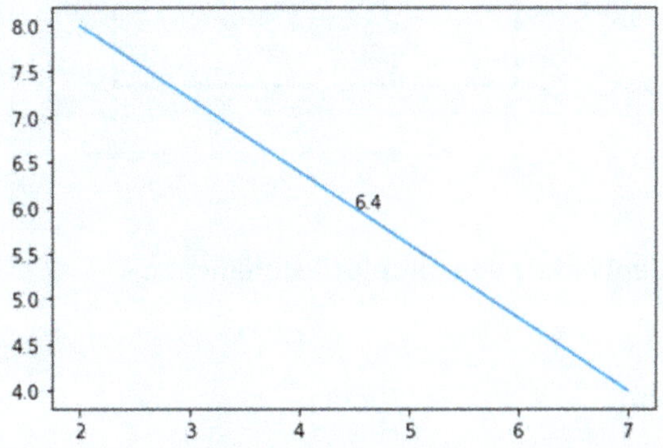

Image 6.33

(Continued)

6. Write a program that asks the user to input one endpoint (*x*1,*y*1) and the midpoint (*m*1,*m*2) then calculates and outputs the coordinates of the other endpoint (*x*2,*y*2).

Sample Output

Point 1? 2,8

Midpoint? 4,12

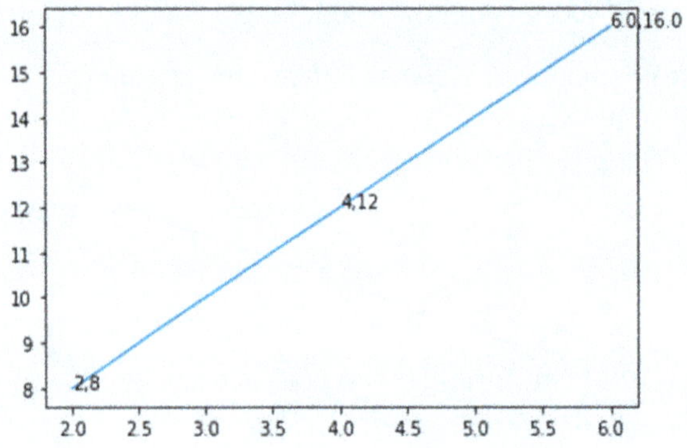

Image 6.34

7. Write a program that draws a square, then connects the midpoints of the square.

Sample Output

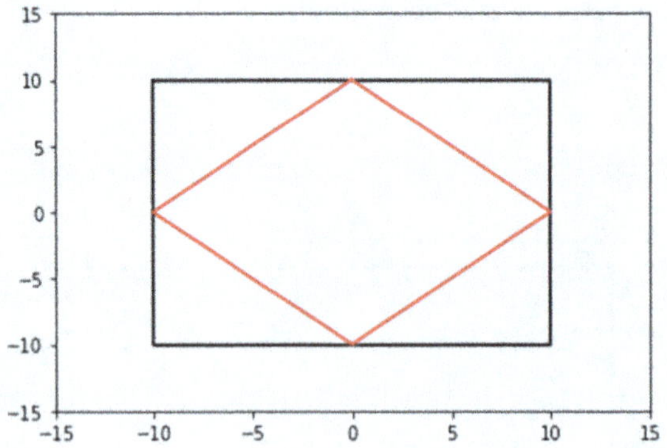

Image 6.35

8. Insert the following statement before your plt.**plot**() statements:

plt.**axis**('equal')

What is the effect on the graph?

9. Continue the program in #7 above by drawing successively smaller squares, each time connecting the midpoints of the previous square.

Sample Output

Image 6.36

3D Midpoint Formula

Just as the 3D distance formula is a logical extension of the 2D distance formula, so is the 3D midpoint formula.

The coordinates of the **midpoint** of the line segment joining (x_1, y_1, z_1) to (x_2, y_2, z_2) are:

$$\left(\frac{x_1 + x_2}{2}, \frac{y_1 + y_2}{2}, \frac{z_1 + z_2}{2}\right)$$

Example

Find the midpoint of the line segment joining (4, 11, −2) and (15, 0, 3) in 3D.

Solution

$$\left(\frac{4+15}{2}, \frac{11+0}{2}, \frac{-2+3}{2}\right) = \left(\frac{19}{2}, \frac{11}{2}, \frac{1}{2}\right)$$

EXERCISE SET 6.4.2

1. Find the midpoint and distance (as a simplified radical) between (−1, 1, 5) and (2, 5, 0).
2. Find the midpoint and distance (as a simplified radical) between (2, −9, 4) and (7, 10, −3).
3. Modify program **3dplot** so that it calculates, plots, and labels the midpoint of the line segment.

   *Hint: Use ax.**text**(x,y,z,text) to output text at the (x, y, z) coordinates.*

6.5 Circles

Another handy derivation based on the Pythagorean theorem is the equation for a circle.

Consider that the circle is simply the set of all points (x, y) that are equidistant from a given point (h, k) (the center).

If you apply the distance formula to points (x,y) and (h,k), you get:

$$r = \sqrt{(x-h)^2 + (y-k)^2}$$

Squaring both sides:

$$r^2 = (x-h)^2 + (y-k)^2$$

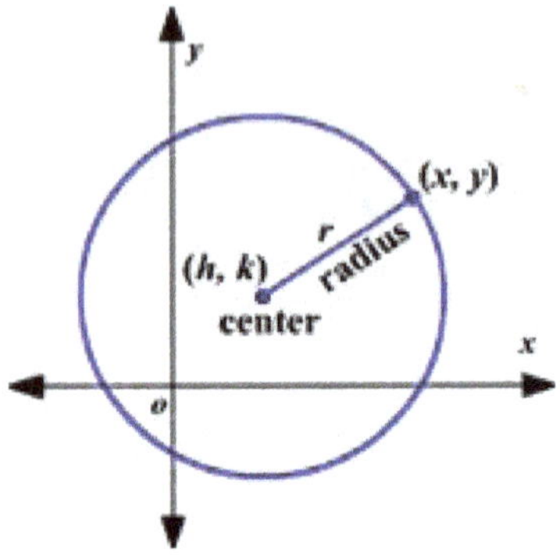

Image 6.37

FORMULA FOR CIRCLE (STANDARD FORM)

The formula for a circle with center (h, k) and radius r is

$$(x-h)^2 + (y-k)^2 = r^2$$

If the center is at (0,0) then the standard form formula simplifies to

$$x^2 + y^2 = r^2$$

General Form

$$x^2 + y^2 + ax + by + c = 0$$

Example 1

Find the general form of the circle with radius $7\sqrt{3}$ and center $(0, -4)$.

Solution

First find the standard form by substituting r and (h, k):

$$(x-0)^2 + (y-(-4))^2 = \left(7\sqrt{3}\right)^2$$

Then expand, simplify, and get 0 on one side:

$$x^2 + (y+4)^2 = 49(3)$$

$$x^2 + y^2 + 8y + 16 = 147$$

$$\therefore\ x^2 + y^2 + 8y - 131 = 0$$

Example 2

Find the center and radius of the circle with general form $2x^2 + 2y^2 + 8x + 7 = 0$.

Solution

First convert to standard form. This will involve completing the square twice—once for the x terms and once for the y terms. So let's apply the commutative and associative laws of addition to reorder the terms:

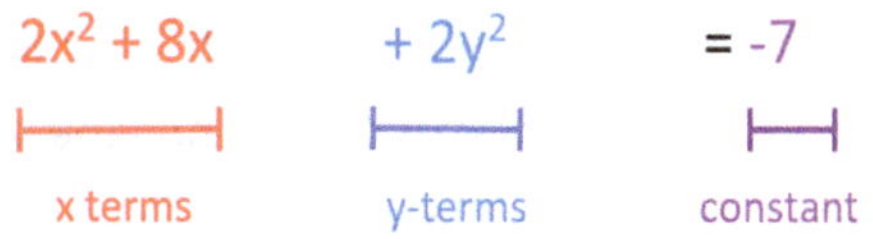

Image 6.38

Factor out the coefficients of the x^2 and y^2 terms:

$2(x^2 + 4x) + 2(y^2) = -7$

Complete the square for the x-terms (y-terms don't need it):

$2(x^2 + 4x + \mathbf{4}) + 2(y^2) = -7 + \mathbf{8}$

$2(x + 2)^2 + 2(y - 0)^2 = 1$

Divide both sides by 2:

$(x + 2)^2 + (y - 0)^2 = \frac{1}{2}$

$\therefore$ The center is $(-2, 0)$ and the radius is $\sqrt{\frac{1}{2}} = \frac{\sqrt{2}}{2}$.

Example 3

Write the standard form of a circle with center (1,0) and containing point (−3, 2).

Solution

We are given $(h, k) = (1, 0)$. We only have to figure out the radius r. Use the distance formula between (1, 0) and (−3, 2):

$$r = \sqrt{(-3-1)^2 + (2-0)^2} = \sqrt{16+4} = \sqrt{20}$$

Note it is not necessary to simplify the radical because we will just end up squaring r:

$$(x-1)^2 + (y-0)^2 = \left(\sqrt{20}\right)^2$$

$$\therefore \; (x-1)^2 + y^2 = 20$$

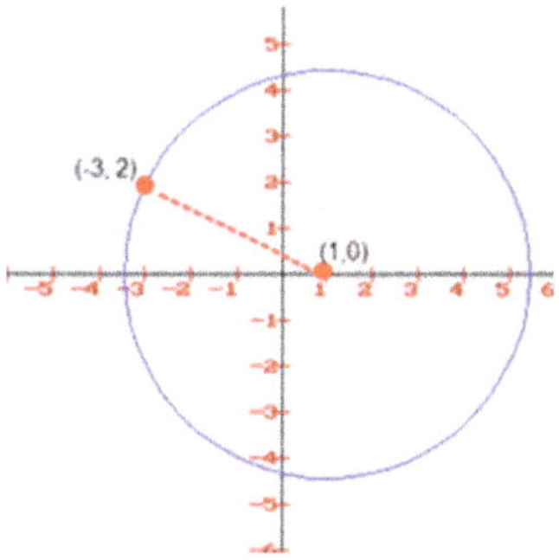

Image 6.39

EXERCISE SET 6.5.1

1. Find the standard form of a circle with center at origin and radius $2\sqrt{3}$.
2. Find the standard form of a circle with center (2, –3) and radius 4.
3. Given circle with standard form $(x-2)^2+(y+7)^2=36$, find the center and radius.
4. Given circle with center at (–5, –2) and $r=7$, find the general form.
5. Given circle with general form $3x^2+3y^2-12y-27=0$, find the center and radius.
6. Given circle with general form $2x^2+2y^2-12x+8y-24=0$, find the standard form.
7. Given circle with endpoints of diameter at (4,3) and (0,1), find the standard form.
8. Given circle with center (2,3) and tangent to the x-axis, find the standard form.
9. Write a program that asks the user to input the center and radius of a circle. Output the equation of the circle in standard form.
10. Write a program that asks the user to input the endpoints of the diameter (as in #7 above); output the equation of the circle in standard form.

Program Goal: Write a program that asks the user to input the center coordinates h, k and radius r, then graphs the circle.

To do this, we need the formula for the circle expressed in $\boldsymbol{y}=$ form.

Start with standard form $(x-h)^2+(y-k)^2=r^2$ and solve for y:

$$(y-k)^2=r^2-(x-h)^2$$

$$y-k=\pm\sqrt{r^2-(x-h)^2}$$

Remember, when you take the square root of both sides, one of the sides needs ±.

$$y=\pm\sqrt{r^2-(x-h)^2}+k$$

A function can only have one output, but this formula provides two outputs, so it is essentially two functions:

$$y=+\sqrt{r^2-(x-h)^2}+k \qquad \text{(top half of circle)}$$

$$y=-\sqrt{r^2-(x-h)^2}+k \qquad \text{(bottom half of circle)}$$

Circle with center (*h*, *k*) and radius *r*

Functional ($y=$) form:

$$y = +\sqrt{r^2 - (x - h)^2} + k$$

$$y = +\sqrt{r^2 - (x - h)^2} + k$$

Domain:

$[h - r, h + r]$

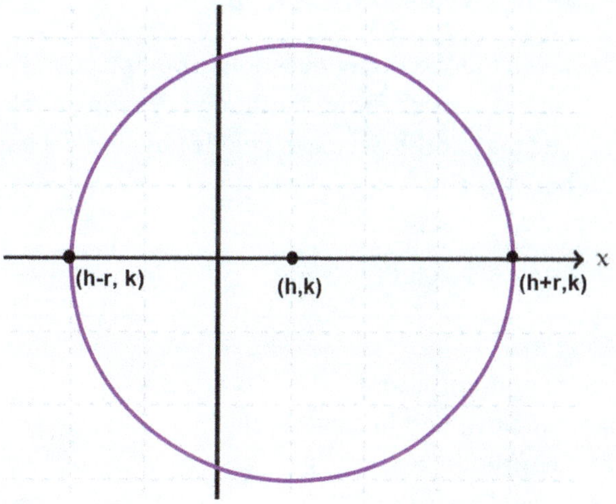

Image 6.40

PROGRAM 6.5.1 CIRCLE1.PY

```
from matplotlib import pyplot as plt
from math import sqrt,floor,ceil
center = input("Coordinates of center h,k: ")
h,k = [float(i) for i in center.split(',') ]
r = float(input("Radius: "))
x_list = range(ceil(h-r),floor(h+r))
y_list1 = [ ]
y_list1 = [sqrt(r**2-(x-h)**2)+k for x in x_list]
plt.plot(x_list,y_list1)
[fill in code to graph lower semicircle]
plt.show()
```

Sample Output

Coordinates of Center *h*, *k*: 200, 300

Radius: 50

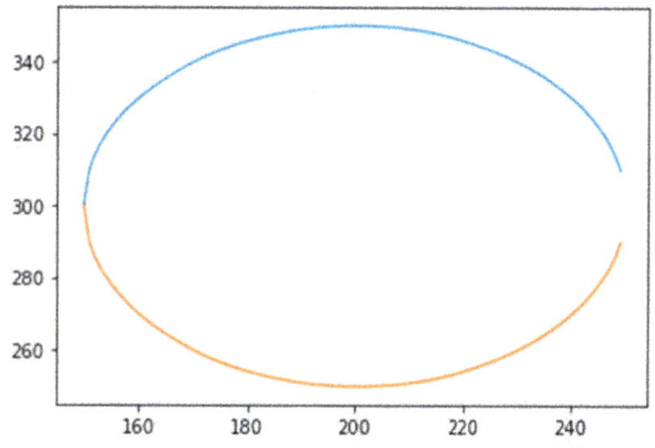

Image 6.41

EXERCISE SET 6.5.2

1. Convert the following circle equations to function ($y=$) form:

   A. $(x + 7)^2 + y^2 = 81$

   B. $x^2 + y^2 - 10x + 2y + 17 = 0$

2. Fill in the yellow highlighted code of program **circle1** to graph the lower semicircle.

Drawing Circles in matplotlib

We haven't asked many questions about this method of drawing circles because, while it has its merits, Python has a much easier method of drawing circles.

The **matplotlib** package includes several packages that draw various geometric shapes that are called **patches**. Try this.

PROGRAM 6.5.2 CIRCLE2.PY

```
from matplotlib import pyplot as plt
center = input("Coordinates of center h,k: ")
h,k = [float(i) for i in center.split(',')]
r = float(input("Radius: "))
circle=plt.Circle((h,k), r)          # creates a Circle object
ax = plt.gca()                       # returns reference to current Axes
ax.add_patch(circle)                 # adds the Circle patch
plt.axis('scaled')                   # automatically scale axes
plt.show()
```

Sample Output

Coordinates of center h, k: 10, −10
Radius: 20

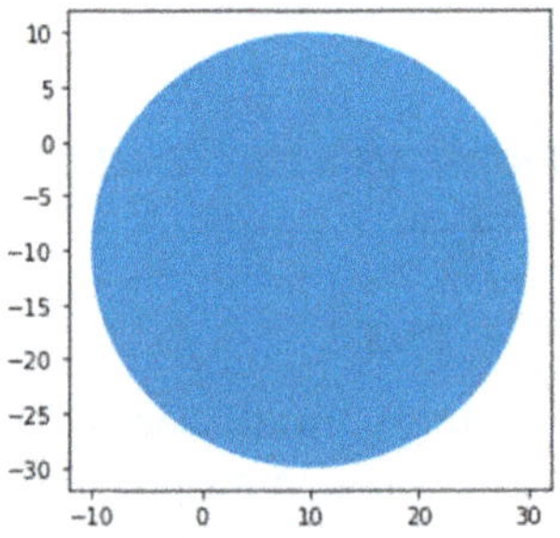

Image 6.42

If the circle on your screen looks more like an ellipse, then add the following statement after *ax* = **plt.gca**():

```
plt.aspect('equal')
```

To specify the edge color and/or face (fill) color, use the *fc* and *ec* parameters to the Circle() function call, for example:

```
circle=plt.Circle((h,k), r,fc='black',ec='pink')
```

EXERCISE SET 6.5.3

1. Write a program that draws a circle inscribed inside a square. It's easiest if the center is at (0, 0).

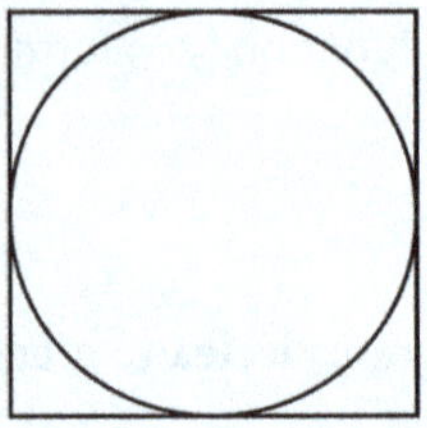
Image 6.43

2. Write a program that draws a square inscribed inside a circle.

   Hint: If the radius of the circle is r, then the length of the side of the square is $\frac{2r}{\sqrt{2}}$.

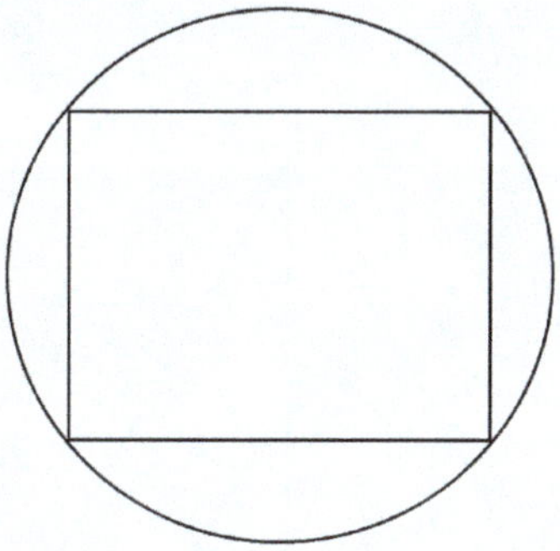

Image 6.44

3. Write a program that generates and draws randomly placed circles in different colors on the screen.

Sample Output

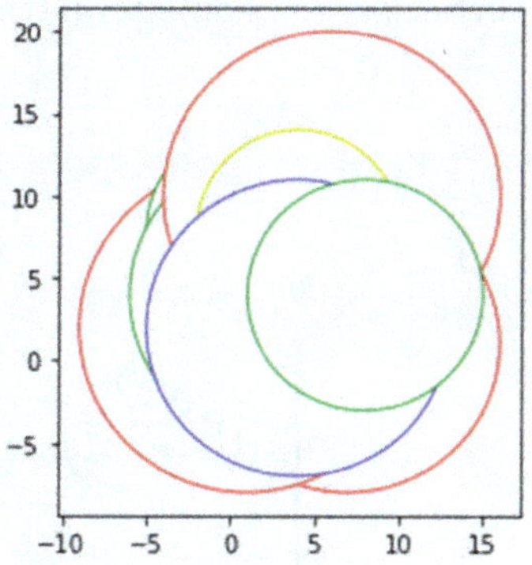

Image 6.45

Hint: To generate random colors, you can use a list:

randcolor = ['b', 'g', 'r', 'c', 'm', 'y', 'k']

4. CHALLENGE. Write a program that will draw this design.

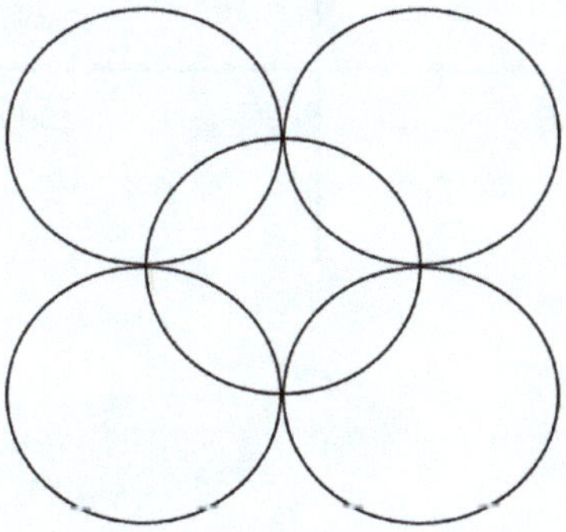

Image 6.46

6.6 More Coordinate Geometry

Drawing an Equilateral Triangle

If the length of each side of an equilateral triangle is D, then the height of the triangle from base to top vertex is $\frac{\sqrt{3}}{2}D$.

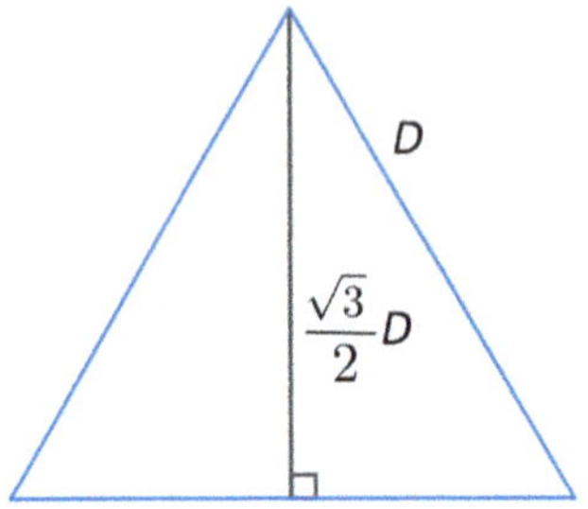

Image 6.47

Example 1

Draw and label the coordinates of the vertices of the equilateral triangle with side length 5, symmetric to the y-axis, and base on the x-axis.

Solution

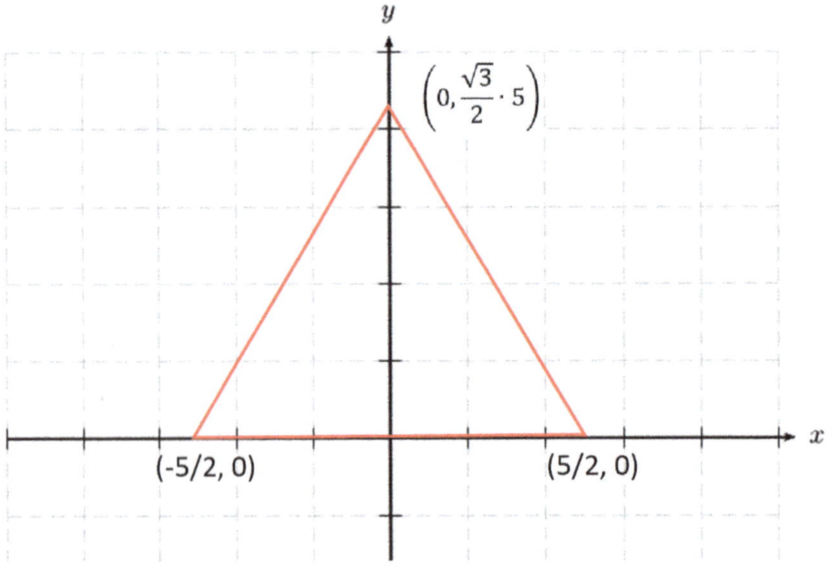

Image 6.48

Example 2

Draw and label the coordinates of the vertices of the equilateral triangle with side length 5, symmetric to the y-axis, and top vertex at (0, 6).

Solution

Because the height of the triangle is $\frac{\sqrt{3}}{2} \cdot 5$, the y-coordinate of the base is $6 - \frac{\sqrt{3}}{2} \cdot 5$.

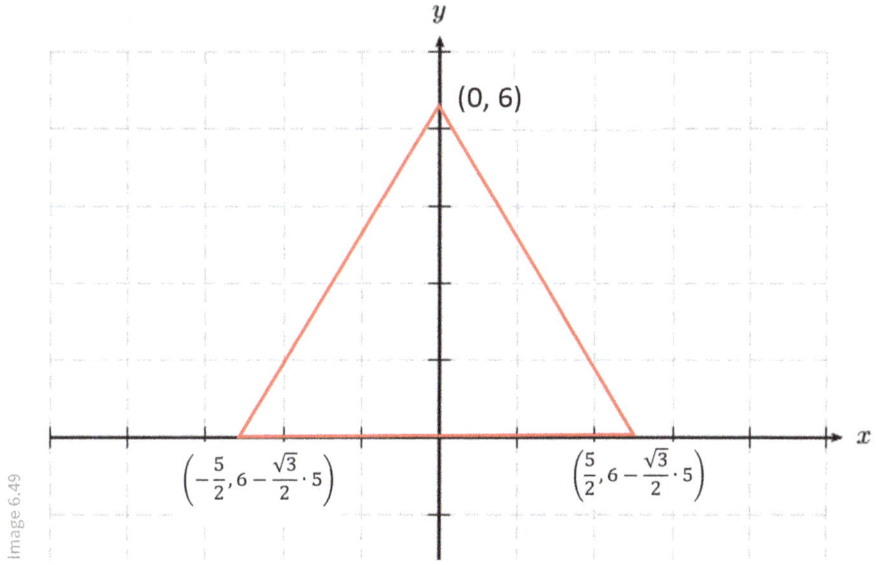

EXERCISE SET 6.6.1

1. A. Draw and identify the coordinates of the vertices of the equilateral triangle symmetric to the y-axis, with base on the x-axis, and length of each side 7.

   B. Calculate the area of the triangle.

2. A. Draw and identify the coordinates of the vertices of the equilateral triangle symmetric to the y-axis, length of each side 7, and top vertex at (0, 10).

   B. Calculate the area of the triangle.

3. Draw and identify the coordinates of the vertices of the equilateral triangle with side length 4 and top vertex at (6, 9).

Program Goal: Write a program that will draw an equilateral triangle with side length D, symmetric to the y-axis, with base on the x-axis.

Here's a basic outline of the program in which the bottom side has been filled in for you. In the Exercise Set below, you will need to fill in the code for the other two sides.

PROGRAM 6.6.1 EQUILATERAL.PY

```
from matplotlib import pyplot as plt
from math import sqrt
d = float(input("Side length? "))
# top vertex is (0,y)
y = d* sqrt(3)/2
# base vertices are (x1,0) and (x2,0)
x1 = -d/2
x2 = d/2
    [fill in code to draw the 3 sides]
plt.show()
```

Sample Output

Side length? 10

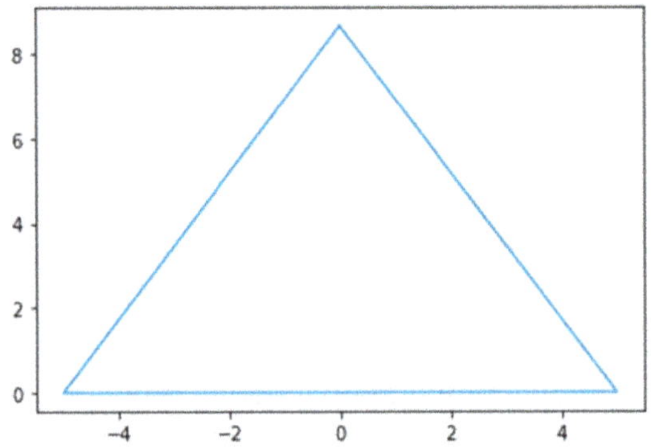

Image 6.50

EXERCISE SET 6.6.2

1. Fill in the yellow highlighted code of the program to draw the 3 sides of the equilateral triangle.
2. Modify the program so that it also calculates and outputs the area at the center of the triangle.
3. Modify the program so that the user is also prompted to enter the y-coordinate of the top vertex.
4. Modify the program so that the user is prompted to enter the side length and coordinates of top vertex (e, f).

Sample Output

Side length? 4

Coordinates of top vertex? 6,9

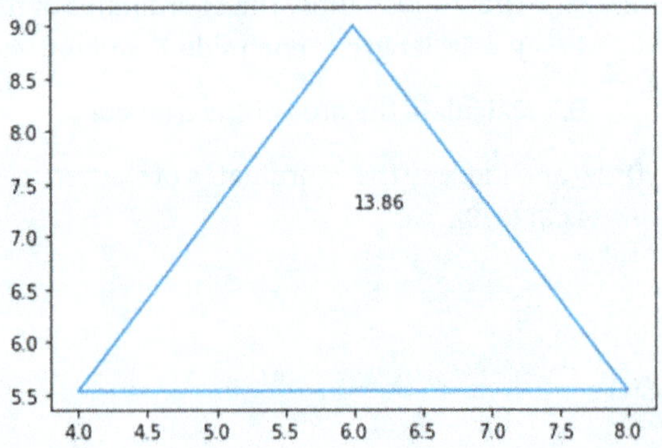

Image 6.51

5. CHALLENGE. Write a program that draws a randomly generated triangle. Draw line segments connecting each vertex to the midpoint of the opposite side. All 3 of these line segments intersect at a special point called the **centroid** of the triangle.

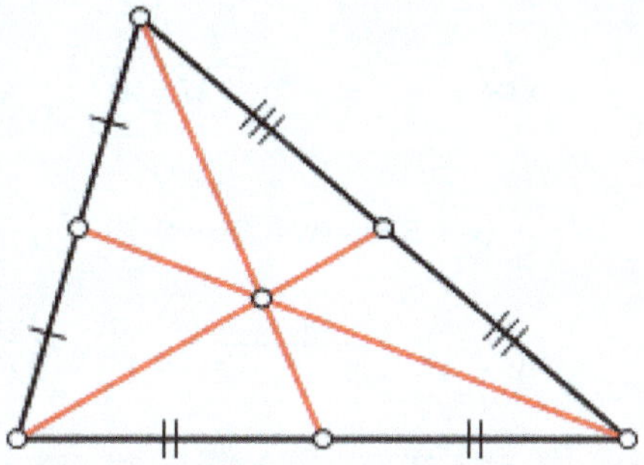

Image 6.52

Congratulations, you can now incorporate fundamental 2D and 3D graphics into your Python programs. Check out Chapter Project XII for an animation program!

Chapter Projects

I. Line Art

Write a program that draws what is sometimes called **String Art**. Here's a basic example:

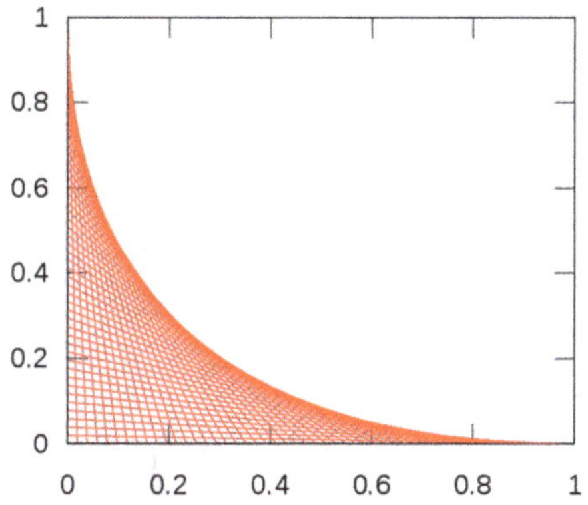

Image 6.53

There are many possibilities limited only by your imagination!

II. Converting Circular Equations

Write a program that will convert the equation of a circle in general form $x^2 + y^2 + ax + by + c = 0$ to standard form $(x - h)^2 + (y - k)^2 = r^2$ and also to functional form $y = \pm\sqrt{r^2 - (x - h)^2} + k$.

Hint: use the pretty print ***pprint()*** *function in* ***SymPy*** *to format the output.*

SAMPLE OUTPUT

Enter the values of a, b, and c in the general form $x^2 + y^2 + ax + by = c$.

a, b, c? −10, 2, 17

The equation in general form is:

$$x^2 + y^2 - 10x + 2y + 17 = 0$$

The equation in standard form is:

$$(x - 5)^2 + (y + 1)^2 = 9$$

The equation in functional form is:

$$Y = +/- \sqrt{9 - (x - 5)^2} - 1$$

III. Area of a Polygon

You can calculate the area of any closed polygon using a formula called the **Shoelace Formula**: Suppose a polygon has vertices (a_1, b_1), (a_2, b_2), ... , (a_n, b_n), listed in clockwise order. Then the area of the polygon is

$$\frac{1}{2}|(a_1b_2 + a_2b_3 + \cdots + a_nb_1) - (b_1a_2 + b_2a_3 + \cdots + b_na_1)|$$

For example, given the polygon:

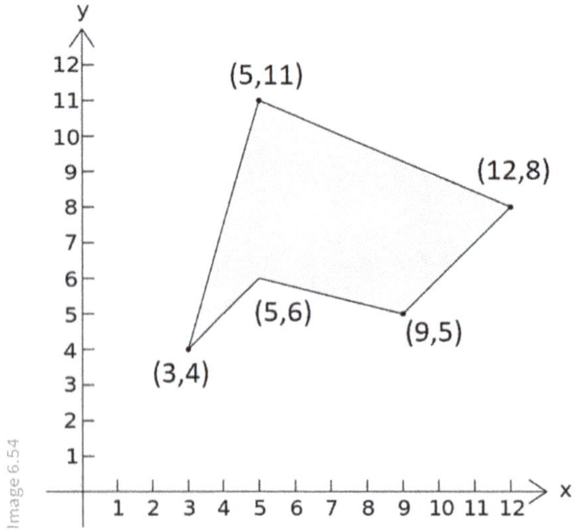

$$A = \frac{1}{2}|(3 \cdot 11 + 5 \cdot 8 + 12 \cdot 5 + 9 \cdot 6 + 5 \cdot 4) - (4 \cdot 5 + 11 \cdot 12 + 8 \cdot 9 + 5 \cdot 5 + 6 \cdot 3)| = 30$$

Write a program that draws a random polygon ($n > 4$) and then calculates its area.

IV. Pythagorean Spiral

The **Pythagorean spiral** is a spiral composed of right triangles, placed edge-to-edge.

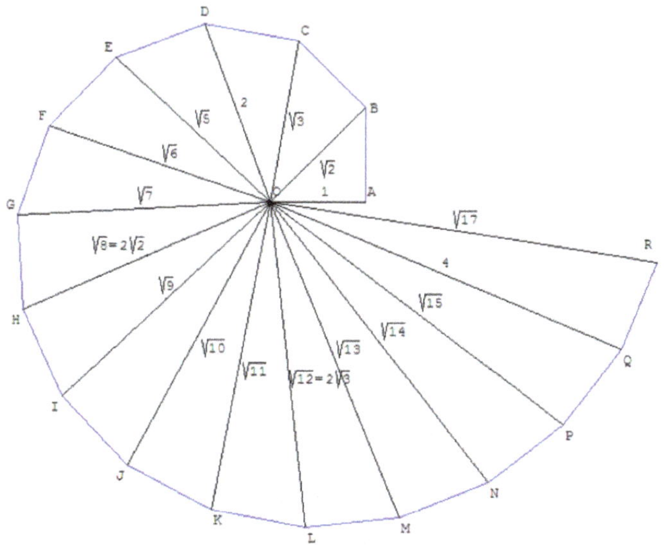

Image 6.55

Draw the Pythagorean spiral shown, with the center at the origin and the initial edge labeled $\sqrt{1}$ on the positive x-axis.

V. Orthocenter of a Triangle

A triangle has three **altitudes**: a line segment perpendicular to each side connected to the opposite vertex. An interesting fact is that the three altitudes intersect at the same point, called the **orthocenter**. Write a program that draws a random triangle and the three altitudes. They should meet at the same point!

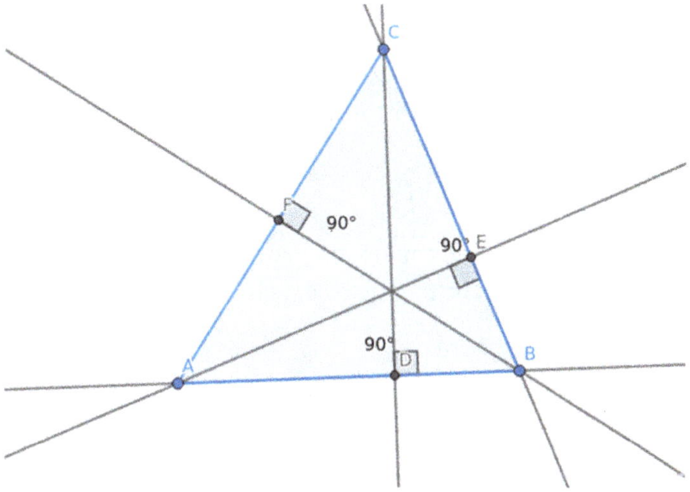

Image 6.56

Calculate and output the (x, y) coordinates of the orthocenter.

VI. Central, Interior, and Exterior Angles of a Regular Polygon

The three types of angles of a regular *n*-gon (polygon with all sides congruent, all angles congruent) can be found as follows:

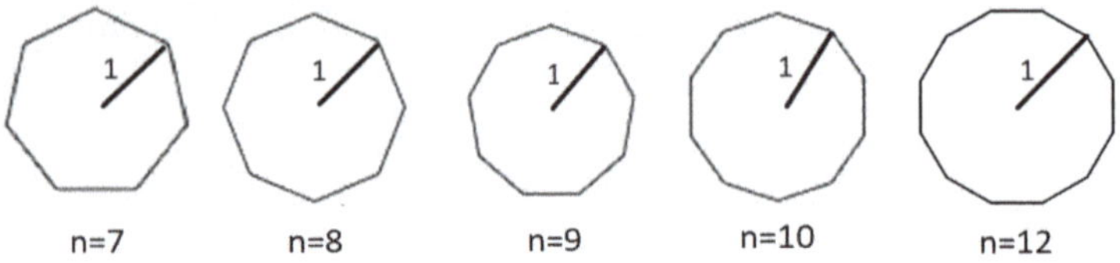

Image 6.57

Each central angle $= \frac{360}{n}$

Sum of interior angles $= (n - 2) \cdot 180$

Each interior angle $= \frac{(n - 2) \cdot 180}{n}$

Each exterior angle $= \frac{360}{n}$

Write a program that asks the user to input the number of sides, *n*. Calculate and output each of the formulas above. Round all measures to 1 decimal point.

There are only three regular polygons that evenly divide 360 and thus can form a tessellation of the plane. Do you know what they are?

VII. Random Convex Polygon

If you were to write a program to randomly generate an *n*-gon (in which the user specifies the number of sides ***n***), most of them will probably be convex or self-intersecting.

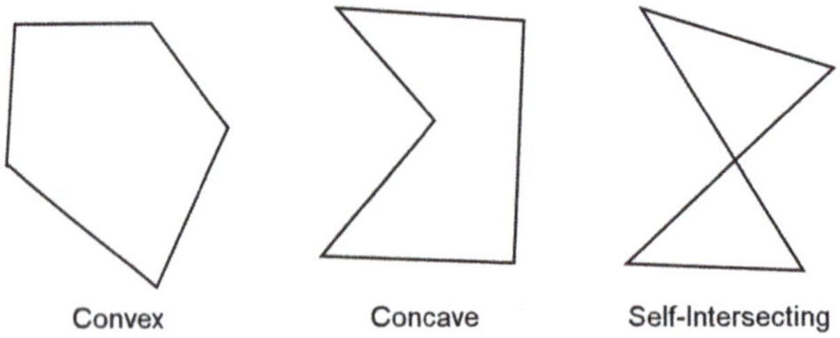

Image 6.58

Write a program that will only draw a convex polygon. There are various articles on the internet about how to do this. Here's one of them: http://cglab.ca/~sander/misc/ConvexGeneration/convex.html.

VIII. Dartboard Probability

Let's take a look at a different kind of dartboard: one in which a circle is inscribed inside a square.

Imagine that a number of random "darts" (n) are generated somewhere within the square.

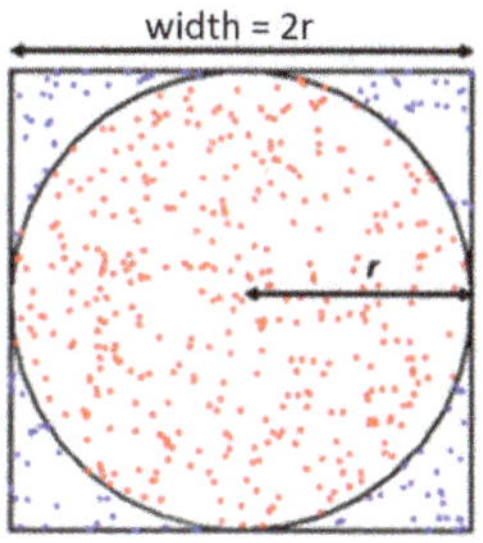

Image 6.59

A. Write a program that keeps track of how many points fall inside the circle and how many points fall outside the circle. Finally, calculate and output 4*(points within circle)/(total points). *Hint: Assume* $r = 200$.

   Run this program several times with increasingly larger values of n. Does the final calculation converge to a specific limit or is it a random value? If it converges, what is the value?

B. Using $2r$ as the length of the side of the square, what is the probability that a randomly thrown "dart" lands within the circle?

IX. Right Triangle Identifier

Given the coordinates of three vertices of a triangle, determine if the triangle is a right triangle. (This can be done using either the slopes of the two shortest sides or confirming that the three side lengths satisfy the Pythagorean theorem.) If it is a right triangle, calculate and output its area. Also plot the triangle.

SAMPLE OUTPUT

Vertex A? (–2,5)

Vertex B? (12,3)

Vertex C? (10, –11)

It is a right triangle.

AB = $10\sqrt{2}$

BC = $10\sqrt{2}$

AC = 20

The area is 100

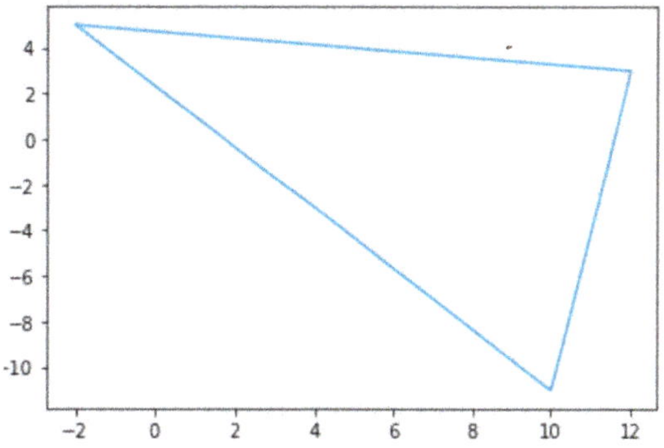

Image 6.60

X. Solving a Linear System

Write a program that asks the user to input two linear functions. Draw the two lines and determine if the linear system is **inconsistent** (parallel lines), **consistent independent** (lines that intersect in exactly one point), or **consistent dependent** (coincident lines, i.e., the same line).

In the case of consistent independent, label the point of intersection with the coordinates of the ordered pair. In this case, also check if the intersecting lines are perpendicular. (Note: the plot may not appear perpendicular due to the aspect ratio of the window. Can you fix this?)

SAMPLE OUTPUT

Line 1? 7*x-4

Line 2? 2*x+1

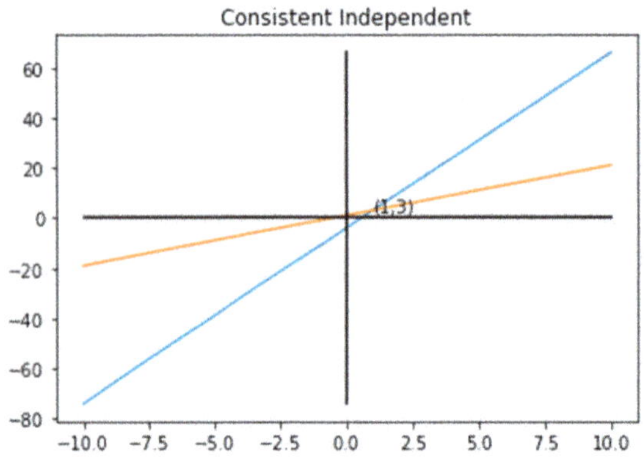

Image 6.61

Line 1? 3*x+4

Line 2? (-1/3)*x−5

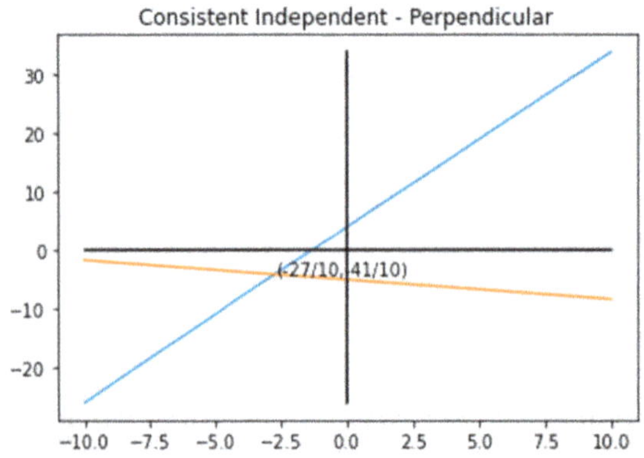

Image 6.62

XI. Solving a Nonlinear System

A nonlinear system involves two functions in which at least one of them is nonlinear. The solution(s) are the points of intersection. Similar to Chapter Project X above, you will write a program that asks the user to input two functions. Plot the two functions and label any point(s) of intersection.

Make sure to check that you are not attempting to plot or label undefined or imaginary values.

SAMPLE OUTPUT

Function 1? sqrt(*x*)

Function 2? sqrt(25-*x***2)

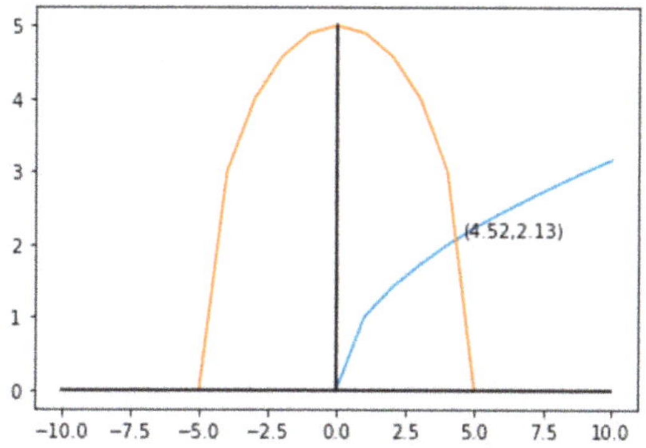

Image 6.63

XII. Animate This

The following program will produce this very simple animation of a dot traveling from one corner of the screen to the other. It accomplishes this by plotting one dot at a time with a brief pause between.

```
from matplotlib import pyplot as plt
x = [ ]
y = [ ]
for i in range(100):
    x.append(i)
    y.append(i)
    plt.xlim(0, 100)
    plt.ylim(0, 100)
    plt.plot(x, y, '.')
    plt.pause(0.01)
plt.show()
```

Output

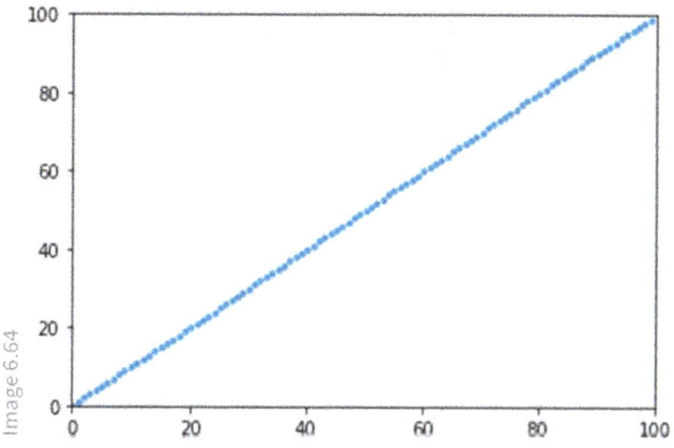

Image 6.64

Your challenge is to modify the program so that the object (whether it is a dot, circle, dash, asterisk, etc.) travels along a different path. Experiment with colors, time delays, and paths. Can you plot an object other than a single character, such as a bouncing ball?

CHAPTER SUMMARY

PYTHON CONCEPTS

matplotlib	Popular graphing package (collection of modules, functions, methods)
pyplot	A plotting module in **matplotlib** To import the **pyplot** module and give it an alias **plt**: **from** matplotlib **import** pyplot **as** plt or **import** matplotlib.pyplot **as** plt
To plot a set of ordered pairs and connect them with blue line segments	plt.**plot**(x_list,y_list) plt.**show**() If x_list is omitted, it defaults to [0,1,2,...**len**(y_values)] *Ex:* *plt.**plot**([0,1,2,3],[0,1,4,9])* *plt.**show**()*
To plot a set of ordered pairs with a marker, not connected with blue line sements	plt.**plot**(x_list,y_list,marker) Choice of markers include 'o' (closed blue dot), '*', 'x', '.' or '+'. Default marker color is blue however they can be preceded by a color 'r' (red), 'b' (blue), 'g' (green). *Ex:* *plt.**plot**([0,−1,−2,−3,−4,−5], 'r*')* *will plot the point (0, 0), (1, −1), (2, −2), (3, −3), (4, −4), (5, −5) using a red asterisk.*
To set the scaling of the axes	plt.**axis(xmin**=xvalue1, **xmax**=xvalue2, **ymin**=yvalue1,**ymax**=yvalue2) or plt.**axis**([xvalue1,xvalue2,yvalue1,yvalue2]) *Ex:* *plt.**axis**(**xmin**=−10,**xmax**=10,**ymin**=−10,**ymax**=10)* *will scale the x-axis from −10 to 10 and the y-axis from −10 to 10.*

To label the axes	plt.**xlabel**(string) plt.**ylabel**(string) *Ex:* *plt.**xlabel**('years')* *plt.**ylabel**('revenue')*
To give the plot a title	plt.**title**(string) *Ex:* *plt.**title**('Revenue Projection')*
To display the plot	plt.**show**()
Specifying color	plt.**plot**(x_list,y_list,**c**='color') or plt.**plot**(x_list,y_list,**color**='color') where 'color' is 'b', 'g', 'r', 'm', 'y', 'k', 'w', an RGB or RGBA string, a tableau color, or a CSS4 color name. *Ex:* *plt.**plot**(x_list,y_list,**color**='cyan')*
Specifying line width and line style	plt.**plot**(x_list,y_list,**linewidth**=thickness, **linestyle**='code') where thickness = 1 (thin line), 2 (medium line), 4 (thick line), etc. and code = ':', '-.', '—', '-' *Ex:* *plt.**plot**(x_list,y_list,**linewidth**=2,**linestyle**=':')*
To add text to the plot	plt.**text**(x,y,text) *Ex:* *plt.**text**(0,0,'hello world')*
To set the aspect ratio of the axes to 1	plt.**axis**('equal')
To draw a circle with center (*h*, *k*) and radius *r*	circle = plt.**Circle**(h,k),**radius**=r) ax = plt.**gca**() ax.**add_patch**(circle) plt.**axis**('scaled')
To specify the edge color and/or fill color of a circle	circle = plt.**Circle**((h,k),**radius**=r,**fc**=color1,**ec**=color2)

PRECALCULUS CONCEPTS

Distance formula in 2D	The distance from (x_1, y_1) to (x_2, y_2) is: $d = \sqrt{(x_2 - x_1)^2 + (y_2 - y_1)^2}$
Distance formula in 3D	The distance from (x_1, y_1, z_1) to (x_2, y_2, z_2) is: $d = \sqrt{(x_2 - x_1)^2 + (y_2 - y_1)^2 + (z_2 - z_1)^2}$
Midpoint in 2D	The midpoint of the line segment joining (x_1, y_1) to (x_2, y_2) is: $\left(\frac{x_1 + x_2}{2}, \frac{y_1 + y_2}{2}\right)$
Midpoint in 3D	The midpoint of the line segment joining (x_1, y_1, z_1) to (x_2, y_2, z_2) is: $\left(\frac{x_1 + x_2}{2}, \frac{y_1 + y_2}{2}, \frac{z_1 + z_2}{2}\right)$
Standard form for a circle with center (h, k) and radius r	$(x - h)^2 + (y - k)^2 = r^2$
Standard form for a circle with center (0, 0) and radius r	$x^2 + y^2 = r^2$
General form of a circle	$x^2 + y^2 + ax + by + c = 0$

Credits

IMG 6.19: Matplotlib Development Team, https://matplotlib.org/3.3.0/gallery/color/named_colors.html. Copyright © 2020 by Matplotlib Development Team.

IMG 6.20: Adapted from Matplotlib Development Team, https://matplotlib.org/3.3.0/gallery/lines_bars_and_markers/linestyles.html?highlight=linestyle. Copyright © 2020 by Matplotlib Development Team.

IMG 6.31: Personline, https://commons.wikimedia.org/wiki/File:Mid_point.svg, 2008.

IMG 6.37: Copyright © by Jack Ver (CC BY-SA 3.0) at https://commons.wikimedia.org/wiki/File:Cirkel.png.

IMG 6.43: Deweirdifier, https://commons.wikimedia.org/wiki/File:Square-circle.svg, 2009.

IMG 6.52: Maksim, https://commons.wikimedia.org/wiki/File:Triangle.Centroid.svg, 2006.

IMG 6.53: Copyright © by IkamusumeFan (CC BY-SA 4.0) at https://commons.wikimedia.org/wiki/File:Envelope_string_art.svg.

IMG 6.54: Adapted from Nat2, https://commons.wikimedia.org/wiki/File:Polygon_area_formula_(English).svg, 2011.

IMG 6.55: Copyright © by Pdebart (CC BY-SA 4.0) at https://commons.wikimedia.org/wiki/File:Escargot_pythagore.png.

IMG 6.56: Copyright © by Mattruffoni (CC BY-SA 4.0) at https://commons.wikimedia.org/wiki/File:Ortocentro_angoli_retti.png.

CHAPTER 7

Graph Features

7.1 Transformations—Stretch and Compress

Our focus in this chapter is to explore the various features of graphs of functions. With the modern-day availability of graphing calculators and online graphing tools, it almost seems unnecessary to put this much analysis into graphs. Granted, many of these concepts are now of only historical interest.

However, the mastery of these concepts can actually make graphing by hand quicker than graphing using technology. Further, we need to be able to interpret the graphs produced by technology and to recognize common functions that will arise often in calculus.

First, let's review a basic graphing program using concepts from Chapter 6.

Program Goal: Write a program that asks a user to input a polynomial function. Graph the function over a domain of [–10,10].

PROGRAM 7.1.1 GRAPH_FUNCTION.PY

```
from matplotlib import pyplot as plt
from sympy import *
x_list = range(−10,11)                    # list of integers
x = Symbol('x')                           between −10 and 10
func1 = sympify(input('Function? '))
# assign y-values
y1_list = [func1.subs({x:a}) for a in x_list]
plt.plot(x_list,y1_list)
plt.show()
```

Sample Output

Function? *x**3*

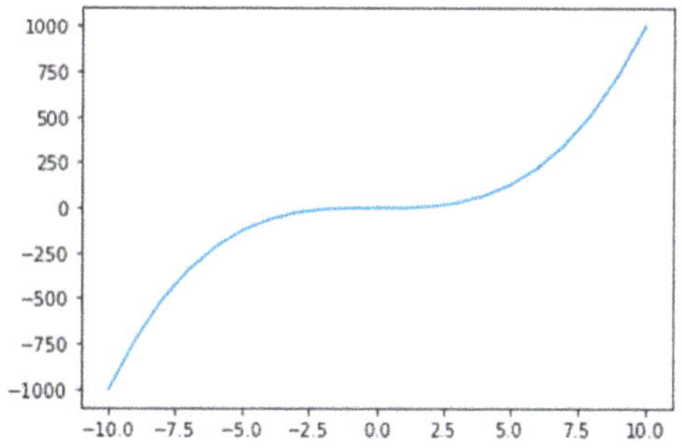

Image 7.1

EXERCISE SET 7.1.1

1. If a graph has domain [–10, 10] and range [–50, 70], write plt.**plot**() statements to draw the *x*- and *y*-axes.

2. Write a function **draw_axes**(x_list,y_list) that will draw the *x*- and *y*-axes using the min and max of x_list to draw the *x*-axis and the min and max of y_list to draw the y_axis. Incorporate the function into program **graph_function**.

Desired Output

Function? *x**3*

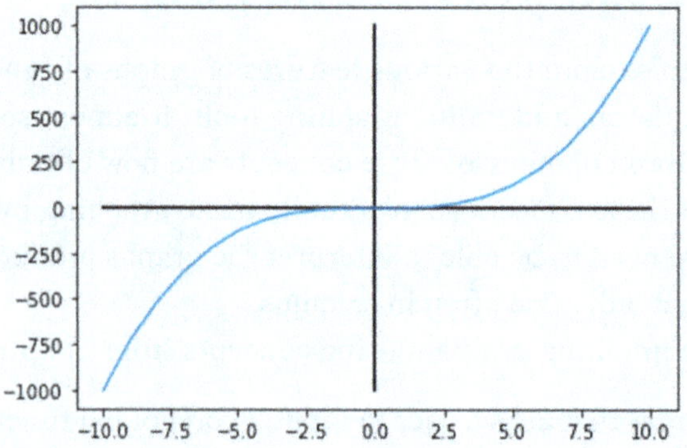

Image 7.2

3. Modify the program so that the *x*-axis is scaled from –20 to 20.

4. Modify the program by adding a title:

   The minimum is [y_min], the maximum is [y_max].

   *(hint: use plt.**title**(text))*

Now let's take a look at our first types of transformation: **stretching** and **compressing**.

Program Goal: Write a program that draws three graphs:

1. $y = x^2$
2. $y = x^2$ stretched by a factor of 2
3. $y = x^2$ compressed by a factor of ½

You should already recognize from Chapter 3 that these are quadratic functions that have parabolic graphs. And from Chapter 6 we'll adapt program 6.2.3 (**function_plot**) to graph our functions.

PROGRAM 7.1.2 STRETCH_COMPRESS.PY

```
## stretch_compress.py
## Draw y=x**2, y=2*x**2, y=(1/2)*x**2
from matplotlib import pyplot as plt
from sympy import Symbol, sympify
x_list = range(−10,11)
y1_list = [ ]
y2_list = [ ]
y3_list = [ ]
x = Symbol('x')
func1 = sympify('x**2')
func2 = sympify('2*x**2')
func3 = sympify('(1/2)*x**2')
y1_list=[func1.subs({x:a}) for a in x_list]
[ insert code to assign values to y2_list and y3_list ]
plt.plot(x_list,y1_list)
[ insert code to plot func2 and func3 ]
plt.show()
```

It's good programming practice to include the filename and a brief description of the program

Desired Output

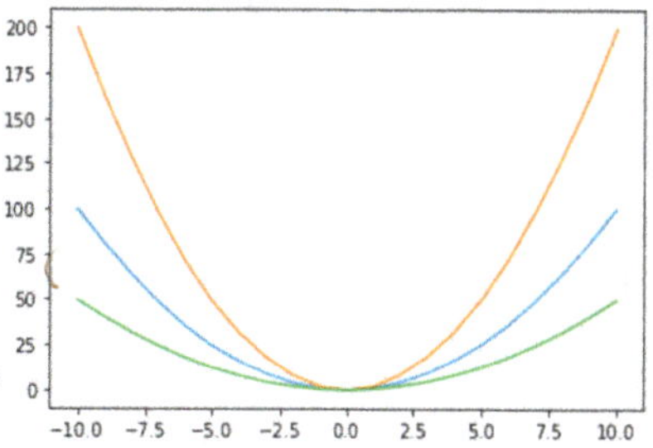

Image 7.3

EXERCISE SET 7.1.2

1. Fill in yellow highlighted code to assign values to y2_list and y3_list.
2. Fill in aqua highlighted code to plot func2 and func3.
3. Add a title to the plot: 'STRETCH AND COMPRESS'. Optional: include x- and y-axes.

We can enhance our drawing by adding a legend. Changes shown below in blue font.

PROGRAM 7.1.2 REVISED

```
## stretch_compress.py
## Draw y=x**2, y=2*x**2, y=(1/2)*x**2
from matplotlib import pyplot as plt
from sympy import Symbol,sympify
x = Symbol('x')
x_list = range(-10,11)
y1_list = [ ]
y2_list = [ ]
y3_list = [ ]
func1 = sympify('x**2')
func2 = sympify('2*x**2')
func3 = sympify('(1/2)*x**2')
y1_list=[func1.subs({x:a}) for x_value in x_list]
[insert code to assign values to y2_list and y3_list ]
plt.plot(x_list,y1_list,label='$y=x^2$')
plt.plot(x_list,y2_list,label='$y=2x^2$')
plt.plot(x_list,y3_list,label='$y=(1/2)x^2$')
ax = plt.subplot()
ax.legend()
plt.show()
```

The $ signs surrounding y = x^2 have the effect of outputting the exponent using a superscript.

Desired Output

Image 7.4

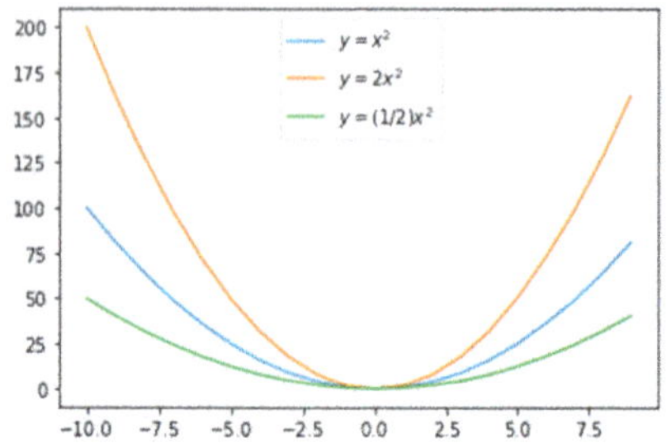

Our algebraic takeaway is as follows:

STRETCH/COMPRESS

If $y = k \cdot f(x)$, k a constant, then

The graph of $y = f(x)$ is **stretched** if $|k| > 1$ (as is $y = 2x^2$ in the graph above)

The graph of $y = f(x)$ is **compressed** if $0 < |k| < 1$ (as is $y = \frac{1}{2}x^2$ in the graph above)

This may be clearer if you observe what happens to the point (1,1) on the graph of $y = x^2$:

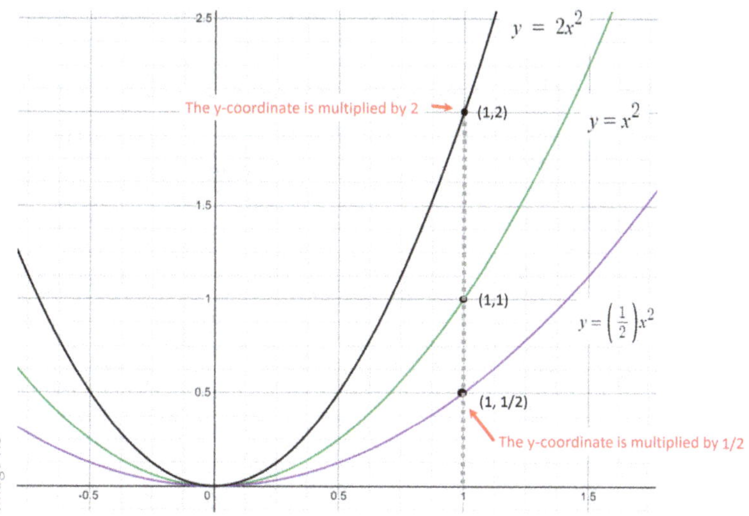

Image 7.5

In fact, the effect of multiplying by a constant is not specific to $y = x^2$. This is true of any function.

It is tempting to assume that stretching makes a graph go "higher" and compressing makes a graph go "lower." But look what happens to the graph of $y = x^3$:

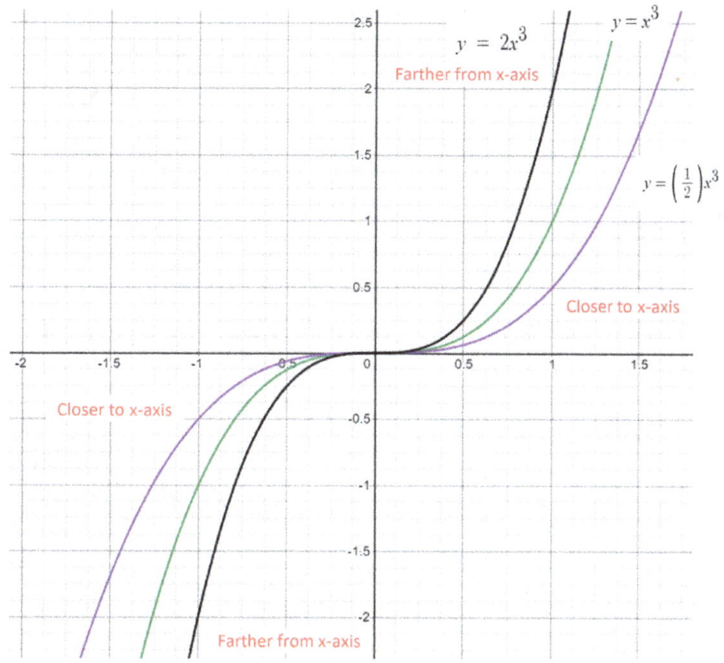

Image 7.6

On the positive side, $y = 2x^3$ is "higher" than $y = x^3$. But on the negative side, $y = 2x^3$ is "lower" than $y = x^3$. So rather than saying that multiplying a function by a constant makes a function "higher" or "lower," think of it as being "farther away" or "closer" to the x-axis.

EXERCISE SET 7.1.3

1. Modify program **stretch_compress** so that the user can input a function that is defined on $[-10, 10]$. Draw the graph of (1) the original function, (2) the function stretched by a factor of 2, and (3) the function compressed by a factor of ½. Include the legend.
2. Sketch the following functions, which we will refer to as **library functions** because they are commonly used in precalculus and calculus. Also identify the domain and range of each function.

   A. **Constant Function:** $y = k$ (k is any constant)

   B. **Identity Function:** $y = x$

   C. **Square Function:** $y = x^2$

   D. **Cube Function:** $y = x^3$

   E. **Absolute Value Function:** $y = |x|$

   F. **Reciprocal Function:** $y = \frac{1}{x}$

   G. **Square Root Function:** $y = \sqrt{x}$

   H. **Cube Root Function:** $y = \sqrt[3]{x}$

3. For each of the library functions in #2, label the given three points.

   A. Constant Function $x = -5, x = 0, x = 2$

   B. Identity Function $x = -5, x = 0, x = 2$

   C. Square Function $x = -2, x = 0, x = 3$

   D. Cube Function: $x = -2, x = 0, x = 2$

   E. Absolute Value Function: $x = -5, x = 0, x = 5$

   F. Reciprocal Function: $x = -1, x = \frac{1}{2}, x = 2$

   G. Square Root Function: $x = 0, x = 4, x = 9$

   H. Cube Root Function: $x = -8, x = 0, x = 1$

7.2 Transformations—Shift

Let's use program **graph_function** to identify some other patterns.

EXERCISE SET 7.2.1

1. Graph the following functions.

   A. $y = x^2 + k$, k any positive constant

   B. $y = x^2 - k$, k any positive constant

   C. $y = (x + k)^2$, k any positive constant

   D. $y = (x - k)^2$, k any positive constant

2. Without using technology, sketch the graphs of the following. Identify the y-intercept of each graph.

   A. $y = x^3$ and $y = x^3 + 2$

   B. $y = \sqrt{x}$ and $y = \sqrt{x + 3}$

   C. $y = |x|$ and $y = 2|x - 1|$

   D. $y = \dfrac{1}{x}$ and $y = \dfrac{1}{2x}$

   E. $y = \sqrt[3]{x}$ and $y = 6 + \sqrt[3]{x}$

   F. $y = \sqrt{x}$ and $y = 2\sqrt{x - 1} + 5$

Our algebraic takeaway is as follows:

SHIFTING (TRANSLATING)

If $y = f(x) + k$, k a positive constant, then the graph of $y = f(x)$ is shifted k units up

If $y = f(x) - k$, k a positive constant, then the graph of $y = f(x)$ is shifted k units down

If $y = f(x + h)$, h a positive constant, then the graph of $y = f(x)$ is shifted h units left

If $y = f(x - h)$, h a positive constant, then the graph of $y = f(x)$ is shifted h units right

Writing Functions from Graphs

Example

Identify the function belonging to this graph.

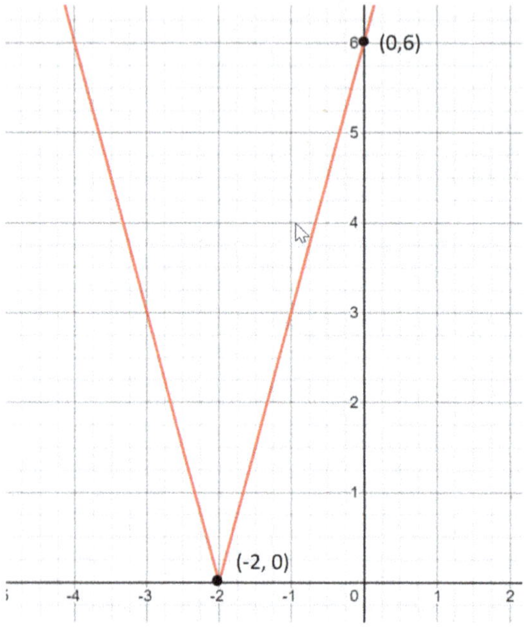

Image 7.7

Solution

On first glance, it appears that this is simply the absolute value function shifted 2 units left:

$$y = |x + 2|$$

However, the y-intercept of that function should be $y(0) = |0 + 2| = 2$, and clearly the y-intercept in the picture is (0, 6). So there is some sort of stretch/compress in action.

One way to determine the stretch/compress is to start with the function $y = a|x + 2|$ (where a is the unknown stretch/compress factor) and then substitute in the y-intercept (0, 6) to get:

$$6 = a|0 + 2|$$
$$6 = 2a$$
$$3 = a$$

Thus, we have a stretch factor of 3 and the equation is:

$$y = 3|x + 2|.$$

EXERCISE SET 7.2.2

1. Write the function that produces each of the following graphs.

A.

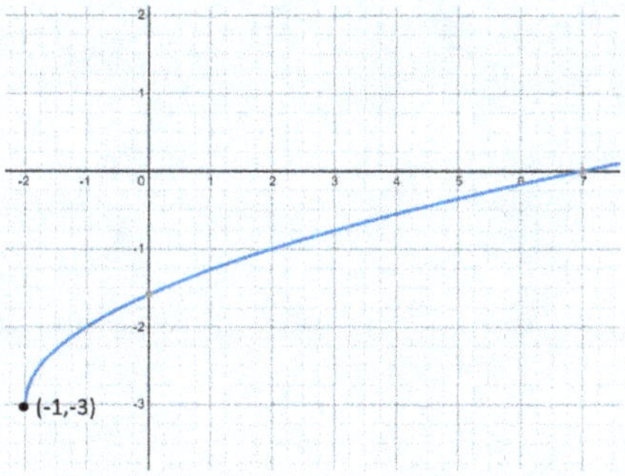

Image 7.8

B.

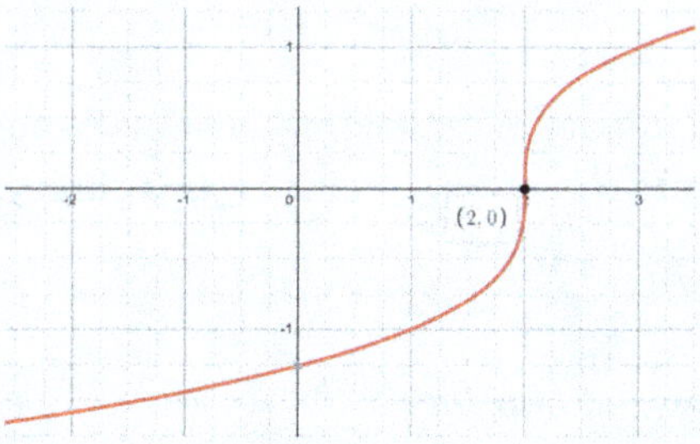

Image 7.9

C.

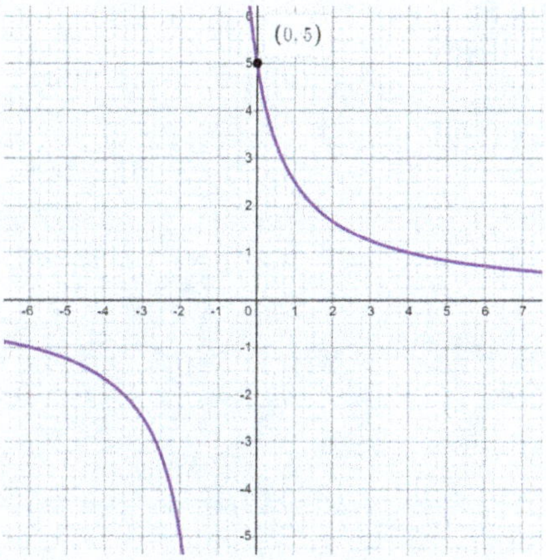

Image 7.10

D.

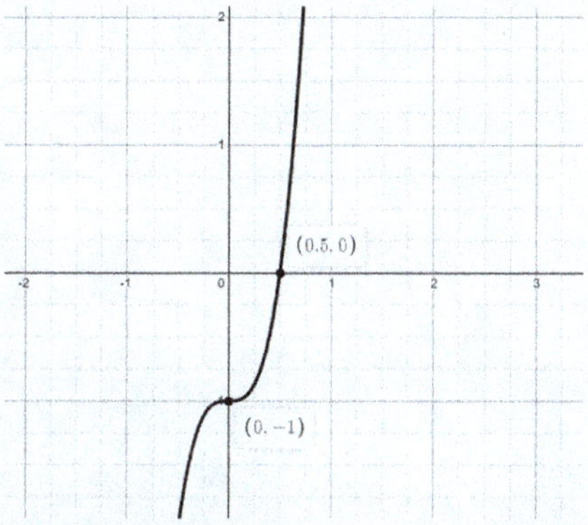

Image 7.11

Handling Difficult Domains

What happens when you try to run program **graph_function** with an input of **sqrt**(x)?

You will probably end up with an empty graph and a series of error messages culminating in:

TypeError: can't convert complex to float

This is because the domain of $y = \sqrt{x}$ is $x \geq 0$, but the program initializes the domain to:

x_list = **range**(–10,11)

When you try to take the square root of a negative number, you end up with an imaginary (complex) number (which we will discuss further in Chapter 8).

An easy fix would be to change the above statement to:

x_list = **range**(0,11)

However, we'd like the program to do the work of recognizing and avoiding imaginary numbers for us. To do so, we'll check if a y_value is imaginary using **SymPy** function y_value.**is_imaginary.**

PROGRAM 7.1.1 GRAPH_FUNCTION.PY REVISED

```
from matplotlib import pyplot as plt
from sympy import *
def draw_axes(x_list,y_list):
    x_min = min(x_list)
    x_max = max(x_list)
    y_min = min(y_list)
    y_max = max(y_list)
    plt.plot([x_min,x_max],[0,0],c='k')
    plt.plot([0,0],[y_min,y_max],c='k')
    return

# main program
x_list = range(-10,11)
x = Symbol('x')
func1 = sympify(input("Function? "))
y1_list = [ ]
domain = [ ]
for a in x_list:
    y_value = func1.subs({x:a})
    if (not y_value.is_imaginary):
        domain.append(a)
        y1_list.append(y_value)
plt.plot(domain,y1_list)
draw_axes(x_list,y1_list)
plt.show()
```

Desired Output

Function? sqrt(*x*)

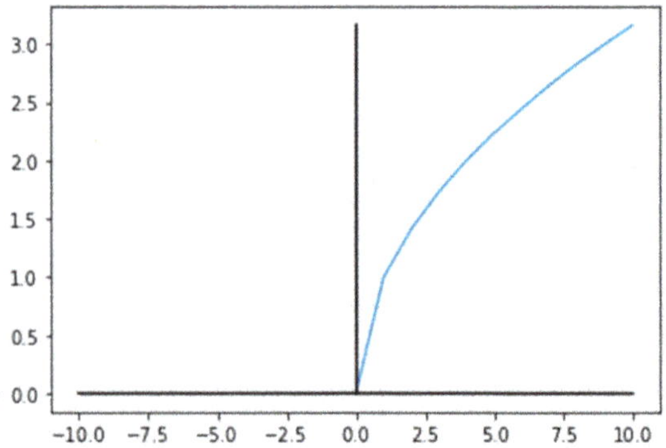

Image 7.12

EXERCISE SET 7.2.3

1. Modify the program so that an additional function can be input and graphed.

   Sample Output

   Function 1? sqrt(*x*)

   Function 2? sqrt(*x* + 3)

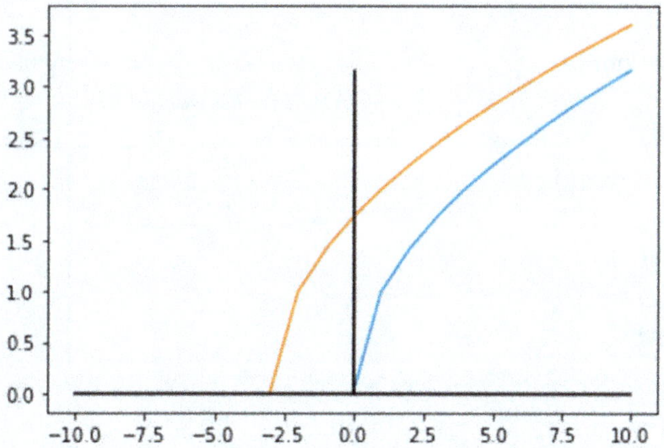

Image 7.13

Writing a program to shift a function up or down is fairly straightforward; one only needs to append "+*k*" or "–*k*" to the function.

Program Goal: Write a program that will ask the user to input a function (any function that can be handled by the **SymPy** library) and a vertical translation. Graph the original function and the translation. Include the legend.

PROGRAM 7.2.1 GRAPH_TRANSLATE.PY

```
## graph_translate.py
## Input: Algebraic Function, Vertical translation
## Output: Graph of original function and translation
from matplotlib import pyplot as plt
from sympy import Symbol,sympify
x = Symbol('x')
x_list = range(−10,11)
func_str1 = input('Function? ')
func1 = sympify(func_str1)
k = input("Vertical Translation? ")
k = int(k)                                    # +k will be converted to k
if k>0:
    func_str2 = func_str1+'+'+str(k)
elif k<0:
    func_str2 = func_str1+str(k)
func2 = sympify(func_str2)
fig = plt.figure()
ax = plt.subplot()
plt.axis([−10,10, −15,15])
y1_list = [ ]
y2_list = [ ]
y1_list=[func1.subs({x:a}) for a in x_list]
y2_list=[func2.subs({x:a}) for a in x_list]
plt.plot(x_list,y1_list,label=func1)
plt.plot(x_list,y2_list,label=func2)
ax.legend()
plt.show()
```

Sample Output

Function? x**3
Vertical Translation? +4

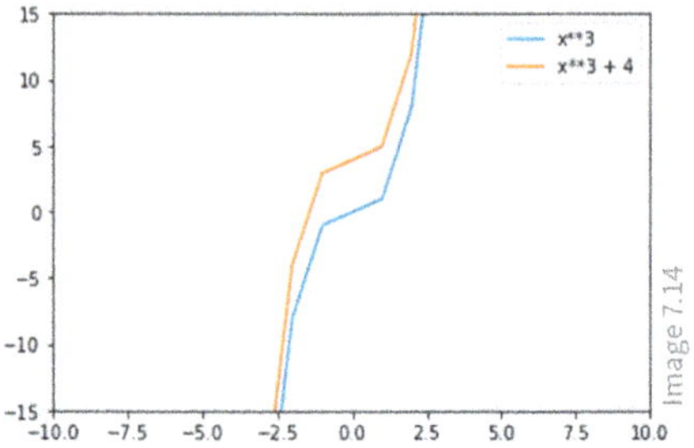

Image 7.14

However, these graphs do not look exactly "smooth." We saw in Chapter 6 that when we try to graph $y = x^3$ over $[-10, 0] \times [-15, 15]$, only 5 points are shown. When Python "connects the dots," we get a function that looks like it's composed of line segments.

Unfortunately the **range**() function does not allow fractional increments such as **range**(−10,10,0.1). But there is a workaround. Replace x_list = **range**(−10,10) with the following list comprehension:

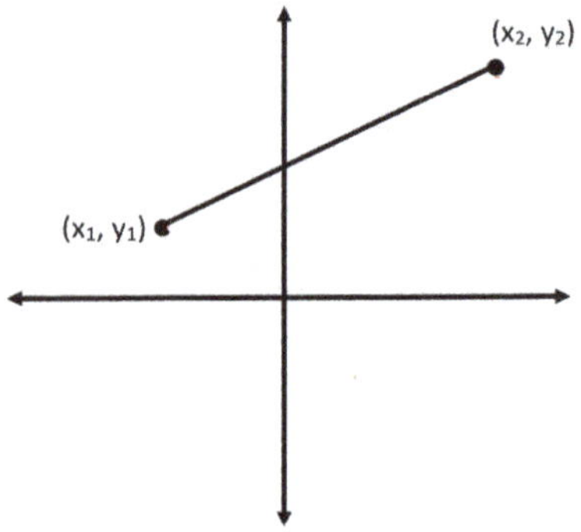

Image 7.15

```
x_list = [i / 10 for i in range(−100,100)]
```

The result is a pair of far "smoother" graphs:

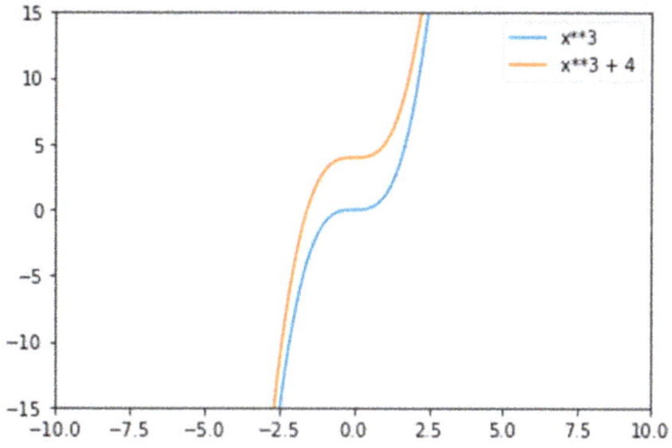

Image 7.16

EXERCISE SET 7.2.4

1. What is the resulting list, and how many elements does it have, when we define x_list using:

   x_list = [*i* / 10 **for** *i* **in range**(−100,100)]

2. When we graph $y = x^3$ using x_list as defined in #1 above, which points are actually plotted if the axes are $[-10, 10, 1] \times [-15, 15, 1]$?
3. Does a vertical shift affect the number of points that are plotted for a function?
4. If you would like to plot points within the domain $[-5, 5]$, then what range should you set for the y-axis?

7.3 Transformations—Reflect

Reflecting Points

Given point (x, y):

Reflection through x-axis	Reflection through y-axis	Reflection through Origin
Reflected point is $(x, -y)$	Reflected point is $(-x, y)$	Reflected point is $(-x, -y)$

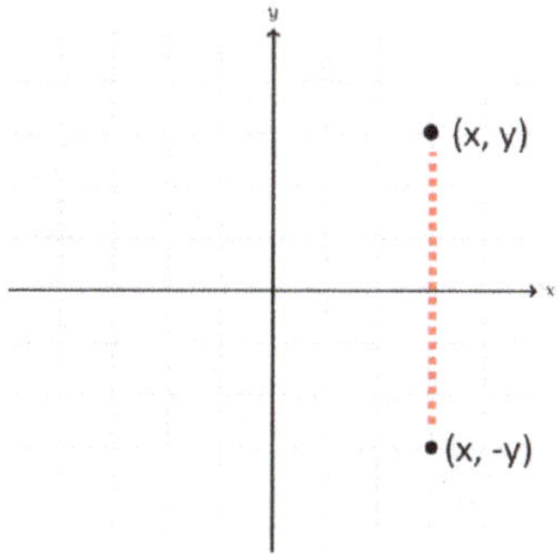

Image 7.17

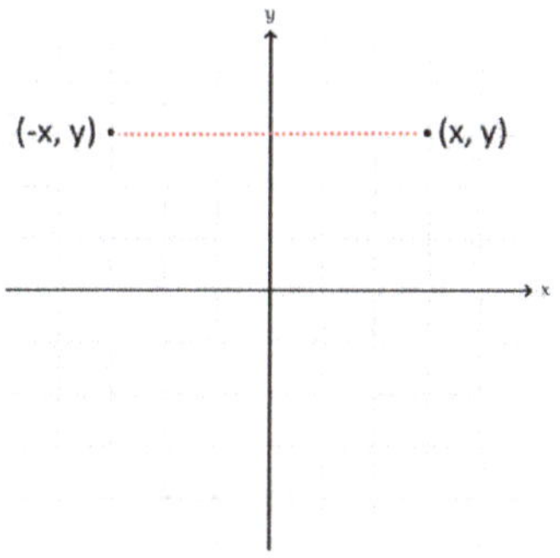

Image 7.18

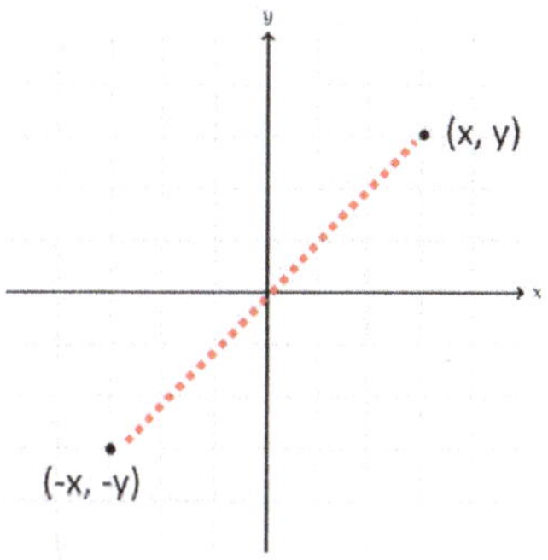

Image 7.19

Program Goal: Write a program that asks the user to input an ordered pair (x, y). Output and identify the coordinates of the 3 reflected points.

SAMPLE OUTPUT

Ordered pair x, y? 3, −2

Reflection through x-axis: 3, 2

Reflection through y-axis: −3, −2

Reflection through origin: −3, 2

PROGRAM 7.3.1 REFLECT_POINT.PY

```
## reflect_point.py
## Input: ordered pair x,y
## Output: coordinates of point reflected through x-axis, y-axis, origin
pair = input('Ordered pair x,y? ')
x,y = [int(i) for i in pair.split(',')]
print('Reflection through x-axis:', x, -y)
    [ Insert code to output the reflections through the y-axis and origin ]
plt.show()
```

EXERCISE SET 7.3.1

1. Fill in the yellow highlighted code to output the reflected point through the y-axis and origin.
2. Given triangle with vertices (−2, 2), (−5, 3) and (−3, 7), draw its reflection:
   A. through the x-axis
   B. through the y-axis
   C. through the origin
3. Write a program that draws the triangle in question 2 along with its three reflections. Use four different colors.

Reflecting Graphs

We will consider two types of reflections:

Reflection through the x-axis $-f(x)$ is the reflection of $f(x)$ through the x-axis. Examples: $y_1 = x^2$ and $y_2 = -x^2$ (shown to right) $y_1 = \sqrt{x}$ and $y_2 = -\sqrt{x}$	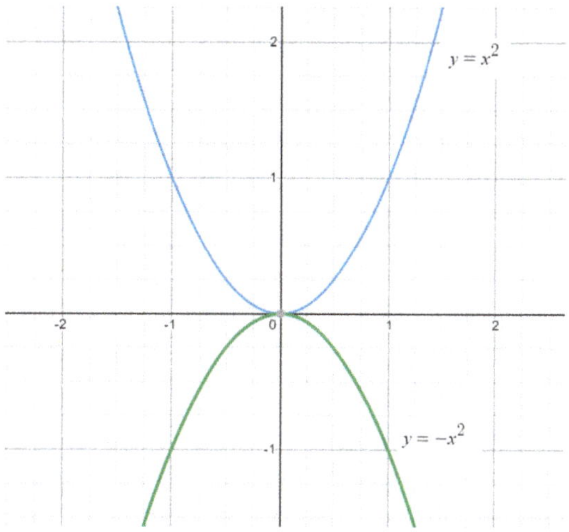 Image 7.20
Reflection through the y-axis $f(-x)$ is the reflection of $f(x)$ through the y-axis. Examples: $y_1 = \sqrt{x}$ and $y_2 = \sqrt{-x}$ (shown to right) $y_1 = \frac{1}{x}$ and $y_2 = -\frac{1}{x}$	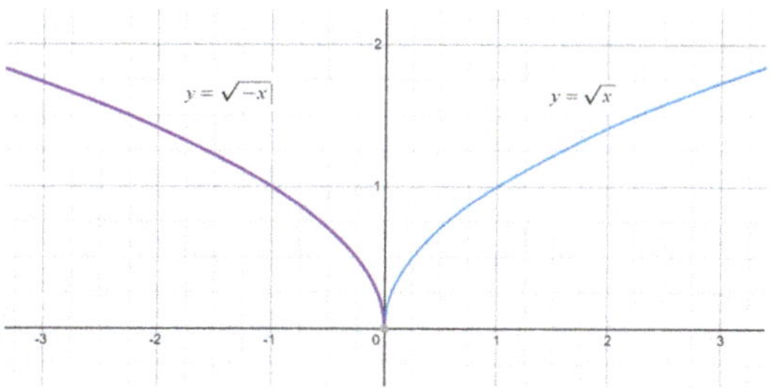 Image 7.21

When combining shifting, reflecting, and stretching/compressing, it is recommended you perform the transformations in this order:

1. Stretch/Compress
2. Reflect
3. Shift

Example

Sketch the graph of $y = \frac{3}{x-2} + 1$

Solution

Step 1: Identify the "starting" function, which is $y = \frac{1}{x}$

Step 2: Rewrite as $y = 3 \cdot \frac{1}{x-2} + 1$

Step 3: The 3 in front of $\frac{1}{x-2}$ indicates that there is a stretch factor of 3.

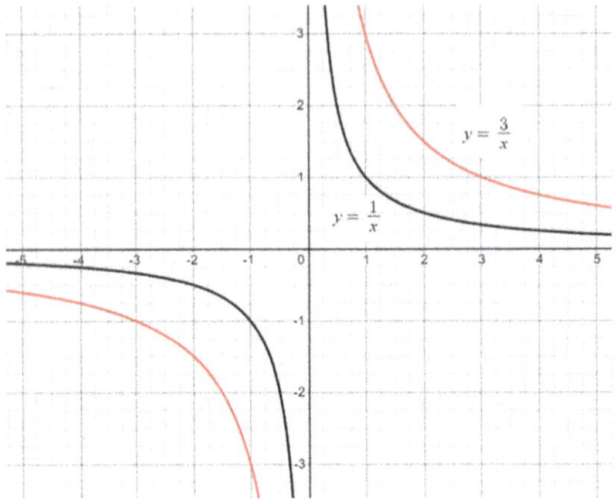

Image 7.22

Step 4: There is no reflection.

Step 5: "$x - 2$" in the denominator indicates you shift 2 units to the right. The "+1" at the end of the function indicates shift 1 unit up. Note the shift right results in the vertical asymptote moving from $x = 0$ to $x = 2$.

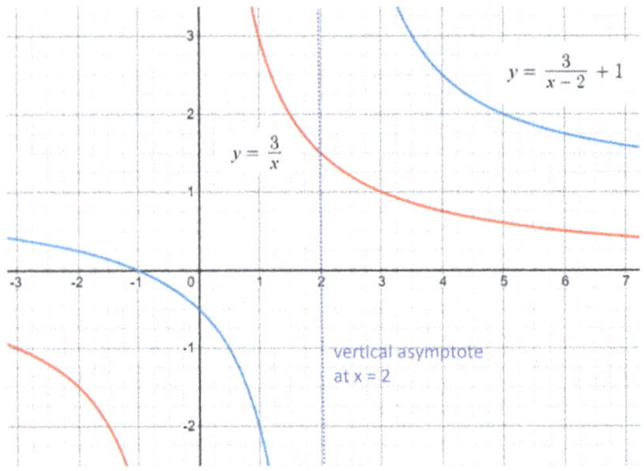

Image 7.23

EXERCISE SET 7.3.2

Sketch the following graphs.

1. $y = -x^3$
2. $y = -|x|$
3. $y = \sqrt{-x} - 5$
4. $y = 2\sqrt[3]{x-4}$
5. $f(x) = -(3 - x)^2 + 2$
6. $y = \sqrt{2 - x}$ (*hint*: convert radicand to $-(x - 2)$)

7. Write the function that produces each of the following graphs. Use the indicated points to determine if there is a stretch/compress.

A.

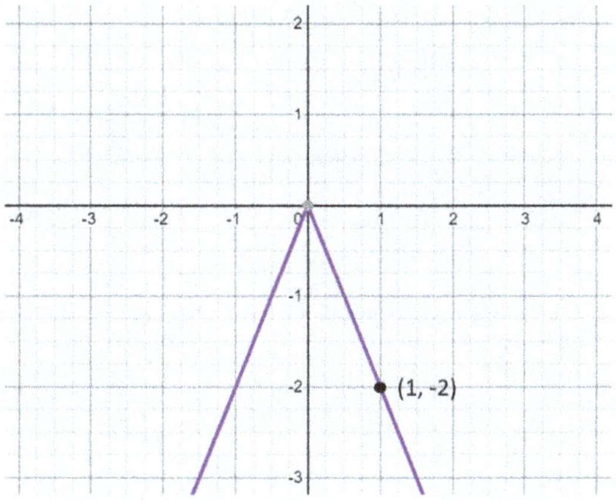

Image 7.24

B.

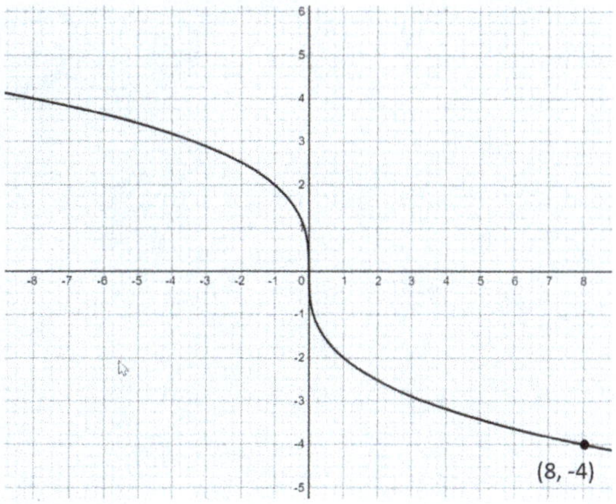

Image 7.25

C.

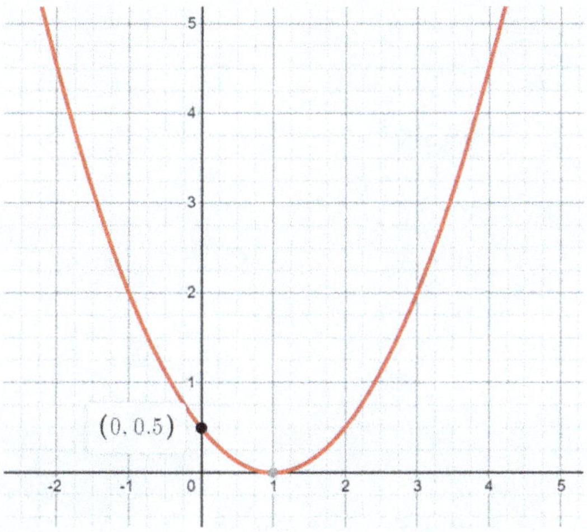

Image 7.26

D.

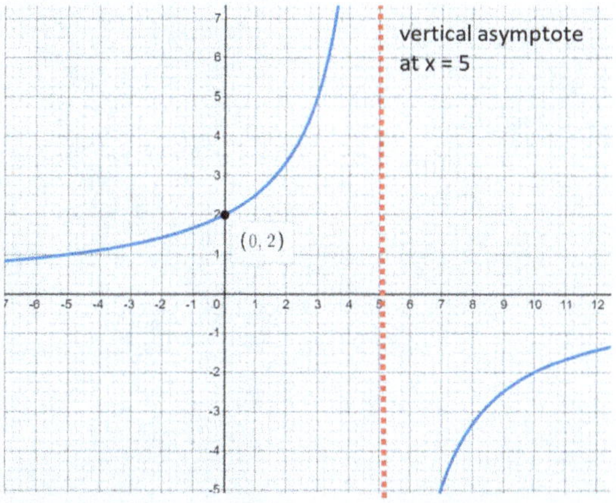

Image 7.27

We can reflect a graph through the x-axis by simply appending a negative sign to the front of the function. Reflect a graph through the y-axis by replacing x with $(-x)$.

For example, to reflect $f(x) = \sqrt{x+1}$ through the y-axis, replace x with $(-x)$ to get $f(x) = \sqrt{(-x)+1}$.

Example

Given $f(x) = -2|x + 4|$:

A. Graph the function

B. Identify and sketch the graph of the function reflected through the y-axis

C. Identify and sketch the graph of the function reflected through the x-axis

Solution

A. $f(x) = -2|x + 4|$ is the absolute value function stretched by a factor of 2, reflected through the x-axis, and translated 4 units left.

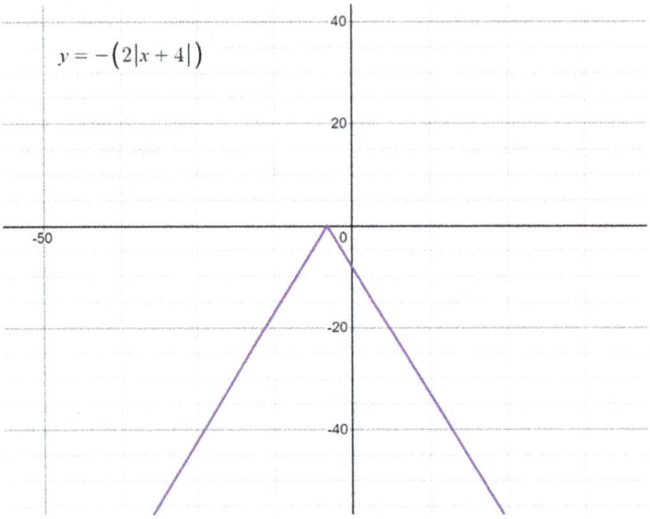

Image 7.28

B. To reflect through the y-axis, replace x with $-x$:

$$g(x) = -2|-x + 4|$$

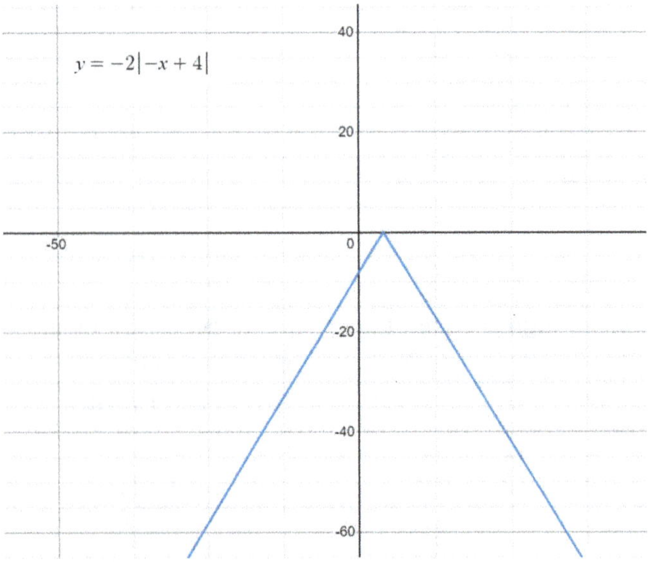

Image 7.29

C. To reflect through the x-axis, append a negative sign to the front of the function:

$$h(x) = -(-2|x + 4|) = 2|x + 4|.$$

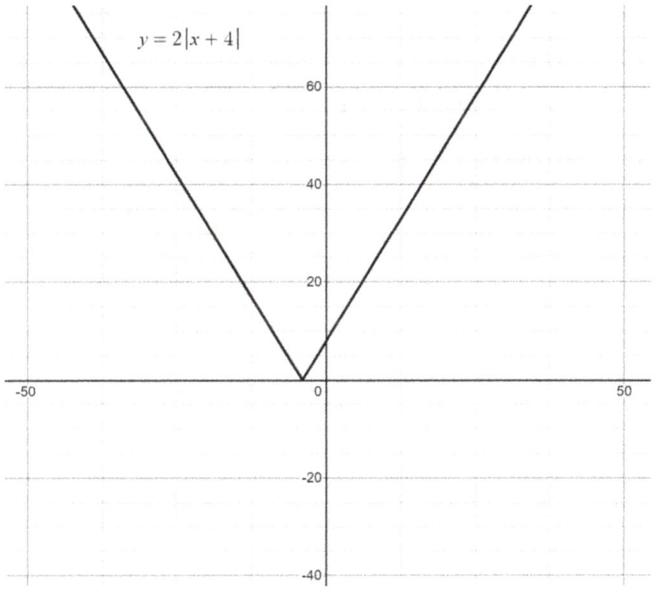

Image 7.30

EXERCISE SET 7.3.3

For each of the functions in #1–4:

A. Graph the function

B. Identify and sketch the graph of the function reflected through the y-axis

C. Identify and sketch the graph of the function reflected through the x-axis

1. $f(x) = (x + 4)^3 - 5$
2. $f(x) = \sqrt[3]{x - 4}$
3. $f(x) = \sqrt{2 - x}$
4. $f(x) = \dfrac{3}{x + 7}$
5. Write a program **graph_reflect** that (A) asks user to input a function and (B) graphs the given function, also its reflections through the x- and y-axes.

   *(Hint: can you adapt program **graph_translate**?)*

7.4 Symmetry

Most of the letters of the English alphabet have either *x*-axis (vertical) symmetry, *y*-axis (horizontal) symmetry, or origin (diagonal) symmetry. Also, corporate logos and much more!

Image 7.31

Likewise, a graph can possibly have three different kinds of symmetry:

y-axis symmetry

For each point (x, y) on the graph, there is also a point $(-x, y)$ on the graph.

Or: $f(-x) = f(x)$

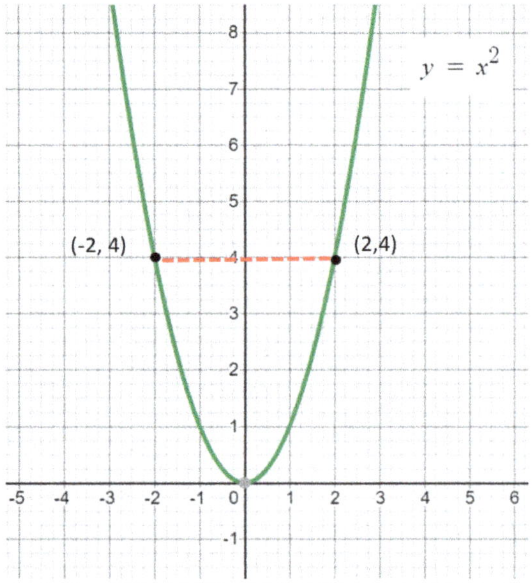

Image 7.32

x-axis symmetry

For each point (x, y) on the graph, there is also a point $(x, -y)$ on the graph.

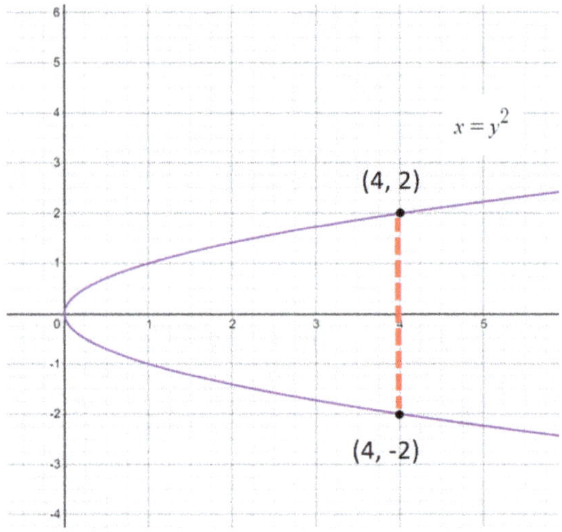

Image 7.33

Origin symmetry For each point (x, y) on the graph, there is also a point $(-x, -y)$ on the graph. Or: $f(-x) = -f(x)$	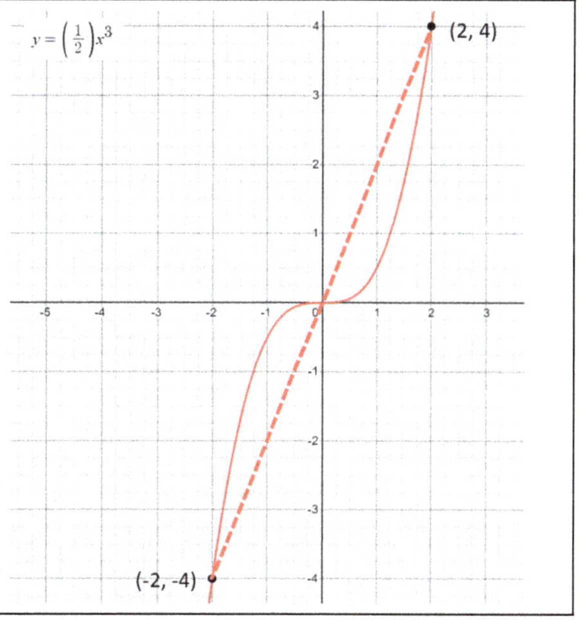 Image 7.34

EXERCISE SET 7.4.1

Identify which type of symmetry each graph has, if any.

1. $y = \sqrt{x}$
2. $y = x$
3. $y = \sqrt{25 - x^2}$
4. $y = \sqrt[3]{x}$
5. $y = |x|$
6. $y = \dfrac{1}{x}$
7. 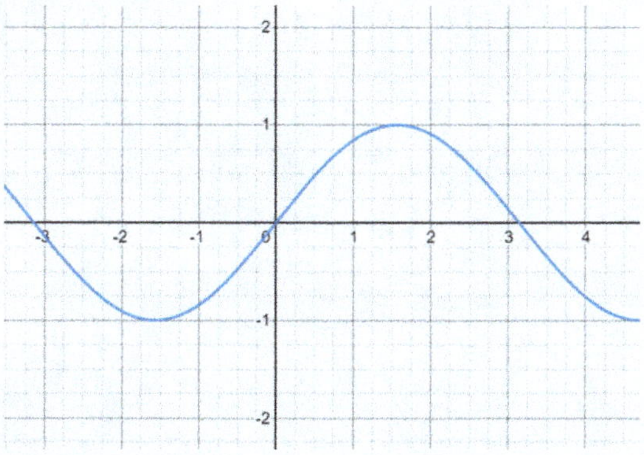

Image 7.35

8. 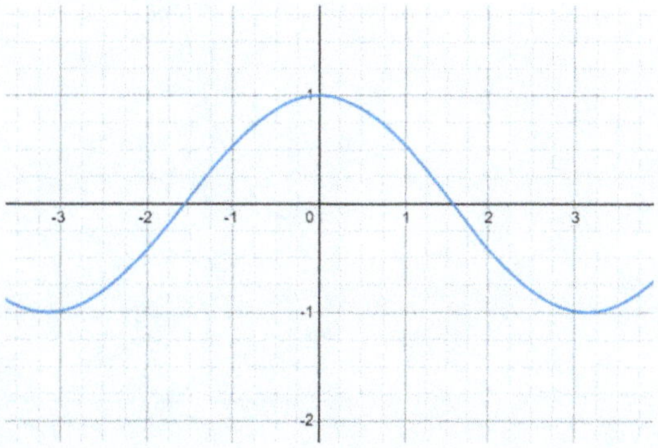

Image 7.36

(*Continued*)

9.

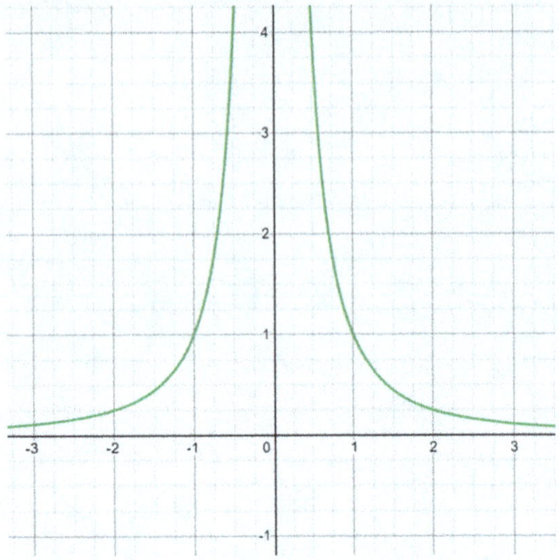

Image 7.37

10.

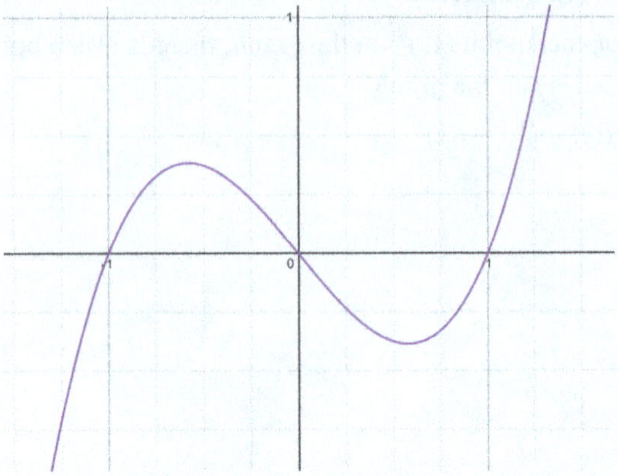

Image 7.38

11.Can a graph with x-axis symmetry be considered a function? Explain.

Identifying Odd and Even Functions

An **even function** has symmetry with respect to the y-axis. If you do not have access to the graph, you can identify whether a function is even by figuring out whether $f(-x) = f(x)$.

Examples of Even Functions

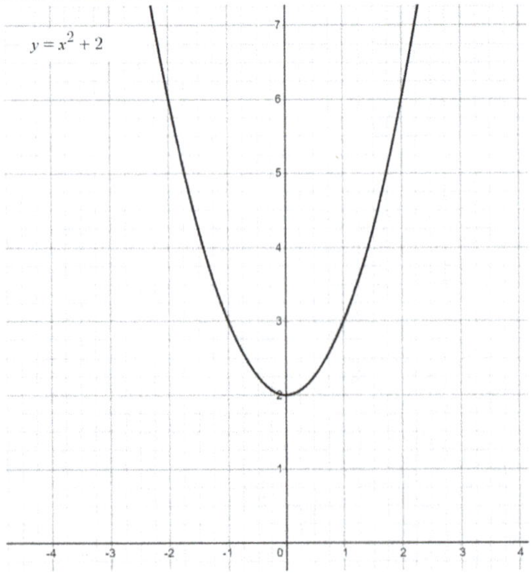

Image 7.39

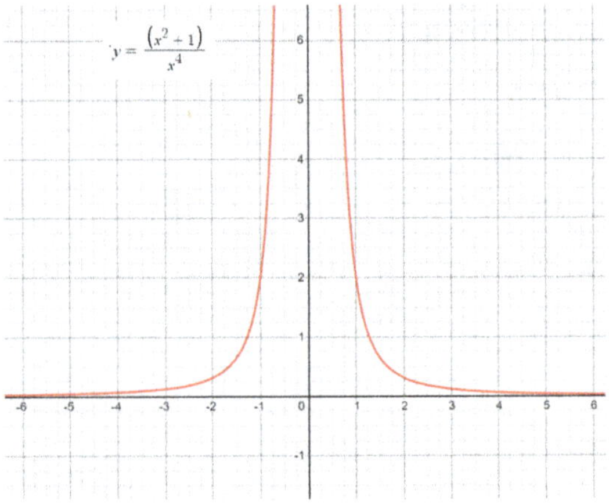

Image 7.40

An **odd function** has symmetry with respect to the origin. If you do not have access to the graph, you can identify whether a function is odd by figuring out whether $f(-x) = -f(x)$.

Examples of Odd Functions

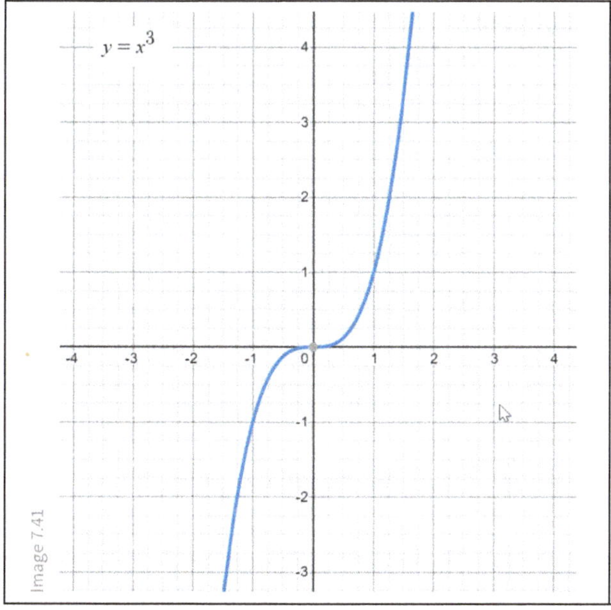

Image 7.41

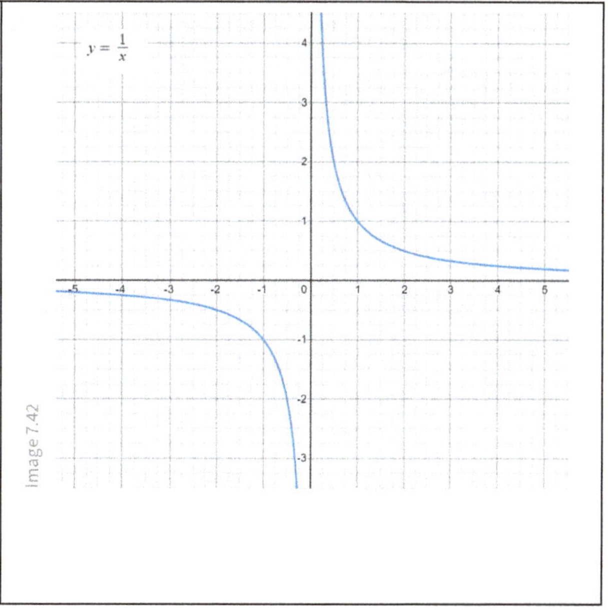

Image 7.42

Example

Identify whether the following functions are odd, even, or neither.

A. $f(x) = 2x^4 - x^2$

B. $f(x) = \dfrac{2x}{|x|}$

C. $f(x) = 3x^3 + 5x^2 + 6$

Solutions

Replace each x with "$-x$":

A. $f(-x) = 2(-x)^4 - (-x)^2 = 2x^4 - x^2$
$= f(x)$
$f(x)$ is even because $f(-x) = f(x)$.

B. $f(x) = \dfrac{2x}{|x|}$

$$f(-x) = \frac{2(-x)}{|-x|} = -\frac{2x}{|x|}$$

$= -f(x)$
$f(x)$ is odd because $f(-x) = -f(x)$.

C. $f(x) = 3x^3 + 5x^2 + 6$
$f(-x) = 3(-x)^3 + 5(-x)^2 + 6 = -3x^3 + 5x^2 + 6$
$f(-x)$ does not equal $f(x)$ or $-f(x)$ so $f(x)$ is neither even nor odd.

EXERCISE SET 7.4.2

Determine algebraically whether the following functions are even, odd, or neither.

1. $f(x) = \dfrac{x^2}{x^4 + 1}$
2. $f(x) = -5x^3 + x$
3. $f(x) = x + |x|$
4. $f(x) = -\dfrac{x^3}{3x^2 - 9}$
5. Is every graph with y-axis symmetry a function? Explain.
6. Is every graph with origin symmetry a function? Explain.

Program Goal: Write a program that asks the user to input a function. Find and output $f(-x)$.

SAMPLE OUTPUT

Function? 3*x**3+sqrt(x)

$f(x) =$

$\sqrt{x} + 3 \cdot x^3$

$f(-x) =$

$-3 \cdot x^3 + \sqrt{-x}$

PROGRAM 7.4.1 F(-X).PY

```
## f(-x).py
## Input: function
## Output: original function f(x) and f(−x)
from sympy import *
x = Symbol('x')
f = input('Function? ')
f2 = f.replace('x','(-x)')
f = sympify(f)
print('f(x)=')
pprint(f)
    [ Insert code to output 'f(-x) =' followed by pretty print of f(−x) ]
```

EXERCISE SET 7.4.3

1. Fill in the yellow highlighted code to output '$f(-x)=$' followed by pretty print of $f(-x)$.
2. Modify the program to determine if the input function is odd, even, or neither by comparing $f(x)$ to $f(-x)$.

7.5 Increasing, Decreasing, Extrema

Let's go back to the roller coaster analogy for graphs of polynomial functions. Recall that a local or absolute maximum occurs when the car is at the peak, ready to plummet downward. A local or absolute minimum occurs when the car is in a valley, ready to zoom upward.

It is natural to extend this analogy and describe a function as **increasing** when the car is traveling upward and **decreasing** when the car is traveling downward.

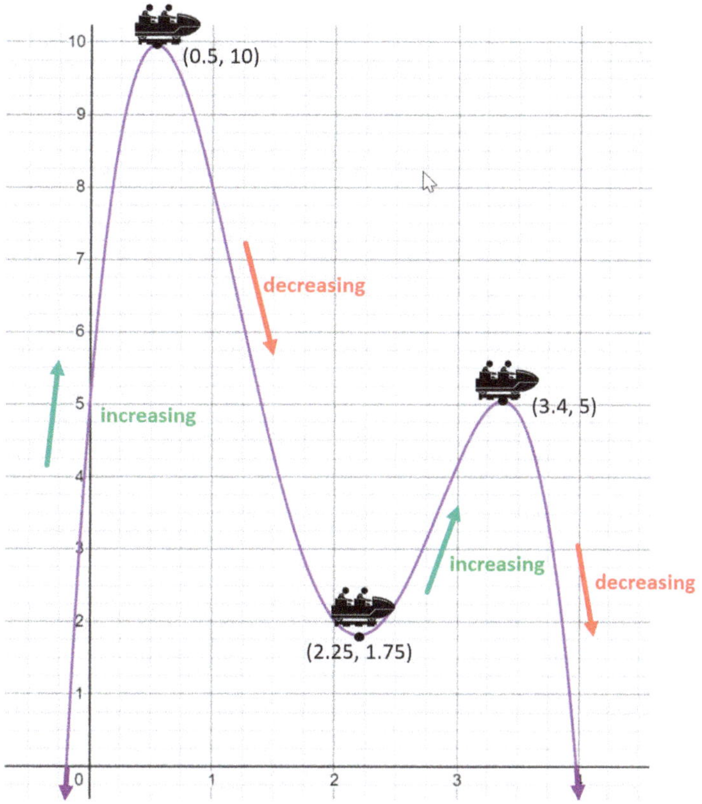

Image 7.43

When the car departs or returns to the station, it is on a level stretch of track that we call **constant**.

Image 7.44

More formally, the definitions are as follows.

A function is **increasing** on interval (a, b) if for every pair of points x_1 and x_2 in (a, b), if $x_1 < x_2$, then $f(x_1) < f(x_2)$.	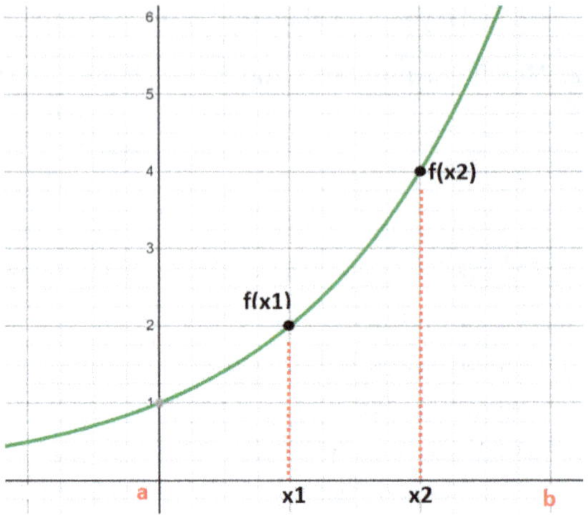 Image 7.45
A function is **decreasing** on interval (a, b) if for every pair of points x_1 and x_2 in (a, b), if $x_1 < x_2$, then $f(x_1) > f(x_2)$.	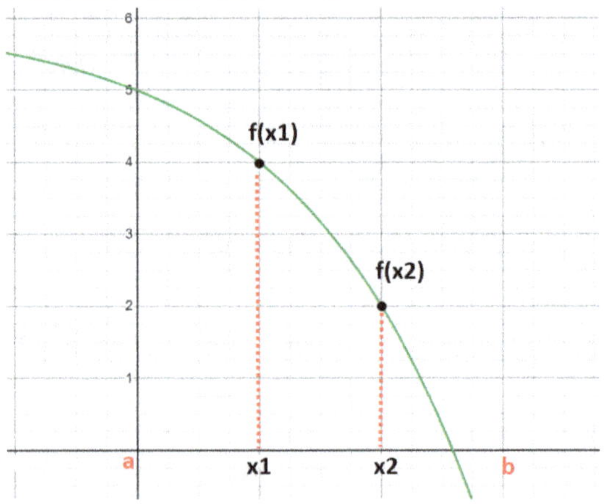 Image 7.46
A function is **constant** on interval (a, b) if for any pair of points x_1 and x_2 in (a, b), $f(x_1) = f(x_2)$.	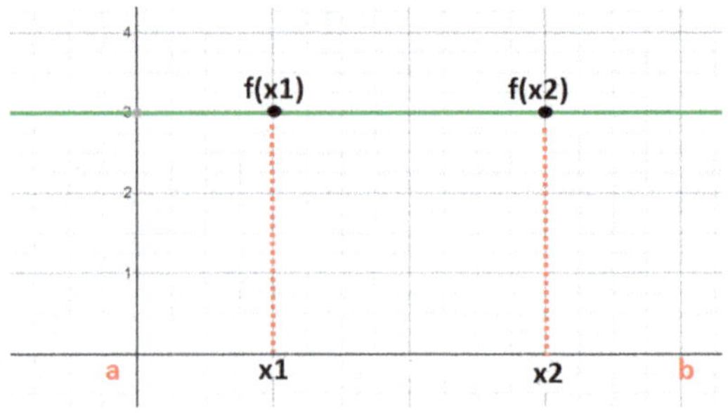 Image 7.47

Absolute and Local Extrema

Recall the discussion of absolute and local extrema from Section 3.9. A local max or local min cannot occur at an endpoint.

We can redefine a **local maximum** as the point $f(M)$ at which a function changes from increasing to decreasing, and a **local minimum** as the point $f(m)$ at which a function changes from decreasing to increasing.

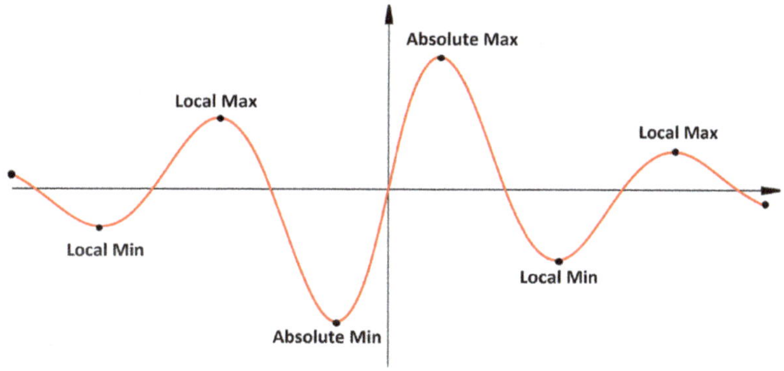

Image 7.48

Example

Identify the intervals on which the function is increasing, decreasing, or constant; and any local maxima or minima.

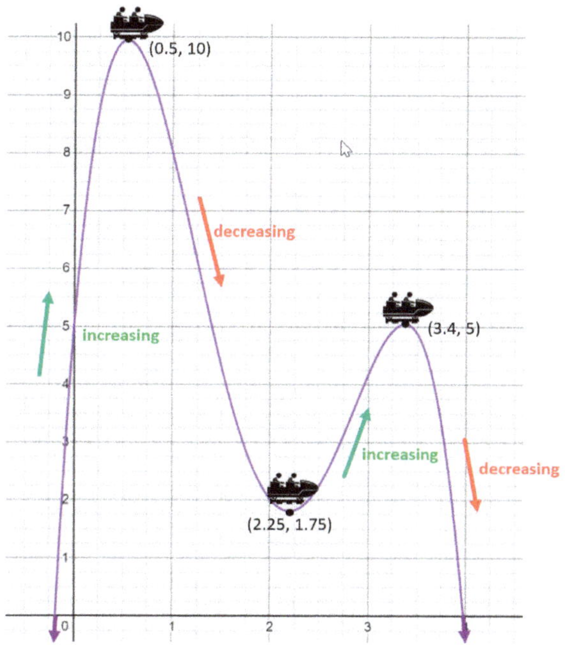

Image 7.43

Solution

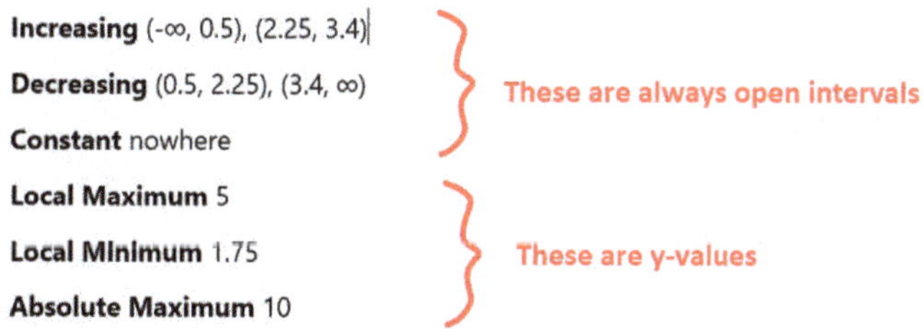

Increasing (-∞, 0.5), (2.25, 3.4)

Decreasing (0.5, 2.25), (3.4, ∞)

Constant nowhere

These are always open intervals

Local Maximum 5

Local Minimum 1.75

Absolute Maximum 10

These are y-values

Image 7.49

EXERCISE SET 7.5.1

For each function in #1–7, identify increasing, decreasing, and constant intervals; also any local maxima and minima.

1.

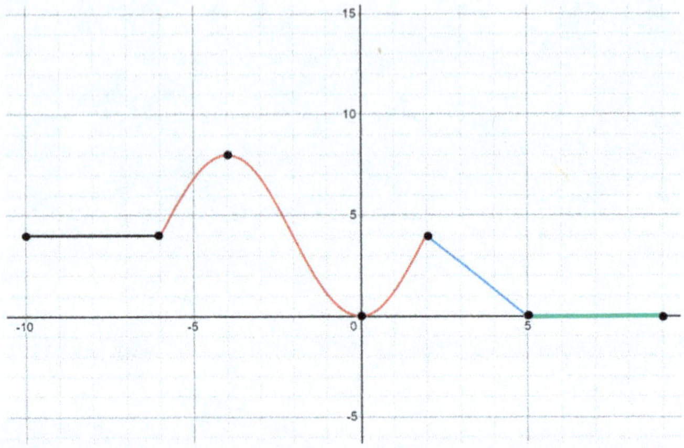

Image 7.50

2.

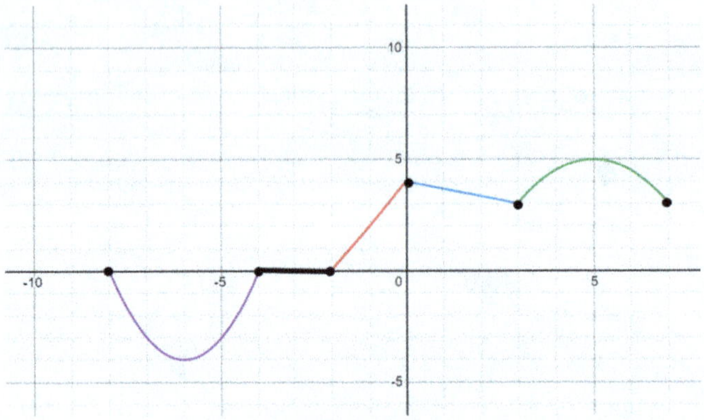

Image 7.51

3.

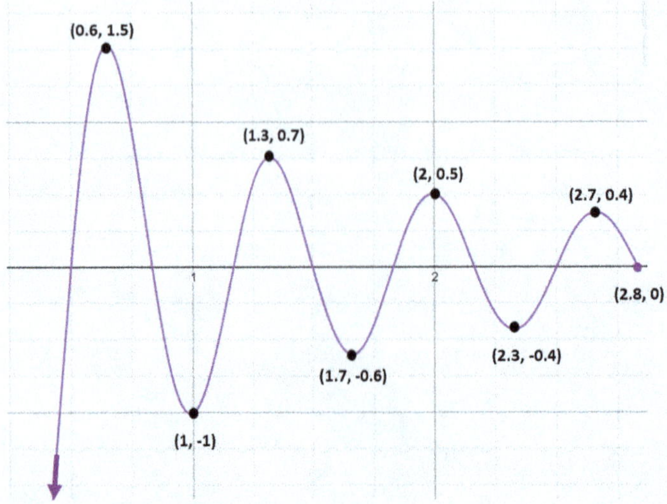

Image 7.52

4. $y = x^3$
5. $y = \frac{1}{x}$
6. $y = \sqrt{x}$
7. $y = (x - 2)^2$
8. True or False:
   A. The graph of a function can have a local maximum or absolute minimum at an endpoint.
   B. The graph of a function can have an absolute maximum or absolute minimum at an endpoint.
   C. A point on a graph can be both an absolute and local maximum.
   D. The graph of a function can have more than one absolute maximum or absolute minimum.
   E. The graph of a function can be neither increasing nor decreasing.
   F. Increasing, decreasing, and constant intervals are open intervals of x-values.
   G. A function is neither increasing nor decreasing at a maximum or minimum point.
   H. The graph of every function has both increasing and decreasing intervals.
9. Write a program that asks the user to input any polynomial function f. Scale the axes at $(-10, 10) \times (-100, 100)$. Generate two random float numbers a and b in the interval $[-10, 10]$. Determine if the function is increasing or decreasing from $(a, f(a))$ to $(b, f(b))$. Is this a reliable way to determine if the function is increasing/decreasing over that entire interval (a, b)?

Program Goal: Calculate the absolute minimum and maximum of a function over domain [–10, 10].

PROGRAM 7.5.1 EXTREMA.PY

```
## extrema.py
## input: function
## output: absolute max and absolute min over the domain [-10, 10].
from sympy import Symbol,sympify
x = Symbol('x')
f = sympify(input('Function? '))
x_list = range(-10,11)
y_list = [f.subs({x:x_value}) for x_value in x_list]
y_min = min(y_list)
print('Absolute minimum:',y_min)
y_max = max(y_list)
print('Absolute maximum:',y_max)
```

The limitation of this program is that it only calculates the absolute max and min of a discrete set of integer values in [−10, 10], which may or may not be the actual absolute extrema. (Chapter Project I uses calculus to calculate the actual extrema.)

EXERCISE SET 7.5.2

1. Run the program using $f(x) = -(x - 0.5)^2$ as input.
   A. What does the program identify as the absolute max?
   B. What is the actual absolute max? (Hint: graph, or calculate $f(0.5)$.)
2. Modify the program so that the user can enter the domain rather than using [−10, 10].
3. Does every function continuous on closed interval $[a, b]$ have both an absolute maximum and absolute minimum in $[a, b]$?

Let's say we want to modify program **extrema** so that it outputs the full ordered pair of the extrema point rather than just the y-value. This involves a "reverse lookup." First, a simple example of a reverse lookup:

```
>>>pets = ['dog', 'cat', 'bird', 'snake']
>>>names = ['Spot', 'Fluffy', 'Squeaky', 'Sneaky']
>>> names.index('Fluffy')
1
>>>pets[1]
'cat'
```

Notice the **.index**() method returns the position of 'Fluffy' in the names list. (Recall that list indices start with 0.) We then use that same position to locate Fluffy in the pets list.

If we were to do the same with x_list and y_list, assume we have ordered pairs (1, 10), (2, 20), (3, 30), (4, 20), and (5, 10), and we want to identify the ordered pair of the maximum.

```
>>>x_list = [1, 2, 3, 4, 5]
>>>y_list = [10, 20, 30, 15, 5]
>>>y_max = max(y_list)            ← This returns 30
>>>k = y_list.index(y_max)        ← This returns 2, because 30 is y_list[2]
>>> x_list[k]
3
```

Thus, the x-coordinate belonging to $y = 30$ is $x = 3$. The ordered pair is (3, 30). If the x_list has more than one x_value corresponding to the given y_value, then it will only return the first instance.

We'll place this code in a function and add it to program **extrema**.

PROGRAM 7.5.1 EXTREMA.PY REVISED

```
## extrema.py
## input: function
## output: absolute max and absolute min over the domain [−10, 10].
from sympy import Symbol,sympify
def get_x(y):
    k = y_list.index(y)
    return x_list[k]
x = Symbol('x')
f = sympify(input('Function? '))
x_list = range(−10,11)
y_list = [f.subs({x:x_value}) for x_value in x_list]
y_min = min(y_list)
x_min = get_x(y_min)
print('Absolute minimum:',x_min, ',', y_min)
    [ insert code to calculate and output ordered pair of absolute maximum ]
```

EXERCISE SET 7.5.3

1. Fill in the yellow highlighted code to calculate and output the ordered pair of the absolute maximum.
2. Use the program to find the ordered pairs of the absolute extrema of the following functions on the domain [−10, 10]:

   A. $y = -3x^3 + 4x - 11$

   B. $y = x^4 + x^3$
3. Modify the program so that it graphs the given function and labels the absolute maximum and absolute minimum on the graph with the ordered pairs.

Sample Output

Function? −x**3

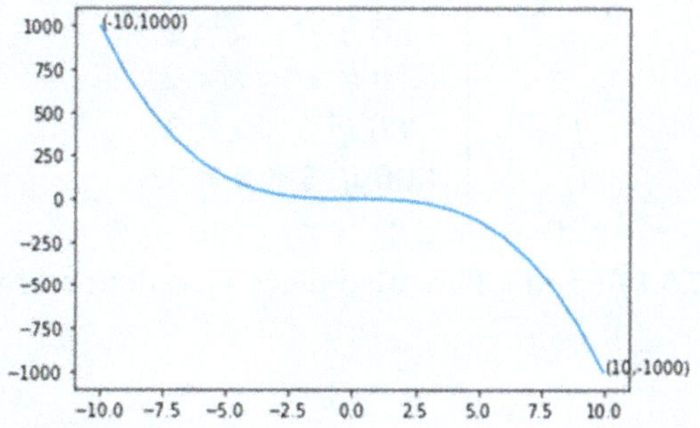

Image 7.53

7.6 Piecewise-Defined Functions

Some businesses (and government agencies) offer staggered pricing. For example, the 2019 postage rates for first-class letters were as follows:

Up to 1 oz.	.55
Between 1 and 2 oz.	.70
Between 2 and 3 oz.	.85
Between 3 and 3.5 oz.	1.00

etc.

The graph looks like this:

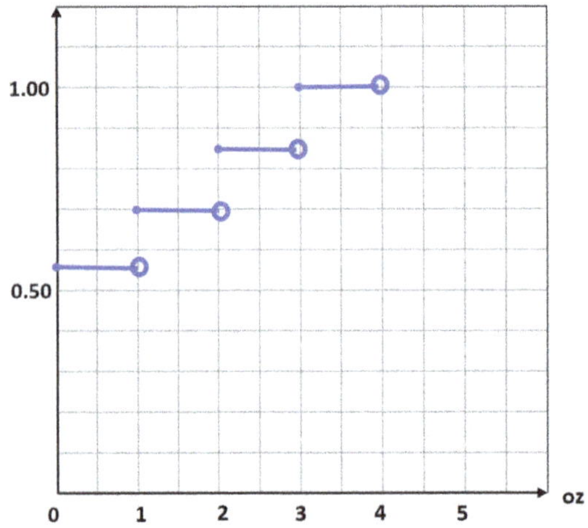

Image 7.54

We have a series of disjoint, discontinuous line segments. Income tax rate tables are also staggered in this way. Using mathematical notation, the **piecewise-defined function** for the above graph is given as follows:

$$f(x) = \begin{cases} .55 \ \textit{if} \ 0 < x < 1 \\ .70 \ \textit{if} \ 1 \le x < 2 \\ .85 \ \textit{if} \ 2 \le x < 3 \\ 1.00 \ \textit{if} \ 3 \le x < 3.5 \end{cases}$$

You can see that Exercise 7.5.1 #1 and #2 are also piecewise-defined functions.

Example 1

Given the following definition of a piecewise-defined function, draw the graph and also evaluate $f(-10)$, $f(-6)$, $f(-2)$, $f(0)$, $f(2)$, and $f(6)$.

$$f(x) = \begin{cases} 4 & \text{if } -10 \le x < -6 \\ -(x+4)^2 + 8 & \text{if } -6 \le x < -2 \\ x^2 & \text{if } -2 \le x < 2 \\ -\frac{4}{3}x + \frac{20}{3} & \text{if } 2 \le x < 5 \end{cases}$$

Solution

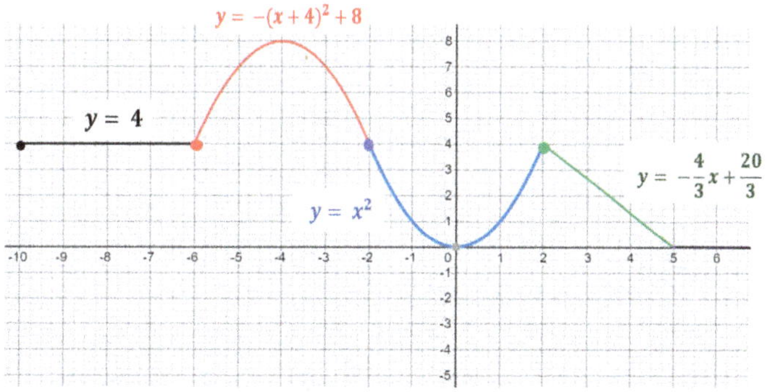

Image 7.55

$f(\mathbf{-10}) = 4$ because $\mathbf{-10}$ is in $[-10, -6)$

$f(-6) = -(-6 + 4)^2 + 8 = 4$ because -6 is in $[-6, -2)$

$f(-2) = (-2)^2 = 4$ because -2 is in $[-2, 2)$

$f(0) = 0^2 = 0$ because 0 is in $[-2, 2)$

$f(2) = -\frac{4}{3}(2) + \frac{20}{3} = 4$ because 2 is in $[2, 5)$

$f(6)$ is undefined because it is not in any of the given intervals.

Example 2

Write the piecewise-defined function for the following graph.

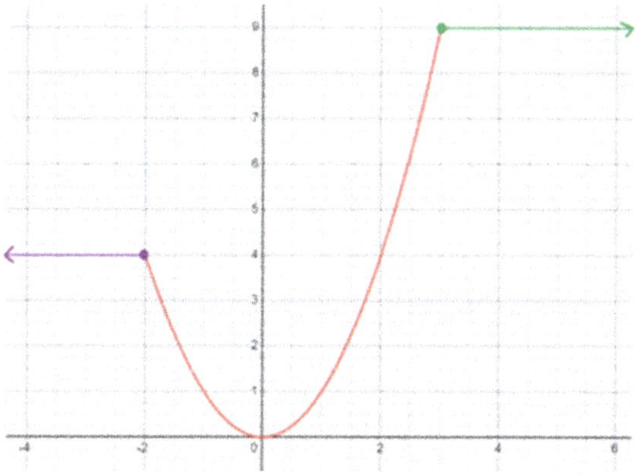

Image 7.56

Solution

The 1st and 3rd pieces are the horizontal lines $y = 4$ and $y = 9$ respectively. The middle piece is the parabola $y = x^2$. Thus:

$$f(x) = \begin{cases} 4 & \text{if } x <= -2 \\ x^2 & \text{if } -2 < x < 3 \\ 9 & \text{if } x \geq 3 \end{cases}$$

This is a **continuous** function in that the endpoints of each piece "join up."

EXERCISE SET 7.6.1

1. In Example 2 above, evaluate the following:
   1. $f(-5)$
   2. $f(-2)$
   3. $f(0)$
   4. $f(3)$
   5. $f(10)$
2. Write the piecewise-defined function for the following graph. (Assume each "piece" begins with a closed dot and ends with an open dot.) Hint: the piece from (0, 4) to (2, 0) is $y = -x^2 + 4$.

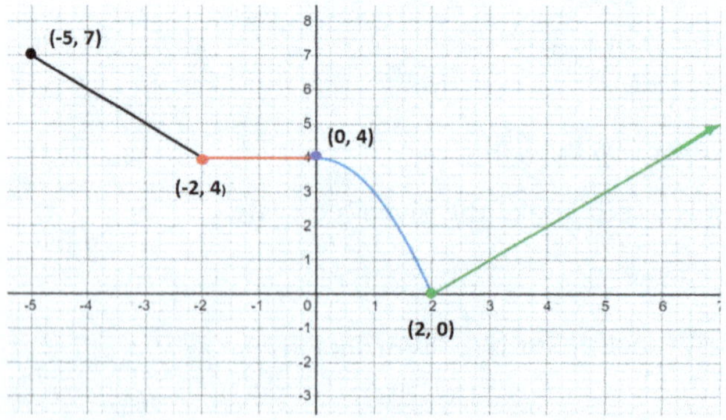

Image 7.57

3. Write the piecewise-defined function for $f(x) = |x|$
4. Write the piecewise-defined function for $f(x) = \text{int}(x)$ (for domain $[-2, 2]$)
5. Draw the graphs for the following piecewise-defined functions and identify whether or not they are continuous.

   A. $f(x) = \begin{cases} 2x + 3 & \text{if } x < 0 \\ 3 - x & \text{if } x \geq 0 \end{cases}$

   B. $f(x) = \begin{cases} \sqrt{4 + x} & \text{if } x < 0 \\ \sqrt{4 - x} & \text{if } x \geq 0 \end{cases}$

C. $f(x) = \begin{cases} 1-(x-1)^2 & \text{if } x \leq 2 \\ \sqrt{x-2} & \text{if } x > 2 \end{cases}$

D. $f(x) = \begin{cases} x+3 & \text{if } x \leq 0 \\ 3 & \text{if } 0 < x \leq 2 \\ 2x-1 & \text{if } x > 2 \end{cases}$

Program Goal: Write a program to plot the piecewise-defined function in Exercise Set 7.6.1 #5D above. Scale the axes at $[-10, 10] \times [-20, 20]$.

Writing a program to graph a piecewise-defined function is relatively straightforward. Simply use a loop to assign y-values for each "piece" of the function.

PROGRAM 7.6.1 DRAW_PIECEWISE.PY

```
## draw_piecewise.py
## Draw hard-coded piecewise-defined function
from matplotlib import pyplot as plt
from sympy import Symbol,sympify
x=Symbol('x')
piece1 = sympify('x+3')
piece2 = sympify('3')
piece3 = sympify('2*x-1')
domain1 = [i/10 for i in range(-100,0)]
domain2 = [i/10 for i in range(0,20)]
domain3 = [i/10 for i in range(20,100)]
fig = plt.figure()
ax = plt.subplot()
plt.axis([-10,11, -20,21])
y1_list = [ ]
y2_list = [ ]
y3_list = [ ]
y1_list=[piece1.subs({x:x_value}) for x_value in domain1]
plt.plot(domain1,y1_list,label=piece1)
    [fill in code to draw the 2nd and 3rd pieces of the graph ]
ax.legend()
plt.show()
```

Desired Output

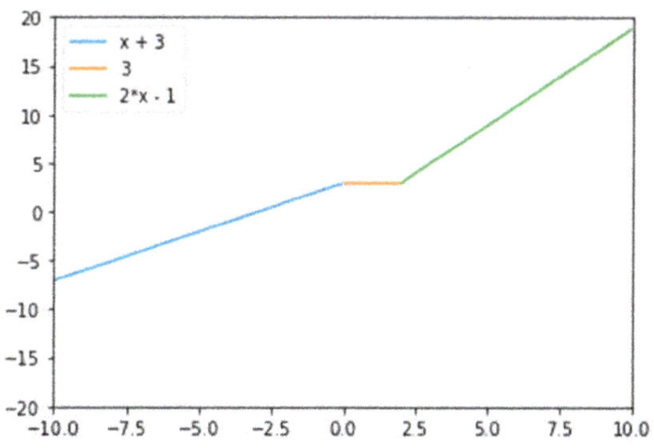

Image 7.58

EXERCISE SET 7.6.2

1. Fill in the yellow highlighted code to draw the 2nd and 3rd pieces.
2. Modify the program so that it graphs the piecewise-defined function in Example 1 above Exercise Set 7.6.1:

$$f(x) = \begin{cases} 4 & \text{if } -10 \le x < -6 \\ -(x+4)^2 + 8 & \text{if } -6 \le x < -2 \\ x^2 & \text{if } -2 \le x < 2 \\ -\dfrac{4}{3}x + \dfrac{20}{3} & \text{if } 2 \le x < 5 \end{cases}$$

3. Modify the program so that the user can input three functions and three corresponding domains.

Sample Output

Piece 1? *x***2

Interval 1? [−10, 1]

Piece 2? 3

Interval 2? (1, 2]

Piece 3? *x*

Interval 3? (2, 10]

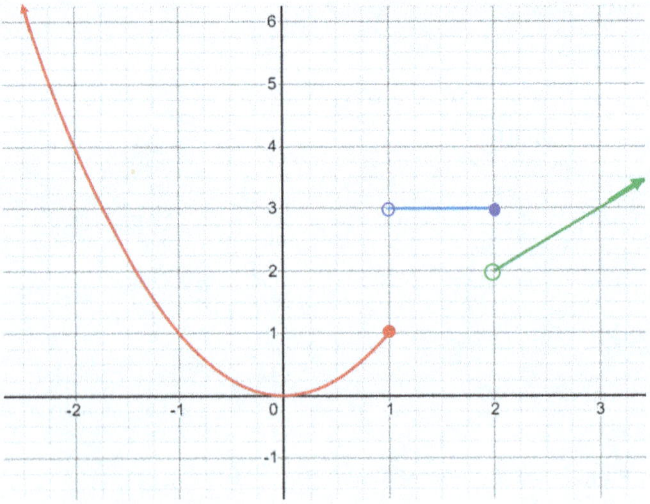

Image 7.59

4. CHALLENGE. Repetitive code is the bane of programming! Rewrite the program so that it makes use of a single function call to generate and graph each piece of the function.

We can make this program even more concise using the **piecewise** function in the **NumPy** library. Before graphing, we'll first demonstrate the use of this function in a very brief program.

Program Goal: Write a program that asks the user to input a value of x. Calculate and output $f(x)$ according to this piecewise-defined function:

$$f(x) = \begin{cases} x+3 & \text{if } x \le 0 \\ 3 & \text{if } 0 < x \le 2 \\ 2x-1 & \text{if } x > 2 \end{cases}$$

PROGRAM 7.6.2 NUMPY_PIECEWISE.PY

```
import numpy as np

x = float(input("Value of x? "))
y = np.piecewise(x, [x>=-10 and x<0, x>=0 and x <2, x >=2 and x <=10],[x+3, 3, 2*x-1])
print(y)
```

list of conditions (domains) · *list of functions* · *variable*

Image 7.60

Sample Output

Value of x? 3

5.0

Note this does not even require the **SymPy** library, which means it can handle various functions, not just polynomials. A flaw is that it returns a value of 0 for undefined values.

EXERCISE SET 7.6.3

1. Replace the piecewise-defined function in program **numpy_piecewise** with the following:

$$f(x) = \begin{cases} x(x+2)^2 & \text{if } x < -2 \\ -x^2+4 & \text{if } 2 \le x < 2 \\ -(x-2) & \text{if } 2 \le x \le 5 \\ x-5 & \text{if } x > 5 \end{cases}$$

2. Use the revised program to answer the following:
   A. $f(-2)$
   B. $f(0)$
   C. $f(2)$
   D. $f(5)$
   E. $f(6)$
3. What variable type is returned by the **piecewise**() function?

The **piecewise()** function however is not a graphing or drawing function in itself. It is a numeric function that returns a number based on the value of x.

Thus, if you would like to employ **piecewise()** to draw the graph of the function, you would need to assign it to a *y_list,* which is passed to the **plot** function.

Program Goal: Modify program 7.6.1 **draw_piecewise()** (or write a new one) to draw a piecewise-defined function using the **piecewise()** function in the **NumPy** library.

PROGRAM 7.6.3 DRAW_NUMPY_PIECEWISE.PY

```
## draw_numpy_piecewise.py
## Draw piecewise-defined function using the NumPy piecewise()
function
from matplotlib import pyplot as plt
import numpy as np
fig = plt.figure()
ax = plt.subplot()
plt.axis([−10,10, −20,20])
x_list = [i/10 for i in range(−100,100)]
y_list = [ ]
for x in x_list:
    y = np.piecewise(x,[x>=−10 and x<0,x>=0 and x<2,x>=2 and
    x<=10],[x+3, 3,2*x−1])
    y_list.append(y)
plt.plot(x_list,y_list)
plt.show()
```

Desired Output

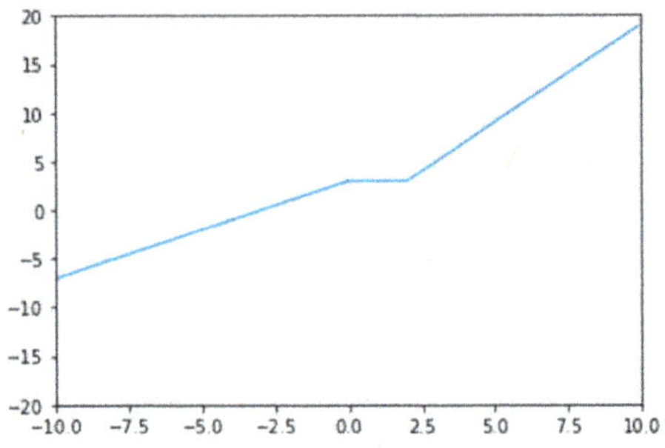

Image 7.61

Compare this to the original program **draw_piecewise.** How much more succinct and elegant!

EXERCISE SET 7.6.4

1. Draw in the *x*- and *y*-axes. Label each piece of the function. Position the label at the initial point of each piece.

   Desired Output

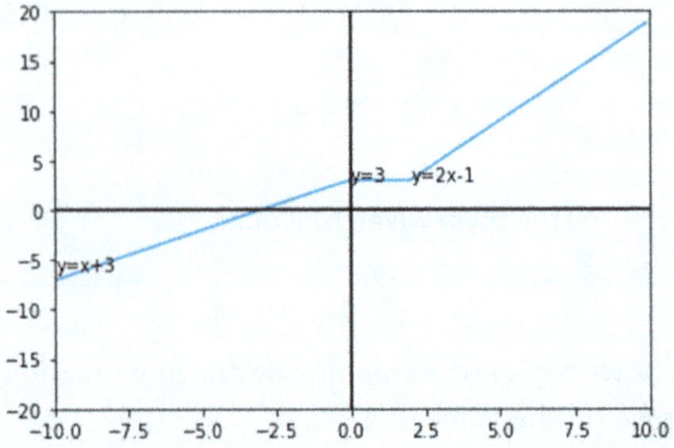

Image 7.62

2. Modify the program to draw the following piecewise-defined function.

$$f(x) = \begin{cases} x & \text{if } -10 \le x < 0 \\ -(x-1)^2 + 1 & \text{if } 0 \le x < 1 \\ 1 & \text{if } 1 \le x < 3 \\ 2(x-3)^2 + 1 & \text{if } x \ge 3 \end{cases}$$

Chapter Projects

I. Identifying Extrema

We know the formula to find the x-coordinate of the vertex of a parabola: $x = -\frac{b}{2a}$. However, finding the extrema of a polynomial function with degree 3 or higher involves the derivative function, which is covered in Calculus. The good news is, we already wrote a program in Chapter 3 that finds the derivative of a polynomial function.

The extrema of a function occur at points where the derivative equals 0; that's because the derivative represents the slope of the tangent line at that point.

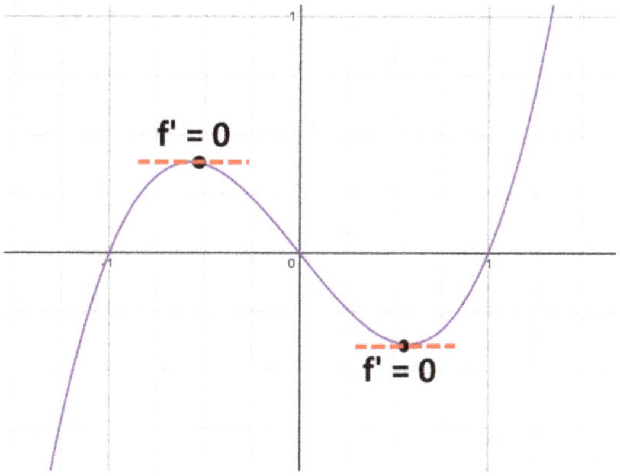

Image 7.63

Example

If $f(x) = \frac{2}{9}x^3 - x^2$, find the coordinates of the extrema.

$$f'(x) = 3\left(\frac{2}{9}\right)x^2 - 2x = \frac{2}{3}x^2 - 2x.$$

Solution

Set $f'(x) = 0$ and solve for x:

$$\frac{2}{3}x^2 - 2x = 0$$

$$\frac{2}{3}x(x - 3) = 0$$

$$x = 0, 3$$

The ordered pairs of the extrema are:
$(0, f(0)) = (0,0)$
$(3, f(3)) = (3, -3)$

Write a program that asks the user to input the coefficients of a polynomial function. Calculate and output (1) the function, (2) the derivative, and (3) the coordinates of the extrema, using simplified fractions instead of decimals.

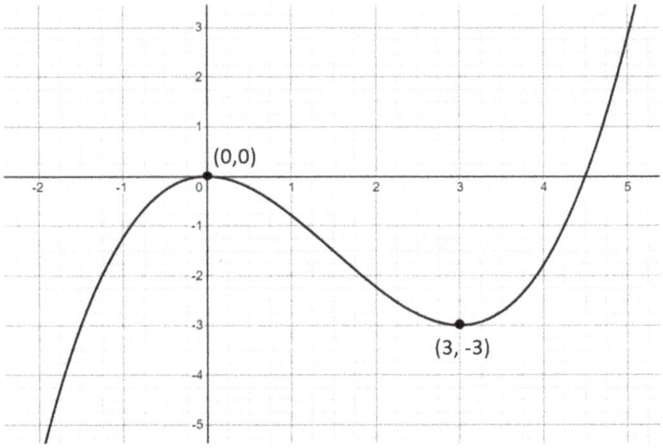

Image 7.64

SAMPLE OUTPUT

Polynomial Function? 2/9, –1, 0, 0

Derivative of $f(x) = (2/9)x^3 - x^2$ is:

$f'(x) = (2/3)x^2 - 2x$

Coordinates of extrema are:

(0, 0)

(3, –3)

II. Classifying Extrema of Polynomial Functions

Refine the program in Chapter Project I above by further classifying each point as either a local maxima or local minima. This requires another calculus feature called the 2nd derivative (notation $f''(x)$). The second derivative is basically the derivative of the first derivative:

$$f''(x) = [f'(x)]'$$

Using the example above where $f(x) = \frac{2}{9}x^3 - x^2$ and $f'(x) = \frac{2}{3}x^2 - 2x$, then:

$$f''(x) = 2\left(\frac{2}{3}\right)x - 2 = \frac{4}{3}x - 2.$$

If $f'(x) = 0$ and $f''(x) > 0$ at a point x, then the function $f(x)$ has a local minimum at that point.
If $f'(x) = 0$ and $f''(x) < 0$ at a point x, then the function $f(x)$ has a local maximum at that point.
Continuing the example above, $f'(x) = 0$ at $x = 0$ and $x = 3$.

$f''(0) = \frac{4}{3}(0) - 2 = -2$, which is negative, therefore the function has a maximum at $x = 0$.

$f''(3) = \frac{4}{3}(3) - 2 = 2$, which is positive, therefore the function has a minimum at $x = 3$.

SAMPLE OUTPUT

Polynomial Function? 2/9, −1, 0, 0

Derivative of $f(x) = (2/9)x^3 - x^2$ is:

$f'(x) = (2/3)x^2 - 2x$

Coordinates of extrema are:

(0, 0) Local Maximum

(3, −3) Local Minimum

III. Continuous Piecewise Function

Given a piecewise-defined function, draw it and determine whether it is continuous over [−10, 10]. The user will need to input the pieces (functions) and corresponding intervals.

SAMPLE OUTPUT

How many pieces? 3

Piece 1? x**2

Interval 1? [−10,1]

Piece 2? 3

Interval 2? (1,2]

Piece 3? x

Interval 3? (2,10]

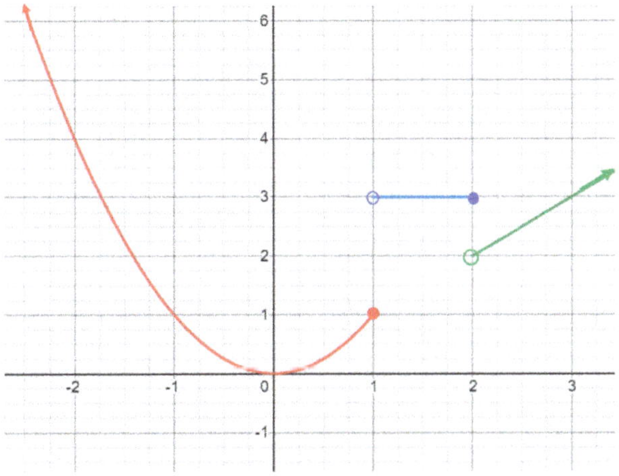

Image 7.59

NOT CONTINUOUS at $x = 1$

NOT CONTINUOUS at $x = 2$

How many pieces? 4

Piece 1? –2*x+5

Interval 1? [–10,1)

Piece 2? 2*x+2

Interval 2? [1,3)

Piece 3? 7

Interval 3? [3,5)

Piece 4? –x+12

Interval 4? [5,10]

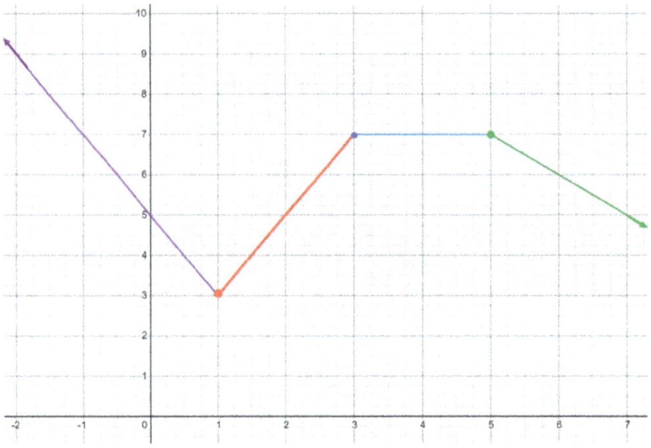

Image 7.65

CONTINUOUS

IV. All You Need

Demonstrate your mastery of the basic library functions and their various transformations with a work of art. Here is an example:

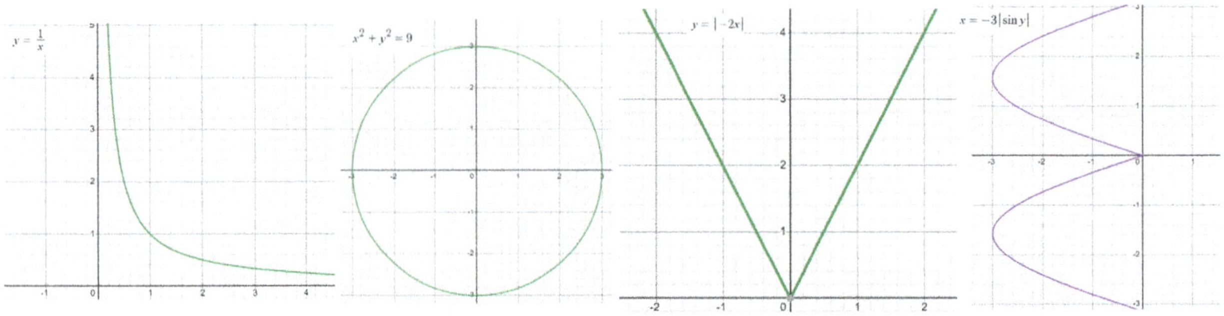

Image 7.66

CHAPTER SUMMARY

PYTHON CONCEPTS

To add a legend to a graph	plt.**plot**(x_list,y_list,**label**=text) Include in program : ax = plt.**subplot**() ax.**legend**()
The piecewise function	Must be imported from **NumPy** library. Evaluates an input based on a piecewise-defined set of functions, returns a single numerical value. np.**piecewise**(input_variable, [domain1, domain2, ...], [function1, function2, ...]) *Ex:* *x = 5* *y = np.**piecewise**(x,[x <=5,x>5],[x−1,x**2])* *will return 4 because 4<=5 so y is evaluated using x−1.*
To identify whether the value of a variable is imaginary	**x.is_imaginary** returns a Boolean value (True/False) based on value of *x*. Useful in determining whether to plot a point. *Ex:* *x = sqrt(−2)* *x.is_imaginary* *will return True because* $\sqrt{-2}$ *is an imaginary number*
To retrieve the index of a list item	list.**index**(item) *Ex:* *list = ['alpha', 'beta', 'gamma', 'delta']* *k = list.**index**('alpha')* *will return 0 because list[0] = 'alpha'*

PRECALCULUS CONCEPTS

Stretching/compressing a graph	If $y = k \cdot f(x)$, k a constant, then: The graph of $y = f(x)$ is **stretched** if $\lvert k \rvert > 1$ The graph of $y = f(x)$ is **compressed** if $0 < \lvert k \rvert < 1$

(Continued)

Shifting (translating) a graph	If $y = f(x) + k$, k a positive constant, then the graph of $y = f(x)$ is shifted k units up If $y = f(x) - k$, k a positive constant, then the graph of $y = f(x)$ is shifted k units down If $y = f(x + h)$, h a positive constant, then the graph of $y = f(x)$ is shifted h units left If $y = f(x - h)$, h a positive constant, then the graph of $y = f(x)$ is shifted h units right
Reflecting a graph	$y = -f(x)$ is the graph of $f(x)$ reflected through the x-axis $y = f(-x)$ is the graph of $f(x)$ reflected through the y-axis
Symmetry	A graph that contains both (x, y) and $(-x, y)$ has symmetry through the **y-axis**. 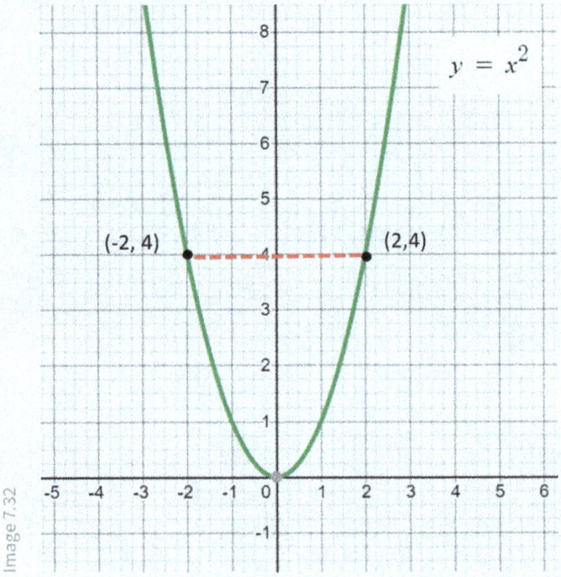

	A graph that contains both (x, y) and $(-x, -y)$ has symmetry through the **origin**. 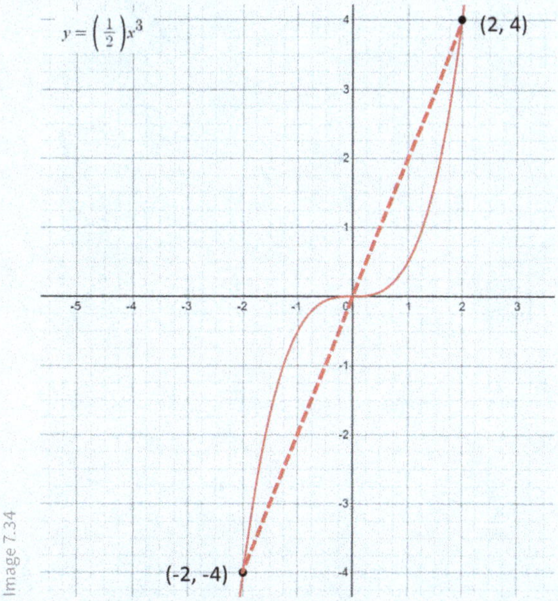 A graph that contains both (x, y) and $(x, -y)$ has symmetry through the ***x*-axis**. 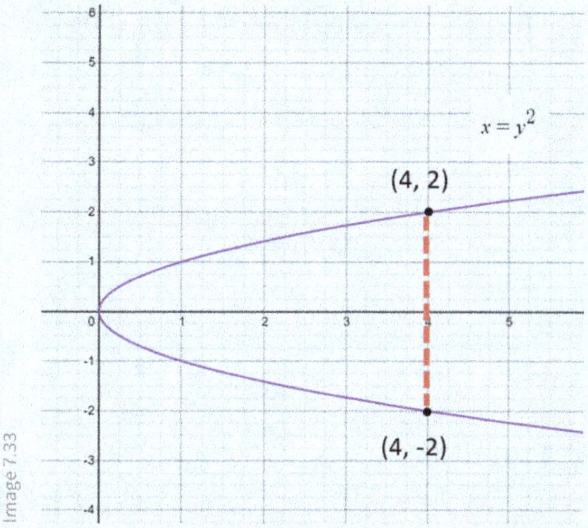

(Continued)

Even function	A function such that $f(-x) = f(x)$; graph has symmetry through the y-axis. *Ex:* 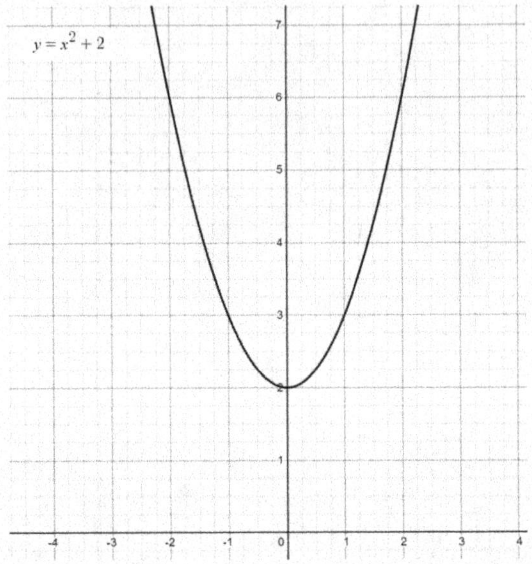 Image 7.39
Odd function	A function such that $f(-x) = -f(x)$; graph has symmetry through the origin. *Ex:* 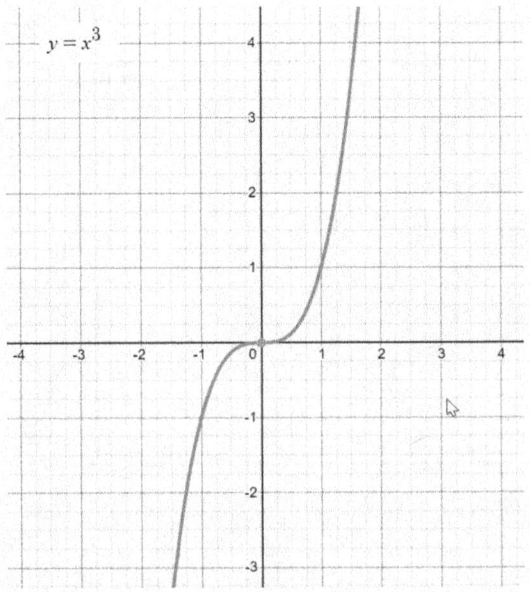 Image 7.41

Increasing on an interval	$f(x)$ is increasing on interval (a, b) if for every pair of points x_1 and x_2 in (a, b), if $x_1 < x_2$ then $f(x_1) < f(x_2)$. *Ex:* 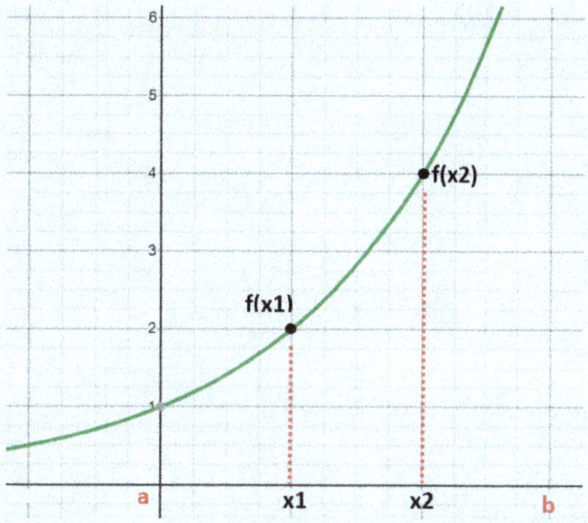 Image 7.45
Decreasing on an interval	$f(x)$ is decreasing on interval (a, b) if for every pair of points x_1 and x_2 in (a, b), if $x_1 < x_2$ then $f(x_1) > f(x_2)$. *Ex:* 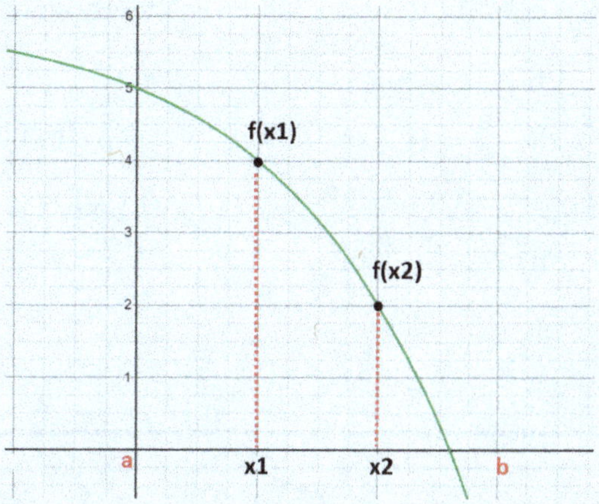 Image 7.46
Constant on an interval	$f(x)$ is constant on interval (a, b) if for any pair of points x_1 and x_2 in (a, b), $f(x_1) = f(x_2)$. *Ex:* 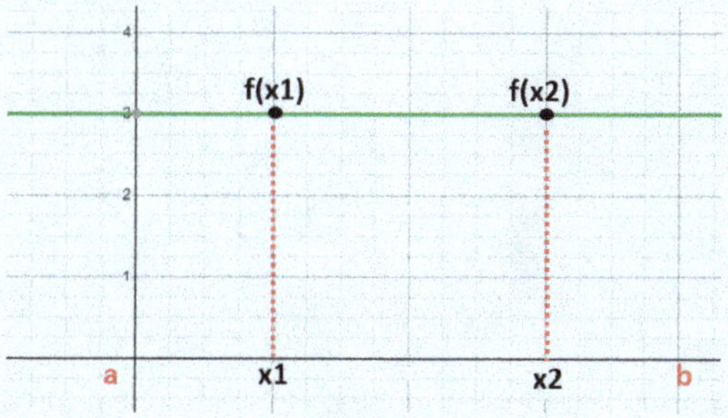 Image 7.47

(*Continued*)

Extrema of a graph	The **absolute maximum** is the y-value of the highest point on the graph. The **absolute minimum** is the y-value of the lowest point on the graph. A **local maximum** is the highest point in a local area of the graph. A **local minimum** is the lowest point in a local area of the graph. Image 7.48 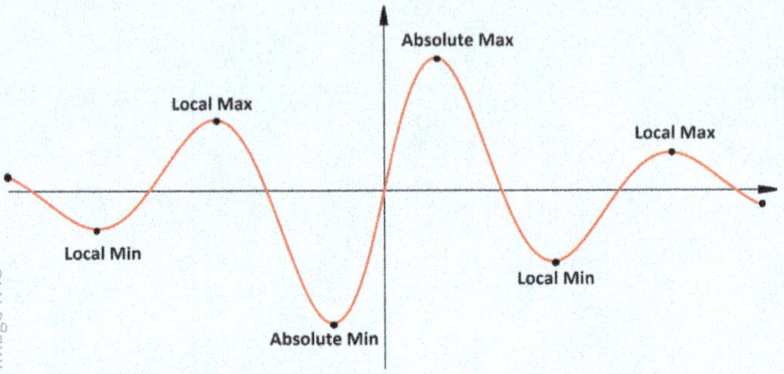
Piecewise-defined function	A function comprised of component functions ("pieces"). $f(x) = \begin{cases} f_1(x) \textit{ if } [\textit{domain of } f_1] \\ f_2(x) \textit{ if } [\textit{domain of } f_2] \\ \vdots \\ f_n(x) \textit{ if } [\textit{domain of } f_3] \end{cases}$ In order to be a function, the domains must not overlap. *Ex:* $f(x) = \begin{cases} x(x+2)^2 \quad \textit{if } x < 2 \\ -x^2 + 4 \quad \textit{if } -2 \leq x < 2 \\ -\sqrt{(x-2)} \quad \textit{if } 2 \leq x \leq 5 \\ x - 5 \quad \textit{if } x > 5 \end{cases}$

Credits

IMG 7.43a: Copyright © by Gan Khoon Lay (CC by 3.0) at https://commons.wikimedia.org/wiki/File:Icon-roller-coaster-881334.svg.

IMG 7.44a: Copyright © by Gan Khoon Lay (CC by 3.0) at https://commons.wikimedia.org/wiki/File:Icon-roller-coaster-881334.svg.

CHAPTER 8

Complex Numbers and the Fundamental Theorem of Algebra

8.1 Zeros and Intercepts

Precalculus might seem like a random collection of topics, but the focus is *the pursuit of the solution of* p(x) = *0*, where $p(x)$ is a polynomial function. Let's take a look at the various types of polynomial functions.

Linear

Equations of the form $ax + b = 0$ are easy to solve: $x = -\frac{b}{a}$ ... so easy that we are not even going to write a program to do this.

Quadratic

We saw in Section 3.6 various methods to solve $ax^2 + bx + c = 0$. One of these is the quadratic formula $x = \dfrac{-b \pm \sqrt{b^2 - 4ac}}{2a}$.

Higher Order Polynomials

The first thing to understand is the connection between the **zeros of a function** and the ***x*-intercepts** of the graph. First, look at this graph of $f(x) = (x + 2)^2(x - 1)$:

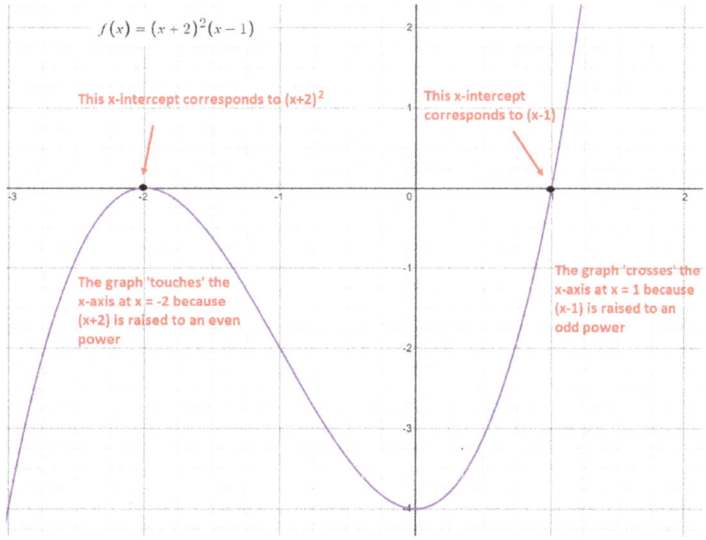

Image 8.1

Relationship between Factors and Zeros

If $f(x) = (x - a)^m(x - b)^n$ ...
Then $f(x)$ has zeros at $x = a, b$, ... and the graph of $f(x)$ has x-intercepts at a, b ...
More generally: if $(x - r)^k$ is a factor of $f(x)$, then:

1. $x = r$ is a zero of $f(x) = 0$
2. $x = r$ is an x-intercept of the graph of $f(x)$

The exponent k is called the **multiplicity** of r. If k is odd, then the graph crosses the x-axis at r. If k is even, then the graph touches the x-axis at r.

Example 1

Identify the zeros and whether the graph touches or crosses at each corresponding x-intercept:

$$f(x) = (x - 3)^3(x + 4)^2(x - 5)$$

Solution

There are three factors:
$(x - 3)$ is raised to the third power, so it has odd multiplicity. The graph crosses the x-axis at $x = 3$.
$(x + 4)$ is raised to the second power, so it has even multiplicity. The graph touches the x-axis at $x = -4$.
$(x - 5)$ is raised to the first power, so it has odd multiplicity. The graph crosses the x-axis at $x = 5$.

Example 2

Identify the zeros and whether the graph touches or crosses at each corresponding x-intercept:

$$y = x^3 + 19x^2 + 80x - 100$$

Solution

First factor:

$$x^3 + 19x^2 + 80x - 100 = (x + 10)^2(x - 1)$$

There are two factors:

$(x + 10)$ has even multiplicity, so the graph touches the x-axis at $x = -10$.
$(x - 1)$ has odd multiplicity, so the graph crosses the x-axis at $x = 1$.

EXERCISE SET 8.1.1

For each function, identify the zeros, multiplicity, and whether the graph touches or crosses at each corresponding x-intercept.

1. $f(x) = (x-3)^3(x+2)^2(x+5)^4$
2. $f(x) = x(x+4)^2$
3. $f(x) = x^2 - 4x + 4$
4. $y = x^3 - x^2 + 9x - 9$

Program Goal: Determine whether the graph of a function touches or crosses at each x-intercept.

Hint: A simple trick for determining whether an integer n is even or odd:

if $n\%2 == 0$ **then** n is even. **if** $n\%2 == 1$ **then** n is odd.

PROGRAM 8.1.1 ZEROS.PY

```
n = int(input("How many zeros? "))
zeros = [ ]
mults = [ ]
for k in range(n):
    z = float(input("Zero "+str(k+1)+": "))
    zeros.append(z)
    m = int(input("Multiplicity of "+str(z)+": "))
    mults.append(m)
for k in range(n):
    [ Fill in code to output 'touches' or 'crosses' ]
```

Sample Output

How many factors? 3
Zero 1? 5
Multiplicity of 5.0: 11
Zero 2? −3
Multiplicity of −3.0: 4
Zero 3? 0
Multiplicity of 0.0: 1
crosses at $z = 5.0$
touches at $z = -3.0$
crosses at $z = 0.0$

EXERCISE SET 8.1.2

1. Fill in the yellow highlighted code to output "touches" or "crosses."
2. Modify the program so that it also outputs the degree of the polynomial. Hint: degree of $(x-5)^{11} \cdot (x+3)^4 \cdot x$ is $11 + 4 + 1 = 16$.

3. Streamline the data entry for the user so that the data can be done with a single input statement. For example:

   Enter Zero,Multiplicity Pairs: 3,5;11,−3;4,0

4. CHALLENGE: Modify the program so that it also outputs the polynomial in factored form. For example, using the sample input above, the program would output:

   crosses at $z = 5.0$
   touches at $z = -3.0$
   crosses at $z = 0.0$
   $f(x) = (x - 5)^{11} \cdot (x + 3)^{4} \cdot x$

Since we will be working with polynomials in this section, let's review (and introduce) some helpful concepts from the **SymPy** library, which simplifies, factors, expands, evaluates, and does a whole lot more.

(Recall >>> indicates a Python statement typed directly into the console so that we can see the result directly below it.)

TABLE 8.1 Symbolic Manipulation (SymPy library)

To define *x* as an algebraic variable (also known as a **symbol**):

```
>>>x = Symbol ('x')
```

To convert a string type to a **SymPy** expression, use **sympify.** Example:

```
>>>poly = sympify(input('polynomial? '))
```

To "pretty print" the output:

```
>>>pprint(poly)
5 · x² + 3 · x + 7
```

Note the exponent is superscripted and the terms are padded with spaces in between.

To evaluate an expression for a specific value of the symbol:

```
>>>value = poly.subs({x:value})
```

To factor a polynomial:

```
>>>factor('x**2−4')
(x + 2)*(x – 2)
```

To expand a polynomial, use **expand**. Example:

```
>>>poly1 =sympify( '(x+2)**2')
>>>poly2 =sympify( '(x–3)')
>>>expand(poly1*poly2)
x**3 + x**2 – 8*x – 12
```

To get a factor list with corresponding multiplicities for a polynomial:

```
>>>factor_list('x**6–30*x**4+20*x**3+225*x**2–108*x–540')
(1, [(x + 5, 1), (x + 2, 2), (x – 3, 3)])
```

which corresponds to $(x + 5)(x + 2)^2(x - 3)^3$

*Note **factor_list()** returns a **tuple** (ordered set), and 1 is always included.*

Program Goal: Write a program that uses the **factor_list** function to factor a polynomial. Output the factored form and then identify the multiplicity of each zero.

PROGRAM 8.1.2 MULTIPLICITY.PY

```
## multiplicity.py
## input: polynomial
## output: list of factors and multiplicities
from sympy import factor_list,factor,pprint,solve
poly_str=input('Polynomial Function? ')
factor_set = factor_list(poly_str)
factored_form = factor(poly_str)
pprint(factored_form)
f_list=list(factor_set[1])
#for each [factor, multiplicity] pair:
for f in f_list:
    fact_str=str(f[0])
    z = solve(fact_str)
    mult=f[1]
    print(z,'has multiplicity',mult)
```

Sample Output

```
Polynomial Function?
x**6–30*x**4+20*x**3+225*x**2–108*x–540
(x – 3) 3 · (x + 2)2 · (x + 5)
[–5] has multiplicity 1
[–2] has multiplicity 2
[3] has multiplicity 3
```

EXERCISE SET 8.1.3

1. Use the program to factor and identify the zeros/multiplicities of the following polynomials:

   A. $x^4 - 2x^3 - 3x^2 + 4x + 4$
   B. $x^3 + 11x^2 + 7x - 147$
   C. $x^6 + 10x^5 + 19x^4 - 68x^3 - 233x^2 - 230x - 75$

2. What is the type and value of each of the following variables when you input x**2+10*x+25?

   A. factor_set
   B. f_list
   C. f
   D. f[0]
   E. fact_str
   F. z

3. Modify the program so that it also outputs whether each zero touches or crosses the x-axis.

4. The output for $x^3 - 1$ is:

   $(x - 1)^2 \cdot (x + x + 1)$

   [1] has multiplicity 1, crosses

   [−1/2 − sqrt(3)*1/2, −1/2 + sqrt(3)*1/2] has multiplicity 1, crosses

   However, the graph has only one x-intercept:

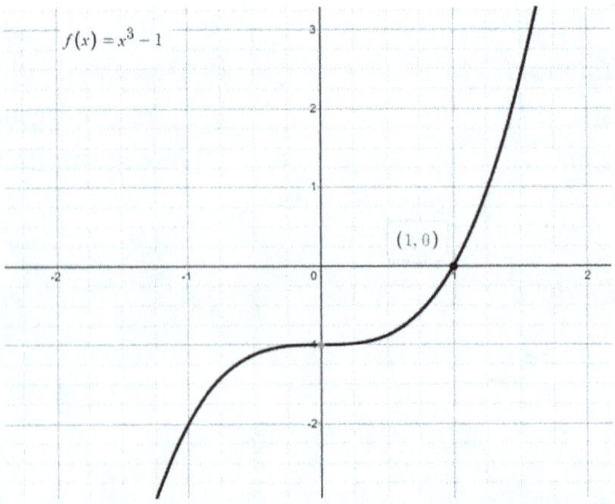

Image 8.2

We will soon find out that $-\frac{1}{2} \pm \frac{\sqrt{3}}{2} i$ is a pair of imaginary numbers that do **not** correspond to x-intercepts. Therefore, it makes no sense to describe their multiplicity and whether they touch or cross the x-axis.

Modify the program so only real numbers are output. *(Hint: use x.**is_imaginary** from the previous chapter to identify whether a number is imaginary.)*

Sample Output

Polynomial Function? x**4 – 4*x**3 + 6*x**2 – 8*x + 8

$(x - 2)^2 \cdot (x^2 + 2)$

[2] has multiplicity 2, touches

5. Use function num_list_to_poly() in **poly_fcn.py** (see chapter 3 or Appendix C) so the user can input coefficients of the polynomial rather than the entire polynomial.

Sample Output

Coefficients of Polynomial Function? 1,–4,6,8,8

$(x - 2)^2 \cdot (x^2 + 2)$

[2] has multiplicity 2, touches

One could argue that **SymPy**'s vast library of polynomial manipulation functions would render most of this chapter useless, however a good programmer needs to understand which tools are applicable, what the tool is accomplishing, and whether the result is valid. Perhaps someday you will be hired to develop a new programming language or hired to teach a Precalculus course. Therefore, it is important to understand the historical, purely mathematical approaches to factoring polynomials.

8.2 Using Synthetic Division to Factor

In Section 3.4 we learned that we could use **synthetic division** to divide a polynomial by a divisor in the form $(x - a)$, and if the remainder was zero, the divisor was a factor. For example, if $2x^4 - 5x^3 - 14x^2 + 47x - 30$ is divided by $(x - 2)$:

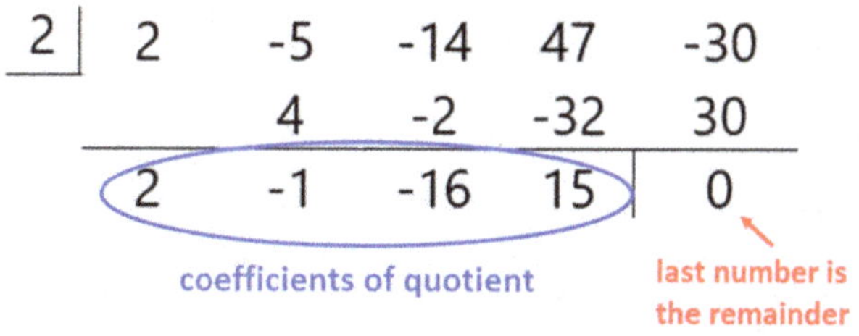

Image 8.3

The quotient is $2x^3 - x^2 - 16x + 15$ with zero remainder, and we have:

$$2x^4 - 5x^3 - 14x^2 + 47x - 30 = (x - 2)(2x^3 - x^2 - 16x + 15)$$

dividend divisor quotient

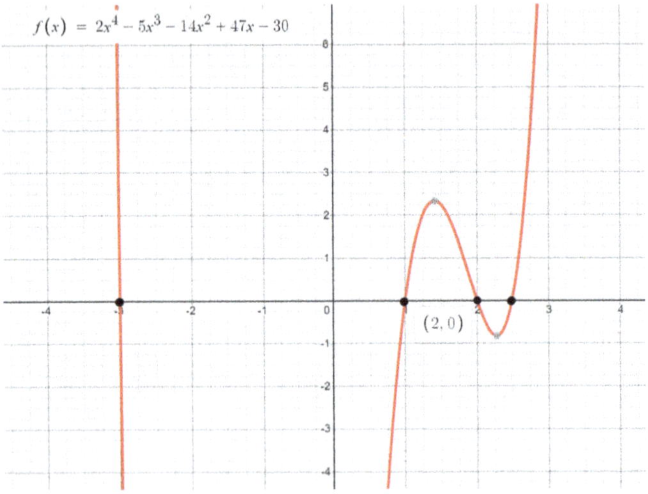

Image 8.4

You can see that 2 is an x-intercept, but there are three other x-intercepts. How do we find them? ... We continue synthetic division on the quotient $2x^3 - x^2 - 16x + 15$.

It appears that 1 is an x-intercept; let's verify that by dividing by $x - 1$.

1	2	-1	-16	15
		2	1	-15
	2	1	-15	0

quotient

Image 8.5

We can see by the zero remainder that $x - 1$ is indeed a divisor, and 1 is an x-intercept.
We can replace $2x^3 - x^2 - 16x + 15$ with
$(x - 1)(2x^2 + x - 15)$

so $2x^4 - 5x^3 - 14x^2 + 47x - 30 = (x - 2)(2x^3 - x^2 - 16x + 15)$ becomes

$(x - 2)(x - 1)(2x^2 + x - 15)$

Now that we have a quadratic $2x^2 + x - 15$, we can factor that:

$2x^2 + x - 15 = (2x - 5)(x + 3)$

thus $2x^4 - 5x^3 - 14x^2 + 47x - 30 = (x - 2)(x - 1)(2x - 5)(x + 3)$

The polynomial is now fully factored, and the x-intercepts are 2, 1, 5/2 and -3.

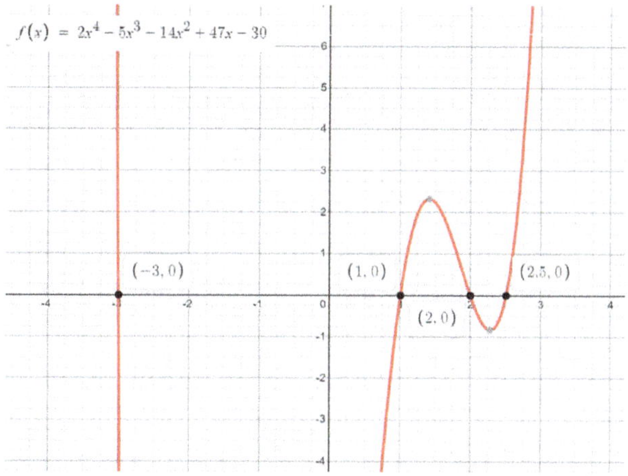

Image 8.6

Example

Factor $4x^4 - 12x^3 - 71x^2 + 120x + 400$ and find its x-intercepts.

Solution

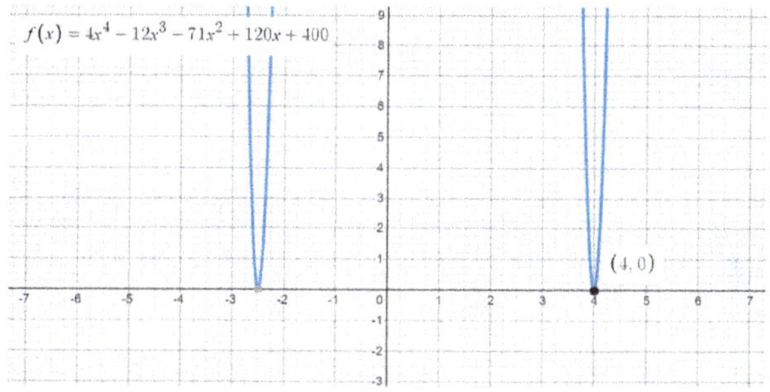

Image 8.7

To do synthetic division, we need a "corner number" to start the ball rolling, so let's try one of the x-intercepts. It appears that 4 might work.

4	4	−12	−71	120	400
		16	16	−220	−400
	4	4	−55	−100	0

Yes, 4 worked. (How do we know?)
We have now factored $4x^4 - 12x^3 - 71x^2 + 120x + 400 =$

$$(x - 4)(4x^3 + 4x^2 - 55x - 100)$$

We need to repeat synthetic division. Note the graph "touches" the x-axis at $x = 4$, which means it has even multiplicity, so a possible factor is $(x - 4)^2$ or $(x - 4)^4$, etc.
Do synthetic division again with 4.

4	4	4	−55	−100
		16	80	100
	4	20	25	0

We now have

$$4x^4 - 12x^3 - 71x^2 + 120x + 400 =$$
$$(x - 4)^2(4x^2 + 20x + 25) =$$
$$(x - 4)^2(2x + 5)^2$$

from which we can conclude that x touches the x-axis at $x = 4$ and $x = -5/2$.

EXERCISE SET 8.2.1

Factor the following polynomials and find the zeros. Use repeated synthetic division.

1. $3x^3 - 50x^2 + 207x - 220$
2. $3x^4 + 10x^3 - 69x^2 - 264x - 80$
3. $8x^5 - 18x^4 - 97x^3 + 129x^2 + 134x + 24$

When the Quadratic Doesn't Factor

In real life, equations describing the world do not always have rational number solutions. The answers might be irrational or even nonreal. (We'll get to imaginary numbers later.)

Fortunately, the quadratic formula will work for any 2nd degree polynomial.

Example 1

Factor $x^2 + 3x + 1$.

Solution

The quadratic doesn't factor using rational numbers, so we apply the quadratic formula to $x^2 + 3x + 1 = 0$:

$$x = \frac{-3 \pm \sqrt{3^2 - 4(1)(1)}}{2(1)} = \frac{-3 \pm \sqrt{5}}{2} = \frac{-3 + \sqrt{5}}{2} \text{ or } \frac{-3 - \sqrt{5}}{2}$$

Recall that if a and b are zeros of $f(x)$, then $f(x) = (x - a)(x - b)$. In the above example, we have that $x^2 + 3x + 1 = \left(x - \frac{-3 \pm \sqrt{5}}{2}\right)\left(x - \frac{-3 - \sqrt{5}}{2}\right)$ or $(x + .382)(x + 2.618)$.

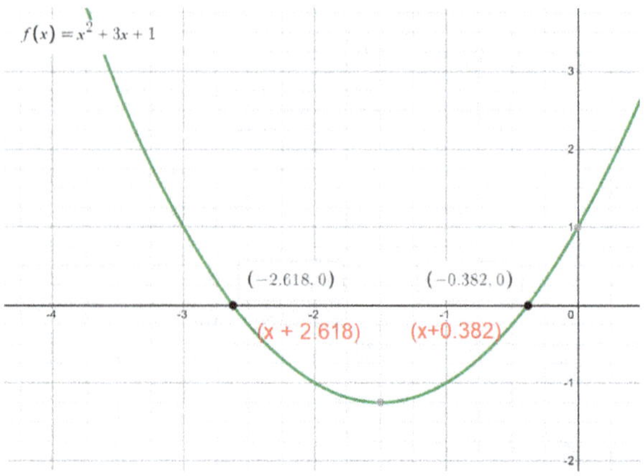

Image 8.8

When the Quadratic Doesn't Have Real Solution(s)

A quadratic such as $x^2 + 5$ doesn't factor because there are no real solutions to $x^2 + 5 = 0$. The quadratic formula yields $x = \frac{\pm\sqrt{-4(1)(5)}}{2(1)} = \frac{\pm\sqrt{-20}}{2}$. At this point we cannot take the square root of a negative number; it is not a real number. No real solutions means no x-intercepts:

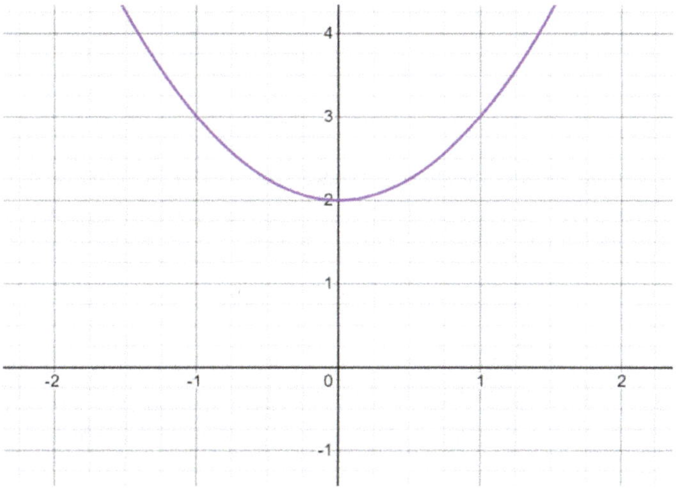

Image 8.9

Example 2

Factor $x^4 + 2x^3 + 7x^2 + 14x$.

Solution

First factor out x and then do factor by grouping to obtain:

$$x^4 + 2x^3 + 7x^2 + 14x = x(x + 2)(x^2 + 7).$$

Setting this equal to 0, we get $x = 0$ or $x = -2$. The quadratic $x^2 + 7$ does not factor over the field of real numbers, thus the quartic only has two x-intercepts.

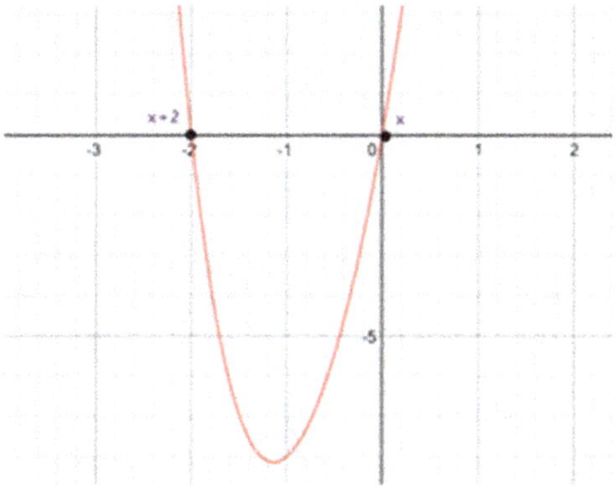
Image 8.10

EXERCISE SET 8.2.2

Factor the following polynomials and find the zeros. Use repeated synthetic division.

1. $3x^4 - 7x^3 + 9x^2 - 35x - 30$
2. $x^5 + 3x^4 - 21x^3 - 71x^2 - 72x - 240$
3. $4x^4 - 3x^3 - 24x^2 - 7x - 6$

Factoring a Quadratic

We are of course heading in the direction of writing a program to perform this tedious factorization for us. We first need a function to factor a quadratic. You might recall from a prerequisite course that this involved a fair amount of guess-and-check, for example:

$$x^2 - 27x + 50 = (x - 25)(x - 2)$$

Given $x^2 + bx + c$, this involved finding a pair of numbers that multiplied to c and added to b. There was a bit more involved when dealing with $ax^2 + bx + c$. Various textbooks tout methods such as the "ac" method, and these are only useful if the quadratic can be factored using rational numbers.

However, we have one foolproof method, which is the quadratic formula. Let's say $x = \dfrac{-b \pm \sqrt{b^2 - 4ac}}{2a}$ yields $x = x = -\frac{1}{2}$ *or* $x = \frac{3}{4}$. This corresponds to the factorization $\left(x + \frac{1}{2}\right)\left(x - \frac{3}{4}\right)$.

An effective program needs to address the various cases in which the solutions are rational, irrational, or nonreal. Use the **discriminant** $b^2 - 4ac$ (the expression under the radical sign) to determine how many and what type of solutions the quadratic formula yields.

SOLUTIONS AS DETERMINED BY DISCRIMINANT

If $b^2 - 4ac$ is ...

	Example	Graph	# and Type of Solutions
Positive perfect square	4	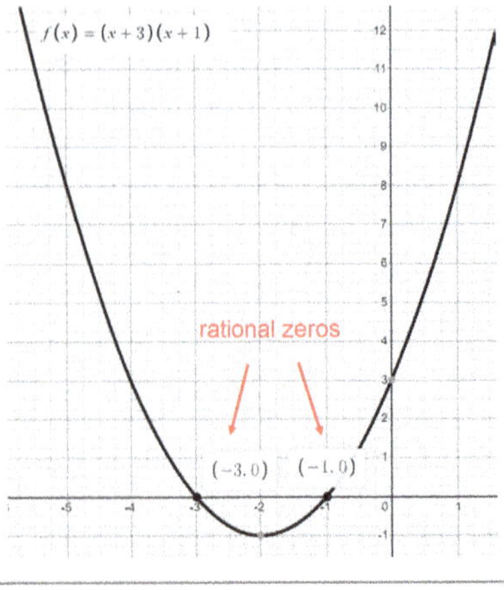 Image 8.11	Two rational solutions $-\dfrac{b}{2a} \pm \dfrac{\sqrt{b^2 - 4ac}}{2a}$

<table>
<tr><td>Positive—not a perfect square</td><td>17</td><td>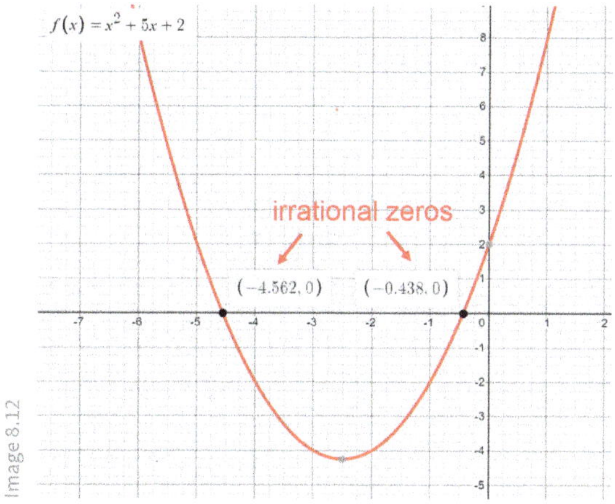
</td><td>Two irrational solutions
$-\frac{b}{2a} \pm \frac{\sqrt{b^2 - 4ac}}{2a}$</td></tr>
<tr><td>Zero</td><td>0</td><td>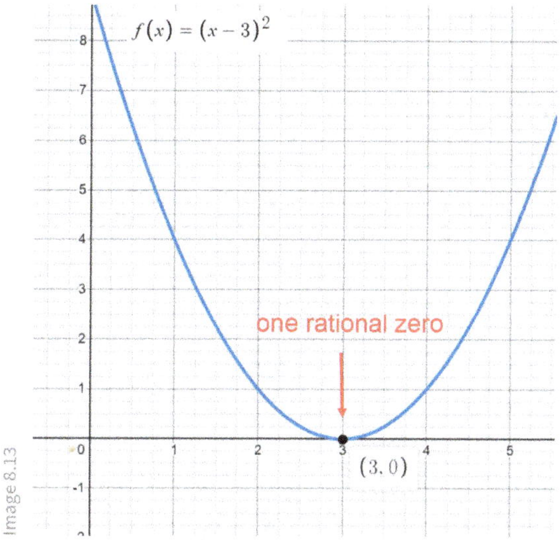
</td><td>One rational solution
$-\frac{b}{2a}$</td></tr>
<tr><td>Negative</td><td>−8</td><td>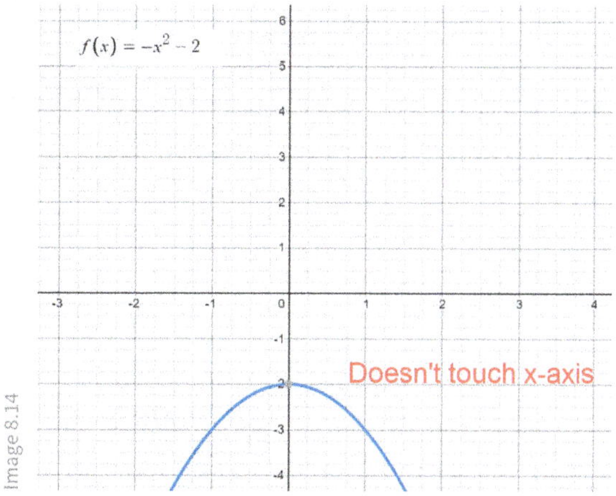
</td><td>No real solutions</td></tr>
</table>

Example

Use the discriminant to find the number and type of solutions of $3x^2 + 4x + 16$.

Solution

The discriminant $= b^2 - 4ac = (4)^2 - 4(3)(16) = -176$.
-176 is a negative number, so there are **no real solutions**.

EXERCISE SET 8.2.3

Use the discriminant to find the number and type of solutions to the following quadratics. Do not solve.

1. $x^2 - 4$
2. $x^2 + 4$
3. $x^2 + 4x + 4$
4. $x^2 + 2x + 5$
5. Write a program that uses the discriminant to identify the number and type of solutions for quadratic equation $ax^2 + bx + c = 0$.

   Hint: Use ***sqrt****(x)==****int****(****sqrt****(x)) to check if x is a perfect square.*

Sample Output

a, b, c? 1,0,1

Discriminant is −4

No real solutions

a,b,c? 1,4,4

Discriminant is 0

One rational solution

We'll start the framework of our quadratic-factoring problem with the most simple case—that in which the solution(s) are rational numbers.
Program Goal: Write a program that accepts as inputs *a*, *b*, and *c*, then calculates and outputs the solutions to $ax^2 + bx + c = 0$ using the quadratic formula. Use fractions, not decimals. Assume all inputs result in rational solutions. Do not use **SymPy**.

PROGRAM 8.2.1 QUAD_ROOT.PY

```
## quad_root.py
## Input: a,b,c (coefficients of ax^2 + bx + c)
## Output: zeros
```

Sample Output

a,b,c? 1,9,14

7, 2

a,b,c? 10,−13,−3

1.5, −0.2

```
from math import sqrt
coeffs = input("a,b,c? ")
a,b,c = coeffs.split(',')
a,b,c = float(a),float(b),float(c)
x1_num = -b + sqrt(b**2-4*a*c)
x2_num = -b - sqrt(b**2-4*a*c)
den = 2*a
x1 = x1_num/den
x2 = x2_num/den
print(x1,x2)
```

a,b,c? 1,4,0
0, −4

a,b,c? 24,1,−3
0.33333, −0.375

We will soon find out the importance of thorough testing, as this program is far from perfect.

EXERCISE SET 8.2.4

1. What is the difference between **Fraction**(1/3), **Fraction**(1,3), and **Fraction**(1.0,3.0)?
2. Modify the program to handle all sample output correctly, i.e.:
   A. Output integers as "5" not "5.0"
   B. Output rational numbers as "1/3" not "0.333333333333" *(hint: section 5.1)*.
   C. Check for nonreal solutions.
3. Does the program handle rational inputs? If not, fix.

Sample Output

a,b,c? 1/2, −9/2, 7

2,7

4. Modify the program to output the factorization instead of the zeros.

Sample Output

a,b,c? 1,2,−15

$(x+5)^*(x-3)$

a,b,c? 24,1,−3

$(x-1/3)^*(x+3/8)$

a,b,c? 1,4,0

$x^*(x+4)$

5. There is a lot of redundancy in this program (i.e., several identical steps to calculate each factor.) If you have not already done so, employ a function to eliminate redundancy.
6. What factorization is returned for 6,19,15? What is the actual factorization?

Now we want to enhance **quad_root** so that it handles irrational numbers. This occurs when the **discriminant** ($b^2 - 4ac$) is a positive number that it not a perfect square, such as 24.

The **SymPy** library has a **sqrt()** function that is superior to that in the **math** library:

math.**sqrt**(24) = 4.898979485566356 (an approximation, because irrational numbers are nonrepeating, nonterminating)

sympy.**sqrt**(24) = 2*sqrt(6)

We prefer the latter. However, when we insert the root into a factor, it looks kind of awkward:

```
factor="(x–"+str(sqrt(24))+")"

print(factor)
>>> (x–2*sqrt(6))
```

It would look much nicer like this: (x – 2√6)

However, there is no square root symbol √ on a standard computer keyboard (which uses ASCII symbols). Never fear, the square root symbol √ can be inserted into strings using "\u221a." This is called a **unicode** symbol and it works like this:

```
factor="(x–2\u221a6)"

print(factor)

>>> (x–2√6)
```

You may be wondering why we don't simply use **SymPy factor**() to factor a quadratic. Try it with $2x^2 + 6x + 3$:

```
>>>factor('2*x**2+6*x+3')

2*x**2+6*x+3
```

But we were expecting something like this:

[x–(–3/2 + √(3)/2)]*[x–(–3/2 – √(3)/2)]

So it appears that **SymPy factor**() is not reliable when the zeros are irrational. This is one reason why we continue to enhance our own **quad_root** program.

The next issue is whether the program will automatically simplify an expression such as $\frac{\sqrt{24}}{4}$ to $\frac{\sqrt{6}}{2}$? I will leave it to the reader to find out through exploration.

EXERCISE SET 8.2.5

1. Write a function that calculates and outputs the simplified square root of a positive integer. Use the radical sign.

Sample Output

n? 24

2√(6)

n? 200

10√(2)

2. Modify program **quad_root** so that it detects irrational zeros and outputs the associated factors in simplified radical form. Be sure to retest previous data with rational zeros. *(Hint: **math.sqrt** is required for rational zeros to effectively simplify fractions. **sympy.sqrt** is required for irrational zeros to effectively simplify radicals.)*

Sample Data

a,b,c? 1,0,–24

[*x*–(2*√(6))]*[*x*–(–2*√(6))]

a,b,c? 2,6,3

[*x*–(–3/2 + √(3)/2)]*[*x*–(–3/2 – √(3)/2)]

3. Modify the program (or write a new one) that uses **SymPy solve()** to find the zeros of the quadratic. This should greatly shorten the program, but we wanted to make sure you could do it on your own first!

   *(Hint: **solve**('a*x**2+b*x+c') will return a list of two elements.)*

4. Modify the program you wrote in #3 above so that repeated roots are expressed using an exponent.

Sample Data

a,b,c? 1,4,4

$(x - 2)^2$

(Hint: Use unicode '\u00b2' to print the superscripted '2'.)

We have a small snafu. When we use **quad_root** to factor $6x^2 + 19x + 15$, it returns: (*x*+3/2)*(*x*+5/3)

But when you multiply that out, you don't get $6x^2 + 19x + 15$, you get $x^2 + \frac{19}{6}x + \frac{5}{2}$. How did that happen? Take a look at the graphs of both:

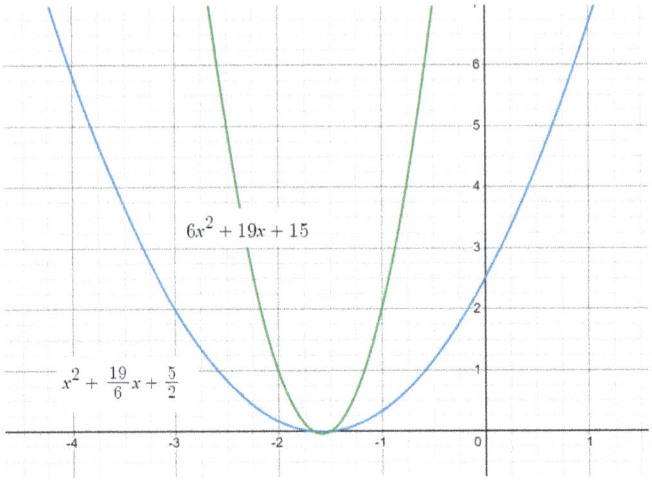

Image 8.15

Both polynomials have the same x-intercepts, but the green one is a "stretch" of the blue one. Thus, two polynomials could share the same zeros, but a set of zeros does not necessarily lead you back to the correct polynomial.

So what can we do to $\left(x + \frac{3}{2}\right)\left(x + \frac{5}{3}\right)$ to transform it to the correct factorization, which is $(2x + 3)(3x + 5)$? It appears the fix would be to multiply the first binomial $\left(x + \frac{3}{2}\right)$ by 2 to eliminate the denominator and likewise multiply the second binomial $\left(x + \frac{5}{3}\right)$ by 3. This gives us $(2x + 3)(3x + 5)$, which indeed multiplies out to $6x^2 + 19x + 15$. I will leave it to you, the intrepid reader, to figure out how to adjust $\left(x - \frac{E}{F}\right)\left(x - \frac{G}{H}\right)$ so that it correctly factors $ax^2 + bx + c$.

EXERCISE SET 8.2.6

1. Modify program **quad_root** so that the correct factorization is output.

Sample Output

a,b,c? 2,7,–15

2*[x–(–5)]*[x–(3/2)]

a,b,c? 6,19,15

6*[x–(–5/3)]*[x–(–3/2)]

2. Given $a = 1, b = 0, c = -4$, give the value and type of each of the following expressions, which are executed sequentially.

   A. quadratic = **str**(a)+'*x**2+'+**str**(b)+'*x+'+**str**(c)
   B. quadratic = **sympify**(quadratic)
   C. quadratic = **str**(quadratic)
   D. quadratic = quadratic.**replace**('**2','\u00b2')

3. Adapt program **quad_root** as a function quad_factor in which the parameters are *a,b,c* (coefficients of the quadratic) and it returns variable **factored_quadratic**.

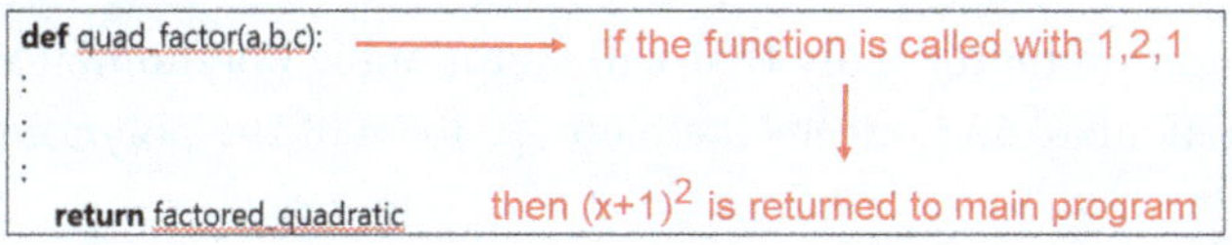

Image 8.16

Return the original quadratic if it cannot be factored using real numbers.

For ease of importing into other programs, you may want to save the function in file **quad_root_fcn.py**.

8.3 Factoring Higher Degree Polynomials

Now that we can factor quadratics using our program **quad_factor** (Exercise Set 8.2.6 #3), we should be ready to write a program that will factor a cubic polynomial. Because the program can't "look" at a graph for a potential zero, we will also need to input one of the known *x*-intercepts in order to get things rolling.

Here's an example of synthetic division to factor $x^3 - 4x^2 - 17x + 60$ using 3 as a zero:

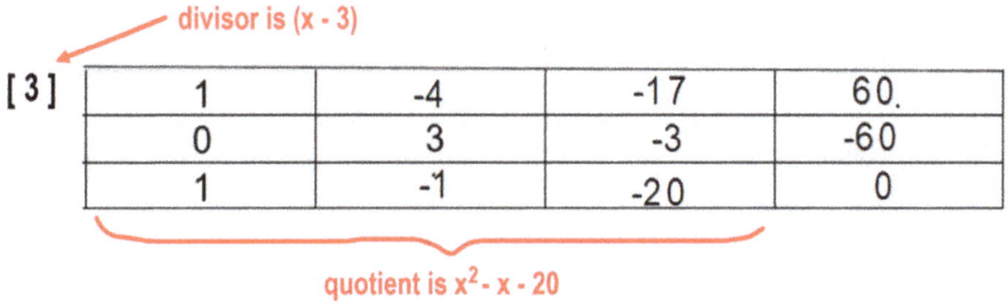

1	-4	-17	60
0	3	-3	-60
1	-1	-20	0

Image 8.17

The next step is to factor the quadratic quotient:

$$x^2 - x - 20 = (x + 4)(x - 5)$$

... giving us a complete factorization of:

$$(x - 3)(x + 4)(x - 5)$$

Using our Python tools, we can replicate the factorization of a cubic polynomial as follows:

F. User inputs coefficients of the cubic a, b, c, d; and a known zero z

G. Form the cubic dividend $ax^3 + bx^2 + cx + d$ and binomial divisor $x - z$

H. Use **SymPy** function **div()** to return the quotient (easier than trying to replicate synthetic division)

I. Use our function **quad_factor()** to factor the quotient

Let's give this a try.

Program Goal: Write a program **factor_poly** that will factor cubic polynomial $x^3 - 4x^2 - 17x + 60$, given that one of its zeros is 3. Calculate and output the factored form of the polynomial.

PROGRAM 8.3.1 FACTOR_POLY.PY

```
## factor_poly.py
## Input: Coefficients of cubic polynomial and one of its zeros
## Output: Factored form of the polynomial
from sympy import *
import quad_root_fcn as qr  # the file that contains function quad_factor()
# main program
a,b,c,d = 1,-4,-17,60          # coefficients of cubic
z = 3                          # corner number
# form first factor (divisor)
factorl = ""
if z>0:
    factorl+="(x-"+str(z)+")"
else:
    factorl+="(x+"+str(abs(z))+")"
# form the cubic dividend
cubic = str(a)+'*x**3+' + str(b)+'*x**2+' + str(c)+'*x+'+str(d)
# perform the division
q,r = div(cubic,factorl)
# get coefficients of quadratic quotient
q = Poly(q)
qa, qb, qc = q.all_coeffs()
# use function quad_factor to factor the quadratic quotient
factored_quad = qr.quad_factor(qa,qb,qc)
# form the full factorization
factored_cubic = factorl+'*'+factored_quad

[ fill in code to output the result ]
```

Desired Output

(x – 3)*[x – (–4)]*[x – (5)]

EXERCISE SET 8.3.1

1. In program **factor_poly**, identify the dividend, divisor, quadratic quotient, and factored quotient using polynomials.
2. Fill in the aqua highlighted code to output the result.
3. Replace a, b, c, d with 6, 7, –14, –8; and z with –2. What is the resulting factorization?
4. Replace the assignment statements to a, b, c, d so that the user can input the coefficients of any cubic and any zero belonging to that cubic.

Sample Output

Coefficients? 1,6,–121,210

Zero? 7

(x–7)*[x–(–15)]*[x–(2)]

Coefficients? 1,–1,–21,45

Zero? –5

$(x+5)$*$[x-(3)]^2$

Coefficients? 2,11,7,10

Zero? –5

2*$(x+5)$*$[2x^2+x+2]$

5. Modify the program so that it also outputs the original polynomial.

Sample Output

Coefficients? 1,6,–121,210

Zero? 7

$x^3+6\bullet x^2-121\bullet x+210$

= (x – 7)*[x – (–15)]*[x – (2)]

6. Modify the program so that it verifies that z is actually a zero of the polynomial.

Sample Output

Coefficients? 1,–2,0,–4

Zero? 3

3 is not a zero of x^3-2x^2-4

Rational Zeros Theorem

One more small improvement will make our factoring program vastly better. What if the user doesn't know or want to enter an initial zero? One thing we could do is start with 1 and just keep looping through the integers until we find a zero. But that would not be helpful for polynomials with factorizations such as:

$$(x - 82)(x + 1{,}000) \text{ or } \left(x - \frac{2}{5}\right)\left(x + \frac{1}{4}\right)$$

Fortunately, there is a theorem to help narrow down the list of possible zeros.

RATIONAL ZEROS THEOREM

If $a_n x^n + a_{n-1}x^{n-1} + \ldots + a_1 x + a_0$ is a polynomial with integer coefficients, and p is a factor of constant term a_0 and q is a factor of leading coefficient a_n, then the list of possible rational zeros is all rational numbers in the form $\frac{p}{q}$.

Example

Given $6x^3 + 7x^2 - 23x - 30$, find the list of possible rational zeros.

Solution

$p =$ factors of $30 = \pm 1, \pm 2, \pm 3, \pm 5, \pm 6, \pm 10, \pm 15, \pm 30$

$q =$ factors of $6 = \pm 1, \pm 2, \pm 3, \pm 6$

$$\frac{p}{q} = \pm 1,\ \pm\frac{1}{2}, \pm\frac{1}{3}, \pm\frac{1}{6}, \pm 2, \pm\frac{2}{3}, \pm 3, \pm\frac{3}{2}, \pm 5, \pm\frac{5}{2}, \pm\frac{5}{3}, \pm\frac{5}{6}, \pm 6, \pm 10, \pm\frac{10}{3}, \pm 15, \pm\frac{15}{2}. \pm 30$$

Recalling the **multiplication principle of counting** from Section 2.1, there are technically $16(8) = 128$ different possibilities for $\frac{p}{q}$; however, many of them are equivalent (for example $\frac{10}{6}$ is the same as $\frac{5}{3}$), so there's no point in checking the same number twice.

EXERCISE SET 8.3.2

For #1–3, find the list of possible rational zeros for each of the following.

1. $6x^3 - 5x^2 - 27x + 14$
2. $12x^3 - 56x^2 + 39x + 35$
3. $2x^4 - 32x^3 + 129x^2 - 16x + 64$
4. Are you guaranteed that at least one of the possible rational zeros is an actual zero?

Program Goal: Write a function **pq** that generates the list of unique possible rational zeros. For testing purposes, call it from a temporary main program that asks the user to input the coefficients of a polynomial. When we're satisfied that it works, we'll call **pq** as a function from program **factor_poly**.

PROGRAM 8.3.2 PQ.PY

```
## function pq
## parameters passed in: first,last (first and last coefficients of an nth
degree polynomial)
## variable returned: possible, a set of possible rational zeros
from fractions import Fraction
def pq(first,last):
    # p = factors of last
    p = [ ]
    for k in range(1,last+1):
        if last % k==0:  # if k is a multiple of last
            p.append(k)
    # q = factors of first
    [ insert code to generate list q of factors of first term ]
    # p_over_q = list of p/q
    p_over_q = [ ]
    for i in range(len(p)):
        for j in range(len(q)):
            p_over_q.append(Fraction(p[i],q[j]))
    possible = p_over_q
    return possible

# main program
coeffs = input("Polynomial coefficients? ")
a = coeffs.split(',')
a = [int(a[i]) for i in range(len(a))]
pq_s = pq(abs(a[0]),abs(a[len(a)-1]))
print(pq_s)
```

Sample Output

```
Polynomial coefficients?
4, -4, -19, -10
[Fraction(1,1), Fraction(1,2),
Fraction(1,4), Fraction(2,1),
Fraction(1,1),
Fraction(1,2), Fraction(5,1),
Fraction(5,2), Fraction(5,4),
Fraction(10,1),
Fraction(5,1), Fraction(5,2)]
```

EXERCISE SET 8.3.3

1. Replace the yellow highlighted code to generate list *q*, factors of the first term.
2. Is it necessary to iterate through **range**(1,*n*+*1*) to check the factors of *n*? How can this search be improved?
3. Modify the program to sort the fractions in the output from least to greatest.
4. Note the use of the **Fraction**() function automatically simplified fractions; for example, (10,4) was automatically simplified to (5,2). However, we still have duplicates. Revise function **pq** so that duplicate factors are not returned.
5. The list is incomplete because it is missing all negative entries. Revise the program to include negative entries.

How to Check If z Is a Zero of a Polynomial

Why do we care so much about factoring polynomials? Recall from the beginning of this chapter that the following four facts go hand in hand:

1. z is a zero of polynomial equation $P(x) = 0$
2. $(x - z)$ is a factor of $P(x)$
3. $P(z) = 0$
4. z is an x-intercept of the graph of P.

Example 1

Given $P(x) = 2x^3 + 5x^2 - 19x - 42$

A. Find the list of possible rational zeros.

B. Is 3 a zero?

C. Is -1 a zero?

D. Is $(2x + 7)$ a factor?

Solution

A. $\pm 1, \pm 1/2, \pm 2, \pm 3, \pm 3/2, \pm 6, \pm 7, \pm 7/2, \pm 14, \pm 21, \pm 21/2, \pm 42$

B. $f(3) = 0$, so, yes, 3 is a zero.

C. $f(-1) = -20$, so, no, -1 is not a zero.

D. $(2x + 7)$ is equivalent to $x = -7/2$. $f(-7/2) = 0$, so, yes, $(2x + 7)$ is a factor.

Example 2

Given $p(x) = 14x^3 - 121x^2 - 368x + 55$

A. Find the list of possible rational zeros.

B. Use the graph to narrow down the list to the most likely possible rational zeros.

C. Which is the least (far left) zero ... is it $-5/2$ or $-15/7$?

D. What is the greatest (far right) zero?

E. Use synthetic division (or program **factor_poly**) to finish factoring the polynomial.

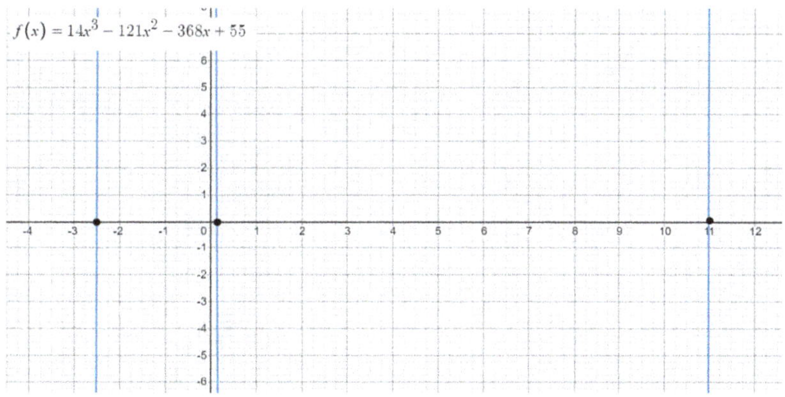

Image 8.18

Solution

A. $\pm 1, \pm 1/2, \pm 1/7, \pm 5, \pm 5/2, \pm 5/7, \pm 11, \pm 11/2, \pm 11/7, \pm 15, \pm 15/2, \pm 15/7$

B. Possibly: $-5/2, -15/7, 1/7, 11$

C. $f(-5/2) = 0$, so it's $-5/2$

D. $f(11) = 0$, so it's 11.

E.

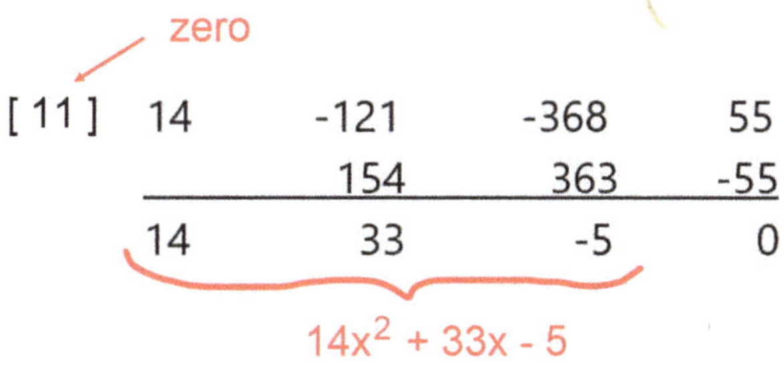

Image 8.19

Use function **quad_factor()** (or factor by hand) to get $14x^2 + 33x - 5 = 14(x + 5 / 2)(x - 1 / 7)$. Thus, the full factorization, including $(x - 11)$, is:

$14(x - 11)(x + 5 / 2)(x - 1 / 7)$ or $(x - 11)(2x + 5)(7x - 1)$.

EXERCISE SET 8.3.4

1. Given $18x^3 + 111x^2 + 17x - 6$:

   A. Find the list of possible rational zeros.
   B. Is 1/3 a zero?
   C. Is 3 a zero?
   D. Is $(3x + 1)$ a factor?

2. Given $10x^4 - 37x^3 + 24x^2 + 23x - 20$:

   A. Find the list of possible rational zeros and then use the graph to narrow down the list.
   B. Is $-1/2$ a zero?
   C. Is $(5x + 4)$ a factor?

3. Write a program that asks the user to input a polynomial p and possible zero. Determine if the zero is an actual zero of $p = 0$.

   Hint: try using:

   q,r = ***div****(f,g)*

   which finds the quotient of f/g. The ***div()*** *function returns a tuple q,r, which is the quotient and remainder.*

We are now ready to modify program 8.3.1 **factor_poly** to incorporate function **pq**. The user will no longer have to input an initial zero *z*. Rather than providing for you the revised **factor_poly**, we will give a hint (in question 1 below).

SAMPLE OUTPUT

Coefficients of *a,b,c,d*? 18,111,17,–6

$18x^3 + 111x^2 + 17x - 6$

18*[*x* – 1 / 6]*[*x* – (–6)]*[*x* – (–1/3)]

Coefficients *a,b,c,d*? 14,–121,–368,55

$14x^3 - 121x^2 - 368x + 55$

14*(*x* – 11)*[*x* – (–5 / 2)]*[*x* – (–1 / 7)]

EXERCISE SET 8.3.5

1. Here's your hint. Figure out where in the main program to insert the following code:

```
# Get the list of possible rational zeros
pq_s = pq(abs(int(a)),abs(int(d)))
found = False
x = Symbol('x')
# Cycle through set elements in pq_s until you find one in which P(z) = 0
while found==False and len(pq_s)>0:
```

```
next = Fraction(pq_s.pop())
if poly.subs({x:next})==0:
z = next
found = True
```

2. Does the program work for a cubic polynomial in which the constant term is 0—for example, $x^3 + 2x^2$? If not, fix. *(Hint: Factor out x.)*
3. Does the program work for a cubic polynomial that has irrational zeros—for example, $x^3 - x^2 - 5x + 5$? If not, fix.
4. Does the program work for a polynomial in which the first or constant term is a fraction—for example, $\frac{1}{3}x^3 + \frac{2}{3}x^2 - \frac{1}{3}x - \frac{2}{3}$? (Hint: Multiply by a temporary constant to transform the first and last numbers to whole numbers. Divide out the constant at the last step.)
5. It's tempting to believe that the program can now handle any cubic polynomial, but try asking it to factor $x^3 + 5$. Does this mean the graph has no real zeros? Take a look:

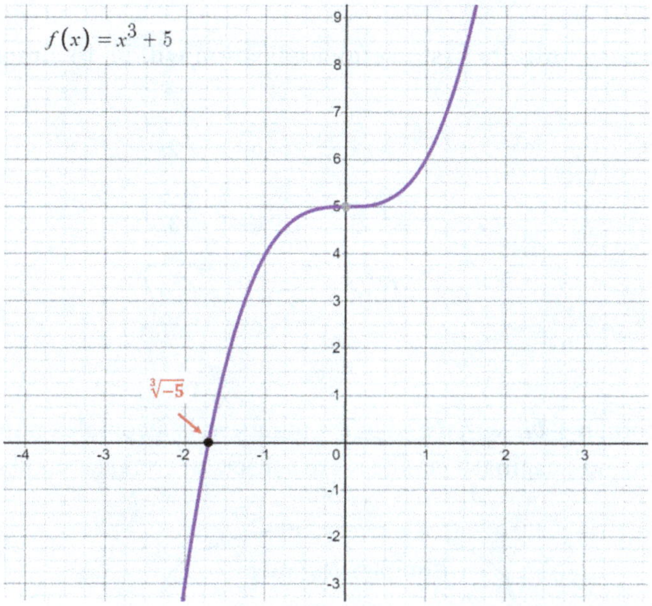

Image 8.20

Even sympy.**factor**('x**3+5') returns the original polynomial without bothering to factor over real numbers because this would involve taking a cube root, which would add another layer of complexity to the program. We will defer that level to Chapter Project II; however, for now, simply output "irreducible."

You've made several changes to program **factor_poly** in Exercise Set 8.3.5. Make sure it still works with all the sample data provided for program 8.3.1. sympy.**factor**() could easily replace most of this program, yet relying exclusively on that function would have provided you no understanding of the underlying iterative process involved in factoring. Further, the current release of sympy.**factor**() does not handle polynomials with radical zeros (such as $x^2 - 7$ or $x^3 + 5$). Take pride that our program handles polynomials with square root zeros.

We are now going to expand our knowledge of number types that will enable us to solve and factor even more polynomials. This new field of numbers is called ...

8.4 Complex Numbers

We know that we can factor $x^2 - 4$ as the difference of squares: $(x + 2)(x - 2)$. However, we have been unable to factor a **sum of squares** such as $x^2 + 4$.

Yet.

If we expand the field of real numbers to the field of **complex numbers**, we can factor $x^2 + 4$ and do much, much more.

The field of complex numbers is predicated on a single definition:

$$\boldsymbol{i} = \sqrt{-1}$$

i is called the **imaginary unit**. Its first useful application will be to simplify the square roots of negative numbers:

$$\sqrt{-4} = \sqrt{4}\sqrt{-1} = 2i$$

$$\sqrt{-9} = \sqrt{9}\sqrt{-1} = 3i$$

$$\sqrt{-16} = \sqrt{16}\sqrt{-1} = 4i$$

etc.

Any number in the form $\boldsymbol{a} + \boldsymbol{bi}$, with a and b real numbers, is a complex number. In the diagram, we can see that real numbers and imaginary numbers are subsets of complex numbers:

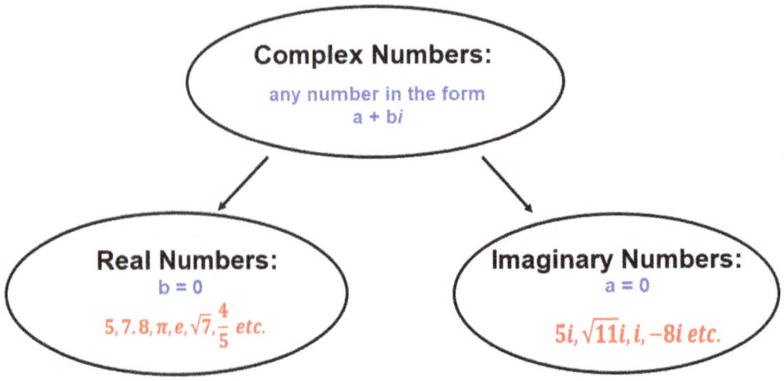

Image 8.21

$a + bi$, in which a and b are both real numbers, is called **standard form** for a complex number. The order is important:

Standard Form	Not Standard Form
$6 + 8i$	$8i + 6$
-5	$6 + i \cdot 8$
$4i$	$2(3 + 4i)$

If the real part $a = 0$, then bi is called a **pure imaginary number**—for example, $3i$ or $-i$.

We can now solve $x^2 + 4 = 0$ using imaginary numbers as follows:

$x^2 = -4$ ← Take square root of both sides

$x = \pm\sqrt{-4} = \pm 2i$

Thus $x^2 + 4$ can be factored as $(x + 2i)(x - 2i)$.

We can now in fact solve *any* quadratic equation because we are no longer intimidated by $b^2 - 4ac < 0$. It will be easier to solve using this slight rearrangement of the quadratic formula:

$$x = -\frac{b}{2a} \pm \frac{\sqrt{b^2 - 4ac}}{2a}$$

Example

Solve $x^2 + 2x + 3 = 0$ and express your answer(s) in standard form $a + bi$.

Solution

$a = 1, b = 2, c = 3$

$$-\frac{b}{2a} \pm \frac{\sqrt{b^2 - 4ac}}{2a} = -\frac{2}{2} \pm \frac{\sqrt{4 - 4(1)(3)}}{2} = -1 \pm \frac{\sqrt{-8}}{2} = -1 \pm \frac{2\sqrt{2}}{2}i = -1 \pm \sqrt{2}i$$

EXERCISE SET 8.4.1

1. Simplify $\sqrt{-120}$
2. Solve $x^2 = -80$
3. Factor $x^2 + 81$
4. Solve $x^2 + 2x = -5$
5. Solve $2x^2 + 3x + 6 = 0$
6. Enter the following into the python console:

(Continued)

```
from sympy import *
solve('x**2+9')
factor('x**2+9')
```

What is the result of **solve**? What is the result of **factor**?

7. CHALLENGE. Modify program **factor_poly** (section 8.3) so that it now factors cubics with nonreal zeros.

Sample Output

Coefficients *a,b,c,d*? 1,0,0,1

$x^3 + 1 =$

$(x + 1)$*[x – (1 / 2 – √(3)*I / 2)]*[x – (1 / 2 + √(3)*I / 2)]

Complex Numbers in Python

Recall from our discussion in Chapter 5 the different types of numbers in Python:

Name	Python Type	Example
integer	int	−4, 0, 3
decimal	float	−7.2, 5.0, 3.14159
fraction	Fraction (must include fractions module)	**Fraction**(3,4) (represents ¾)

Python also has a **complex** type number, which can be expressed in the following ways:

$a + bj$ (where the j represents i)
complex(a,b)

The real (a) and imaginary (b) parts of a complex number can be represented as shown in the following example.

```
>>>z = −4 + 5j
>>>z.real
−4.0
>>>z.imag
5.0
```

To input a complex number:

z = **complex(input**("z?"))

Adding and subtracting complex numbers is as simple as adding and subtracting polynomials. The imaginary unit i can be treated like any other variable. For example, $3 + 9i - (7 - 2i) = (3 - 7) + (9i + 2i) = -4 + 11i$. In general:

ADDING COMPLEX NUMBERS

$(a + b\,i) + (c + d\,i) = (a + c) + (b + d)\,i$

EXERCISE SET 8.4.2

1. Write a program that asks the user to input two complex numbers $a + bi$ and $c + di$, then calculates and outputs their sum.

Sample Output

z1? 3−5j

z2? 4.5+6j

(3−5j) + (4.5+6j) =

(7.5+1j)

2. What is the type of answer returned when you add a float and complex number—for example, 7.0 + complex(4,3)?

Oddly, **complex(input())** does not enable the user to input a square root such as **sqrt**(4) or **sqrt**(−9). Here's the workaround that requires the **simplify()** function in the **SymPy** library:

>>>z = **simplify**(**input**("z?")) ← User enters **sqrt**(−9)

z is converted to a **SymPy** complex type number in the form 3**I*.

However, if you wish to do Python arithmetic calculations, you need to convert it to a Python complex type variable:

>>>z = **complex**(z) ← converts 3**I* to complex number which is 3*j*

Yet another representation of a complex number by **SymPy**:

>>>**pprint**(4+sqrt(5)*I)
$4 + \sqrt{5}\cdot i$

Keep in mind that a **SymPy** complex number type is not the same as the built-in Python complex number type. To identify the real and imaginary parts of a **SymPy** complex number type, import the **re**() and **im**() functions from the **SymPy** library and use:

re(z)

im(z)

EXERCISE SET 8.4.3

1. Modify the program you wrote in Exercise Set 8.4.2 #1 so that the user can input square root expressions. Express answer as a complex number with a,b as simplified radicals.

Sample Output

z1? 2+sqrt(−9)

z2? sqrt(11)+sqrt(25)*I

$2 + \sqrt{11} + 8{\cdot}i$

z1? sqrt(3)+10*I

z2? 14+sqrt(12)*I

$\sqrt{3} + 14 + 2{\cdot}\sqrt{3}{\cdot}i + 10{\cdot}i$

z1? 9-sqrt(−3)

z2? 6+2j

$15 - \sqrt{3}{\cdot}i + 2{\cdot}i$

2. Let's improve the program even more. Technically, a complex number should be expressed in the form $\boldsymbol{a} + \boldsymbol{bi}$. As you can see in the sample output above, the **SymPy simplify()** function does not combine coefficients of i if one is a square root and the other is not. Fix this.

Sample Output

z1? sqrt(3)+10*I

z2? 14+sqrt(−12)

sqrt(3) + 14 + (2*sqrt(3) + 10)i

z1? 9−sqrt(−3)

z2? 6+2j

15 + (2 − sqrt(3))i

*Hint: As of this writing, **SymPy** does not correctly simplify the sum of complex numbers, so you will need to add the real and imaginary parts separately.*

*If you'd like to output that really cool italicized i as used by **SymPy's pprint**(), use unicode '\u2148'*

8.5 Powers of *i*

Recall the definition of i is:

$$i = \sqrt{-1}$$

Squaring both sides, you get:

$$i^2 = -1$$

Building upon these two identities, we can derive the powers of i are as follows:

$i^0 = 1$	$i^4 = (i^2)^2 = (-1)^2 = 1$	$i^8 = (i^4)^2 = (-1)^8 = 1$
$i^1 = i$	$i^5 = i^4 \cdot i^1 = (1)(i) = i$	$i^9 = i^8 \cdot i^1 = (1)(i) = i$
$i^2 = i \cdot i = -1$	$i^6 = i^4 \cdot i^2 = (1)(-1) = -1$	$i^{10} = i^8 \cdot i^2 = (1)(-1) = -1$
$i^3 = i^2 \cdot i = (-1)\, i = -i$	$i^7 = i^4 \cdot i^3 = (1)(-i) = -i$	$i^{11} = i^8 \cdot i^3 = (1)(-i) = -i$

And so on. Can you see the pattern?

EXERCISE SET 8.5.1

Express your answer(s) in standard form $a + bi$.

1. Simplify i^{93}
2. Simplify $11 - i^{11}$
3. Simplify $12i^2 - 7i + 4$
4. What is the result of entering the following code into the console:

   ```
   >>>i = complex(0,1)
   >>>i**101
   ```

   Is the result accurate? What would the simplest answer be?
5. Write a program that will accept positive integer n as input and then simplify and output i^n.

Sample Output

```
n? 101
i^101 = i
```

(Hint: This relates to the modulo % operator.)

8.6 Multiplying Complex Numbers

Use the distributive property to multiply a real number by a complex number, or a pure imaginary number by a complex number:

$$2(11 - 6i) = 22 - 12i$$

$$3i(7 + 4i) = 21i + 12i^2 = 21i + 12(-1) = -12 + 21i$$

Multiplying a pair of complex numbers is similar to the FOIL (First-Outer-Inner-Last) method of multiplying two binomials. For example:

$$(3 - 10i)(16 + i) = 48 + 3i - 160i - 10i^2$$

The only difference is the addition of a final step, which is to substitute i^2 with -1 and then combine like terms.

$$48 + 3i - 160i - 10i^2$$

$$= 48 - 157i - 10(-1)$$

$$= 58 - 157i$$

(Remember to express your complex number answer in the form $a + bi$.)

EXERCISE SET 8.6.1

Express your answer(s) in standard form $a + bi$.

1. Simplify $(4 - 5i) \bullet 6$
2. Simplify $2i(-5 - 7i)$
3. Simplify $(7 + 6i)(8 - 3i)$
4. Simplify $(5 - 2i)(5 + 2i)$
5. Simplify $(14 + 2i)^2$
6. Simplify $(3 - i)^3$
7. Which Python statement will find the product of $6 + 5i$ and $5 + 7i$, and what is the value returned?
8. Which **SymPy** function call will find the product of $6 + 5i$ and $5 + 7i$, and what is the value returned?
9. Write a program that asks the user to input two complex numbers, and find their product by manipulating their real and imaginary parts. Do not use any **SymPy** functions.

Sum of Squares

In a previous math course, you learned to factor a **difference of squares**:

$$a^2 - b^2 = (a + b)(a - b)$$

A special feature of complex numbers is that we can now factor a **sum of squares**. Look at Exercise Set 8.6.1 # 4 above: $(5 - 2i)(5 + 2i)$. Notice the answer was simply $5^2 + 2^2 = 29$. In general:

FACTORING A SUM OF SQUARES

$$a^2 + b^2 = (a + bi)(a - bi)$$

Very similar to difference of squares, just stick the i in there.
The **conjugate** of $a + bi$ is $a - bi$ and vice-versa. As you can see, the product of conjugate pairs is $a^2 + b^2$.

Example

Find the product: $(7 + i)(7 - i)$

Solution

In this example, $a = 7$ and $b = 1$.

$$(7 + i)(7 - i) = (7)^2 + (1)^2 = 49 + 1 = 50$$

To find the conjugate of a Python complex type number z, use z.**conjugate**().

To find the conjugate of a **SymPy** complex type number z, use **conjugate**(z).

EXERCISE SET 8.6.2

1. Multiply: $(7 + 8i)(7 - 8i)$
2. Multiply: $(1 - 10i)(1 + 10i)$
3. Factor $x^2 + 81$
4. Factor $121x^2 + 49$
5. Factor $45x^2 + 50$
6. One of the shortcomings of the **SymPy factor()** function is that it does not factor polynomials with either imaginary or irrational roots. What is returned by the following? (Remember to declare $x =$ **Symbol**('x') first).
   A. **factor**(x**2–25)
   B. **factor**(x**2–24)
   C. **factor**(x**2+25)
7. Write a program that will factor a sum of squares.

Sample Output

Sum of squares in the form x**2 + a? x**2+50

(x – 5*√(2) i)(x + 5*√(2) i)

(Continued)

8. Write a brief program that finds the conjugate of a built-in complex number.

Sample Output

Complex Number? 16–92*j*

The conjugate is (16+92*j*)

9. Write a brief program that finds the conjugate of a **SymPy** complex number.

Sample Output

Complex Number? 16+92**I*

The conjugate is 16–92**I*

8.7 Quotients of Complex Numbers

Recall from a previous math course how to simplify fractions with radicals in the denominator, examples:

A. $\frac{1}{\sqrt{2}} = \frac{1}{\sqrt{2}} \cdot \frac{\sqrt{2}}{\sqrt{2}} = \frac{\sqrt{2}}{2}$

B. $\frac{3}{\sqrt[3]{7}} \cdot \frac{\sqrt[3]{7^2}}{\sqrt[3]{7^2}} = \frac{3\sqrt[3]{49}}{\sqrt[3]{7^3}} = \frac{3\sqrt[3]{49}}{7}$

The goal was to "rationalize" the denominator by multiplying by whatever expression would turn it into an integer.

We will take a similar tack in simplifying quotients with imaginary numbers in the denominator, such as $\frac{4}{i}$, $\frac{\sqrt{2}}{3-7i}$, etc.

We saw when adding, subtracting, and multiplying complex numbers that i can be handled as any other variable, i.e., combining like terms, rules for exponents, etc.

We only need three tools to convert an imaginary expression to a real one:

4. $i^2 = -1$

5. $i^4 = 1$

6. $(a + bi)(a - bi) = a^2 + b^2$

Our goal in simplifying is to eliminate imaginary numbers in the denominator.

Example 1

Simplify $\frac{4i}{i^3}$

Solution

$$\frac{4i}{i^3} = \frac{4}{i^2} = \frac{4}{-1} = -4$$

Example 2

Simplify $\frac{6\sqrt{3}}{i}$

Solution

$$\frac{6\sqrt{3}}{i} = \frac{6\sqrt{3}}{i} \bullet \frac{i}{i} = \frac{6i\sqrt{3}}{i^2} = \frac{6i\sqrt{3}}{-1} = -6i\sqrt{3}$$

Example 3

Simplify $\frac{-15}{i+1}$

Solution

$$\frac{-15}{1+i} = \frac{-15}{1+i} \cdot \frac{1-i}{1-i} = \frac{-15+15i}{1+1} = \frac{-15+15i}{2} = -\frac{15}{2} + \frac{15}{2}i$$

Multiply by conjugate of 1 + i to form a sum of squares

Image 8.22

Example 4

Simplify $\frac{2+7i}{5-2i}$

Solution

$$\frac{2+7i}{5-2i} = \frac{2+7i}{5-2i} \cdot \frac{5+2i}{5+2i} = \frac{10+4i+35i+14i^2}{5^2+2^2} = \frac{10+39i+14(-1)}{29} = \frac{-4+39i}{29} = \frac{-4}{29} + \frac{39}{29}i$$

Multiply by conjugate of 5 - 2i to form a sum of squares

Image 8.23

EXERCISE SET 8.7.1

Express your answers in standard complex form $a + bi$ using rational numbers for a and b.

1. Simplify $\frac{3}{7i}$
2. Simplify $\frac{1-3i}{1+2i}$
3. Simplify $\frac{8}{i-3}$
4. Simplify $\frac{7+4i}{-3-i}$
5. Simplify $\frac{-5+5i}{-5-5i}$
6. Finding the quotient of complex numbers is relatively straightforward using the **SymPy simplify**() function: **simplify**(dividend/divisor) returns the quotient as a complex number. Write a program that asks the user to input a dividend and divisor, then output the quotient. Use it to verify your answers to #1–5 above.

Sample Output

Dividend? 3+2*I

Divisor? −4*I

$-\frac{1}{2}+\frac{3 \cdot i}{4}$

7. Will this program work with polynomials as well? For example will it find $x^2 + 25$ divided by $x + 5i$?

8.8 Fundamental Theorem of Algebra

Most of precalculus, especially the factoring techniques, lead up to what is considered its most important concept:

THE FUNDAMENTAL THEOREM OF ALGEBRA

The theorem of equations was proved by Carl Friedrich Gauss in 1799. It states that every polynomial equation of degree $\boldsymbol{n}$ with complex coefficients has exactly $\boldsymbol{n}$ complex zeros.

Thus, a 1st-degree polynomial $ax + b$ has one zero, a 2nd-degree polynomial $ax^2 + bx + c$ has two zeros, a 3rd-degree has 3 zeros, etc. For degree $n \geq 2$, this would be some combination of real and imaginary zeros.

A zero with multiplicity ≥ 2 would be considered a repeated zero. For example, $f(x) = (x + 2)^2$ has two (identical) zeros: $x = -2$, because the multiplicity of -2 is 2.

Imaginary zeros always occur in conjugate pairs—for example, if one of the zeros is $2 + 5i$, there is another zero $2 - 5i$.

Example 1

Use the given zero to find the remaining zeros of the polynomial, and then multiply it out so that it is in standard form.

Degree: 2; zero: $-6 - 7i$

Solution

Because imaginary zeros always occur in conjugate pairs, the other zero is $-6 + 7i$. Substituting into $(x - z_1)(x - z_2)$, we get:

$$(x - (-6 - 7i))(x - (-6 + 7i))$$

Here's a handy shortcut: $(x - (a + bi))(x - (a - bi)) = x^2 - 2ax + (a^2 + b^2)$. In this example, $a = -6$ and $b = 7$. Thus:

$$(x - (-6 - 7i))(x - (-6 + 7i)) = x^2 - 2(-6)x + ((-6)^2 + 7^2)$$

$$= x^2 + 12x + 85$$

Example 2

Use the given zeros to find the remaining zeros of the polynomial, and then multiply it out so that it is in standard form.

Degree: 4; zeros: $2 - i$, $-i$.

Solution

A 4th-degree polynomial has four complex zeros. We are only given two of them. Because imaginary zeros occur in pairs, if we are given that $2 - i$ and $-i$ are zeros, then so are $2 + i$ and i. Substituting these into $(x - z_1)(x - z_2)(x - z_3)(x - z_4)$, we get:

$$(x - (2 - i))(x - (2 + i))(x + i)(x - i)$$

$$= (x^2 - 4x + 5)(x^2 + 1)$$

$$\therefore\ x^4 - 4x^3 + 6x^2 - 4x + 5$$

EXERCISE SET 8.8.1

In #1–6, use the given zero(s) to find the remaining zero(s) of the polynomial function, and find the standard form $a_n x^n + a_{n-1} x^{n-1} + \ldots + a_1 x + a_0$.

1. Degree 2: zero: $7 + 2i$
2. Degree 3: zeros: $4, 3 + i$

(Continued)

3. Degree 5; zeros: 0, 1, 2, i
4. Degree 4: zeros: i, $1+i$
5. Degree 3: zeros: 3, $4-i$
6. Degree 4: zeros: $\pm 3i$; 5 with multiplicity 2
7. Prove the formula used in Example 1: $(x-(a+bi))(x-(a-bi)) = x^2 - 2ax + (a^2+b^2)$.
8. Write a program that uses the formula in Example 1 to expand $(x-(a+bi))(x-(a-bi))$. Ask the user to input values for a and b. Output the corresponding quadratic polynomial $x^2 - 2ax + (a^2+b^2)$.

Sample Output

a? −3

b? 7

$(x-(-3+7i))(x-(-3-7i))=$

x**2 + 6x + 58

Program Goal: Write a program that asks the user to input a complex number. Find the conjugate of the input and then find the quadratic polynomial having that conjugate pair as its zeros.

SAMPLE OUTPUT

complex number? 2+3*I
$(x-(2+3^*I))(x-(2-3^*I)) =$
$x^2 - 4.0x + 13.0$

This program performs the same operations as Exercise Set 8.8.1 #8.

PROGRAM 8.8.1 FORM_POLY_FROM_ZEROS.PY

```
## conjugate.py
## Input: complex number
## Output: conjugate and factorization
from sympy import *
z1 = simplify(input("Complex Number? "))
z2 = conjugate(z1)
x = Symbol('x')
factor1 = '(x-('+str(z1)+'))'
factor2 = '(x-('+str(z2)+'))'
print(factor1,factor2,'=')
[ fill in code to calculate and output the expanded quadratic ]
```

Note: This program makes use of **SymPy** complex type numbers, not Python's built-in complex type numbers.

EXERCISE SET 8.8.2

1. Fill in the yellow highlighted code to calculate and output the expanded quadratic. You may need to import additional functions from **SymPy**.
2. CHALLENGE. Modify the program so that the user can enter any additional zeros of the function.

Sample Output

complex number? 3−4j

other zeros? 7,−10

$x^4 - 3.0 \cdot x^3 - 63.0 \cdot x^2 + 495.0 \cdot x - 1750.0$

complex number? 6+7j

other zeros? ← leave blank if none

$x^2 - 12{\cdot}0x + 85.0$

Now that you know how to handle polynomial factors that involve complex zeros, you are ready to completely factor a polynomial function and find its zeros. By *completely* we mean both the real and imaginary zeros.

Quick recap of all the techniques we've learned thus far:

If It's a Quadratic

1. Try to factor.

   Examples:

   $x^2 - 13x + 42 = (x - 7)(x - 6);$

   $10x^2 + 11x - 6 = (2x + 3)(5x - 2).$

2. If it does not factor using integers, use any of the quadratic solving techniques in Section 3.6.

   Example: $x^2 - 7x + 3 = \left(x - \frac{7+\sqrt{37}}{2}\right)\left(x - \frac{7-\sqrt{37}}{2}\right) = (x - 6.541)(x - 0.459)$

3. If it does not factor using real numbers, use the quadratic formula with imaginary numbers.

   Example: $x^2 + 3x + 5 = \left(x - \frac{-3+i\sqrt{11}}{2}\right)\left(x - \frac{-3-i\sqrt{11}}{2}\right)$

If It's a Higher Degree Polynomial

1. Try doing synthetic division using the list of possible rational zeros.

   Example: $x^3 + 2x^2 - 3x - 10 = (x - 2)(x^2 + 4x + 5)$.

2. Iterate through synthetic division until the quotient is a quadratic. Then use any of the steps above under the heading "If It's a Quadratic."

3. If synthetic division does not produce any integer or rational zeros, then use an approximation technique such as the **midpoint method** (Chapter 3 Chapter Project VIII) to find possible irrational zeros.

4. If that doesn't work, you might not have any real zeros.

   Example: $x^4 + 1$ has no real zeros. As you can see, it does not touch the x-axis:

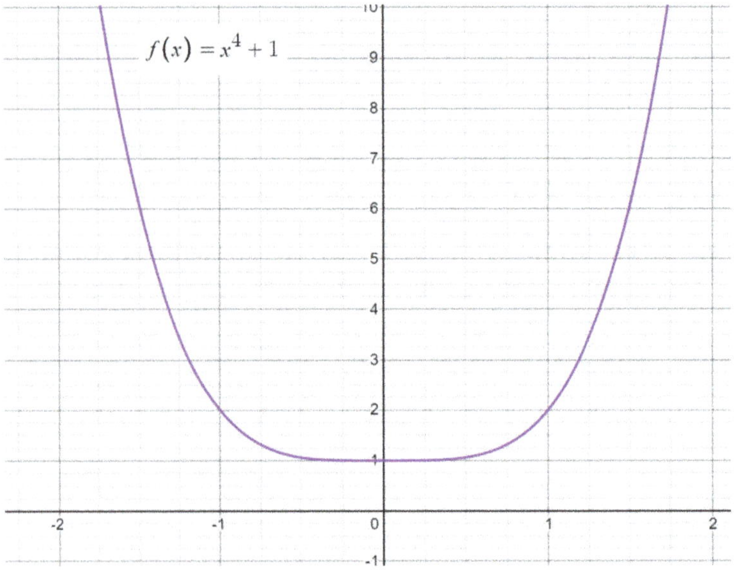

Image 8.24

The following examples use purely algebraic techniques, although you may have written programs earlier in the chapter to help with any or all of the steps.

Example 1

Factor $P(x)$ completely, and identify the zeros of $P(x) = 0$.

$$P(x) = -x^2 + 4x - 11$$

Solution

The quadratic cannot be factored. Use the quadratic formula to find the zeros:

$$x = \frac{-4 \pm \sqrt{(4)^2 - 4(-1)(-11)}}{2(-1)} = 2 \pm \sqrt{7}i$$

The factorization is thus $(x - (2 + \sqrt{7}i))(x - (2 - \sqrt{7}i))$.

Example 2

Factor $P(x)$ completely, and identify the zeros of $P(x) = 0$.

$$P(x) = 6x^3 + 25x^2 + 3x - 4$$

Solution

The list of possible rational zeros is $\{\pm 1, \pm 2, \pm 4, \pm\frac{1}{2}, \pm\frac{1}{3}, \pm\frac{1}{6}, \pm\frac{2}{3}, \pm\frac{4}{3}\}$. An examination of the graph shows that $x = -4$ appears to be an x-intercept:

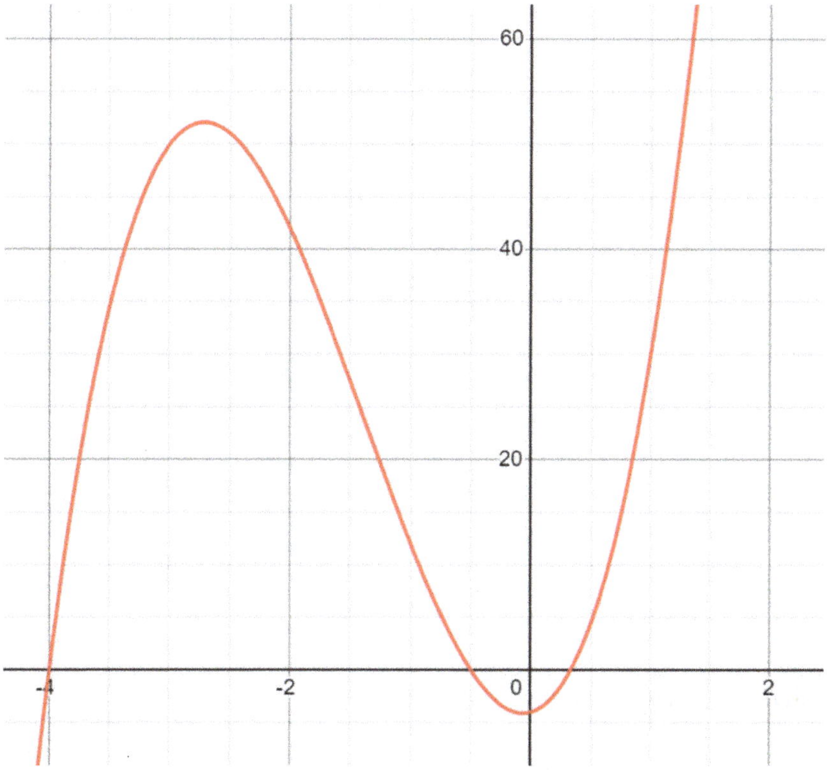

Image 8.25

Try synthetic division using -4 as the “corner number”:

-4	6	25	3	-4
		-24	-4	4
	6	1	-1	0

The remainder of zero verifies that -4 is a zero. The factorization thus far is:

$$(x + 4)(6x^2 + x - 1)$$

The quadratic factors nicely, giving us a complete factorization of:

$$(x + 4)(2x + 1)(3x - 1)$$

∴ The 3 corresponding zeros are -4, $-1/2$, $1/3$ (all rational).

Example 3

Factor $P(x)$ completely, and identify the zeros of $P(x) = 0$.

$$P(x) = 2x^4 + 3x^3 - 2x^2 + 27x - 180$$

Solution

This list of possible rational zeros is quite long, but an examination of the graph shows that $x = -4$ appears to be an x-intercept:

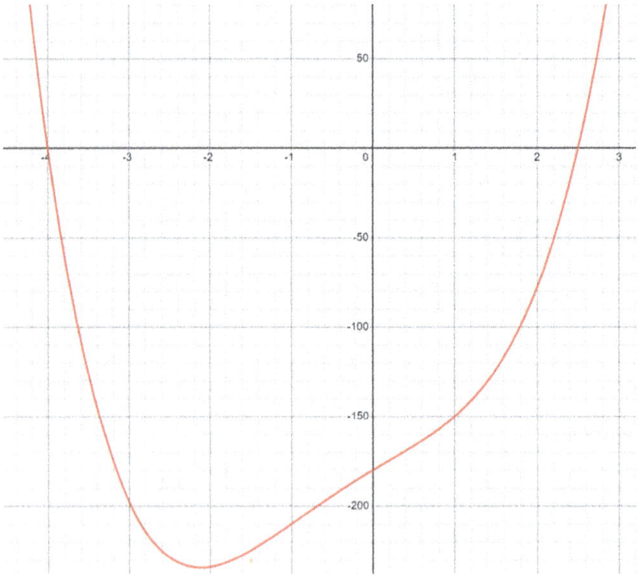

Image 8.26

Try synthetic division using -4 as the "corner number":

$\boxed{-4}$	2	3	-2	27	-180
		-8	20	-72	180
	2	-5	18	-45	0

The remainder of zero verifies that -4 is a zero. The factorization thus far is:

$$(x + 4)(2x^3 - 5x^2 + 18x - 45)$$

Another glance at the graph shows that 5/2 is probably the other real zero, so confirm using synthetic division again:

5/2	2	−5	18	−45
		5	0	45
	2	0	18	0

This results in a factorization of:

$$(x + 4)(x - 5/2)(2x^2 + 18)$$

$= (x + 4)(x - 5 / 2) \cdot 2(x^2 + 9)$ ← Factor 2 out of $(2x^2 + 18)$

$= (x + 4)(2x - 5)(x^2 + 9)$ ← Distribute 2 to $(x - 5 / 2)$

$= (x + 4)(2x - 5)(x + 3i)(x - 3i)$ ← Factor $(x^2 + 9)$ using Sum of Squares

∴ The corresponding zeros are -4, $5/2$ (rational), $\pm 3i$ (imaginary).

Example 4

Factor $P(x)$ completely, and identify the zeros of $P(x) = 0$.

$$P(x) = x^4 + 50x^2 + 49$$

Solution

The set of possible rational zeros is $\{ \pm1, \pm7, \pm49 \}$; however, the graph reveals there are no real zeros (no x-intercepts):

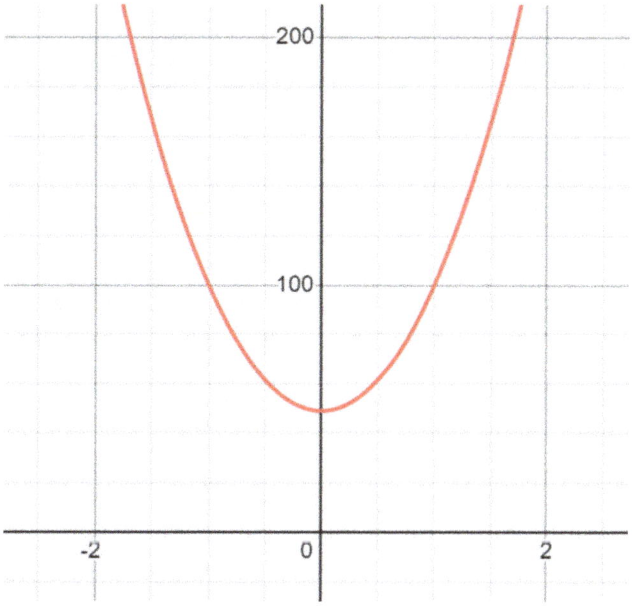

Image 8.27

We therefore have no rational numbers with which to attempt synthetic division. However, notice that the quartic can be factored as a quadratic:

$$x^4 + 50x^2 + 49$$
$$= (x^2 + 1)(x^2 + 49)$$
$$= (x + i)(x - i)\,(x + 7i)(x - 7i)$$

$\therefore$ The corresponding zeros are $\pm\, i$, $\pm 7i$ (all imaginary)

EXERCISE SET 8.8.3

Find the zeros of the following polynomial functions and use them to completely factor the polynomial.

1. $x^2 + 72$
2. $x^2 + 7x + 13$
3. $x^3 - 6x^2 - 31x + 36$
4. $x^3 + 27$
5. $2x^3 - 3x^2 - 12x + 20$
6. $x^4 + x^3 - 10x^2 - 4x + 24$
7. $x^4 + x^3 - 2x^2 + 4x - 24$
8. $x^4 + 4$

Is there life beyond polynomial functions? Yes, in the next two chapters we will explore exponential and logarithmic functions.

Chapter Projects

I. Plotting Complex Numbers

Complex numbers are easy to graph. Relabel the horizontal axis as the **real** axis and the vertical axis as the **imaginary** axis. The point representing complex number $a + bi$ is in the same position as ordered pair (a,b).

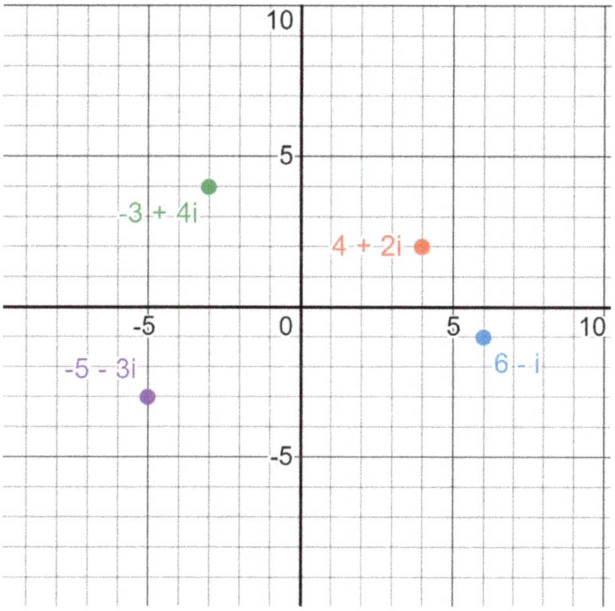

Image 8.28

Complex numbers are typically named with the variable ***z***; i.e., $z = a + bi$. The **modulus** of a complex number denoted $|z|$ refers to its distance from the origin.

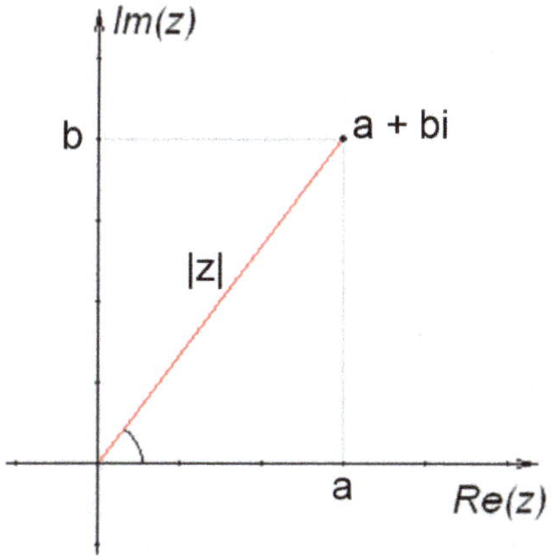

Image 8.29

An easy derivation of the distance formula $\sqrt{(x_2 - x_1)^2 + (y_2 - y)^2}$ yields the following.

MODULUS OF A COMPLEX NUMBER

If $z = a + bi$, then $|z| = \sqrt{a^2 + b^2}$

Using Python:

```
abs(z)
```

Write a program that asks the user to enter a complex number $a + bj$. Draw the axes, labeled “Real” (x-axis) and “Imaginary” (y-axis). Draw a line segment from the origin to (a, b). Calculate the modulus and output the value of the modulus at the approximate midpoint of the line segment. Express the modulus as a simplified radical, not a decimal number.

SAMPLE OUTPUT

$a + bj$? $-2+8j$

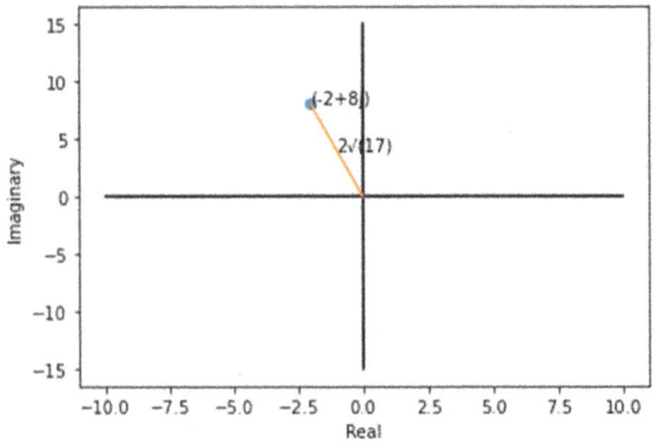

Image 8.30

Plotting complex numbers plays an important role in the stunning graphs of fractals. It is left to the motivated reader to explore this fascinating field.

II. Factoring a Cubic with Irrational Zeros

Modify program 8.3.1 **factor_poly** (which factors cubic polynomials) so that it also factors cubics with irrational roots such as $x^3 + 5$.

III. Binomial Expansion Using Complex Numbers

In Chapter 3 Chapter Project III, you wrote a program that did the binomial expansion of $(a + b)^n$. Modify this program (or write a new one) so that it performs binomial expansion of complex numbers in the form $(a + bi)^n$. Express result as both an nth degree polynomial and a simplified complex number in standard form $a + bi$.

SAMPLE OUTPUT

A? 3
B? 4
N? 3
(3 + 4*i*)^3 =
64*I**3 + 144*I**2 + 108*i**I+ 27 =
−117 + 44*I

IV. A Robust Polynomial Factorization Program

Modify program **factor_poly** so that it factors any degree polynomial (not just cubics) over the field of complex numbers.

Make sure your program handles the following:

A. polynomials of any degree

B. polynomials with irrational zeros

C. polynomials with complex zeros

D. polynomials in which the leading coefficient is not necessarily 1

SAMPLE OUTPUT

Coefficients a,b,c,d? 18,111,17,–6

18*x**3 + 111*x**2 + 17*x – 6 =

18*(x – 1 / 6)*[x – (–6)]*[x – (–1/3)]

Coefficients a,b,c,d? 1,–1,–5,5

x**3 – x**2 – 5*x + 5 =

(x – 1)*[x – (–√(5))]*[x – (√(5))]

Coefficients? 1,–2,1,3,–10

$x^4 - 2x^3 + x^2 + 3x - 10 =$

$(x - 2)*(x^3 + x + 5)$

Coefficients? 1,–7,0,4,–13

$x^5 - 7x^4 + 4x^2 - 13 =$

irreducible

Coefficients? 6,–31,–14,89,30

$6x^4 - 31x^3 - 14x^2 + 89x + 30$

6*(x – 2)*(x + 3 / 2)*(x – 5)*(x + 1/3)

Use the existing code in this chapter. Do not copy and paste a factorization program from another source, and do not use the **SymPy factor()** function.

CHAPTER SUMMARY

PYTHON CONCEPTS

Assume x has been declared as a **SymPy** symbol using x = **Symbol**('x')

Factor a polynomial	**factor**(sympy_expression) *Ex:* *>>>**factor**('x**2+4*x+4)')* *(x + 2)**2*
Expand a polynomial	**expand**(sympy_expression) *Ex:* *>>> factor1 = **sympify**('(x–1)**2')* *>>> factor2 = **sympify**('7*x')* *>>>**expand**(factor1*factor2)* *7*x**3 – 14*x**2 + 7*x*
Factor list of a polynomial	**factor_list**(sympy_expression) *Ex:* *>>> **factor_list**('x**3+7*x**2+8*x–16')* *(1, [(x – 1, 1), (x + 4, 2)])* *which corresponds to* $(x-1)(x+4)^2$
Divide polynomials	q,r = **div**(f,g) **SymPy** function that divides dividend f by divisor g and returns a tuple (quotient,remainder). *Ex:* *>>>q,r = div('x**2+2*x+1','x+1')* *>>>q,r* *(x+1,0)*
Output the radical symbol √	Use \u221a (a Unicode) *Ex:* *>>>**print**('x–2\u221a6')* *x–2√6*
Output a superscript	Precede the desired superscript with \u00b *Ex:* *>>>**print**('(x–2)\u00b2')* *(x–2)²*

"Built-in" Python complex type number	A complex number $a + bi$ can be expressed as either: $a + bj$ or **complex**(a,b) *Ex:* *2 + 3i can be represented using either:* *2 + 3j* *or* ***complex****(2,3)*
Real and imaginary parts of a Python complex number	If z is a complex number $a + bi$ then **z.real** is the real part a **z.imag** is the imaginary part b *Ex:* *>>>z =* ***complex****(4,−5)* *>>>z.****real*** *4* *>>>z.****imag*** *−5*
Conjugate of a Python complex number	**z.conjugate()** *Ex:* *>>>z = 99 − 8j* *>>>z.****conjugate()*** *99 + 8j*
Product of complex numbers	$(a + bj)*(a - bj)$ *Ex:* *>>>(2+3j)*(4−6j)* *(26 + 0j)*
Complex number in SymPy	$a + b*I$ or a + sqrt($-b$) *Ex:* *>>>****simplify****('4+10*I')* *4 + 10*I* *>>>****simplify****('sqrt(16)+sqrt(−25)')* *4 + 5*I*

(*Continued*)

Real and imaginary parts of a SymPy complex number	**re**(z) **im**(z) *Ex:* *>>>z = **simplify**('sqrt(16)+sqrt(−25)')* *>>>**re**(z)* *4* *>>>**im**(z)* *5*
Conjugate of a SymPy complex number	**conjugate**(z) *Ex:* *>>>**conjugate**(4+19*I)* *4 – 19*I*
Expanded product of two SymPy expressions	**expand**(z1*z2) *Ex:* *>>>z1=**simplify**('2+3j')* *>>>z2=**simplify**('5−11j')* *>>>z1*z2* *(2 + 3*I)*(5 – 11*I)* *>>>**expand**(z1*z2)* *43 – 7*I*
Quotient of two SymPy complex numbers	**simplify**(z1/z2) *Ex:* *>>>z1 = **sympify**('2+4*I')* *>>>z2 = **sympify**('3−5*I')* *>>> **simplify**(z1/z2)* *−7/17 + 11*I/17*

PRECALCULUS CONCEPTS

Zero of a function	$x = r$ is a zero of function $f(x)$ if $f(x)$ has an x-intercept at $x = r$ and a factor $(x - r)^k$. *Ex:* *2 and 3 are zeros of* $f(x) = (x-2)^2(x-3)$; *the graph of* $f(x)$ *has x-intercepts at* $x = 2$ *and* $x = 3$.

Multiplicity of a zero	If $(x - r)^k$ is a factor of $f(x)$: If k is **even,** then the graph touches the x-axis at $x = r$. If k is **odd,** then the graph crosses the x-axis at $x = r$. *Ex:* *Given $f(x) = (x-2)^2(x-3)$, the graph touches the x-axis at $x = 2$ and crosses the x-axis at $x = 3$.*
Discriminant of a quadratic	The discriminant $b^2 - 4ac$ is used to determine the number and type of solutions of quadratic equation $ax^2 + bx + c = 0$. If $b^2 - 4ac$ is... A positive perfect square — Two rational solutions A positive non-perfect square — Two irrational solutions Zero — One rational solution Negative — No real solutions *Ex:* *Given $f(x) = 6x^2 + 12x + 5$, $b^2 - 4ac = 24$, which is a positive non-perfect square, so the equation $f(x) = 0$ has* ***two irrational solutions****.*
Rational zeros theorem	If $a_nx^n + a_{n-1}x^{n-1} + \ldots + a_1x + a_0$ is a polynomial with integer coefficients and p is a factor of constant term a_0 and q is a factor of leading coefficient a_n, then the list of possible rational zeros is all rational numbers in the form $\frac{p}{q}$. *Ex:* *Given $f(x) = 4x^3 - 4x^2 - 19x - 10$:* *$P$ = factors of $-10 = \pm 1, \pm 2, \pm 5, \pm 10$* *$Q$ = factors of $4 = \pm 1, \pm 2, \pm 4$* *Possible rational zeros = $P/Q = \pm 1, \pm 2, \pm 5, \pm 10, \pm 1/2, \pm 5/2, \pm 1/4, \pm 5/4$*
Imaginary unit	$i = \sqrt{-1}$
Complex number	$a + bi$ where a and b are real numbers and i is the imaginary unit
Square root of a negative number	$\sqrt{-a^2} = ai$ *Ex:* $\sqrt{-81} = 9i$

(*Continued*)

Factorization of $a^2 + b^2$	$x^2 + a^2 = (x + ai)(x - ai)$ *Ex:* $x^2 + 121 = (x + 11i)(x - 11i)$
Conjugate of a complex number	The conjugate of $a + bi$ is $a - bi$ and vice versa. *Ex:* *The conjugate of* $3 - 8i$ *is* $3 + 8i$ *The conjugate of* $4i$ *is* $-4i$
Product of conjugates	$(a + bi)(a - bi) = a^2 + b^2$ *Ex:* $(11 + 12i)(11 - 12i) = 121 + 144 = 265$

Credit

IMG 8.28: Brigban, https://commons.wikimedia.org/wiki/File:Plotting_complex_numbers.png, 2021.

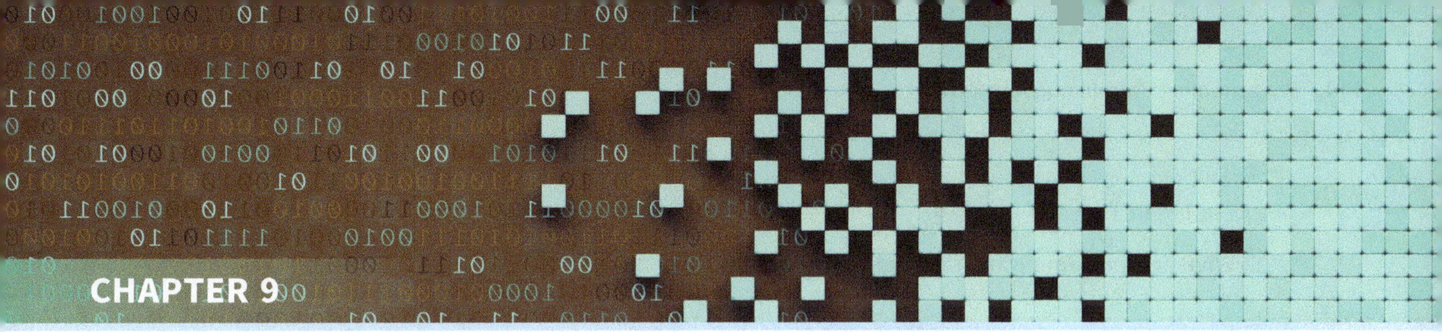

CHAPTER 9

Rational Functions

9.1 Intro to Rational Functions: Domain

Recall a **rational number** is any number that can be expressed as a ratio of two integers such as $\frac{1}{2}$, -4, 5.2, 7.3333333…, $\sqrt{4}$.

A **rational function** is a ratio of two polynomials in which the denominator cannot equal zero (as any number divided by a zero produces an undefined value).

Examples:

$$f(x) = \frac{1}{x}$$

$$g(x) = \frac{x^2 + 2}{x - 1}$$

$$h(x) = \frac{x - 3}{x^2 + 2x - 5}$$

Let's start with some basic rational functions. Compare the graph of $f(x) = \frac{1}{x}$ to $f(x) = \frac{1}{x^2}$:

$$f(x) = \frac{1}{x}$$

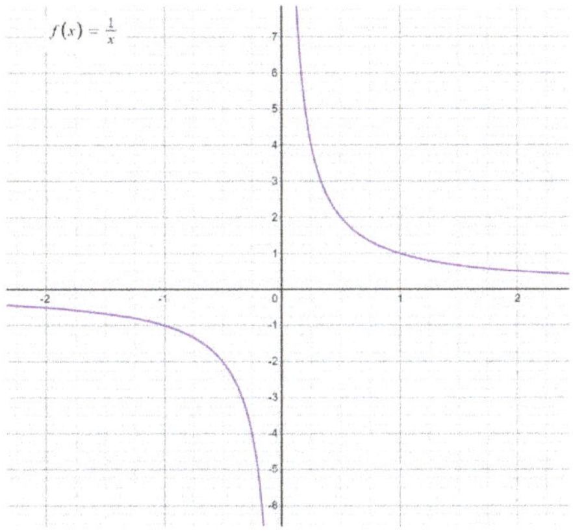

Image 9.1

$$f(x) = \frac{1}{x^2}$$

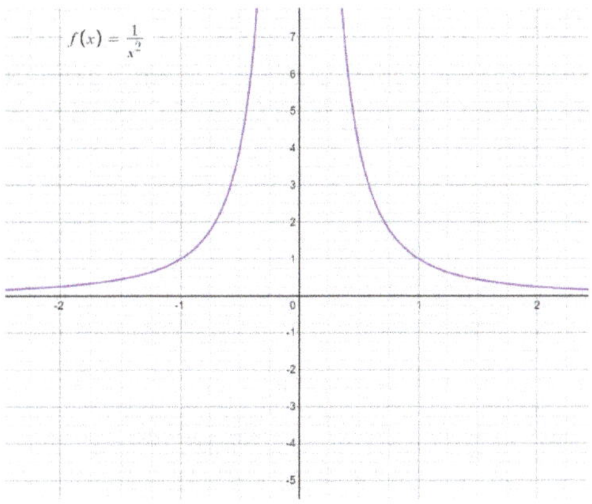

Image 9.2

EXERCISE SET 9.1.1

1. What are the domain and range of $f(x) = \frac{1}{x}$?
2. What type of symmetry does the graph of $f(x) = \frac{1}{x}$ have?
3. What are the domain and range of $f(x) = \frac{1}{x^2}$?
4. What type of symmetry does the graph of $f(x) = \frac{1}{x^2}$ have?
5. What happens to the values of $\frac{1}{x}$ as x gets larger and larger?

The graph of $f(x) = \frac{1}{x}$ also has a form of symmetry seldom mentioned in precalculus textbooks: **rotational symmetry**. This means that if the graph is rotated around the origin, it will at some point coincide with itself. We say that $f(x) = \frac{1}{x}$ has **two-fold rotational symmetry** because in one full rotation of 360°, it coincides with itself two times.

Domain

You probably determined the domain of $f(x) = \frac{1}{x}$ and $f(x) = \frac{1}{x^2}$ by looking at the graphs. However, "eyeballing" is not a reliable method; we need an algebraic method to determine the domain of a rational function.

Because a rational function is undefined when the denominator equals 0, we can find its domain by setting the denominator NOT equal to 0 and solving.

Example

Find the domain of $\frac{3x+5}{7x^2-13x-2}$.

Solution

Ignore the numerator. Set the denominator NOT equal to zero:

$$7x^2 - 13x - 2 \neq 0$$

$$(7x+1)(x-2) \neq 0$$

$$\therefore x \neq -\frac{1}{7}, x \neq 2$$

In interval notation, the domain is $(-\infty, -\frac{1}{7}) \cup (-\frac{1}{7}, 2) \cup (2, \infty)$.

I'm going to coin a new term, which is **nonzero** (as opposed to a zero). This is a number that causes the denominator to equal zero and hence where the function is undefined. In the above example, $-\frac{1}{7}$ and 2 are the nonzeros.

EXERCISE SET 9.1.2

Find the domain of the following rational functions. You might want to use one of the programs you wrote in Chapter 8 to find the zeros of quadratic or cubic functions.

1. $f(x) = \dfrac{2x}{2x+4}$
2. $f(x) = \dfrac{7}{x^2+4}$
3. $f(x) = \dfrac{x-4}{x^2-4}$
4. $y = \dfrac{24x^2-11}{2x^3+13x^2+13x-10}$
5. $f(x) = \dfrac{x^2-4}{x+2}$ *(Note: Do not simplify. Use the denominator as is.)*

Let's brush up our **SymPy** skills by writing a program to find the nonzeros of a rational function.

Program Goal: Write a program to find the nonzeros of a rational function. Ask the user to input only the denominator.

PROGRAM 9.1.1

```
## input: polynomial denominator of a rational function
## output: list of nonzeros (undefined values)
from sympy import *
x = Symbol('x')
poly = sympify(input("Denominator? "))
# solve poly=0 and return complex solutions in a list
nonzeros=solve(poly)
print("Nonzeros:")
print(nonzeros)
```

Sample Output

```
Denominator? x**2-1
Nonzeros:
[-1,1]
Denominator? x**3 + 10*x**2 - 17*x - 66
Nonzeros:
[-11, -2, 3]
```

However, **solve** returns imaginary zeros as well, which have no bearing on the domain of a function. For example, the denominator of $f(x) = \dfrac{1}{x^2+1}$ has complex roots $\pm\, i$ (result of $x^2 + 1 \neq 0$), but the graph has domain $(-\infty, \infty)$:

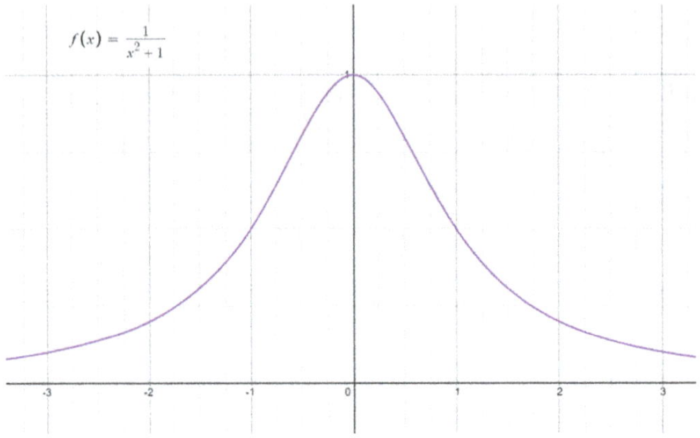

Image 9.3

To exclude imaginary zeros from our output:
Replace nonzeros=**solve**(poly) with nonzeros = **solveset**(poly,domain=S.Reals).

EXERCISE SET 9.1.3

1. What is the output of Program 9.1.1 using Method 1 (**solve**) and input x**2+4?
2. What is the output of Program 9.1.1 using Method 2 (**solveset**) and input x**2+4?
3. What is the "type" returned by **solveset**? Modify the program so that the type returned is an ordered list, least to greatest.

Sample Output

Denominator? x**3 – 8*x**2 – 133*x – 340

Nonzeros:

[–5, –4, 17]

4. Modify the program so that if there are no real nonzeros, it outputs "none."

Let's give the program a filename (**domain.py**) so that in addition to outputting nonzeros, it also outputs the domain in interval notation. The domain is constructed by iterating through the list of nonzeros.

Example of Program Flow:

Denominator? *x***2 – 4

nonzeros =	[–2, 2] *(an ordered list)*
domain =	(-∞
–2 is a nonzero: concatenate:	,–2) ∪ (–2
2 is a nonzero: concatenate:	,2) ∪ (2,
No more nonzeros: concatenate:	,∞)
End result: domain =	(–∞, –2) ∪ (–2, 2) ∪ (2, ∞)

PROGRAM 9.1.1 DOMAIN.PY REVISED

```
## domain.py
## input: polynomial denominator of a rational function
## output: domain in interval notation
from sympy import *
x = Symbol('x')
poly = sympify(input("Denominator?"))
# solve poly=0 and return complex solutions in a list
nonzeros=solveset(poly,domain=S.Reals)
print("Nonzeros:")
print(nonzeros)
print("Domain:")
# initialize domain string; Unicode symbol for infinity is \u221E
domain = "(-\u221E "
for nz in nonzeros:
    [ insert code to form interior of domain string ]
domain += ", \u221E)"
print(domain)
```

Sample Output

```
Denominator? x**2 – 4
Nonzeros:
[–2, 2]
Domain:
(–∞,–2) ∪ (–2,2) ∪ (2,∞)

Denominator? x**3 – 8*x**2 – 133*x – 340
Nonzeros:
[–5, –4, 17]
Domain:
(–∞,–5) ∪ (–5, –4) ∪ (–4, 17) ∪ (17, ∞)

Denominator? x**2 + 49
Nonzeros:
[ ]
Domain:
(–∞, ∞)

Denominator? x+4
Nonzeros:
[–4]
Domain:
(–∞,–4) & (–4,∞)
```

EXERCISE SET 9.1.4

1. Fill in the yellow highlighted code to form the interior of the domain string.
2. Modify the program so that the factorization of the denominator is also output.
3. Use the program to find the nonzeros, domain, and factorization of the denominator of $f(x) = \dfrac{2x}{4x^3 - 3x^2 - 25x - 6}$.

Distinguishing Nonzeros

The graph of a rational function exhibits either of two features at these nonzeros:

5. A **vertical asymptote** as in $f(x) = \dfrac{2x}{2x+4}$:

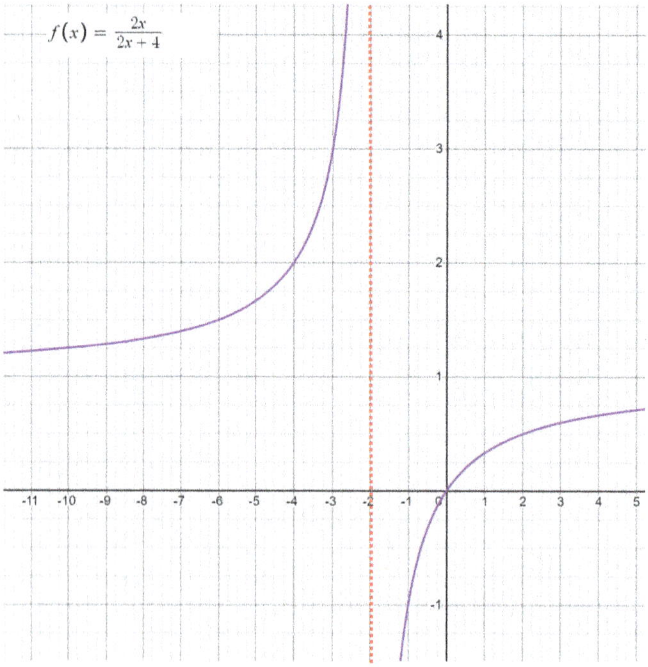

Image 9.4

Notice the nonzero is -2, so the vertical asymptote is at $x = -2$. A vertical asymptote is a vertical line that the graph never crosses but approaches from one or both sides.

6. A **removable discontinuity** as in $f(x) = \frac{x^2 - 4}{x + 2}$:

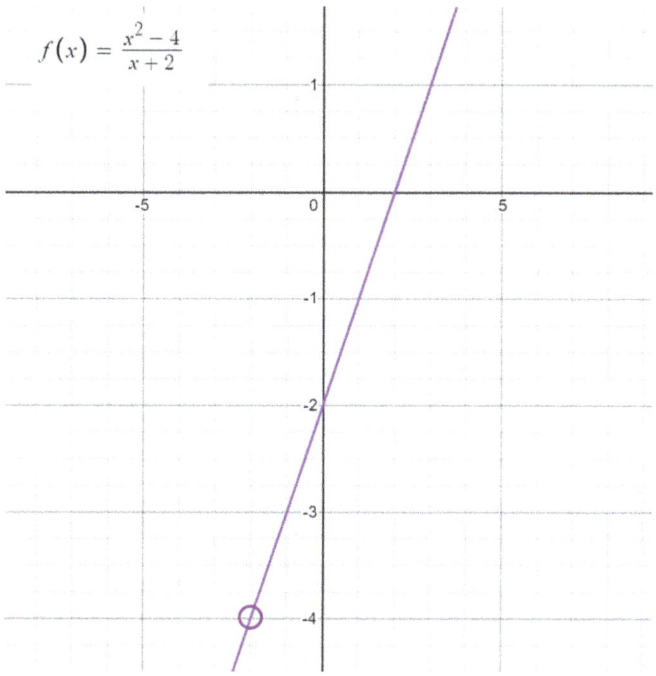

Image 9.5

The removable discontinuity is at the nonzero −2. It indicates the function is undefined at that point. It is called **removable** because if it were "filled in," the function would be continuous. We will informally refer to a removable discontinuity as a "hole" because it has fewer letters, is easier to say, and is more descriptive.

Let's take a look at $f(x) = \frac{2x^2 + 11x + 12}{2x^2 - 7x - 15}$.

$$\frac{2x^2 + 11x + 12}{2x^2 - 7x - 15} = \frac{(2x + 3)(x + 4)}{(2x + 3)(x - 5)} = \frac{x + 4}{x - 5}$$

It is tempting to deduce that the only nonzero is $x = 5$, looking at the fully simplified version. But there are actually two nonzeros: 5 AND $-\frac{3}{2}$. Remember I said to ignore the numerator? That's right. To determine the domain, IGNORE THE NUMERATOR. Don't simplify, don't cancel. Stick with the original denominator.

However, the numerator comes into play when distinguishing between the two types of nonzeros generated by the original denominator. Confused? Let's look again at the example.

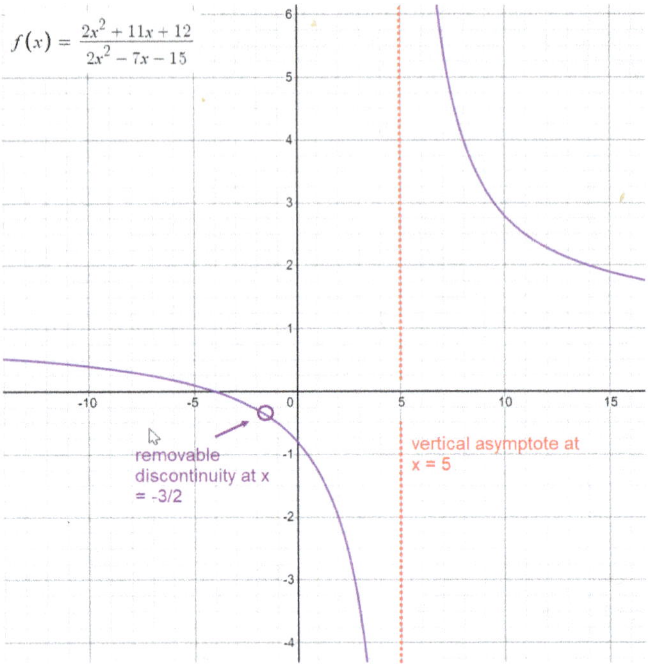

Image 9.6

The nonzero that "survived" the simplifying and cancelling, $x = 5$, resulted in a **vertical asymptote**.

The nonzero that was cancelled out, $x = -\frac{3}{2}$, resulted in a **hole**.

More formally:

If a rational function $\frac{p(x)}{q(x)}$ is fully simplified—i.e., $p(x)$ and $q(x)$ have no common factors—then the nonzero (the result of setting $q(x) = 0$) results in a **vertical asymptote**. For example, if $q(x) = (x - a)$, the vertical asymptote is at $x = a$.

If a rational function $\frac{p(x)}{q(x)}$ is not fully simplified—i.e., $p(x)$ and $q(x)$ have common factor(s)—then the common factor(s) result in a **removable discontinuity** (aka "hole"). For example, if $p(x)$ and $q(x)$ have a common factor $(x - b)$, then the graph has a removable discontinuity at $x = b$.

Example

Find the nonzeros of $f(x) = \frac{x^2 - x - 6}{x^2 + x - 2}$

Solution

Factor and simplify:

$$\frac{x^2 - x - 6}{x^2 + x - 2} = \frac{(x + 2)(x - 3)}{(x + 2)(x - 1)} = \frac{\cancel{(x + 2)}(x - 3)}{\cancel{(x + 2)}(x - 1)} = \frac{x - 3}{x - 1}$$

Image 9.7

The factor $(x + 2)$ was cancelled, so a hole occurs at $x = -2$.
The factor $(x - 1)$ remains in the denominator, so a vertical asymptote occurs at $x = 1$.

EXERCISE SET 9.1.5

Find the "nonzeros" of the following rational functions and indicate whether they result in a vertical asymptote or a removable discontinuity.

1. $f(x) = \dfrac{7x+2}{x-11}$
2. $f(x) = \dfrac{x+2}{10-x}$
3. $f(x) = \dfrac{(x+3)(x-2)}{(x-2)(2x+5)}$
4. $f(x) = \dfrac{x^2+6x+8}{x^2-x-20}$
5. $f(x) = \dfrac{x^3-8}{x^3-6x^2+12x-8}$
6. $f(x) = \dfrac{6x^3+5x^2-2x-1}{3x^2-14x-5}$
7. Find a rational function with vertical asymptotes $x = 1$ and $x = -7$; and removable discontinuity at $x = 5$.

Let's summarize a few of the **SymPy** methods we've used so far and add a few new ones that will be useful with rational expressions.

TABLE 9.1 Symbolic Manipulation (SymPy library)

To evaluate an expression for a specific value of the symbol:

```
>>>value = poly.subs({x:value})                    # returns a number
```

To factor a polynomial:

```
>>>factor('x**2–4')
(x + 2)*(x – 2)
```

To expand a polynomial:

```
>>> expand('(x+2)*(x–3)')
x**2–x–6
```

(continued)

To solve an algebraic equation in the form expression = 0 over the field of complex numbers:

```
>>>solve('x**4-12*x**2-64')                         # returns a list
[-4, 4, -2*I, 2*I]
```

To solve an algebraic equation in the form expression = 0 over the field of real numbers:

```
>>>solveset('x**4-12*x**2-64', domain=S.Reals)      # returns a set
{-4, 4}
```

To cancel common factors in a rational expression:

```
>>>cancel('(x**2-1)/(x-1)')
x + 1
```

To extract the numerator and denominator:

```
>>>f = '(x**2+10*x+16)/(x-4)'
>>>numer(f)
x**2+10*x+16
>>>denom(f)
x-4
```

For simplicity, we will only consider factorizations involving rational numbers. For example:

$\frac{x^2+4x+3}{x+1}$ will be factored and simplified to $x+3$.

$\frac{x^2-11}{x-\sqrt{11}}$ will **not** be factored to $x+\sqrt{11}$.

EXERCISE SET 9.1.6

Given $f(x) = \frac{2x^3+3x^2-3x-2}{x^4+2x^3-21x^2-22x+40}$:

Give the result of the following commands in the Python console (or in a program). Perform the commands sequentially.

1. Assign the function to string variable **rational_fcn**. What is the result of:

   >>>num,den = **numer**(rational_fcn),**denom**(rational_fcn)?

2. What is the result of:

   >>>**factor**(den)

3. What is the result of:

   >>>**cancel**(rational_fcn)

4. What is the factorization of the denominator of the expression returned in #3 above?
5. Are the results of #2 and #4 the same? Why or why not?
6. Identify the vertical asymptotes and removable discontinuities of $f(x)$.
7. Find the y-coordinates of the removable discontinuities.

Program Goal: Write a program in which the user inputs a rational function. Output the simplified rational expression and find the nonzeros. Classify each as either a vertical asymptote or a hole (removable discontinuity).

Hint: Testing will be easier if we initially hardcode the rational function rather than ask for input. Use $f(x) = \dfrac{2x^3 + 3x^2 - 3x - 2}{x^4 + 2x^3 - 21x^2 - 22x + 40}$ from Exercise Set 9.1.6 above.

PROGRAM 9.1.2 NONZEROS.PY

```
## nonzeros.py
## input: rational function
## output: simplified rational expression;
## list of nonzeros classified as either vertical asymptote or removable discontinuity
## Assume numerator and denominator are expanded, not factored
from sympy import *
x=Symbol('x')
# hard-coded rational function (for testing purposes)
rational_fcn='(2*x**3+3*x**2-3*x-2)/(x**4+2*x**3-21*x**2-22*x+40)'
# parse the numerator and denominator
num,den = numer(rational_fcn),denom(rational_fcn)
# find the nonzeros of the denominator and convert to a sorted list
original_nonzeros = solveset(den,domain=S.Reals)
original_nonzeros = sorted(list(original_nonzeros))
# simplify and reduce the rational function
reduced_form = cancel(rational_fcn)
print("Reduced form:")
```

Desired Output

```
Reduced form:
(2*x + 1)/(x**2 + x - 20)
Nonzeros:
-5, vertical asymptote
-2, removable discontinuity
1, removable discontinuity
4, vertical asymptote
```

```
print(reduced_form)
# parse the numerator and denominator of reduced form
num2,den2 = numer(reduced_form),denom(reduced_form)
# find nonzeros of denominator of reduced form
reduced_nonzeros = solveset(den2,domain=S.Reals)
reduced_nonzeros = list(reduced_nonzeros)
# print and classify the list of original nonzeros
print("Nonzeros:")
for nz in original_nonzeros:
    [fill in code to print and classify the list of original nonzeros]
```

EXERCISE SET 9.1.7

1. Fill in the yellow highlighted code to print and classify the list of original nonzeros.

   Hint: compare the list of original_nonzeros to the list of reduced_nonzeros.

2. Modify the output as follows:

   Reduced form:

   (2*x + 1)/(x**2 + x − 20)

   Nonzeros:

   vertical asymptote at $x = -5$

   removable discontinuity at (−2, 1/6)

   removable discontinuity at (1, −1/6)

   vertical asymptote at $x = 4$

   *Hint: find the y-coordinates of the removable discontinuities using reduced_form.**subs**().*

3. Replace the hard-coded assignment statement to **rational_fcn** with an input statement.

4. Use the modified program to verify your answers to Exercise Set 9.1.5.

If we want to modify the program to draw the graph, we add the following statements:

```
from matplotlib import pyplot as plt
:
:
x_list=list(range(-20,20))
y_list = [reduced_form.subs({x:x_value}) for x_value in x_list]
plt.plot(x_list,y_list)
plt.title(reduced_form)
```

SAMPLE OUTPUT

Rational function? (x**2–2*x+7)/(x**2+4)

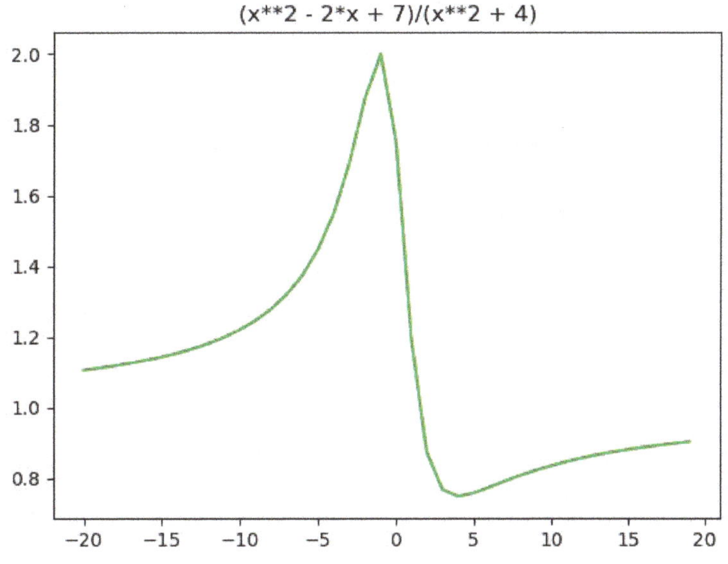

Image 9.8

This works great if the function does not have vertical asymptotes. (Holes are okay.) However, anything with vertical asymptotes or removable discontinuities results in a slew of error messages culminating in:

TypeError: can't convert complex to float

The best way to handle this is to remove the values of the vertical asymptotes from x_list. Changes to program **nonzeros** are in blue:

PROGRAM 9.1.2 NONZEROS.PY REVISED

```
## nonzeros.py
## input: rational function
## output: simplified rational expression;
## list of nonzeros classified as either vertical asymp. or removable discontinuity
## Assume numerator and denominator are expanded, not factored
from sympy import *
from matplotlib import pyplot as plt
x=Symbol('x')
rational_fcn='(2*x**3+3*x**2–3*x–2)/(x**4+2*x**3–21*x**2–22*x+40)'
num,den = numer(rational_fcn),denom(rational_fcn)
original_nonzeros = solveset(den,domain=S.Reals)
original_nonzeros = sorted(list(original_nonzeros))
reduced_form = cancel(rational_fcn)
va = [ ]                          # initialize list of vertical asymptotes
rd = [ ]                          # initialize list of removable discontinuities
# find nonzeros of reduced form
num2,den2 = numer(reduced_form),denom(reduced_form)
reduced_nonzeros = [ ]
reduced_nonzeros = solveset(den2,domain=S.Reals)
reduced_nonzeros = list(reduced_nonzeros)
for nz in original_nonzeros:
    if nz in reduced_nonzeros:
        va.append(nz)
    else:
        rd.append(nz)
x_list=list(range(–20,20))
# remove vertical asymptotes from domain
for x_value in x_list:
    if x_value in va:
        x_list.remove(x_value)
y_list = [reduced_form.subs({x:x_value}) for x_value in x_list]
plt.plot(x_list,y_list)
plt.title(str(reduced_form))
```

DESIRED OUTPUT

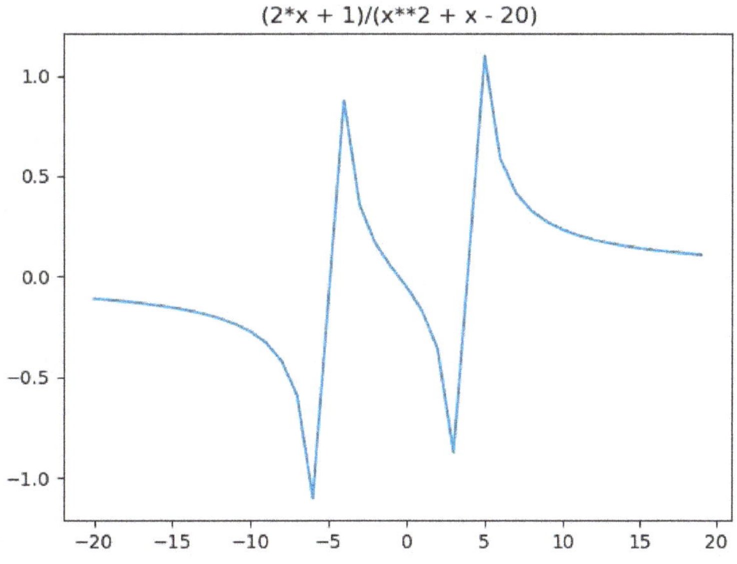

Image 9.9

However, this graph is not desirable at all. It makes it look as though the function is continuous, which we know it is not. Recall, **plot** will "connect the dots" whether you want it to or not—in this case connecting points from the left side of each vertical asymptote to the right side. A workaround is proposed in the online Supplementary Material for this text.

9.2 Graph Features

Intercepts of a Rational Function

Finding the intercepts of a rational function is basically the same process as any other function.

Given simplified rational function $f(x) = \dfrac{p(x)}{q(x)}$:

The ***y*-intercept** is $f(0)$, provided $q(0) \neq 0$.
To find the ***x*-intercept**, set $p(x) = 0$ and solve for x.

Example 1

Find the x- and y-intercepts of $f(x) = \dfrac{x^2 - 1}{x^2 - x - 6}$.

Solution

y-intercept:

$$f(0) = \frac{0^2 - 1}{0^2 - 0 - 6} = \frac{-1}{-6} = \frac{1}{6}$$

Ordered pair is $(0, \frac{1}{6})$

x-intercept:

Solve $x^2 - 1 = 0$

$(x + 1)(x - 1) = 0$

$x = -1, x = 1$

Ordered pairs: (–1,0), (1,0)

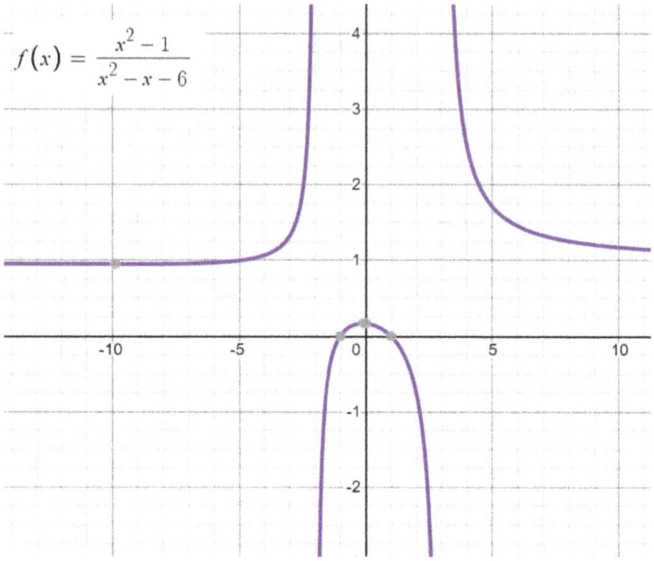

Image 9.10

Example 2

Find the x- and y-intercepts of $f(x) = \frac{x^2 - x - 20}{x^2 + 2x - 8}$.

Solution

y-intercept:

$$f(0) = \frac{0^2 - 0 - 20}{0^2 + 2(0) - 8} = \frac{-20}{-8} = \frac{5}{2}$$

Ordered pair is (0,5/2)

x-intercept:

Solve $x^2 - x - 20 = 0$

$(x + 4)(x - 5) = 0$

$x = -4, x = 5$

Ordered pairs: (–4,0), (5,0)

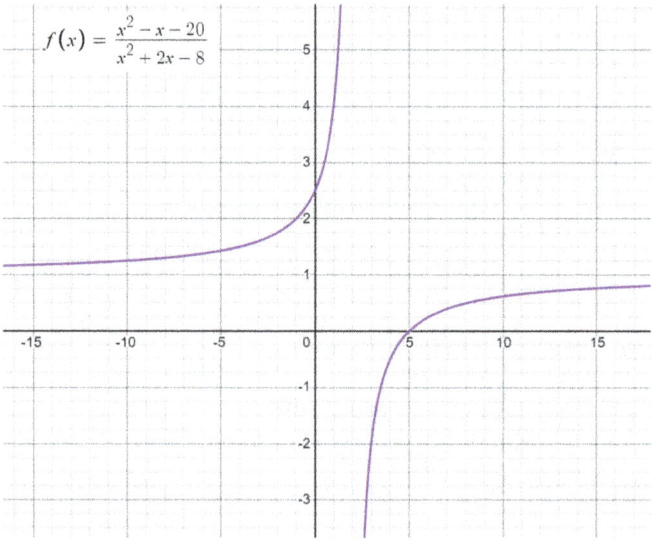

Image 9.11

But! Notice the domain of $f(x)$ is found by setting $q(x) \neq 0$:

$x^2 + 2x - 8 \neq 0$

$(x + 4)(x - 2) \neq 0$

$x \neq -4, x \neq 2$

This means (–4,0) is not an x-intercept after all, because –4 is in fact a removable discontinuity. (–4,0) is not even a point on the graph.

We can "head off at the pass" this type of issue by first factoring and simplifying. In Example 2 above, the function factors to $f(x) = \dfrac{(x+4)(x-5)}{(x+4)(x-2)} = \dfrac{x-5}{x-2}$ which clearly only has an x-intercept at $x = 5$.

Example 3

Find the x- and y-intercepts of $f(x) = \dfrac{x^3 + x^2 - 9x - 9}{x^3 - 9x}$.

Solution

First factor and simplify:

$$\frac{x^3 + x^2 - 9x - 9}{x^3 - 9x} = \frac{(x+1)(x+3)(x-3)}{x(x+3)(x-3)}$$

$$= \frac{x+1}{x}$$

y-intercept:

$$f(0) = \frac{0+1}{0} = \text{undefined. None.}$$

x-intercept:

Solve $x + 1 = 0$

$x = -1$

Ordered pair: $(-1,0)$

Image 9.12

EXERCISE SET 9.2.1

Simplify if possible. Then identify the x- and y-intercepts of the following rational functions.

1. $f(x) = \dfrac{x}{x^2 - 25}$

2. $g(x) = \dfrac{x^3 - 27}{x^2 - 9}$

3. $f(x) = \dfrac{x^2 - 16}{x^4 - 81}$

4. $y = \dfrac{x^3 - 2x^2 - 25x + 50}{x^2 + 5x}$

5. $f(x) = \dfrac{7x - 49}{x^2 + 5x - 14}$

6. Write a program that simplifies a rational function, then identifies the x- and y-intercepts of the graph.

Sample Output

Rational function? (7*x**2–42x–49)/(x**2–5*x–14)

Simplified function is 7*(x + 1)/(x + 2)

(Continued)

x-intercepts:

[−1]

y-intercept:

7/2

rational expression? (x**2+9)/(x**2-9)

Simplified function is (x**2 + 9)/(x**2 − 9)

x-intercepts:

none

y-intercept:

-1

Vertical Asymptotes

Return to this pair of graphs our most basic rational functions: $f(x) = \frac{1}{x}$ and $f(x) = \frac{1}{x^2}$. Note they share the same vertical ($x = 0$, or y-axis) and horizontal ($y = 0$, or x-axis) asymptotes.

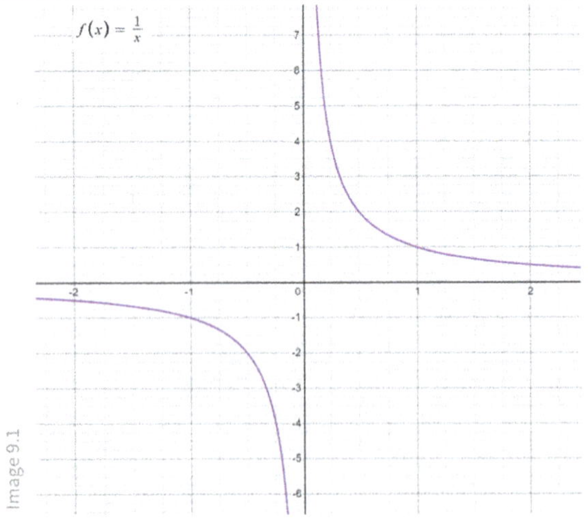

The graph approaches $-\infty$ on the left side and $+\infty$ on the right side.

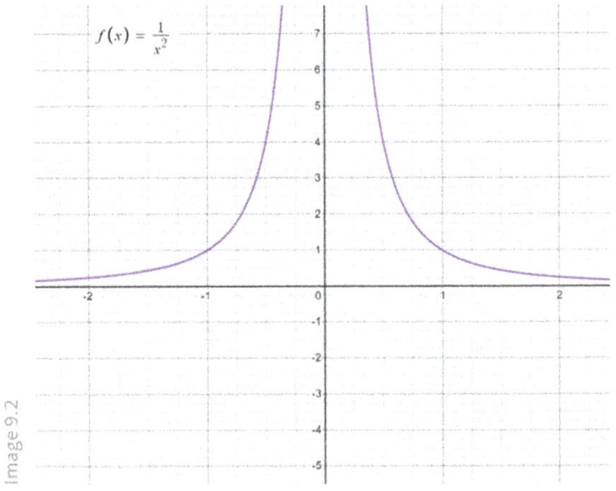

The graph approaches $+\infty$ on the left side and $+\infty$ on the right side.

We can graph many rational functions if we recognize that they are transformations of $f(x) = \frac{1}{x}$ and $f(x) = \frac{1}{x^2}$. Recall the graph transformations from Chapter 7:

TABLE 9.2

$F(x) = f(x) + a$ is the graph of $f(x)$ shifted **up** a units
$F(x) = f(x) - a$ is the graph of $f(x)$ shifted **down** a units
$F(x) = f(x + a)$ is the graph of $f(x)$ shifted **left** a units
$F(x) = f(x - a)$ is the graph of $f(x)$ shifted **right** a units
$F(x) = -f(x)$ is the graph of $f(x)$ reflected through the **x-axis**
$F(x) = f(-x)$ is the graph of $f(x)$ reflected through the **y-axis**
$F(x) = a \cdot f(x)$ is the graph of $f(x)$ **stretched** (pushed away from x-axis) by a factor of a if $\lvert a\rvert > 1$
$F(x) = a \cdot f(x)$ is the graph of $f(x)$ **compressed** (pulled in toward the x-axis) by a factor of a if $0 < \lvert a\rvert < 1$

These transformations can be applied to rational functions as well. Note that shifting left or right also shifts the vertical asymptote(s). Shifting up or down also shifts the horizontal asymptote.

Example

State the type of transformation, and then sketch the following transformations of $f(x) = \frac{1}{x}$ or $f(x) = \frac{1}{x^2}$ by hand. Label any x- or y-intercepts.

$$f(x) = -\frac{x+2}{(x+2)(x+3)}$$

Solution

First, simplify to get $f(x) = -\frac{1}{(x+3)}$. This is a transformation of $f(x) = \frac{1}{x}$ reflected through the x-axis and shifted 3 units left.

The vertical asymptote has been shifted left to $x = 3$.

The y-intercept is $f(0) = -\frac{1}{0+3} = -\frac{1}{3}$

Ordered pair: $\left(0, -\frac{1}{3}\right)$

The x-intercept is found by solving $f(x) = 0$:

$$-\frac{1}{(x+3)} = 0$$

No solution because the x-axis is a horizontal asymptote.

Note there is a hole at $x = -2$ because the factor $(x + 2)$ was cancelled.

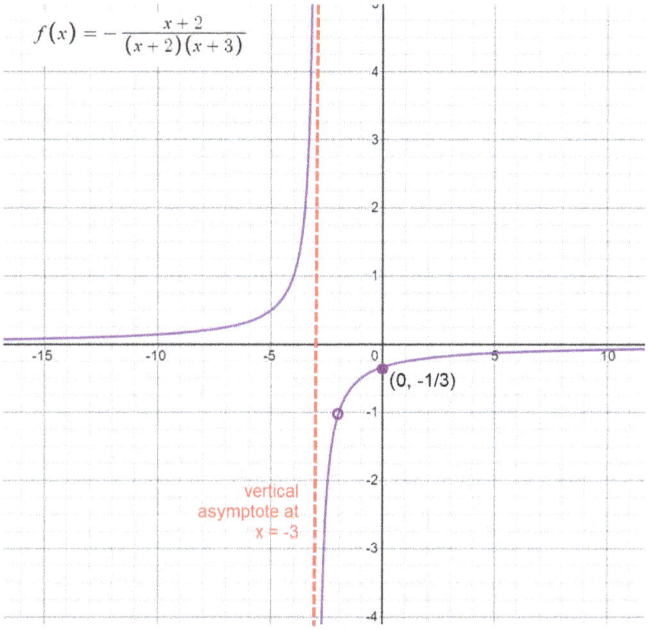

Image 9.13

EXERCISE SET 9.2.2

State the type of transformation, and then sketch the following transformations of $f(x) = \frac{1}{x}$ or $f(x) = \frac{1}{x^2}$ by hand, including any vertical asymptotes or removable discontinuities. Label any *x*- or *y*-intercepts.

1. $f(x) = \dfrac{1}{x-2}$
2. $f(x) = 2 + \dfrac{1}{x^2}$
3. $f(x) = \dfrac{x-1}{x^2-1}$
4. $f(x) = -\dfrac{2}{x+1}$
5. $f(x) = \dfrac{1}{-x^2-4x-4}$
6. $y = \dfrac{x-4}{x}$
7. Write the equation of the following graph:

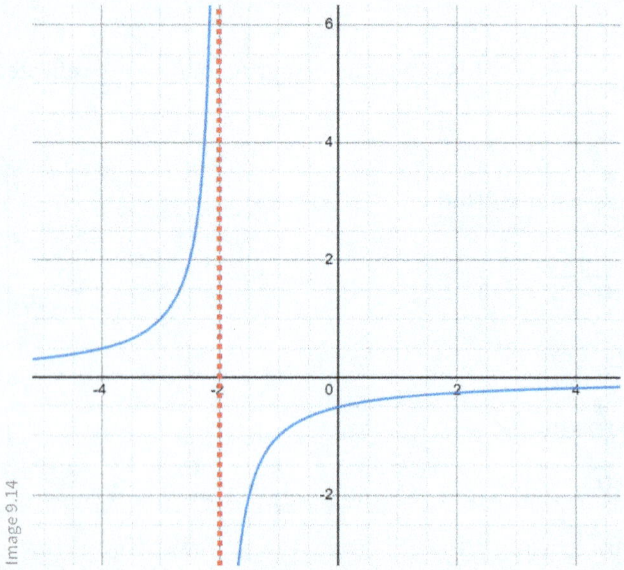

8. Write the equation of the following graph:

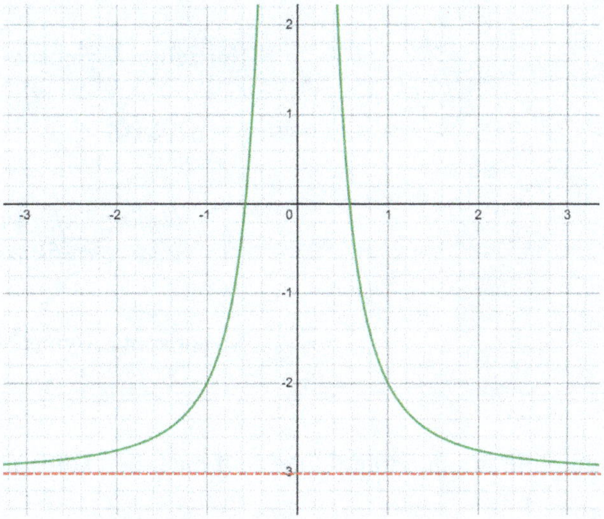

Image 9.15

Horizontal Asymptotes

As you have noticed, graphs of rational functions may also have horizontal asymptotes. Similar to vertical asymptotes, horizontal asymptotes are "invisible" lines that the graph approaches but does not cross as it approaches positive or negative infinity.

But! The graph could possibly cross the horizontal asymptote somewhere in between positive and negative infinity as shown here:

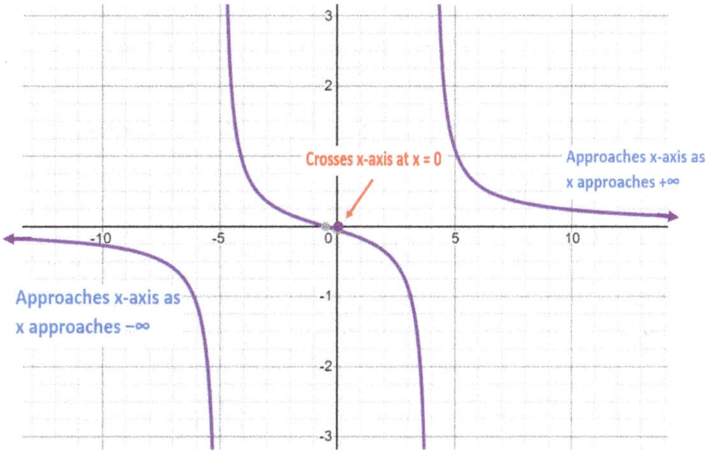

Image 9.16

There is a very simple test to determine if the graph of a rational function has a horizontal asymptote, and its equation. It's based on comparing the degree of the numerator and degree of denominator.

Rational Function	Horizontal Asymptote
$f(x) = \dfrac{\textit{smaller degree}}{\textit{larger degree}}$	$y = 0$ (x-axis)
$f(x) = \dfrac{\textit{same degree}}{\textit{same degree}}$	$y = \dfrac{\textit{leading coefficient}}{\textit{leading coefficient}}$
$f(x) = \dfrac{\textit{larger degree}}{\textit{smaller degree}}$	No horizontal asymptote

Note the horizontal asymptote is a horizontal line, so it must be expressed as the equation of a line: $y = b$, not just b.

Example 1

Find the vertical and/or horizontal asymptotes.

$$f(x) = \frac{2x^2 + 5}{x^2 - 25}$$

Solution

The degree of the numerator equals the degree of the denominator, so the horizontal asymptote is the ratio of the leading coefficients:

$$y = 2$$

The simplified rational function is $f(x) = \dfrac{2x^2 + 5}{(x + 5)(x - 5)}$, so the vertical asymptotes are determined by setting the denominator equal to zero:

$$(x + 5)(x - 5) = 0$$
$$x = -5, x = 5$$

Here is the graph with the asymptotes drawn in:

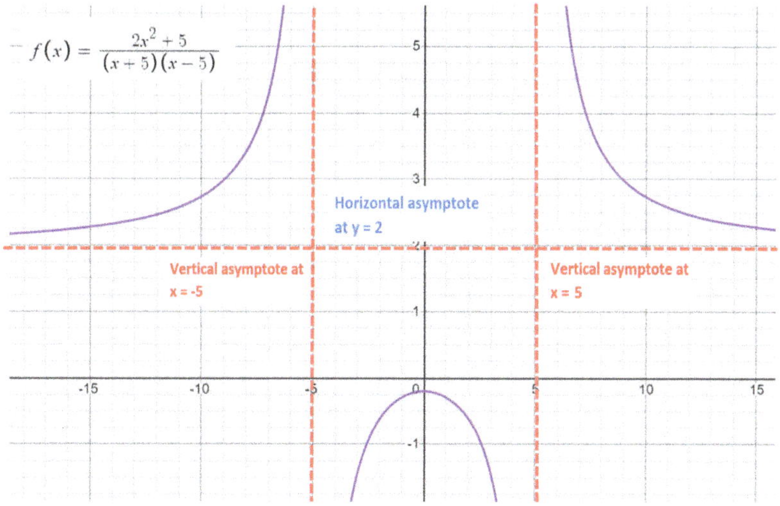

Image 9.17

Notice it is an even function (symmetric to the y-axis) because $f(-x) = f(x)$.

The following **SymPy** method and attributes will come in helpful when writing a program to identify horizontal asymptotes:

To return the degree and coefficients of a polynomial with variable x:
polynomial = **Poly**(string expression,x)
polynomial.**degree**()
polynomial.**all_coeffs**()

EXERCISE SET 9.2.3

For #1–4, find the vertical and horizontal asymptotes of the following rational functions.

1. $f(x) = \dfrac{x^2}{(x+3)(x-2)}$

2. $f(x) = \dfrac{x}{x^2-25}$

3. $f(x) = \dfrac{x^2-16}{x^4-81}$

4. $f(x) = \dfrac{x^3-27}{x^2-9}$

5. Modify program **nonzeros**, or write a new one, that calculates whether a rational function has a horizontal asymptote. If so, draw and label the horizontal asymptote on the graph.

(Continued)

Sample Output

Rational function? (2*x**2+5)/(x**2–25)

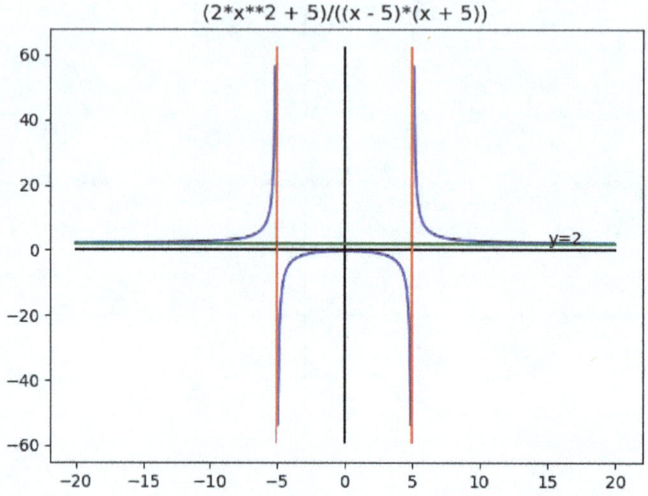

Image 9.18

The axes were added to the diagram.

Oblique Asymptotes

Oblique asymptotes (aka slant asymptotes) are "invisible" lines that are neither horizontal nor vertical, therefore they have an equation in the form $y = mx + b$.

Consider this function with an oblique asymptote:

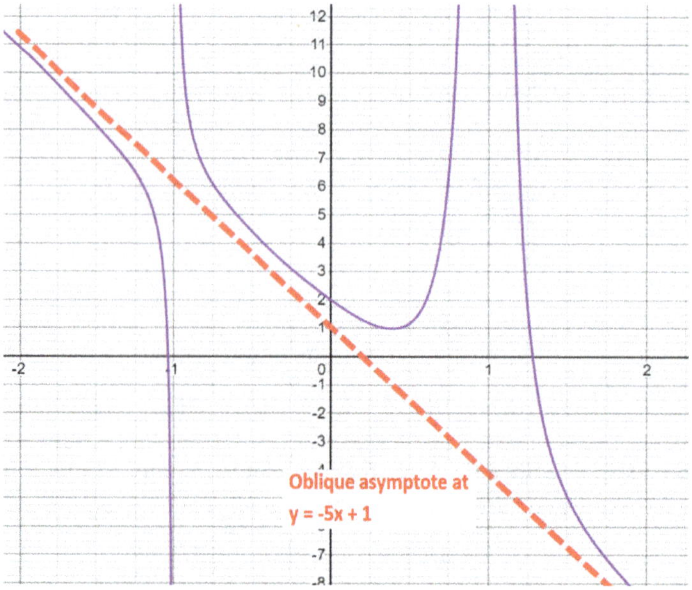

Image 9.19

An oblique asymptote only occurs in the special case:

$$f(x) = \frac{smaller\ degree + 1}{smaller\ degree}$$

i.e., the degree of the numerator is exactly 1 more than the degree of the denominator, such as $f(x) = \frac{2x^3 - 3x^2 + 4x - 7}{x^2 - 5x + 11}$.

The equation of an oblique asymptote is the quotient of the rational function (without the remainder).

Example

Find the oblique asymptote of $f(x) = \frac{3x^2 - 2x + 1}{x - 1}$.

Solution

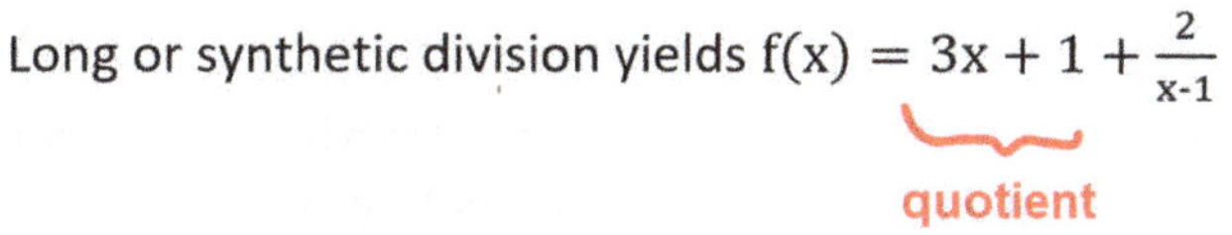

Image 9.20

Thus, the oblique asymptote is $y = 3x + 1$ as shown here:

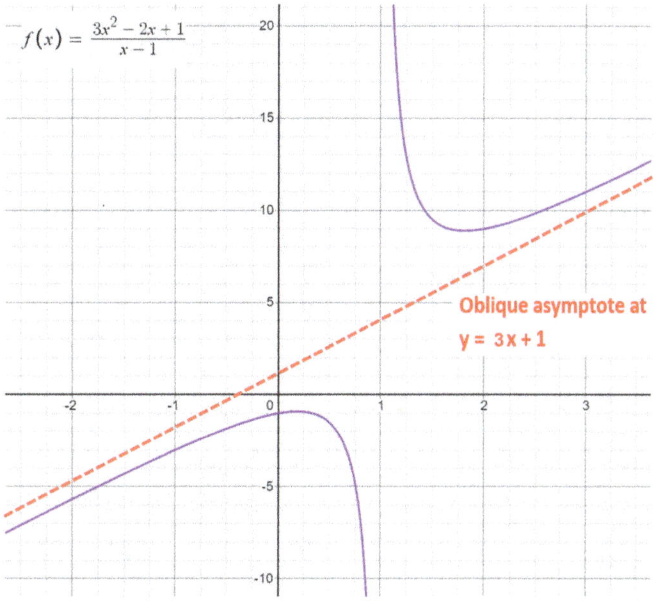

Image 9.21

EXERCISE SET 9.2.4

For #1–6, identify any vertical, horizontal, or oblique asymptotes, any removable discontinuities, *x*- or *y*-intercepts:

1. $f(x) = \dfrac{3x}{x^2 + 2x - 8}$

2. $f(x) = \dfrac{4 - 3x - x^2}{x - 2}$

3. $f(x) = \dfrac{x^2 - x - 6}{x^2 - 1}$

4. $f(x) = \dfrac{4x + 2}{x^2 + 4x - 5}$

5. $f(x) = \dfrac{3x^2 + 2x}{4x^3 - 5x + 7}$

6. $f(x) = \dfrac{5x^3 - 8}{x^2 + 3x - 1}$

7. Is it possible for a rational function to have both horizontal and oblique asymptotes? Explain.
8. Modify program **nonzeros**, or write a new one, that calculates whether a rational function has an oblique asymptote. If so, draw and label the oblique asymptote on the graph.

   *Hint from Chapter 3: quotient,remainder=**div**(dividend,divisor)*

Sample Output

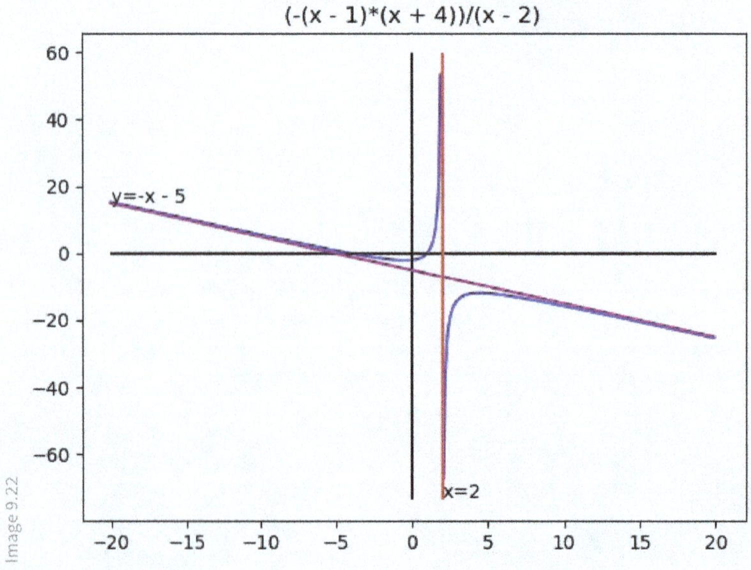

The axes were added to the diagram.

9.3 Application Problems

Example 1

The following function approximates the probability that, in a room of n people, at least two people share the same birthday:

$$P(n) = 100 - \frac{113.3198}{1 + 0.115e^{0.0912n}}$$

(e is an irrational constant equal to approximately 2.71828, which you will learn more about in Chapter 10.)

Image 9.23

A. What is the approximate probability that at least two people will share a birthday in a room of 15 people? Of 50 people?

B. Graph the function. Does it appear to have a limit?

C. What is the fewest number of people in a room that will provide a 75% probability that at least two of them share a birthday?

Solutions

A. $P(15) \approx 21.9\%$, $P(50) \approx 90.6\%$

B. The graph demonstrates that the limit (horizontal asymptote) is 100%, which makes sense in the context of probability.

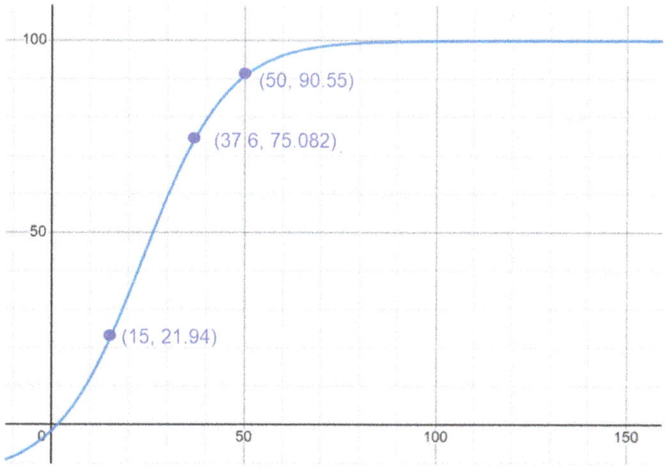

Image 9.24

C. The fewest number of people in a room that will yield a 75% probability is 37.6, which needs to be rounded to the nearest whole person, or 38.

Example 2

An aluminum can has a capacity of 500 cm^3. The top and bottom of the can are made of an aluminum alloy that costs 0.05 /cm^2. The sides of the can are made of a material that costs .02/cm^2.

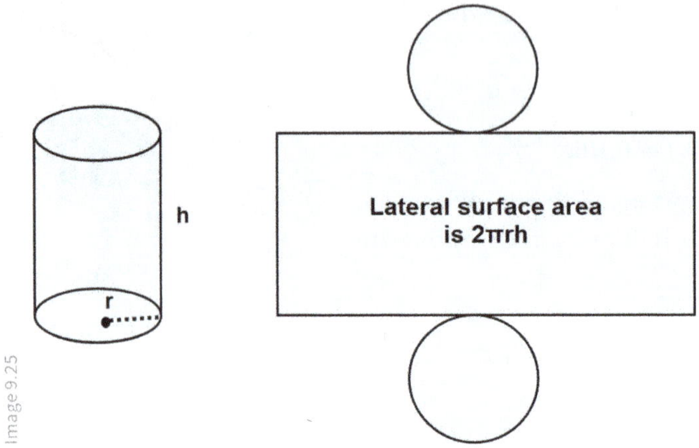

A. Express the cost of material for the can as a function of the radius r of the can. (Hint: The lateral surface area of a cylinder is $2\pi rh$ and the top and bottom are each πr^2. The volume of a cylinder is $\pi r^2 h$.)

B. Graph the cost function.

C. Use the graph to estimate the radius of the cylinder with the least cost.

Solution

$$\boxed{\text{A.}}\ C = \boxed{\text{Cost of top and bottom}} + \boxed{\text{Cost of side}}$$

$$= .05(2\pi r^2) \qquad + .02\,(2\pi rh)$$

$$= 0.1\,\pi r^2 \qquad + .04\,\pi rh$$

However, this is a function of two variables: r and h. We want a function of a single variable: r, so we must find a formula that expresses h in terms of r. This is derived from the fact that the volume of the can is 500 cm^3:

$$\text{If } V = \pi r^2 h = 500, \text{ then } h = \frac{500}{\pi r^2}.$$

Substitute this into $C = 0.1\,\pi r^2 + .04\,\pi rh$:

$$C(r) = 0.1\,\pi r^2 + .04\,\pi r\left(\frac{500}{\pi r^2}\right) = \frac{0.1\pi r^3}{r} + \frac{20}{r} = \frac{0.10\pi r^3 + 20}{r}$$

A. If you use program **nonzeros** to graph the cost function, you will need to enter the cost function like this: (.314*x**3 + 20) / x, i.e., use .314 instead of 0.1π.

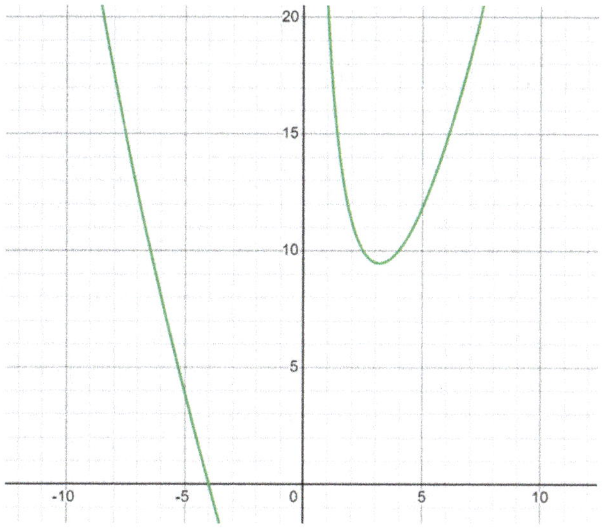

Image 9.26

B. It appears the cost function has a minimum at about $r = 3.17$. We are not interested in negative values of r.

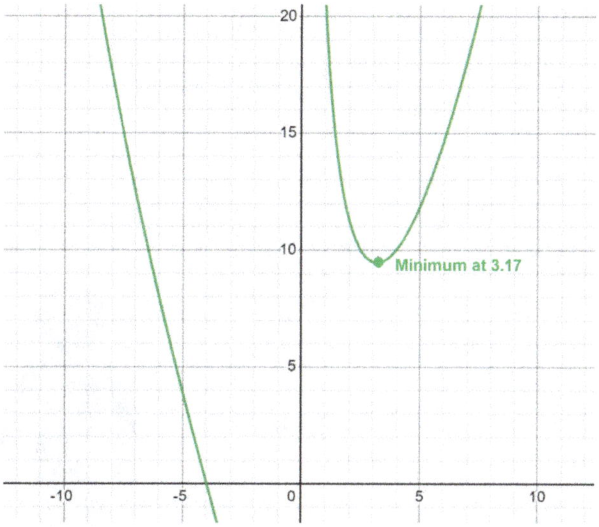

Image 9.27

EXERCISE SET 9.3.1

1. In physics, the acceleration due to gravity g (in m/sec²) at height h meters above sea level is given by $g(h) = \dfrac{3.99 \cdot 10^{14}}{(6.374 \cdot 10^{6} + h)^{2}}$, where $6.374 \cdot 10^{6}$ is the radius of Earth in meters.

   A. What is the acceleration due to gravity at sea level?

   B. Find the horizontal asymptote.

   C. As of this writing, the tallest building in the world (the Burj Khalifa in Dubai) is about 828 meters tall. What is the acceleration due to gravity at the top of this building?

Image 9.28

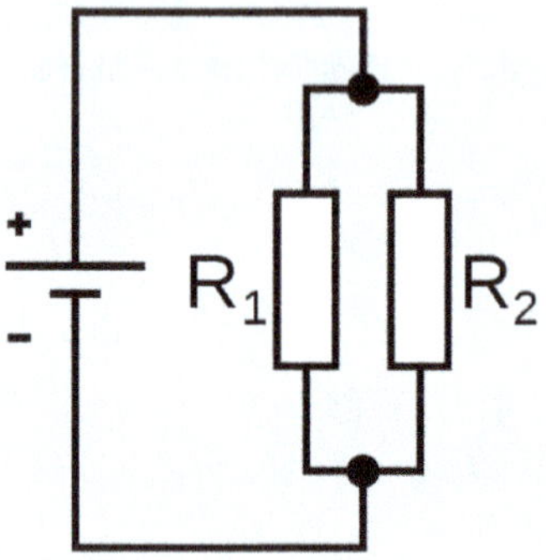

Image 9.29

2. The total resistance R_{tot} of two components hooked in a parallel circuit is given by Ohm's Law: $R_{tot} = \dfrac{R_1 R_2}{R_1 + R_2}$, where R_1 and R_2 are the individual resistances.

   A. Let $R_1 = 10$ ohms and graph R_{tot} as a function of R_2.

   B. Find and interpret any asymptotes of the graph in part A.

3. The concentration C of a certain drug in a patient's bloodstream t minutes after injection is given by $C(t) = \dfrac{50t}{t^2 + 25}$ mg/L.

   A. Find the horizontal asymptote. What happens to the concentration of the drug as t increases?

   B. Graph the function.

   C. Determine the time at which the concentration is highest.

   D. What is the max concentration at that time?

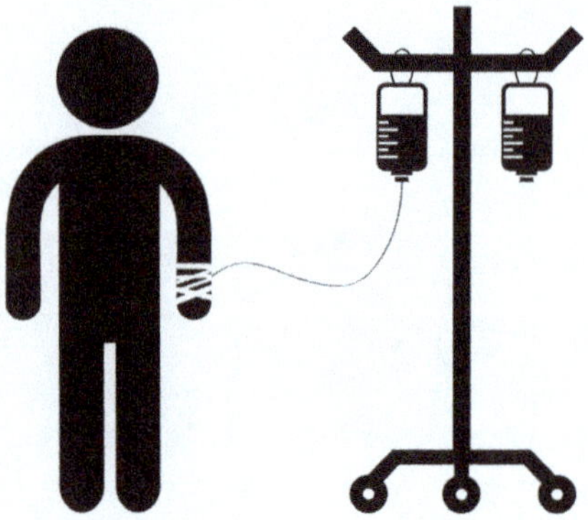

Image 9.30

Image 9.31

4. A delivery service is designing a closed box with a square base that has a volume of 1,000 in^3.

   A. If the length of the edge of the square base is x, express the surface area S of the box as a function of x.

   B. Graph the function.

   C. Based on your graph, estimate the minimum amount of cardboard that can be used to construct the box.

   D. If the cost of the cardboard used to make the square bases is .32/in^2 and the cost of the cardboard that forms the four sides is .25/in^2, write a cost function and minimize the cost for the given volume.

5. A steel drum in the shape of a right circular cylinder is required to have a volume of 100 ft^3.

   A. Express the amount A of material required to make the drum as a function of the radius r of the cylinder.

   B. How much material is required if the drum's radius is 3 ft.?

   C. Graph the function. Use the graph to estimate the value of r that would result in a minimum value of A.

Image 9.32

Chapter Project V uses a calculus technique called the **derivative** to find the actual max/min of a function without having to estimate from the graph.

Chapter Projects

I. Optimal Plotting Window

In previous programs (such as **nonzeros**), we have explicitly stated (or "hard-coded") [xmin, xmax], usually with [−20, 20]. However, this generic window does not always show the key features of the graph. For example, $f(x) = \dfrac{9x - 900}{x - 50}$, shown below, has a vertical asymptote at $x = 50$ and a horizontal asymptote at $y = 9$.

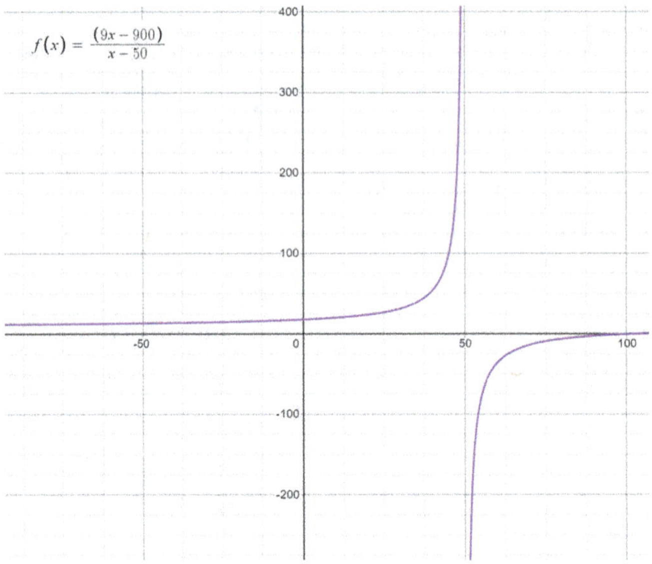

Image 9.33

However, a standard viewing window of [−10,10] × [−10,10] would look like this:

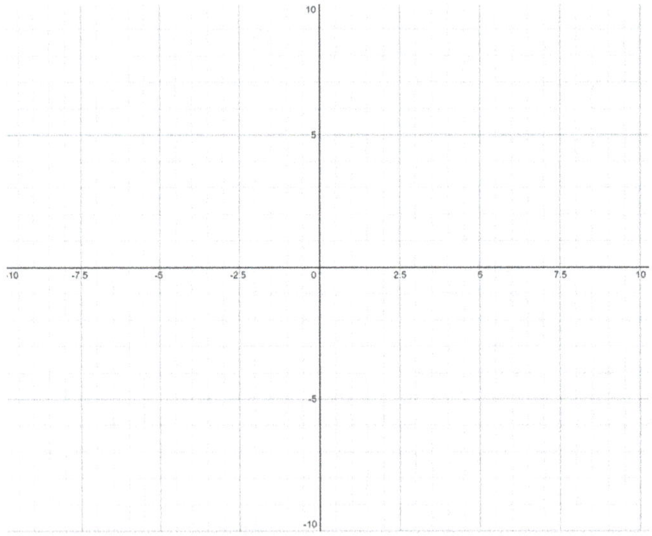

Image 9.34

Modify program **nonzeros** in this chapter so that it scales the x- and y-axes to show the extrema, asymptotes, and zeros.

II. Engineering a Rational Function

Construct a rational function given the vertical asymptote(s) and removable discontinuities. The output should include a graph and the x- and y-axes.

SAMPLE OUTPUT

Are there vertical asymptotes? Y
V.A.? 6

Are there removable discontinuities? Y
R.D.? −5

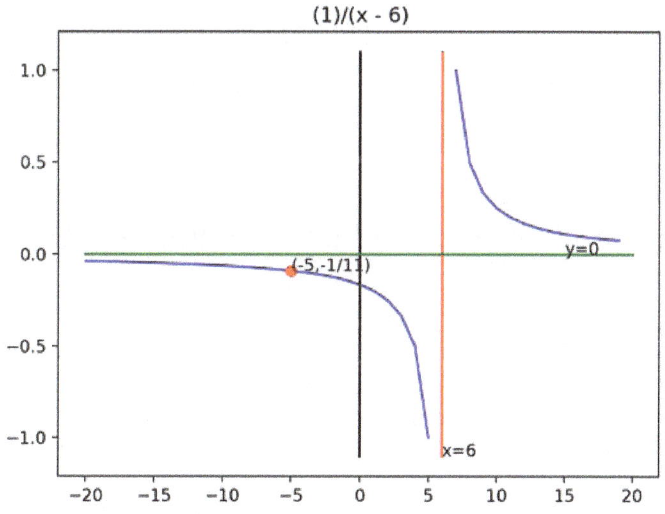

Image 9.35

III. Evaluating Derivative of a Rational Function

Recall that the derivative of a function is useful in identifying extrema (we found the derivative of polynomial functions in Chapter 7 Chapter Project I). The formula to find the derivative of a rational function $f(x) = \frac{P(x)}{Q(x)}$ is:

$$f'(x) = \frac{P'(x) \cdot Q(x) - Q'(x) \cdot P(x)}{[Q(x)]^2}$$

Example: If $f(x) = \frac{3x^2 - 2x + 4}{5x - 11}$

$$P(x) = 3x^2 - 2x + 4,\ P'(x) = 6x - 2.$$

$$Q(x) = 5x - 11,\ Q'(x) = 5.$$

$$f'(x) = \frac{P'(x) \cdot Q(x) - Q'(x) \cdot P(x)}{[Q(x)]^2} = \frac{(6x - 2)(5x - 11) - (5)(3x^2 - 2x + 4)}{(5x - 11)^2}$$

To evaluate at a given value such as $x = 1$:

$$f'(1) = \frac{(4)(-6) - (5)(5)}{(-6)^2} = -\frac{49}{36}$$

Write a program that will ask the user to input the rational function and a given value of x, then calculate and output $f'(x)$ (in the form of a simplified fraction). Do NOT use **SymPy** method **deriv**().

SAMPLE OUTPUT (USING ABOVE EXAMPLE)

Rational Function? (3*x**2-2*x+4)/(5*x–11)
x? 1

$$f'(x) = \frac{-15 \cdot x^2 + 10 \cdot x + (5 \cdot x - 11) \cdot (6 \cdot x - 2) - 20}{(5 \cdot x - 11)^2}$$

$$f'(1) = -49/36$$

Note, when you sympify f', it may rearrange the terms; however, it is still equivalent to the correct derivative.

IV. Crossing Asymptotes

It is possible for a rational function to "cross" a horizontal asymptote, as in this example:

$$f(x) = \frac{2x + 1}{x^2 + x - 20}$$

The horizontal asymptote is $y = 0$.

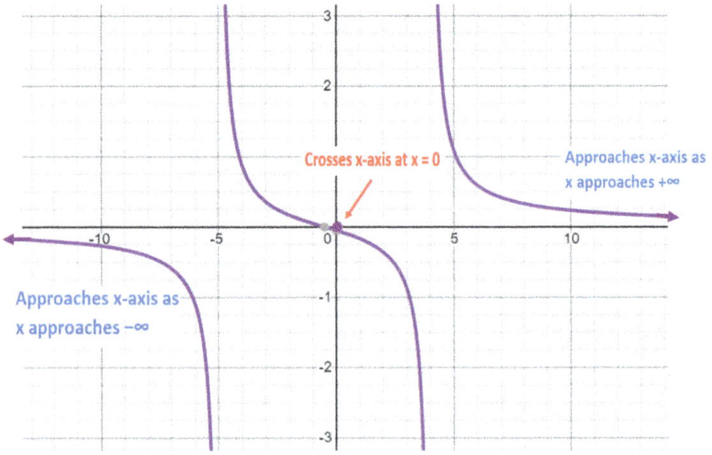

Image 9.16

It is also possible to cross an oblique asymptote as in this example:

$$f(x) = \frac{x^3 - 8}{x^2 - 81}$$

The oblique asymptote is $y = x$.

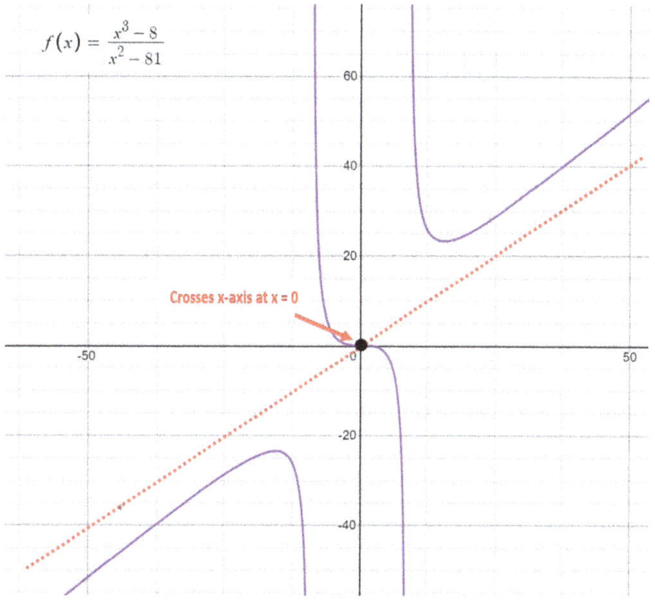

Image 9.36

Write a program that identifies whether the graph of a rational function intersects the horizontal or oblique asymptote and, if so, where it crosses.

V. Relating Horizontal Asymptotes to Limits

As we saw in Section 9.2, horizontal asymptotes of rational functions can be identified by comparing the degree of the numerator to the degree of the denominator.

If we were to again invoke a fundamental concept of calculus, we can identify the horizontal asymptote of a rational function by examining the **limit** of y-values as x approaches infinity.

The calculus notation for this is:

$$\lim_{x \to \infty} f(x) = 2.$$

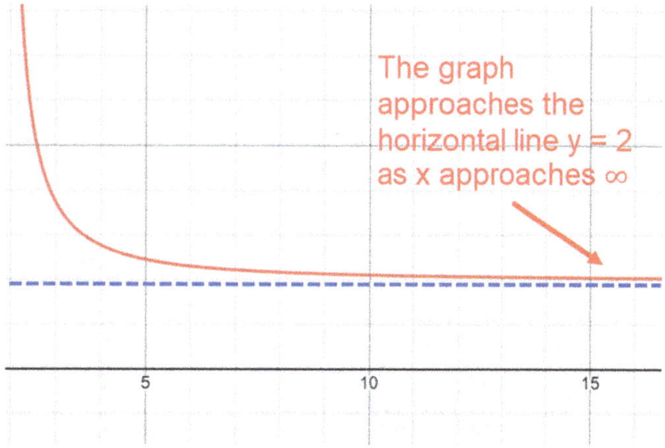

Image 9.37

The limit of function f as x approaches infinity can be found using the **SymPy** function **limit()**: **limit**(f,x,oo).

Notice that you use two lower-case letter os to designate infinity.

Write a program that uses the limit to calculate the horizontal asymptote of a rational function.

SAMPLE OUTPUT

Rational function? (2*x**3−4*x+5)/(4*x**3+7*x)
The horizontal asymptote is $y = ½$

VI. The Work Problem Workaround

You may recall from a prerequisite algebra course a problem that reads something like this:

Working alone, Jack can paint a room in 5 hours. Working alone, Jill can paint the same room in 4 hours. How long will it take them to paint the room if they work together?

Image 9.38

Your instructor undoubtedly showed you how to set up a table and/or equation leading to an equation that looked something like this:

[Jack's rate][# of hours] + [Jill's rate][# of hours] = [1 whole job]

$$\frac{1}{5}x + \frac{1}{4}x = 1$$

Solving this yields $x = \frac{20}{9}$ (or about 2.2 hours).

But notice $\frac{20}{9}$ is the same as $\frac{5 \cdot 4}{5 + 4}$. Does this shortcut work all the time?

Yes, it can be proven that solving $\frac{1}{a}x + \frac{1}{b}x = 1$ always results in $x = \frac{a \cdot b}{a + b}$.

Write a program that asks the user to input two hourly rates, ***a*** and ***b***. Calculate the amount of time it would take two people working at these individual rates to complete the job working together.

SAMPLE OUTPUT
Hourly rate of Jack? 20
Hourly rate of Jill? 5
Working together, Jack and Jill can complete the job in:
4 hours

VII. Partial Fraction Decomposition

When you get to Calculus, a useful integration technique is called **partial fraction decomposition,** in which a rational expression is "decomposed" into the sum of fractions with denominators that are factors of the original denominator. For example:

$$\frac{5x-3}{x^2-2x-3}=\frac{2}{x+1}+\frac{3}{x-3}$$

We begin with what is called a **proper fraction,** i.e., a ratio of polynomials in which the degree of the numerator is less than the degree of the denominator. If done "by hand," the mathematician begins by factoring the original denominator and then setting the fraction equal to a sum of fractions as so:

$$\frac{5x-3}{x^2-2x-3}=\frac{5x-3}{(x+1)(x-3)}=\frac{A}{x+1}+\frac{B}{x-3}$$

The challenge is to find the numerators A and B. This involves several lengthy steps that include solving a system of equations (Chapter 4).

However, I will share with you some formulas that provide rapid partial fraction decomposition of a rational function in the form $\frac{ax+d}{(x-b)(x-c)}$:

$$\frac{ax+d}{(x-b)(x-c)}=\frac{A}{x-b}+\frac{B}{x-c}$$

where $A=\frac{d+ab}{b-c}$, $B=\frac{ac+d}{c-b}$.

You can verify the above example by setting $a=5$, $d=-3$, $b=-1$, $c=3$:

$$A=\frac{-3+(5)(-1)}{-1-3}=\frac{-8}{-4}=2$$

$$B=\frac{(5)(3)+(-3)}{3-(-1)}=\frac{12}{4}=3$$

Thus $\frac{5x-3}{(x+1)(x-3)}=\frac{2}{x+1}+\frac{3}{x-3}$.

Your programming goal is as follows:

A. Ask user to input a proper fraction in the form (*a***x* + *d*) / ((*x* – *b*)*(*x* – *c*)).

B. Extract the values of *a*, *b*, *c*, *d* from the string input.

C. Calculate and output the partial fraction decomposition using the formulas for *A* and *B* above. If *A* and *B* are not integers, use fractions instead of floats. Use pretty print.

SAMPLE OUTPUT

Proper fraction? (2*x)/((x–2)*(x+1)

$$\frac{2 \cdot x}{(x-2) \cdot (x+1)}$$

$$\frac{2}{3 \cdot (x+1)} + \frac{4}{3 \cdot (x-2)}$$

CHAPTER SUMMARY

PYTHON CONCEPTS

Assume *x* has been declared as a **SymPy** symbol using *x* = **Symbol**('x')

To convert a string type to a SymPy expression	poly = **sympify**(string_expression) *Ex:* *>>>f_str = '3*x**3+1'* *>>>f = sympify(f_str)*
To evaluate a SymPy expression for a specific value of the symbol	y_value = poly.subs({x:x_value}) *Ex:* *>>>poly = **sympify**('x**2+4')* *>>>y_value = poly.**subs**({x:2})* *>>>y_value* *8*
To factor a SymPy polynomial	**factor**(sympy_expression) *Ex:* *>>>**factor**('3*x**3+3')* *3*(x + 1)*(x**2 – x + 1)*

To expand a SymPy expression	**expand**(sympy_expression) *Ex:* *>>>**expand**('(x+2)*(x–3)')* *x**2–x–6*
To solve an algebraic equation in the form expression = 0 over the field of complex numbers	**solve**(expression) *returns a list* *Ex:* *>>>solve('x**4–12*x**2–64')* *[-4, 4, -2*I, 2*I]*
To solve an algebraic equation in the form expression = 0 over the field of real numbers	**solveset**(expression, **domain**=S.Reals) *returns a set* *Ex:* *>>>solveset('x**4–12*x**2–64',**domain**=S.Reals)* *(–4, 4)*
To cancel common factors in a rational expression	**cancel**(rational_expression) *Ex:* *>>>f = '(x**2–4)/(x–2)'* *>>>**cancel**(f)* *x+2*
To extract the numerator and denominator of a rational expression	**numer**(rational_expression) **denom**(rational_expression) *Ex:* *>>>f = '(x**3–x**2+6*x–12)/(x**3+1)'* *>>>**numer**(f)* *X**3–x**2+6*x–12* *>>>**denom**(f)* *x**3+1*
To append an item to a list	list.**append**(item) *Ex:* *>>>odds=[1,3,5]* *>>>odds.**append**(7)* *>>>odds* *[1, 3, 5, 7]*
To create a list of integers from *a* to *b*	**range**(a, b+1) *Ex:* *>>>**for** k **in range**(1,9):* *>>> **print**(k)* 1,2,3,4,5,6,7,8

(*Continued*)

To create a clone of a list	list2 = [*k* **for** *k* **in** list1]
To return the degree and coefficients of a polynomial with variable *x*	polynomial = **Poly**(string expression,x) polynomial.**degree**() polynomial.**all_coeffs**() *Ex:* *>>>poly1 = **Poly**('x**3 – 4*x**2 + 10*x – 12')* *>>>poly1.**degree**()* *3* *>>>poly1.**all_coeffs**()* *[1, –4, 10, –12]*

PRECALCULUS CONCEPTS

To find the domain of a rational function	Set the denominator *not* equal to 0. *Ex:* *If* $f(x) = \dfrac{2x+1}{x-7}$ *:* *Set* $x - 7 \neq 0$ $x \neq 7$ *Interval notation:* $(-\infty, 7)\ U\ (7, \infty)$
Vertical asymptote	A vertical line that the graph never crosses but approaches from one or both sides. *Ex:* 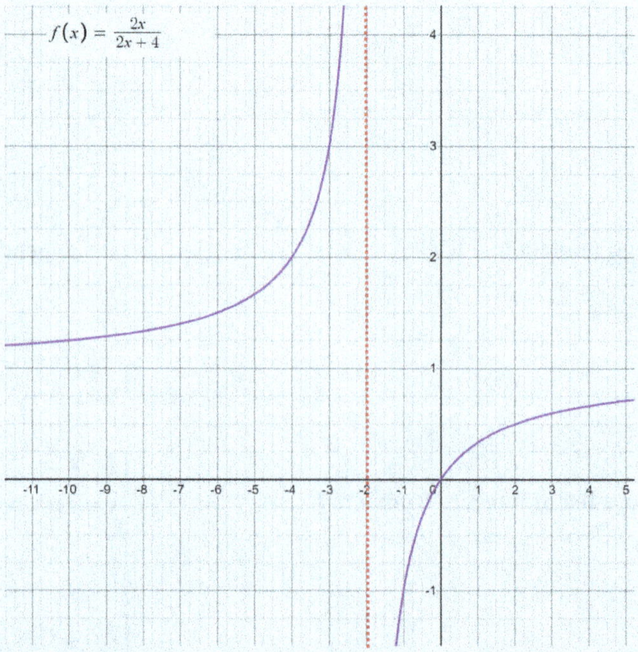 Image 9.4

<table>
<tr>
<td>Horizontal asymptote</td>
<td>A horizontal line that the graph approaches but does not cross when heading to the left or right.

Ex:

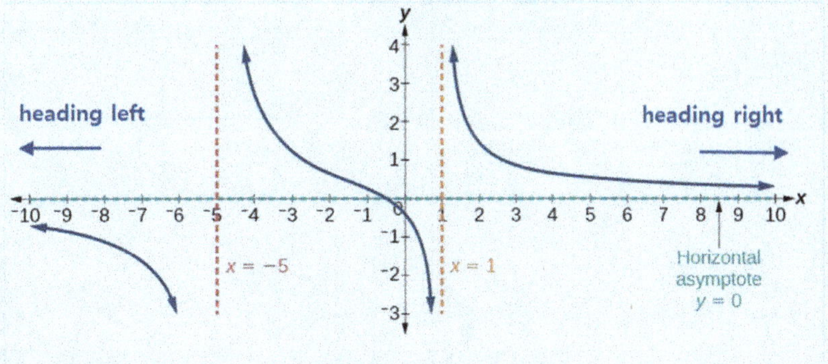

Image 9.39
</td>
</tr>
<tr>
<td>Oblique asymptote</td>
<td>A diagonal (not horizontal or vertical) line that the graph approaches heading to the left or right.

Ex:

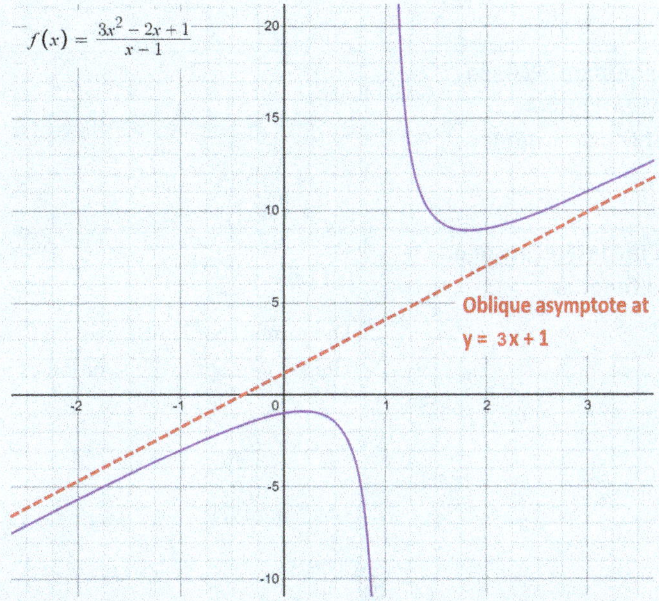

Image 9.21
</td>
</tr>
</table>

(Continued)

Removable discontinuity	An undefined value ("hole") that is not a vertical asymptote. *Ex:* 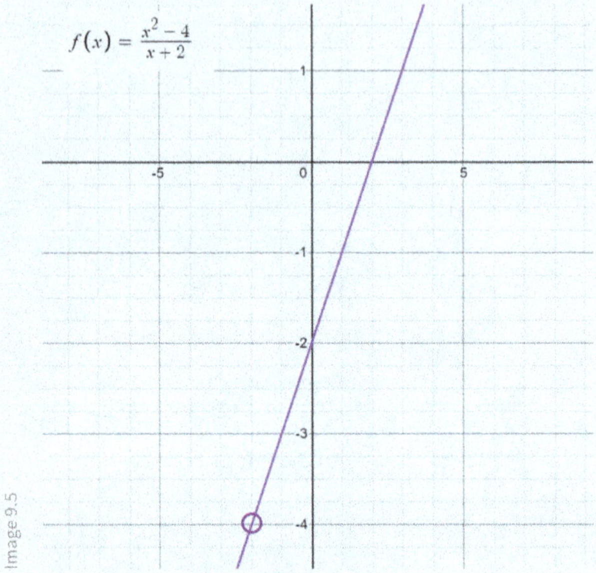
To discriminate between a vertical asymptote and a removable discontinuity	*Ex:* $f(x) = \frac{x^2-6x+8}{3x^2-5x-2} = \frac{(x-2)(x-4)}{(x-2)(3x+1)} = \frac{x-4}{3x+1}$ *then* $x = 2$ *is a "hole" because (x–2) gets cancelled;* $x = -\frac{1}{3}$ *is a vertical asymptote because (3x+1) does not get cancelled.*
To find the intercepts of a rational function	If $f(x) = \frac{p(x)}{q(x)}$: The y-intercept is $f(0)$, provided $q(0) \neq 0$ The x-intercept is the solution of $p(x) = 0$. *Ex:* *If* $f(x) = \frac{x^2-1}{2x-4}$ *The y-intercept is* $f(0) = \frac{0^2-1}{2(0)-4} = \frac{-1}{-4} = \frac{1}{4}$ *The x-intercept is the solution of* $x^2 - 1 = 0$: $x = \pm 1$ ∴ *The intercepts are (0, ¼), (1, 0), and (–1,0).*

Credits

IMG 9.23: Copyright © 2011 Depositphotos/Krisdog.

IMG 9.28: Rafikas, https://commons.wikimedia.org/wiki/File:Burj.jpg, 2007.

IMG 9.29: Copyright © by MikeRun (CC BY-SA 4.0) at https://commons.wikimedia.org/wiki/File:Two-resistors-in-parallel-vert.svg.

IMG 9.30: Copyright © 2014 Depositphotos/tackgalich.

IMG 9.31: Copyright © 2019 Depositphotos/nikiteev.

IMG 9.32: Copyright © 2016 Depositphotos/valdum.

IMG 9.38: Copyright © 2020 Depositphotos/vectorlab.

CHAPTER 10

Exponential and Inverse Functions

10.1 Exponential Functions

Back in Chapter 1 we looked at:

Arithmetic Sequences, in which we keep adding the same number over and over again. For example, 2, 4, 6, 8, 10, 12, ... The functional analogue of an arithmetic sequence is a linear function (i.e., we extend the domain from positive integers to all real numbers).

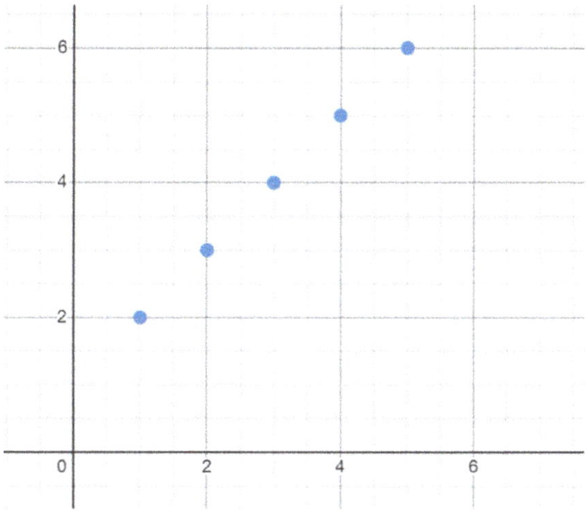

Arithmetic Sequence
Domain is Positive Integers
$a_n = a_1 + (n-1)d$

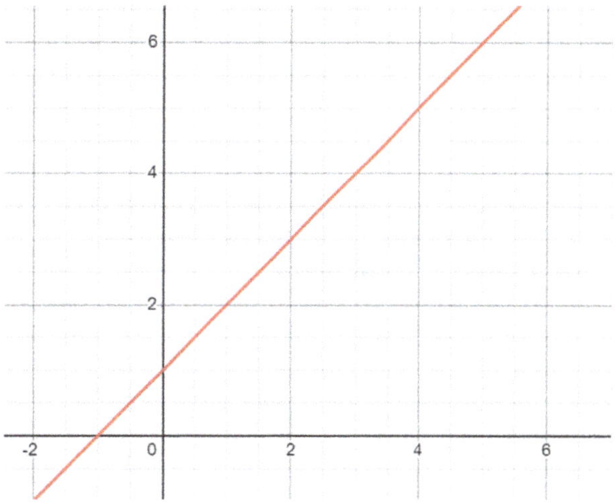

Linear Function
Domain is Real Numbers
$y = mx + b$

Image 10.1

In Chapter 1 we also looked at:

Geometric Sequences, in which we keep multiplying by the same number over and over again. For example 2, 4, 8, 16, 32, ... The functional analogue of a geometric sequence is the heart of this chapter: an **exponential function**.

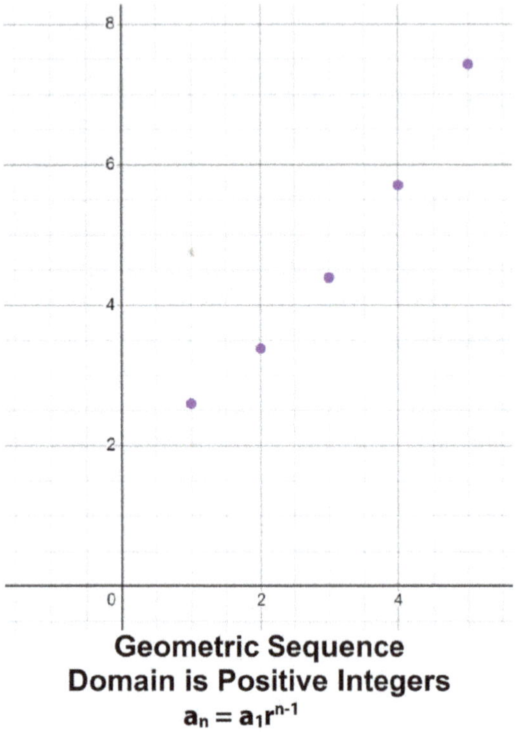

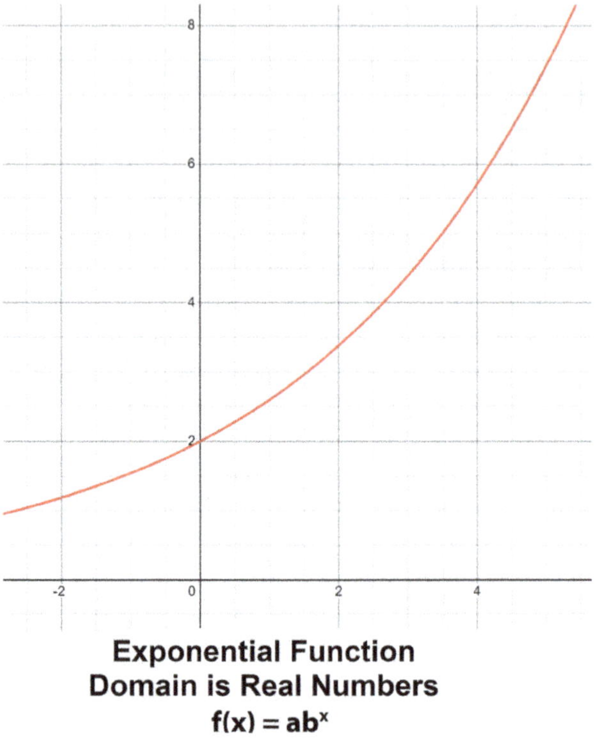

Image 10.2

Typically, in math textbooks we use:

$$f(x) = ab^x$$

because it doesn't matter whether the exponent is x or $x - 1$, as long as the exponent contains a variable. a is any constant nonzero real number, and b (called the **base**) is any real number constant that is neither 0 nor 1.

Examples of exponential functions:

$$f(x) = 2^x,\ y = 3(4)^x,\ f(x) = 12^x$$

Vertical and horizontal shifts of $f(x) = ab^x$ are also exponential functions because they do not change the shape of the graph. Example: $f(x) = 5^{x+1} + 4$ is a shift of $y = 5^x$ one unit right and 4 units up.

EXERCISE SET 10.1.1

1. Graph on the same axes the following functions: $y = 2^x, y = 3^x, y = 5^x$. What is the y-intercept of each?
2. Graph on the same axes the following functions: $y = (1/2)^x, y = (1/3)^x, y = (1/5)^x$. What is the y-intercept of each?

3. Considering your answers to #1 above, sketch a graph of $y = b^x$ in which $|b| > 1$. Identify the domain, range, *y*-intercept, and any asymptotes.
4. Considering your answers to #2 above, sketch a graph of $y = b^x$ in which $0 < |b| < 1$. Identify the domain, range, *y*-intercept, and any asymptotes.
5. Why do you think there is a restriction that given $y = a \cdot b^x$, that *b* is neither 0 nor 1? Why is there a restriction that *a* is not 0?

Graph the following pairs of functions without using your calculator, using your knowledge of shifting and reflecting.

6. $y_1 = 7^x$ and $y_2 = 7^x + 5$
7. $y_1 = 3^x$ and $y_2 = 3^{x+4}$
8. $y_1 = 4^x$ and $y_2 = -4x$
9. $y_1 = \left(\frac{3}{4}\right)^x$ and $y_2 = \left(\frac{3}{4}\right)^{x+1} - 1$
10. $y_1 = \left(\frac{2}{3}\right)^x$ and $y_2 = \left(\frac{2}{3}\right)^{-x}$
11. $y_1 = e^x$ and $y_2 = -e^{-x}$
12. Evaluate $f(x) = 6 \cdot \left(\frac{1}{2}\right)^x$ at $x = -3$. Express your answer as a fraction and a decimal.
13. Evaluate $h(x) = \frac{1}{2} \cdot \left(\frac{1}{3}\right)^x$ at $x = 2$. Express your answer as a fraction and a decimal.
14. The formula for continuously compounded interest is $P(t) = P_0 e^{rt}$ where P_0 is the initial amount, *r* is the periodic rate (as a decimal), and *t* is the amount of time. Find the amount of interest earned on a $100,000 investment at 6.5% for 5 years.
15. You are given .01 on the first day of the month, .02 on the second day, .04 on the third day, and it keeps doubling every day. How much will you receive on the 30th day of the month?
16. Write a program that asks the user to input values for *a*, *b*, and *x*. Calculate and output $a \cdot b^x$. Make sure the user does not input $a = 0$, $b = 0$, or $b = 1$. Express answer as a simplified fraction if possible.
17. Write a program that asks the user to input values for *a* and *b*. Graph $y = a \cdot b^x$. Make sure the user does not input $a = 0$, $b = 0$, or $b = 1$.

Your answers to #1 and #2 above should have yielded the following results:

Graph of $y = b^x$, $\lvert b\rvert > 1$	Graph of $y = b^x$, $\lvert b\rvert < 1$

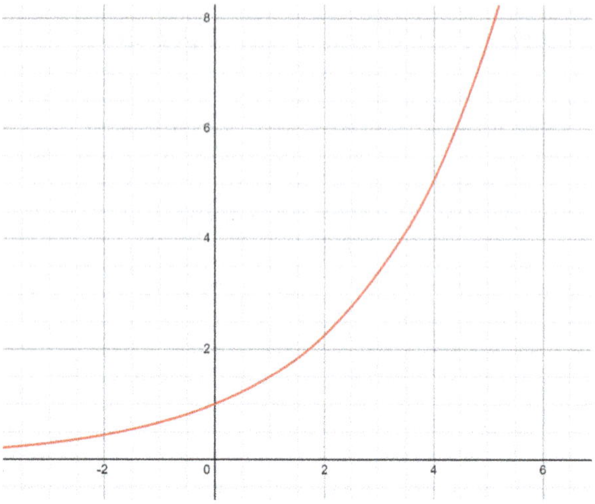

Image 10.3

Domain: $(-\infty, \infty)$

Range: $(0, \infty)$

Horizontal asymptote: $y = 0$

y-intercept: (0, 1)

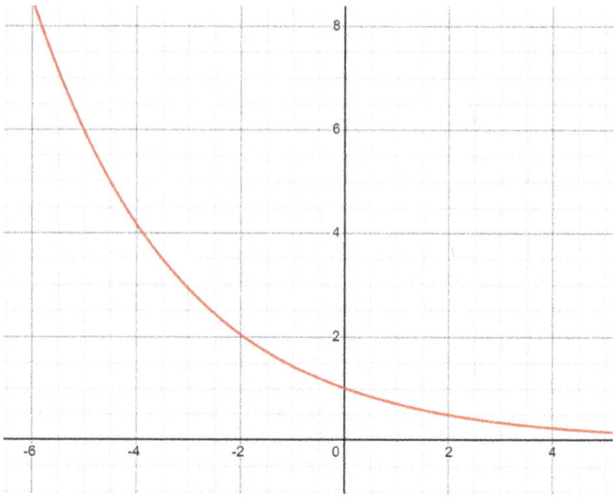

Image 10.4

Domain: $(-\infty, \infty)$

Range: $(0, \infty)$

Horizontal asymptote: $y = 0$

y-intercept: (0, 1)

Rules for Exponents

As a review of prerequisite coursework, here are the rules for exponents.

Rule	Example
$a^m \cdot a^n = a^{m+n}$	$10^4 \cdot 10^2 = 10^{4+2} = 10^6 = 1{,}000{,}000$
$a^m \cdot b^m = (ab)^m$	$5^5 \cdot 2^5 = (5 \cdot 2)^5 = 10^5 = 100{,}000$
$\frac{a^m}{a^n} = a^{m-n}$	$\frac{3^{11}}{3^8} = 3^{11-8} = 3^3 = 27$
$(a^m)^n = a^{m \cdot n}$	$(4^5)^2 = 4^{5 \cdot 2} = 4^{10} = 1{,}048{,}576$
$a^{1/n} = \sqrt[n]{a}$	$(125)^{1/3} = \sqrt[3]{125} = 5$
$a^0 = 1$	$(-100)^0 = 1$
$\left(\frac{a}{b}\right)^n = \frac{a^n}{b^n}$	$\left(\frac{3}{5}\right)^3 = \frac{3^3}{5^3} = \frac{27}{125}$
$a^{-n} = \frac{1}{a^n}$	$4^{-3} = \frac{1}{4^3} = \frac{1}{64}$
$\frac{1}{a^{-n}} = a^n$	$\frac{1}{6^{-3}} = 6^3 = 216$

Examples

Simplify the following. Express answers using positive exponents or simplified radicals.

A. $(p^2)^{-2}$

B. $\dfrac{-15x^4}{3x}$

C. $(s^2t)^3(st)$

D. $(5x)^{-2} \cdot x^3$

E. $\left(\dfrac{x^3}{y^2}\right)^{-3}$

F. $9a^{-1/2}$

G. $\left(x^{\frac{2}{3}}\right)^{12}$

H. $\sqrt[3]{\dfrac{64x^9}{343}}$

Solutions

1. $(p^2)^{-2}$

   p^{-4}

   $\dfrac{1}{p^4}$

2. $\dfrac{-15x^4}{3x}$

   $-5x^3$

3. $(s^2t)^3(st)$

   $s^6t^3 \cdot st$

   s^7t^4

4. $(5x)^{-2} \cdot x^3$

   $5^{-2}x^{-2} \cdot x^3$

   $5^{-2}x$

   $\dfrac{x}{5^2}$

   $\dfrac{x}{25}$

5. $\left(\dfrac{x^3}{y^2}\right)^{-3}$

   $\dfrac{x^{-9}}{y^{-6}}$

   $\dfrac{y^6}{x^9}$

6. $9a^{-1/2}$

   $\dfrac{9}{a^{1/2}}$

   $\dfrac{9}{\sqrt{a}}$

7. $\left(x^{\frac{2}{3}}\right)^{12}$

   $x^{\frac{2}{3} \cdot 12}$

   x^8

8. $\sqrt[3]{\dfrac{64x^9}{343}}$

   $\left(\dfrac{64x^9}{343}\right)^{1/3}$

   $\dfrac{64^{1/3}(x^9)^{1/3}}{343^{1/3}}$

   $\dfrac{\sqrt[3]{64} \cdot x^3}{\sqrt[3]{343}}$

   $\dfrac{4x^3}{7}$

EXERCISE SET 10.1.2

Simplify using the rules for exponents. Express rational numbers as simplified fractions with positive exponents.

1. $6^{3/2} \cdot 6^2$
2. $\dfrac{2^{-4}}{2^3}$
3. $(10^3)^{1/6}$
4. $\left(\dfrac{x^2}{y^3}\right)^{-7}$
5. $\left(\dfrac{\omega^4}{\omega^5}\right)^{-1}$
6. $(a^4)^3 + (a^{1/2})^{24}$
7. $\left(\dfrac{6s}{3s^5}\right)^2$
8. $\dfrac{2n^{-2}}{5n^6b^{-4}}$

Program Goal: Write a program in which the user inputs two integer exponents, m and n, corresponding to a^m and a^n. Simplify the expression $a^m \cdot a^n$.

SAMPLE OUTPUT

```
m? 4

n? -2

(a**4)*(a**-2)

= a**2

m? 3

n? -5

(a**3)*(a**-5)

= a**-2
```

PROGRAM 10.1.1 ADD_EXPONENTS.PY

```
m = int(input('m? '))
n = int(input('n? '))
p = m + n
print('(a**', m, ')*(a**', n, ')')
print('= a**', p)
```

EXERCISE SET 10.1.3

1. Modify the program so that if the result is *a***1, it just outputs *a*.

Sample Output

m? 3

n? -2

(*a***3)*(*a***-2)

= *a*

2. Modify the program so that if the result is a negative exponent, it outputs the result as a fraction, and if the result is a 0 exponent, then the output is simply 1.

Sample Output

m? 1

n? −3

(*a***1)*(*a***−3)

= 1/(*a***2)

m? 3

n? −3

(*a***3)*(*a***−3)

= 1

3. Modify the program so that the output is formatted using **SymPy**'s **pprint()** function.

Sample Output

m? 1

n? 4

(*a***1)*(*a***4)=

a^5

Can **SymPy** simplify exponential expressions? You know by now that is a rhetorical question because of course the answer is yes.

SymPy Functions to Simplify Exponential Expressions

simplify(expression)—will attempt to combine all exponents of the same base

Example: $\frac{a^m a^n}{a^5}$ *will be simplified to* a^{m+n-5}

expand_power_base(expression)—will attempt to apply a constant exponent to each variable/constant in the base

Example: $(ab)^5$ *will be simplified to* a^5b^5, *but* $(ab)^n$ *will not be simplified to* a^nb^n

expand_power_exp(expression)—will attempt to apply each term in an exponent to the single variable base

Example: a^{m+n} *will be simplified to* a^ma^n *but* $(ab)^{m+n}$ *will* ***not*** *be simplified to* $a^{m+n}b^{m+n}$

EXERCISE SET 10.1.4

For #1–10, write a program, or use the Python console, to print the results of the following expressions. Begin with:

>>>**from** sympy **import** *

>>>*a*,*b*,*m*,*n* = **symbols**('a b m n')

1. **simplify**(a**m*a**n)
2. **simplify**((a**m)/(a**n))
3. **simplify**((a*b)**n)
4. **expand_power_base**((a*b)**3)
5. **simplify**(a**(m+n))
6. **expand_power_exp**(a**(m+n))
7. **simplify**(a**–1)
8. **simplify**(a**3/a**–2)
9. **simplify**((a**(3/2))**6)
10. **expand_power_exp**(a**(m–n))

10.2 Solving Exponential Equations

If you were asked to solve $2^x = 512$, you'd probably recognize that this involved converting 512 to a power of 2.

$$2^x = 2^9$$

Therefore, $x = 9$. This leads to a fairly obvious principle:

LAW OF EXPONENTS

If $a^m = a^n$, then $m = n$.

In other words, if the bases are the same, then the exponents are the same.

Sometimes you might need to simplify one or both sides of the equation using the rules for exponents.

Example

Solve $\frac{3^x}{3^5} = 81$

Solution

Simplify the left side, and convert the right side to a power of 3:

$$3^{x-5} = 3^4$$

Now that the bases are the same, set the exponents equal and solve:

$$x - 5 = 4$$

$$x = 9$$

EXERCISE SET 10.2.1

Solve.

1. $4^{x+1} = 64$
2. $5^{2x-1} = 5^{x+3}$
3. $3^{x-9} = 9^3$
4. $e^{3x} = e^{2-x}$
5. $4^{-x} = \frac{1}{64}$
6. $e^{-x^2} = (e^x)^2 \cdot \frac{1}{e^3}$ (Hint: there are two solutions)
7. $\sqrt{225} = 3^{x-1}5^{x-1}$

As a quick Python refresher, let's look at a program that will solve an exponential equation.

Recall that **SymPy solve**(expr) will solve expr = 0, so to solve exprl = expr2, we must first convert the equation to exprl − expr2 = 0.

PROGRAM 10.2.1 SOLVE_EXPONENTIAL.PY

```
from sympy import Symbol,solve
x=Symbol('x')
exp_eqn = input('Exponential Equation? ')
left,right = exp_eqn.split('=')
exp_eqn2 = left + '–('+right + ')'
result = solve(exp_eqn2)
print(result)
```

SAMPLE OUTPUT

Exponential Equation? 2**x=64

[6]

EXERCISE SET 10.2.2

1. When you run the program with input 2**x=64:
   A. What values are assigned to variables **left** and **right**?
   B. What value is assigned to variable **exp_eqn2**?
2. The program will also provide unwanted complex solutions to certain exponential equations. Run the program with input 4**x=64. Which is the real solution? Which is the complex solution?
3. Replace result=**solve**(exp_eqn2) with the following:

   *result=**solveset**(exp_eqn2,**domain**=S.Reals)*

   What effect does this have on the output? *(Note: the **domain=S.Reals** parameter does not appear to work in all Python interpreters.)*
4. Use the program to check your answers in Exercise Set 10.2.1 #1–5. Ignore complex solutions. Does the program handle square root expressions such as in #7?

10.3 Application Problems

Exponential functions (often with base e) occur frequently in the fields of medicine, economics, probability, nature, and other fields of science.

Example

The normal healing of wounds can be modeled by an exponential function. If A_0 represents the original area of the wound and if $A(t)$ equals the area of the wound at time t, then $A(t) = A_0 e^{-0.35t}$ provided there is no infection. Suppose that a wound initially had an area of 100 square millimeters.

If healing is taking place, how large will the area of the wound be after 3 days? Round to the nearest millimeter.

Solution

Substitute $A_0 = 100$ and $t = 3$ into the function:

$$A(3) = 100e^{-0.35(3)} = 100e^{-1.05} \approx 35 \text{ mm}^2.$$

EXERCISE SET 10.3.1

1. The atmospheric pressure P (in pounds per square inch) decreases exponentially with altitude h (in miles above sea level) as given by the equation:

   $$P(h) = 14.7e^{-0.21h}$$

   Find the pressure at 1.2 miles above sea level.

Image 10.5

2. The price p in dollars of a car that is x years old is modeled by $p(x) = 22{,}265(0.9)^x$. How much would a 3-year-old car cost?

Image 10.6

(Continued)

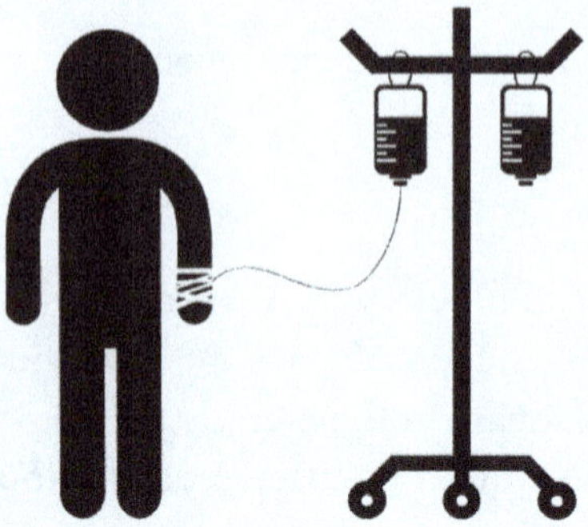

Image 10.7

3. The function $D(h) = 5e^{-0.4h}$ can be used to find the number of milligrams D of a certain drug that is in a patient's bloodstream h hours after the drug has been administered. How many mg will be present after 1 hour?

4. A model for the number N of people in a college community who have heard a certain rumor is $N(d) = P(1 - e^{-0.15}d)$ where P is the total population of the community and d is the number of days that have elapsed since the rumor began. In a community of 1,000 students, how many students will have heard the rumor after 3 days?

Image 10.8

Image 10.9

5. Pierre de Fermat (1601–1665) conjectured that the function

$$P(x) = 2^{2^x} + 1$$

would always produce a prime number.

Find the values for $x = 1, 2, 3, 4,$ and 5. Determine if these are in fact prime numbers.

Exponential Growth/Decay

Questions 1 and 3 in Exercise Set 10.3.1 above are special types of a model called **exponential growth/decay**.

A notable example is described in the children's fable about the king's chessboard (mentioned in Chapter 1) in which 1 piece of rice is placed on the first square, 2 pieces on the second, 4 pieces on the third, resulting in the geometric sequence 1, 2, 4, 8, 16, ... This can be modeled using $A(t) = e^{0.6931t}$. In fact, any geometric sequence (or function, if continuous) can be modeled using:

EXPONENTIAL GROWTH AND DECAY

If a quantity follows a geometric progression, then its quantity (or population) at time t is:

$$P(t) = P_0 e^{kt},$$

where P_0 is the initial quantity (population) and k is a constant called the growth factor (or growth rate). If $k > 0$, the quantity experiences **exponential growth**. If $k < 0$, the quantity follows **exponential decay**.

This can be visualized with the graphs:

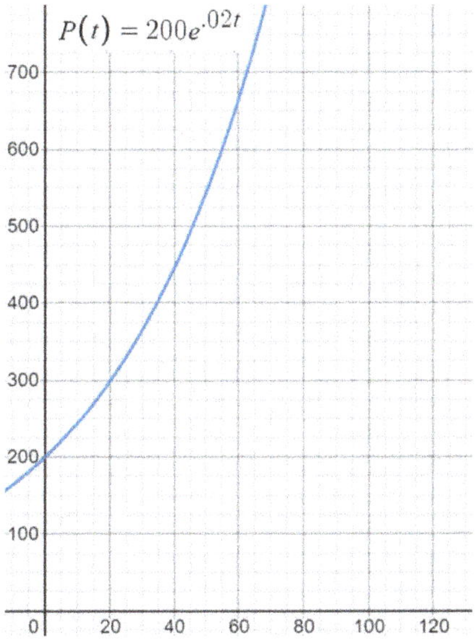

Image 10.10

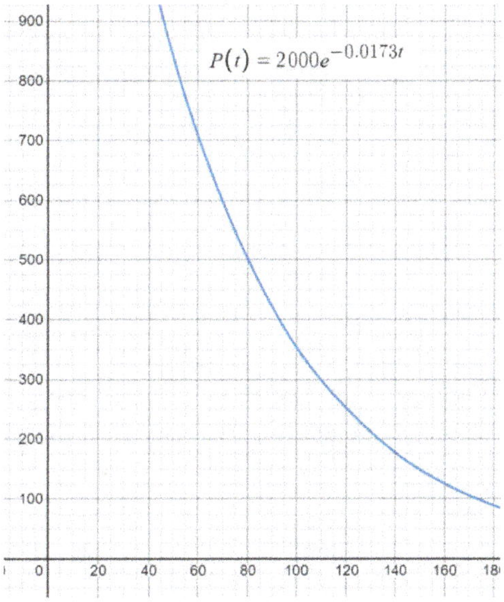

Image 10.11

EXERCISE SET 10.3.2

1. What is the y-intercept (or in this case the P-intercept) of the graph of an exponential function?
2. Because time and quantities are always positive, what should Xmin and Ymin be when graphing exponential functions?
3. In the case of exponential growth, if you would like Xmax = 25, what should you set Ymax to be?
4. In the case of exponential decay, if you would like Xmax = 25, what should you set Ymax to be?

Populations that experience rapid growth include bacterial cultures, continuously compounded interest, human population, pandemics (spread of disease), wildfire, the zombie apocalypse, etc. In the case of continuously compounded interest, the constant k is equal to the annual interest rate expressed as a decimal.

Example 1

A 25-year-old student would like to invest some money in a retirement account that pays 5% annual interest compounded continuously. How much does the student need to invest today to have $1 million when she retires at age 65?

Solution

In this problem we are trying to find P_0, the initial amount. The time $t = 40$ years and the interest rate $k = .05$. The amount we want after 40 years is $P(t) = 1{,}000{,}000$.
Substituting into $P(t) = P_0e^{kt}$ we get:

$$1{,}000{,}000 = P_0e^{0.05(40)}$$

$$1{,}000{,}000 = 7.38906P_0$$

$$P_0 = \$135{,}335.28$$

This may seem like a lot of money, but it is less than 1/7th the amount of desired savings.

Examples of exponential decay include radioactive substances, the dampening of an oscillating object, etc.

Example 2

Carbon-14 is an element that decays using the following model:

$$P(t) = P_0e^{-0.000121t}$$

Scientists can approximate the age of an organic substance by observing the amount of original carbon-14 that remains. If we have a preserved plant that originally had 10 mcg of carbon-14, how much of the substance remains after 5,730 years?

Solution

We are trying to find $P(t)$ with $t = 5{,}730$. The original amount is $P_0 = 10$.
Substituting into $P(t) = P_0e^{kt}$ we get:

$$P(5{,}730) = 10e^{-0.000121(5730)}$$

$$\approx 5 \text{ mcg}$$

We end up with half the original substance. (This is why 5,730 is the half-life of carbon-14.)

EXERCISE SET 10.3.3

1. A colony of bacteria grows rapidly according to the exponential model $P(t) = 100e^{0.045}t$, where P is measured in grams and t is measured in days.

   A. What is the population after 5 days?

   B. If the population after 10 days is 200 g, what was the original amount?

   C. If the population starts at 200 g, how much is present after 5 days?

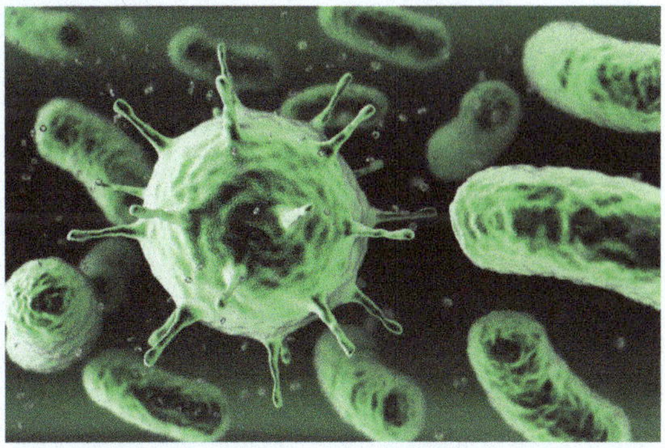

Image 10.12

Image 10.13

2. The size of a certain insect population at time t (in days) is modeled by the exponential model $P(t) = 500e^{0.02t}$.

   A. Identify the initial amount and the growth rate.

   B. What is the population after 10 days?

   C. If the population starts at 750, what is the population after 10 days?

   D. If the population after 10 days is 1,100, what was the initial population?

3. Carbon-14 is an element that decays using the model $P(t) = P_0e^{kt}$ where $k = -\dfrac{ln2}{5730}$.

   A. How much of a substance remains after 33,830 years? Express your answer as a percentage of the initial amount.

   B. If 140 g were found after 30,000 years, what was the initial amount?

Image 10.14

4. Write a program that asks the user to input the initial population, time t, and growth rate k. Calculate and output $P(t)$ rounded to the nearest integer. (Hint: Use **exp**(x) to represent e^x. The **exp**() function should be imported from the **math** library.)

Sample Output

Initial Population? 1,000

Growth Rate? 1.05

Time? 12

$P(12) = 296558565$

(Continued)

5. The **half-life** of an element is the amount of time it takes for that element to decay to half its initial size. If you know the half-life h of an element and its initial population P_0, you can calculate the population at time t using the following formula:

$$P(t) = P_0 \cdot \left(\frac{1}{2}\right)^{t/h}$$

Write a program that asks the user to input the initial population P_0, half-life h, and amount of time t. Calculate and output $P(t)$ rounded to the nearest integer.

Sample Output

Initial Population? 1,000

half-life? 47

Time? 40

$P(40) = 554$

6. Investors are often interested in the **doubling time** for a given interest rate. The **Rule of 72** estimates the doubling time as $t = \frac{72}{r}$, where r is the annual interest rate (expressed as a percent, not decimal.)

Write a program that asks that user to input the annual interest rate r and amount of initial investment A. Calculate the doubling time and the amount.

Sample Output

Annual interest rate? 5.5

Amount invested? 15,000

You will have 30,000.00 in about 13.09 years.

BONUS: Express the amount of time in years and months.

Could you calculate the amount of time it would take for an initial population to reach a given amount? The answer is ... not yet. To calculate a value in the exponent, we need something called a logarithm, which we look forward to learning in Chapter 11.

10.4 Composition of Functions

You probably already have an intuitive understanding of how the **composition** of functions works. For example, **math.sqrt(–25)** returns a domain error because the domain (acceptable parameters) of math.**sqrt** is $[0, \infty)$. However, math.**sqrt(abs**(–25)) will work because we have converted –25 to a nonnegative number. This is a composition of functions.

f* composed with *g is designated **f(g(x))** or **(f °g)(x).**

The domain of $f\circ g$ is the set of all x's in the domain of g for which $g(x)$ is in the domain of f.

Note, the ° is **not** a multiplication symbol. It is a composition symbol.

The inner function is evaluated first.

Example 1

If $f(x)=\sqrt{x}$ and $g(x)=\frac{1}{x-1}$, evaluate $f(g(10)), f(g(-3)), f(g(1)), g(f(25)), g(f(0))$. Identify the domains of $f\circ g$ and $g\circ f$.

Solution

$f(g(10)) = f\left(\frac{1}{10-1}\right) = f\left(\frac{1}{9}\right) = \sqrt{\frac{1}{9}} = \frac{1}{3}$

$f(g(-3)) = f\left(\frac{1}{-3-1}\right) = f\left(-\frac{1}{4}\right) = \sqrt{-\frac{1}{4}}$ = not a real number because $g(-3)$ is not in domain of f

$f(g(1)) = f\left(\frac{1}{1-1}\right) = f\left(\frac{1}{0}\right)$ = undefined because 1 is not in domain of g

$g(f(25)) = g(\sqrt{25}) = g(5) = \frac{1}{5-1} = \frac{1}{4}$

$g(f(1)) = g(\sqrt{1}) = g(1) = \frac{1}{1-1} = \frac{1}{0}$ = undefined because $f(1)$ is not in domain of g

The **domain of $f\circ g$** is the set of all x's in the domain of g for which $g(x)$ is *in* the domain of f.
The domain of g is $\{ x \mid x \neq 1 \}$
The domain of f is $\{ x \mid x \geq 0 \}$. We must therefore make sure $g(x) \geq 0$:
$\frac{1}{x-1} \geq 0$
$x-1 \geq 0$ (so that both numerator and denominator are positive)
$x \geq 1$
Taking the intersection of $x \neq 1$ and $x \geq 1$, we get simply $\mathbf{x > 1}$.
$\therefore$ Domain of $f\circ g = \{ x \mid x \geq 1 \}$

The **domain of $g\circ f$** is the set of all x's in the domain of f for which $f(x)$ is in the domain of g.
The domain of f is $\{ x \mid x \geq 0 \}$
The domain of g is $\{ x \mid x \neq 1 \}$. We must therefore make sure $f(x) \neq 1$:
$\sqrt{x} \neq 1 \rightarrow x \neq 1$
Taking the intersection of $x \geq 0$ and $x \neq 1$, we get $0 \leq x < 1$ or $x > 1$.
$\therefore$ Domain of $g\circ f = \{ x \mid 0 \leq x < 1 \text{ or } x > 1 \}$

Example 2

Given $f(x)=\frac{1}{\sqrt{x}}$ and $g(x)=-x$, express $f\circ g$ and $g\circ f$ as functions of x, and identify the domain of each composition.

Solution

$(f°g)(x) = f(g(x)) = f(-x) = \frac{1}{\sqrt{-x}}$

$(g°f)(x) = g(f(x)) = g\left(\frac{1}{\sqrt{x}}\right) = -\frac{1}{\sqrt{x}}$.

The **domain of $f°g$** is the set of all xs in the domain of g for which $g(x)$ is in the domain of f.
The domain of g is $(-\infty, \infty)$.
The domain of f is $\{ x \mid x > 0 \}$. We must therefore make sure $g(x) > 0$:
$-x > 0 \rightarrow x < 0$
The intersection of $(-\infty, \infty)$ and $x < 0$ is $\mathbf{x < 0}$.
$\therefore$ Domain of $f°g = \{ x \mid x < 0 \}$

The **domain of $g°f$** is the set of all xs in the domain of f for which $f(x)$ is in the domain of g.
The domain of f is $\{ x \mid x > 0 \}$
The domain of g is $(-\infty, \infty)$. Trivially, all $f(x) \in (-\infty, \infty)$.
The intersection of $x > 0$ and $(-\infty, \infty)$ is $\mathbf{x > 0}$.
$\therefore$ Domain of $g°f = \{ x \mid x > 0 \}$

Example 3

Each of the following represents a composition of functions $(f°g)(x)$. "Decompose" each function into the outer function $f(x)$ and inner function $g(x)$.

A. $(f°g)(x) = |x^2 + 1|$

B. $(f°g)(x) = \sqrt[3]{1 - x^2}$

C. $(f°g)(x) = \frac{2}{x^2 + 7}$

Solutions

A. $f(x) = |x|$, $g(x) = x^2 + 1$ OR $f(x) = |x + 1|$, $g(x) = x^2$

B. $f(x) = \sqrt[3]{x}$, $g(x) = 1 - x^2$ OR $f(x) = \sqrt[3]{1 - x}$, $g(x) = x^2$

C. $f(x) = \frac{2}{x}$, $g(x) = x^2 + 7$ OR $f(x) = \frac{2}{x + 7}$, $g(x) = x^2$

EXERCISE SET 10.4.1

Express answers as simplified fractions if possible.

1. If $f(x) = \frac{x}{x+3}$ and $g(x) = \frac{2}{x}$, find the following:

A. $f(g(0))$

B. $f\left(g\left(-\frac{2}{3}\right)\right)$

C. $g(f(1))$

D. $f(g(1))$

E. $g(f(-3))$

F. $(f°g)(x)$

G. $(g°f)(x)$

H. Domain of $f°g$

I. Domain of $g°f$

2. If $f(x) = \frac{1}{x^3}$ and $g(x) = \sqrt{x}$, find the following if possible:

   A. $f(g(-1)$
   B. $f(g(1))$
   C. $f(g(0))$
   D. $g(f(4)$
   E. $g(f(0))$
   F. $(f°g)(x)$
   G. $(g°f)(x)$
   H. Domain of $f°g$
   I. Domain of $g°f$

3. If $f(x) = x^2$ and $g(x) = = \sqrt{x}$, find the following if possible:

   A. $f(g(-1)$
   B. $f(g(1))$
   C. $f(g(0))$
   D. $g(f(4)$
   E. $g(f(-x))$
   F. $(f°g)(x)$
   G. $(g°f)(x)$
   H. Domain of $f°g$
   I. Domain of $g°f$

4. Each of the following represents a composition of functions $(f°g)(x)$. "Decompose" each function into the outer function $f(x)$ and inner function $g(x)$. More than one answer is possible.

   A. $(f°g)(x) = \frac{5 + x^2}{4 - x^2}$

   B. $(f°g)(x) = 2 + \sqrt{x^3 - 4}$

   C. $(f°g)(x) = e^{\sqrt{x}}$

Why do we care so much about domains? They play an important role in program error-checking, as we shall soon see.

Program Goal: Write a program that will ask the user to input a value x. Given $f(x) = \frac{x}{x+3}$ and $g(x) = \frac{2}{x}$, calculate and output $f(g(x))$ and $g(f(x))$.

PROGRAM 10.4.1 COMPOSE_FUNCTIONS.PY

```
def f(a):
    return a/(a+3)
def g(a):
    return 2/a
# main program
x1 = float(input("x? "))
print ("f(g(",x1,"))=")
print (f(g(x1)))
print ("g(f(",x1,"))=")
print(g(f(x1)))
```

SAMPLE OUTPUT

x? 1

f(g(1.0))=

0.4

g(f(1.0))=

8.0

However, what happens when you input 0 or −2/3 or −3? To have a truly robust program, let's (1) do an error check and (2) input and output as a simplified fraction.

We need to have each function return a value indicating that the function was undefined at a given parameter. Because we do not expect any results to be equal to ∞, we will use float('inf') as an error flag.

PROGRAM 10.4.1 COMPOSE_FUNCTIONS.PY V2

```
from fractions import Fraction
def f(a):
    if a!=-3:                    # make sure a is in domain of f
        return a/(a+3)
    else:
        return float('inf')
def g(a):
    if a!=0:                     # make sure a is in domain of g
        return 2/a
    else:
        return float('inf')
# main program
x1 = Fraction(input("x? "))
result = 'undefined'             # default value
print ("f(g(",x1,"))=")
if g(x1)!=float('inf'):
    if f(g(x1))!=float('inf'):
        result = Fraction(f(g(x1))).limit_denominator(100)
print(result)
[insert code to calculate g(f(x1))]
```

SAMPLE OUTPUT

x? 0
$f(g($ 0 $))=$
undefined
$g(f($ 0 $)) =$
undefined

x? 1
$f(g($ 1 $))=$
2/5
$g(f($ 1 $))=$
8

x? −2/3
$f(g($ −2/3 $))=$
undefined
$g(f($ −2/3 $))=$
−7

EXERCISE SET 10.4.2

1. Fill in the yellow highlighted code so that the program also outputs $g(f(x1))$.
2. Modify the program so that the user can input a list of x-values, and the program outputs $f(g(x))$ and $g(f(x))$ for each x-value.

Sample Output

x list? −3, −1, 1, 7

$f(g(x))$ = [−0.29, −2.0, 0.4, 0.09]

$g(f(x))$ = ['undefined', −4.0, 8.0, 2.86]

3. Modify the program so that $f(x) = \frac{1}{x^3}$ and $g(x) = \sqrt{x}$.

Sample Output

x list? 0, 1, 4, −3

$f(g(x))$ = ['undefined', 1.0, 0.12, 'undefined']

$g(f(x))$ = ['undefined', 1.0, 0.12, 'undefined']

To make this program truly versatile, is it possible to enable the user to input functions for f and g? The answer is yes (!) using a method in the **SymPy** library called **lambdify**.

PROGRAM 10.4.1 COMPOSE_FUNCTIONS.PY V3

```
from fractions import Fraction
from sympy import Symbol, lambdify
x = Symbol('x')
f_str = input("function f?")
f = lambdify(x,f_str)
[insert code to input function g and lambdify it]
x1 = Fraction(input("x? "))
print ("f(g(",x1,"))=")
result = Fraction(f(g(x1))).limit_denominator(100)
print(result)
[insert code to calculate and output g(f(x1))]
```

SAMPLE OUTPUT

function *f*? *x***2

function *g*? *x*–1

x? 10

f(*g*(10))=

81

g(*f*(10))=

99

EXERCISE SET 10.4.3

1. Fill in the yellow highlighted code to input function *g* and lambdify it.
2. Fill in the aqua highlighted code to calculate and output $g(f(x1))$.
3. Use the program to calculate $f(g(x))$ and $g(f(x))$ for the following:

   A. $f(x) = \dfrac{1}{x^2+1}, g(x) = 2x - 5, x = 1$

   B. $f(x) = 100(x^2 - 1), g(x) = \dfrac{1}{x}, x = 5$

   C. $f(x) = \dfrac{x}{x^2+11}, g(x) = 1 - x, x = -2$

In general, to define and invoke a function using **lambdify**:

from sympy **import** Symbol, lambdify	*# import the methods from SymPy library*
x = **Symbol**('x')	*# identify x as a SymPy variable*
func = **lambdify**(x, expression)	*# define the function of x using a string expression*
y = **func**(x_value)	*# assign to y the function value evaluated at x_value*

Handling Errors

Now that we are allowing users to input their own functions, how can we anticipate what kind of errors might occur? For example, if they input $f = 1/x$, how do we know to check for $x = 0$?

Since Python already does the error check (as we have seen every time we get a runtime error), we just need a way to circumvent Python's stream of messy tracebacks, etc., with a much more succinct conclusion. This is called **exception handling** and it works something like this:

```
try:
    result = Fraction(f(g(x1))).limit_denominator(100)
except:
    result="Undefined"
print(result)
```

In other words, the program will try to execute the code in the **try** block. If that results in a Python runtime error (an **"exception"**), then the code in the **except** block will be executed.

It is possible to raise different types of exceptions (**ValueError**, **TypeError**, **ZeroDivisionError**, etc.) to make your error messages more specific.

EXERCISE SET 10.4.4

1. Modify Program **compose_functions** so that it incorporates exception handling for both $f(g(x))$ and $g(f(x))$.

Sample Output

function *f*? *x*+2

function *g*? *x***2

x? 1

(Continued)

$f(g(1))=$

3

$g(f(1))=$

9

function f? 1/x

function g? x−4

x? 4

$f(g(4))=$

Undefined

$g(f(4))=$

−15/4

2. Modify the program so that it allows the user to input a list of values.

Sample Output

function f? 1/(x−3)

function g? 4*x**2+5*x−11

x list? −2,3,5,10

$f(g(x))$ = [−0.12, 0.03, 0.01, 0.0]

$g(f(x))$ = [−11.84, 'undefined', −7.5, −10.2]

3. What happens with the following input and why?

function f? x**3

function g? x**(1/3)

x list? −8

10.5 One-to-One Functions

A **relation** is a correspondence between two sets X and Y that pairs elements; i.e., a set of ordered pairs (x,y).

Example: $x^2 + y^2 = 25$ describes the set of ordered pairs on the circle centered at (0,0) with radius 5.

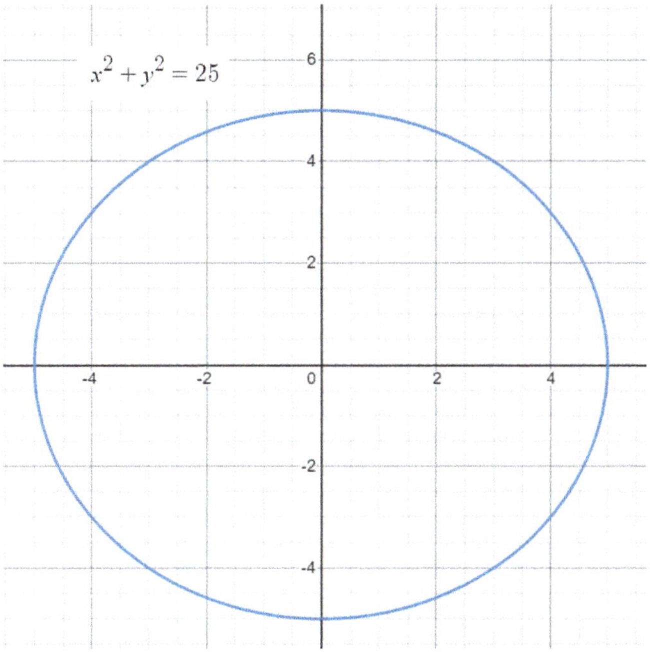

Image 10.15

A **function** is a relation in which each x is paired to exactly one y (not necessarily vice versa). We use the **vertical line test** to determine if a graph represents a function. If any vertical line crosses more than one point on the graph, it is not a function.

Example: $y = x^2$ is a function. $x = y^2$ is not a function.

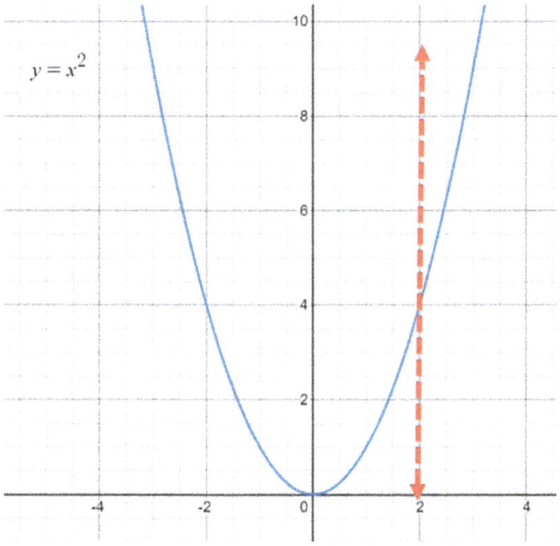

Image 10.16

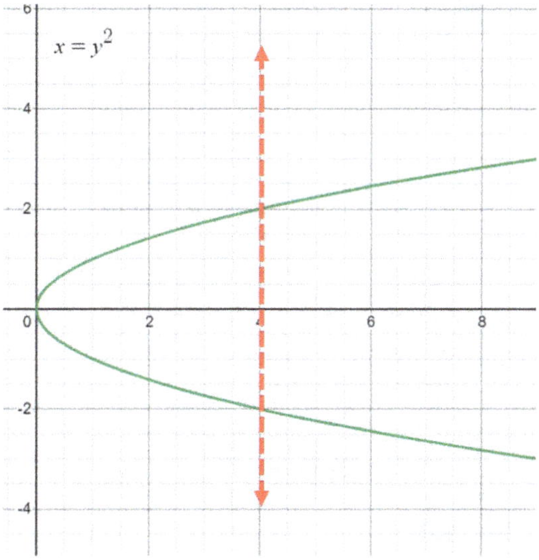

Image 10.17

A **one-to-one function** is a function in which each y is paired to exactly one x. We use the **horizontal line test** to determine if a graph represents a function. If any horizontal line crosses more than one point on the graph, it is not a function.

Example: $y = \sqrt[3]{x}$ is one-to-one. $y = x^2$ is not one-to-one.

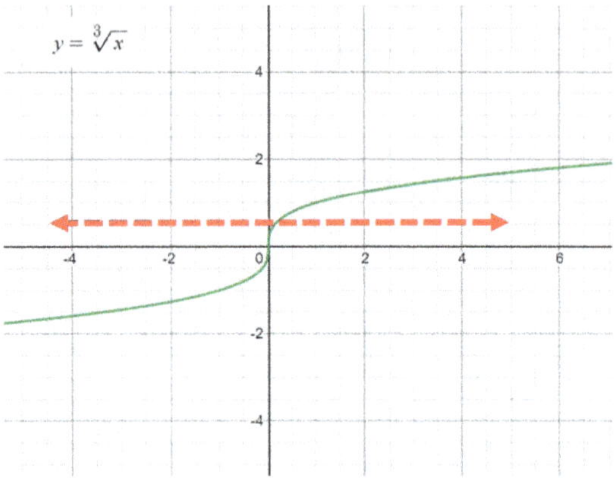

Image 10.18

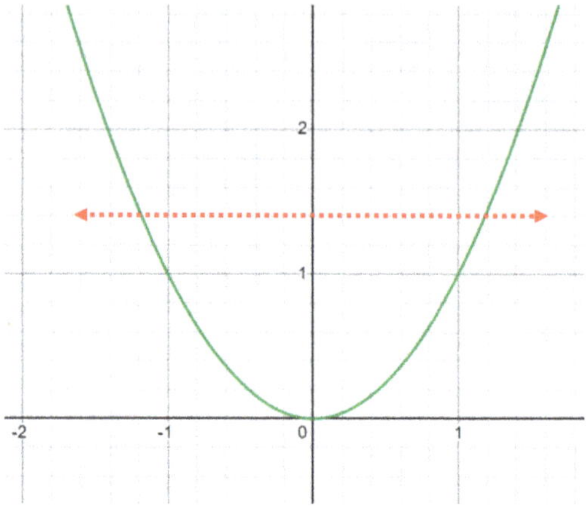

Image 10.19

Another characteristic of functions is that y is **not** raised to an even power. For example, $x + y^2 = 52$ is not a function because y is raised to an even power.

EXERCISE SET 10.5.1

Label each with all that apply: Relation (R), Function (F), One-to-One Function (O).

1. $y = x$
2. $y = 10$
3. $y = \frac{1}{x}$
4. $y = x^3$
5. $y = \sqrt{x}$
6. $y = |x|$
7. $y = \text{int}(x)$
8. $2x + 3y = 12$
9. $x^2 - y^2 = 1$
10. $\{ (-8, 5), (-4,3), (-4,4), (7,5) \}$
11. $\{ (-100, 90), (79, 90), (11, -32), (0,0) \}$
12. $\{ (6, 7), (7, 8), (8,9) \}$

13.

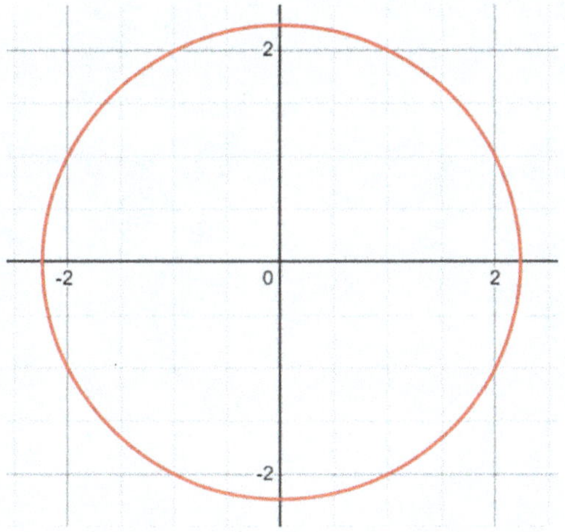

Image 10.20

14.

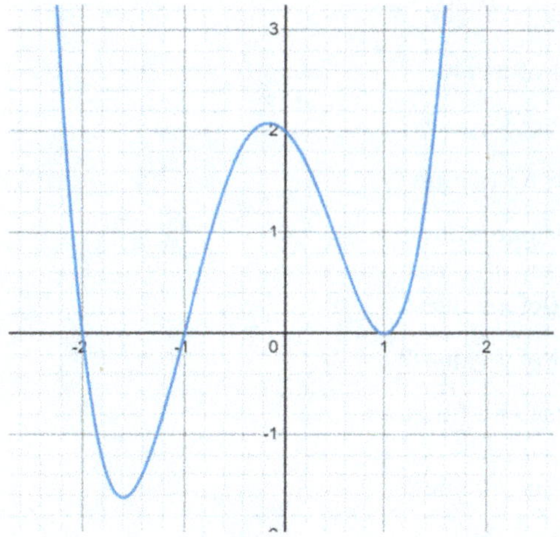

Image 10.21

15. Can an even function be one-to-one? Explain.

The following program will handle questions #10, 11, and 12 above; and provide practice with Python list functions.

Program Goal: Write a program that asks the user to input a set of ordered pairs. Determine if the relation is a function, and if so, if it is one-to-one.

PROGRAM 10.5.1 IDENTIFY_FUNCTIONS.PY

```
x_str = input('set of x-values? ')
x_values = [float(x) for x in x_str.split(',')]
y_str = input('set of y-values? ')
y_values = [float(y) for y in y_str.split(',')]
if len(x_values) != len(y_values):
    print("Your x- and y-values do not match up")
else:
    function = True
    for x in x_values:
        if x_values.count(x) > 1:
            function =False
    if function==False:
        print("It is not a function")
    else:
        print("it is a function")
        [Insert code to check if it is a one-to-one function]
```

SAMPLE OUTPUT

Set of x-values? 1,2,3,4,5

Set of y-values? 2,3,4,5,6

It is a function

It is one-to-one

Set of x-values? 1,2,3,2,1

Set of y-values? 1,4,9,4,1

It is not a function

Set of x-values? 1,2,3,4,5

Set of y-values? 3,4,3,4,5

It is a function

It is not one-to-one

EXERCISE SET 10.5.2

1. Replace the yellow highlighted code to check if the function is one-to-one.
2. Modify the program so that it outputs the set of ordered pairs.

Sample Output

Set of x-values? 1,2,3,4,5

Set of y-values? 1,4,9,16,25

{(1,1), (2,4), (3,9), (4,16), (5,25)}

It is a function

It is one-to-one

10.6 Inverse Functions

The reason we're interested in one-to-one functions is that they play a part in the definition of **inverse functions**. A pair of one-to-one functions are inverses to each other if their sets of xs and ys are swapped.

Example: If $f = \{(1,2), (3,4), (5,6), (7,8), (9,10)\}$ then its inverse (denoted f^{-1}) is $\{ (2,1), (4,3), (6,5), (8,7), (10,9) \}$

Another way to think of inverse functions is that they "reverse" or "undo" one another. For example, if $f(x) = x + 1$, then $f^{-1}(x) = x - 1$.

x	$f(x) = x + 1$	x	$f^{-1}(x) = x - 1$
−2	−1	−1	−2
−1	0	0	−1
0	1	1	0
1	2	2	1
2	3	3	2

Or, if $f(x) = x^3$, then $f^{-1}(x) = \sqrt[3]{x}$.

x	$f(x) = x^3$	x	$f^{-1}(x) = \sqrt[3]{x}$
−2	−8	−8	−2
−1	−1	−1	−1
0	0	0	0
1	1	1	1
2	8	8	2

Example 1

If $f(x) = 3x - 11$, what is $f^{-1}(x)$?

Solution

Think of $f(x)$ in words: "Multiply by 3 then subtract 11."

The inverse would do the opposite operations in reverse order: "Add 11 then divide by 3."

$$\therefore f^{-1}(x) = \frac{x + 11}{3}$$

Example 2

If $f(x) = \frac{5 + x}{4}$ find $f^{-1}(7)$.

Solution

Think of $f(x)$ as: "Add 5, then divide by 4."

The inverse would be: "Multiply by 4 then subtract 5," which is $f^{-1}(x) = 4x - 5$.

$$f^{-1}(7) = 4(7) - 5 = 23$$

We will find out later in this section how to derive the inverse algebraically.

EXERCISE SET 10.6.1

1. If $f(x) = 2x$, what is $f^{-1}(x)$?
2. If $f(x) = 3x + 1$, what is $f^{-1}(x)$?
3. If $f(x) = \frac{1}{x}$, what is $f^{-1}(x)$?
4. If $f^{-1}(x) = \frac{x-2}{3}$, what is $f(x)$?
5. If $f(3) = 11$ for some function f, what is $f^{-1}(11)$?
6. If $f^{-1}(24) = 17$ for some function f, what is $f(17)$?
7. If $f(x) = 5x - 2$, what is $f^{-1}(8)$?
8. If $f(x) = 2x^2$, what is $f^{-1}(64)$?

Refer to the graph of $f(x)$ shown to answer #9–12.

9. What is $f(-2)$?
10. What is $f^{-1}(2)$?
11. What is $f^{-1}(6)$?
12. What is the set of ordered pairs for f^{-1}?
13. Write a program that asks the user to input a function *f* and a value *a*. Calculate and output $f(a)$, then evaluate f^{-1} at that value.

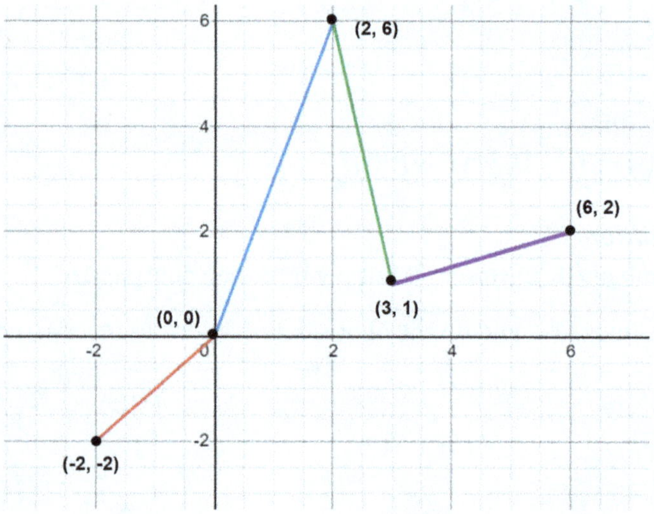

Image 10.22

Sample Output

function? x**2 + 1

a? 3

f(3) = 10

$f^{-1}(10) = 3$

Consider the inverse of $f(x) = x^2$. The logical choice would be $f^{-1}(x) = \sqrt{x}$. But look at the table of values.

x	$f(x) = x^2$	x	$y = f^{-1}(x)$
−2	4	4	−2 or 2
−1	1	1	−1 or 1
0	0	0	0
1	1	1	1 or −1
2	4	4	2 or −2

It appears that the inverse is not a function because every nonzero x is mapped to *two* different y-values. The reason is that $f(x) = x^2$ is not one-to-one. Recall it fails the horizontal line test:

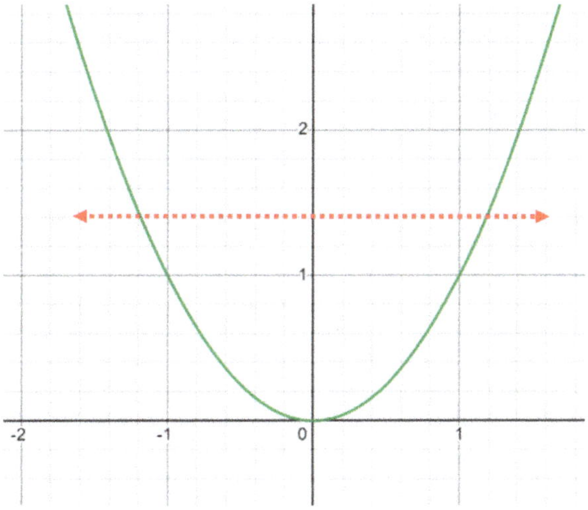

Image 10.19

In order to guarantee that a function has an inverse that is also a function, they must both be one-to-one. The workaround is to restrict the domain of $y = x^2$ to an interval on which it *is* one-to-one. In this case we restrict the domain of $y = x^2$ to $[0, \infty)$.

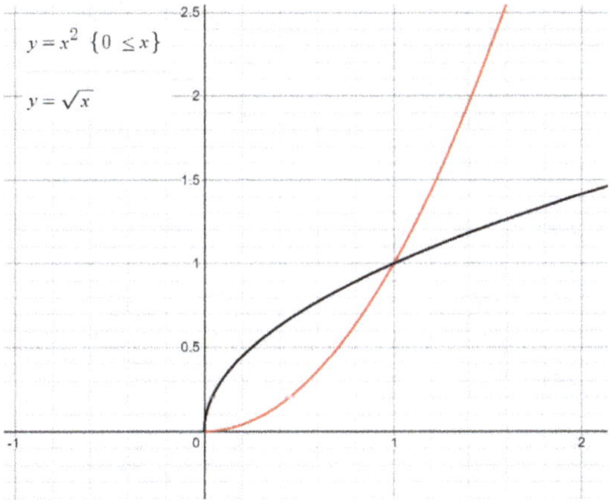

Image 10.23

EXERCISE SET 10.6.2

Graph each function and its inverse on the same set of axes.

1. $f(x) = 2x$
2. $f(x) = 3x + 1$
3. $f(x) = \dfrac{1}{x}$
4. $f(x) = \dfrac{x^3 + 1}{2}$

(Hint: Think of inverse as being the reverse of "cube, add 1, divide by 2.")

Notice that each pair of functions f and f^{-1} is symmetric to the line $y = x$.

Examples:

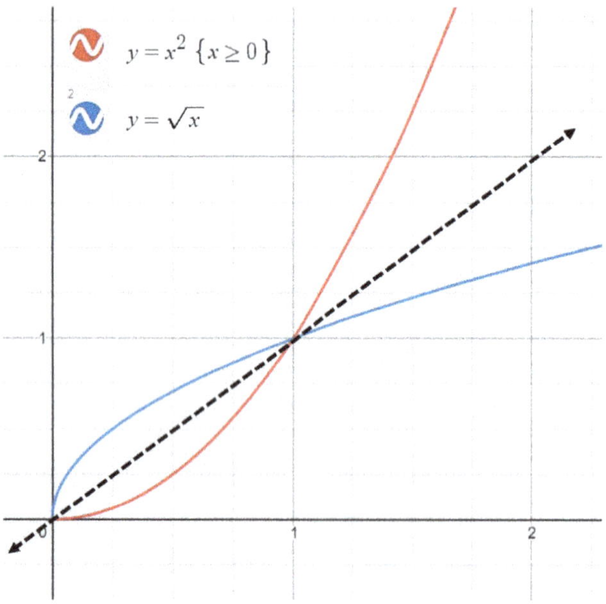

Image 10.24

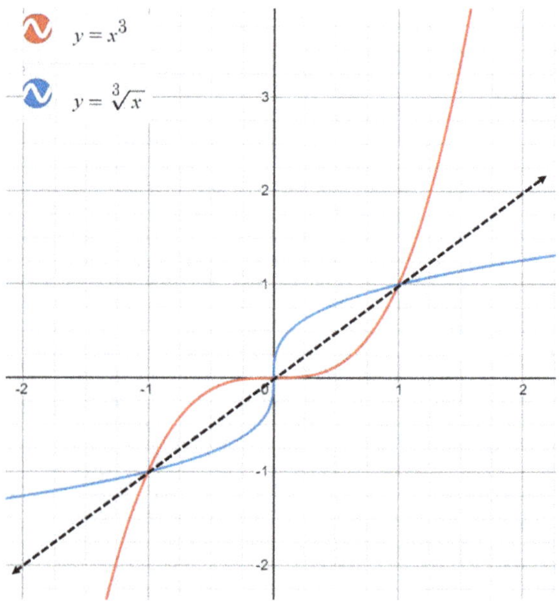

Image 10.25

Finding an Inverse Function Algebraically

Some inverse functions are easy to identify, such as $f(x) = x^3$ and $f^{-1}(x) = \sqrt[3]{x}$. Other functions such as $f(x) = \dfrac{3x + 1}{2 - x}$, it is not so easy to guess.

Fortunately, there's a fairly straightforward method to find the inverse.

Example 1

Find the inverse function of $f(x) = 3x + 10$.

Solution

Step 1—Replace $f(x)$ with y	$y = 3x + 10$
Step 2—"Swap" x and y	$x = 3y + 10$
Step 3—Solve for y	$x - 10 = 3y$ $\frac{x-10}{3} = y$
Step 4—Replace y with $f^{-1}(x)$	$f^{-1}(x) = \frac{x-10}{3}$

Example 2

Find the inverse of $f(x) = \frac{x+2}{2x-5}$

Solution

Step 1—Replace $f(x)$ with y	$y = \frac{x+2}{2x-5}$
Step 2—"Swap" x and y	$x = \frac{y+2}{2y-5}$
Step 3—Solve for y	$x(2y - 5) = y + 2$ $2xy - 5x = y + 2$ $2xy - y = 5x + 2$ $(2x - 1)y = 5x + 2$ $y = \frac{5x+2}{2x-1}$
Step 4—Replace y with $f^{-1}(x)$	$f^{-1}(x) = \frac{5x+2}{2x-1}$

EXERCISE SET 10.6.3

Find the inverse of each function.

1. $f(x) = x^3 - 1$
2. $f(x) = \frac{4}{2-x}$
3. $f(x) = -\frac{3x+1}{x}$
4. $f(x) = \frac{2x-3}{x+4}$
5. $f(x) = \sqrt{x-4}$

Program Goal: Write a program that asks user to input a function f and its inverse f^{-1}. Draw the functions; also draw the x- and y-axes and the line $y = x$.

Note the use of **SymPy** attribute **is_real,** which identifies if the value of a **SymPy** variable is real (as opposed to complex) value, hence graphable.

Therefore num.**is_real** will return a Boolean value of True or False if num is a **SymPy** variable. **is_real**(num) will not work unless you define your own function called **is_real**.

PROGRAM 10.6.2 DRAW_INVERSE.PY

```
## draw_inverse.py
## input: function f and its inverse f-1
## output: graphs of both functions and line y=x
from sympy import *
from matplotlib import pyplot as plt
from numpy import arange
x=Symbol('x')
# input the function/inverse and save in list functions:
# functions[0] = function
# functions[1] = inverse
functions=[ ]
```

```
fun=input("Function? ")
functions.append(fun)
inv=input("Inverse? ")
functions.append(inv)
# initialize subplot and set the axes
ax = plt.subplot()
plt.axis([−20,20, −20,20])
# Graph each function in functions[]
for f in functions:
    f_exp = sympify(f)                # convert to SymPy expression
    y_list = [ ]
    x_list = [ ]
    # initialize x_list as the interval [−20,20] with increment 0.2
    x_list = list(arange(−20,20,0.2))
    x_list2 = [k for k in x_list]     # x_list2 is a temporary clone of x_list
    for x_value in x_list2:
        y_value = f_exp.subs({x:x_value})
        if not y_value.is_real:
            # remove x_value from x_list if y_value is not real
            [ fill in code to remove x_value from x_list ]
        else:
            # add y_value to y_list if y_value is real
            [ fill in code to append y_value to y_list ]
    plot_label='$y='+str(f)+'$'
    plt.plot(x_list,y_list,label=plot_label)
ax.legend()
[Fill in code to draw line of symmetry y = x ]
```

SAMPLE OUTPUT

Function? 3*x+10

Inverse? (x–10)/3

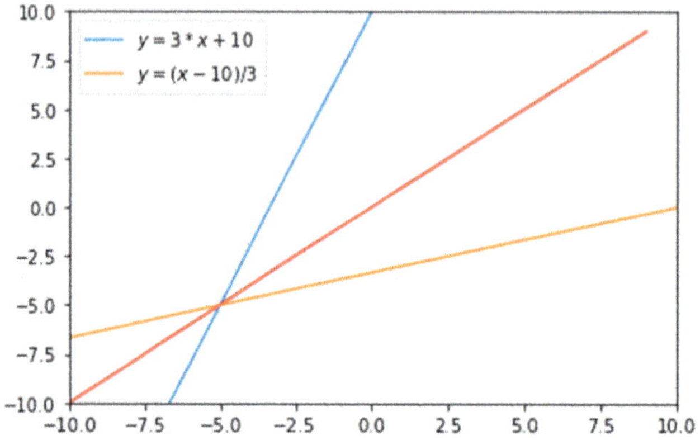

Image 10.26

Function? (x+2)/(2*x–5)

Inverse? (5*x+2)/(2*x–1)

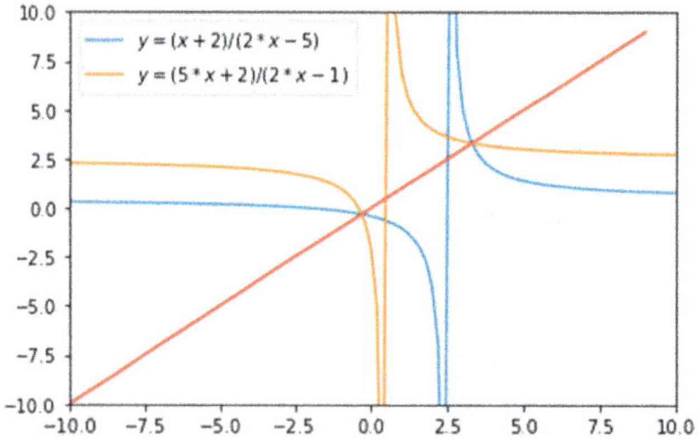

Image 10.27

EXERCISE SET 10.6.4

1. Fill in the aqua and pink highlighted code.
2. Fill in the yellow highlighted code to draw the line of symmetry.
3. It's not necessary for the user to input f^{-1}. Modify the program so that it draws the graph of the inverse by simply reversing every (*x*, *y*) in *f*.

Sample Output

Function? x**3+sqrt(x)

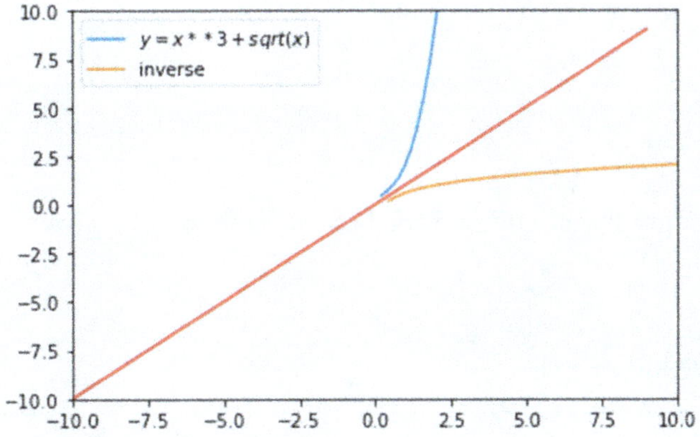

Image 10.28

Is it possible to have Python calculate the formula for the inverse of a given function?

Yes, **SymPy** can solve an expression for a given variable. For example, here's how we would find the inverse of $f(x) = \sqrt{2x-5}+1$ by hand (left) and using **SymPy** (right).

Step 1—Replace $f(x)$ with y: $y = \sqrt{2x-5}+1$	Identify both x and y as **SymPy** variables. User inputs sqrt(2*x – 5) + 1
Step 2—Swap x and y: $x = \sqrt{2y-5}+1$	Replace all occurrences of x with y: sqrt(2*y – 5) + 1
Step 3—Solve for x: $x-1 = \sqrt{2y-5}$ $(x-1)^2 = 2y-5$ $(x-1)^2+5 = 2y$ $\frac{(x-1)^2+5}{2} = y$	Solving x = sqrt(2*y – 5) + 1 for y is the same as solving: sqrt(2*y – 5) + 1 – x = 0 Solve the equation in terms of y: solve(2*y – 5) + 1 – x, y)

Step 4—Replace y with $f^{-1}(x)$:

$$f^{-1}(x) = \frac{(x-1)^2+5}{2} \quad or \quad \frac{(x-1)^2}{2}+\frac{5}{2}$$

Inverse function is:

[(x – 1)**2/2 + 5/2]

Recalling order of operations (exponent before divide), this is equivalent to:

$$f^{-1}(x) = \frac{(x-1)^2}{2}+\frac{5}{2}$$

Good job, **SymPy**!

Program Goal: Write a program that will find the inverse function of a given function.

PROGRAM 10.6.3 FIND_INVERSE.PY

```
## find_inverse.py
## Input: a function that can be handled by SymPy
## Output: inverse function
from sympy import *
from math import sqrt
x= Symbol('x')
y = Symbol('y')
fcn_x = sympify(input("Function? "))
fcn_y = fcn_x.subs({x:y})     # swap x and y, forming g(y)
expr = expand(fcn_y−x)       # form g(y) − x
[insert solve and print statements]
```

SAMPLE OUTPUT

Function? (x + 2) / (2*x – 5)

inverse fcn is g(x) =

[(5*x + 2)/(2*x – 1)]

Function? (3*sqrt(x – 6)) / 4

inverse fcn is g(x) =

[16*x**2/9 + 6]

EXERCISE SET 10.6.5

1. Fill in the yellow highlighted code to solve and print.
2. Use the program to find the inverse functions of the following:

   A. $f(x) = \sqrt{x+4}$

   B. $f(x) = \sqrt{x} + 4$

   C. $f(x) = x^2 - 3$

   D. $f(x) = e^{x-8}$

   (hint: use ***exp****(x−8))*

3. What variable type is returned by **solve**?
4. Modify the program so that it also draws the graph of the function and its inverse including the line of symmetry. If there are two inverse functions, graph both of them.

   Hint: refer to program ***draw_inverse****.*

Sample Output

Function? 2*sqrt(x + 7) − 1

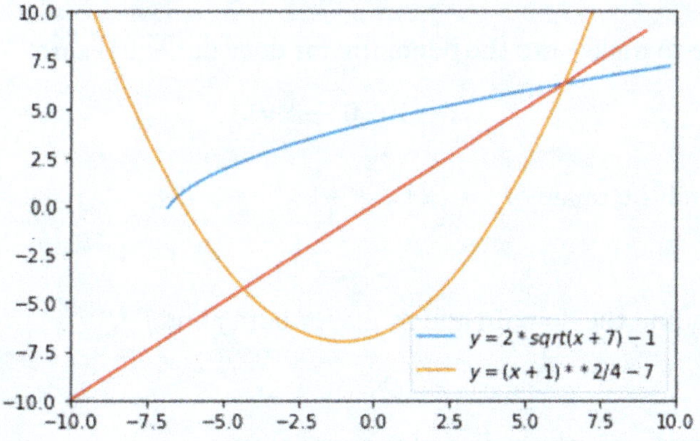

Image 10.29

Function? x**2 − 3

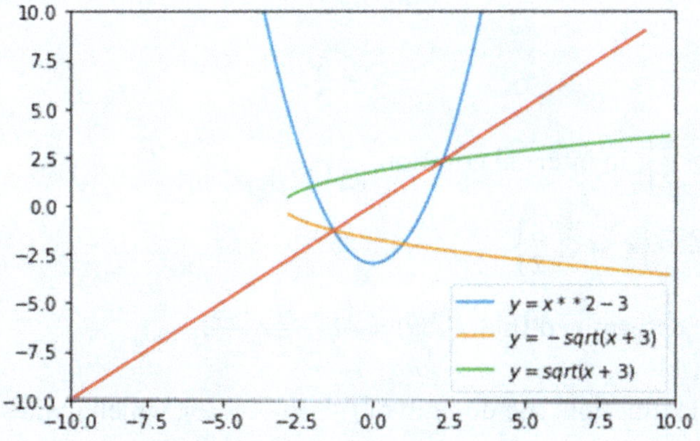

Image 10.30

Note this program does not always return a succinct elegant solution but it's pretty effective for square roots, quadratics, and rational functions.

Identifying the Domain and Range of an Inverse Function

Because inverse functions have their *x*s and *y*s swapped, their domains and ranges are swapped.

Example 1

Given $f(x) = \frac{x+2}{3x-5}$, find the domain and range for both f and f^{-1}.

Solution

Because $f(x) = \frac{x+2}{3x-5}$, we have to make sure the denominator does not equal zero:

$$3x - 5 \neq 0 \rightarrow x \neq \frac{5}{3}$$

The domain of f is $\left\{x \mid x \neq \frac{5}{3}\right\}$; in interval notation: $\left(-\infty, \frac{5}{3}\right) \cup \left(\frac{5}{3}, \infty\right)$.

Thus, the range of f^{-1} is also $\left\{x \mid x \neq \frac{5}{3}\right\}$.

The range of f might be difficult to ascertain without graphing, so it's easier to find the domain of f^{-1}.

If $f^{-1}(x) = \frac{5x+2}{3x-1}$, we have to make sure the denominator does not equal zero:

$$3x - 1 \neq 0 \rightarrow x \neq \frac{1}{3}$$

Thus, the domain of f^{-1} and the range of $f = \left\{x \mid x \neq \frac{1}{3}\right\}$.

Example 2

Given $f(x) = \sqrt{2x+5} - 1$, find the domain and range for both f and f^{-1}.

Solution

To find the domain of f, make sure the radicand is nonnegative:

$$2x - 5 \geq 0$$

$$x \geq \frac{5}{2}$$

The domain of f is $\left\{x \mid x \geq \frac{5}{2}\right\}$; in interval notation: $\left[\frac{5}{2}, \infty\right)$.

Thus, the range of f^{-1} is also $\left\{x \mid x \geq \frac{5}{2}\right\}$.

In a previous example we determined that $f^{-1}(x) = \frac{(x-1)^2 + 5}{2}$.

Because this is a quadratic function, the domain of f^{-1} is $(-\infty,\infty)$, which makes it also the range of f.

EXERCISE SET 10.6.6

For each of the functions in Exercise Set 10.6.3 (copied below), find the domain and range of both the function and its inverse.

1. $f(x) = x^3 - 1$
2. $f(x) = \dfrac{4}{2 - x}$
3. $f(x) = -\dfrac{3x + 1}{x}$
4. $f(x) = \dfrac{2x - 3}{x + 4}$
5. $f(x) = \sqrt{x - 4}$

Composition of a Function with Its Inverse

In Exercise Set 10.4.4 #2 you modified program **compose_functions** to allow the user to input a list of x-values. Run the program now, but input for f and g a pair of functions that are **inverses**.

SAMPLE OUTPUT

function f? 2*x–1

function g? (x+1)/2

x list? –3, –1, 1, 3

$f(g(x))$ = [–3.0, –1.0, 1.0, 3.0]

$g(f(x))$ = [–3.0, –1.0, 1.0, 3.0]

Is it a coincidence the $f(g(x)) = g(f(x)) = x$ every time?
No, it is not.

For any pair of inverse functions f and f^{-1}:

$f(f^{-1}(x)) = x$ and $f^{-1}(f(x)) = x$

Example

Prove that $f(x) = \dfrac{2x + 3}{x + 4}$ and $f^{-1}(x) = \dfrac{4x - 3}{2 - x}$ are inverses by showing $f(f^{-1}(x)) = x$ and $f^{-1}(f(x)) = x$.

Solution

$$f(f^{-1}(x)) = f\left(\frac{4x-3}{2-x}\right) = \frac{2\left(\frac{4x-3}{2-x}\right)+3}{\left(\frac{4x-3}{2-x}\right)+4} = \frac{\frac{8x-6}{2-x}+3\left(\frac{2-x}{2-x}\right)}{\frac{4x-3}{2-x}+4\left(\frac{2-x}{2-x}\right)} = \frac{\frac{8x-6+6-3x}{2-x}}{\frac{4x-3+8-4x}{2-x}} = \frac{\frac{5x}{2-x}}{\frac{5}{2-x}} = \frac{5x}{5} = x$$

$$f^{-1}(f(x)) = f^{-1}\left(\frac{2x+3}{x+4}\right) = \frac{4\left(\frac{2x+3}{x+4}\right)-3}{2-\left(\frac{2x+3}{x+4}\right)} = \frac{\frac{8x+12}{x+4}-3\left(\frac{x+4}{x+4}\right)}{2\left(\frac{x+4}{x+4}\right)-\left(\frac{2x+3}{x+4}\right)} = \frac{\frac{8x+12}{x+4}+\frac{-3x-12}{x+4}}{\frac{2x+8}{x+4}+\frac{-2x-3}{x+4}}$$

$$= \frac{\frac{8x+12-3x-12}{x+4}}{\frac{2x+8-2x-3}{x+4}} = \frac{\frac{5x}{x+4}}{\frac{5}{x+4}} = \frac{5x}{5} = x$$

EXERCISE SET 10.6.7

For each pair of inverse functions below, prove that f and f^{-1} are inverses by showing $f(f^{-1}(x)) = x$ and $f^{-1}(f(x)) = x$.

(Remember that program ***compose_functions*** *will not work when trying to raise a negative number to a fractional exponent.)*

1. $f(x) = 3x + 4, f^{-1}(x) = \frac{1}{3}(x - 4)$
2. $f(x) = x^3 - 8, f^{-1}(x) = \sqrt[3]{x + 8}$
3. $f(x) = (x - 2)^2, f^{-1}(x) = \sqrt{x} + 2$
4. $f(x) = \frac{x - 5}{2x + 3}, f^{-1}(x) = \frac{3x + 5}{1 - 2x}$
5. Can you think of a function that is its own inverse?

In the next chapter, we will define the inverse of an exponential function.

Chapter Projects

I. Generating Ordered Pairs of the Inverse Function

Write a program that asks the user to input a set of ordered pairs (representing function f). Output the inverse function f^{-1} as a set of ordered pairs.

SAMPLE OUTPUT

Ordered pairs? (3,1), (2,4), (3,6), (7,1)

Inverse:

(1,3),(4,2),(6,3),(1,7)

II. Mortgage Payments and Amortization

You're ready to buy your first house. Given the price and interest rate, what will your payment be?

The following formula will calculate the fixed monthly payment (***P***) required to amortize (pay off) a loan of ***A*** dollars over a term of N months at an annual interest rate of r. ***i*** is the interest rate per period (typically month), not an imaginary number. (How we wish our payments were imaginary.) If the quoted annual rate is 6%, for example, i is .06/12 or .005.

$$P = \frac{iA}{1-(1+i)^{-N}}$$

Example 1

You are buying a $250,000 house, with 10% down, on a 30-year mortgage at a fixed rate of 7.8%. What is the monthly payment?

Solution

30 years $\rightarrow N = 360$ months, and

Annual interest rate 7.8% $\rightarrow$ the monthly interest rate is $I = \frac{.078}{12}$, or 0.0065.

Loan amount is 90% of \$250,000 $\rightarrow A =$ \$225,000.

$$\therefore\ P = \frac{0.0065(225000)}{1-(1+.0065)^{-360}} = \$1{,}619.71 \text{ monthly payment}$$

A more useful number while house hunting is your target price range based on what monthly payment you can afford, with A, P, I, and N as defined above:

$$A = \frac{P}{i}\left(1-(1+i)^{-N}\right)$$

Example 2

You can afford to pay $2,250 per month. If you want a 30-year loan at 4.5% fixed rate, what price house could you afford?

Solution

$$P = 2{,}250,\ I = \frac{.045}{12} = 0.00375,\ N = 360 \text{ months.}$$

$$A = \frac{2250}{.00375}\left(1-(1+.00375)^{-360}\right) = \$440{,}062.61.$$

Write a program that will ask the user whether they want to calculate a payment or a home price. Based on their answer, prompt for the appropriate variables (*A* or *P*, annual interest rate *R*, and term of loan in months.) Calculate and output the answer.

SAMPLE OUTPUT

Calculate payment (*P*) or cost (*C*)? C

Payment you can afford? 2,250

I? (annual rate as a percent) 4.5

T? (number of years) 30

You can afford up to $440,062.61

Calculate payment (*P*) or cost (*C*)? P

Cost of house? 250,000

Amount of down payment (as a percent)? 10

I? (annual rate as a percent) 7.8

T? (number of years) 30

With $25,000.00 down, your monthly payment will be $1,619.71

III. Logistic Growth—"Leveling the Curve"

The first edition of this textbook was written in 2020, the year in which the world suffered through the COVID-19 virus. Scientists and politicians discussed a goal of "leveling the curve." Whether they knew it or not, they were referring to a mathematical model in which the virus initially escalates exponentially but then eventually levels off.

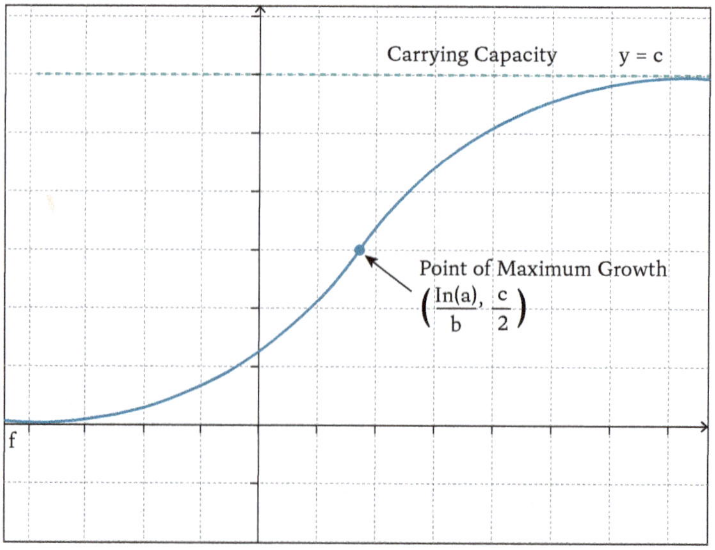

Image 10.31

The logistic growth model is:

$$f(x) = \frac{c}{1 + ae^{-bx}}$$

where

- $\frac{c}{1+a}$ is the initial value
- c is the *carrying capacity*, or *limiting value*
- b is a constant determined by the rate of growth.

Write a program in which the user inputs the values of initial population P_0, b, and c. The value of a is calculated using $\frac{c}{1+a} = P_0$. Draw the model, including the horizontal asymptote $y = c$. The graph title should be the calculated model. Also indicate the point of maximum growth on the graph.

If possible, predict when the population reaches a point within 1% of the carrying capacity.

SAMPLE OUTPUT

Initial population? 1

b? 0.6030

c? 1000

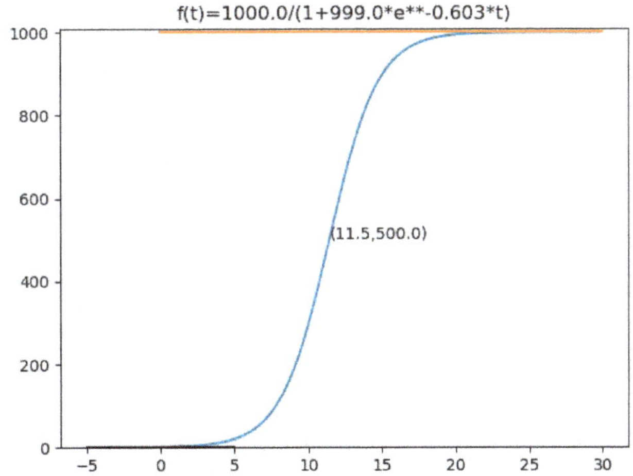

Image 10.32

CHAPTER SUMMARY

PYTHON CONCEPTS

Assume x has been declared as a **SymPy** symbol using x = **Symbol**('x')

Simplify an exponential expression	**expand_power_base**(expression)—will attempt to apply a constant exponent to each variable/constant in the base Note: $(ab)^5$ will be simplified to a^5b^5, but $(ab)^n$ will **not** be simplified to a^nb^n *Ex:* *>>>expand_power_base((5*a)**3)* *125*a**3*
Simplify an exponential expression	**expand_power_exp**(expression)—will attempt to apply each term in an exponent to the single variable base Note: a^{m+n} will be simplified to a^ma^n but $(ab)^{m+n}$ will **not** be simplified to $a^{m+n}b^{m+n}$ *Ex:* *>>>expand_power_exp(3**(4+m+n))* *81*3**m*3**n*
Count the occurrences of a list element in a list	list.**count**(x) will count the occurrences (frequency) of x in list *Ex:* *>>>list1 = [1,1,2,2,3,3,3,3]* *>>>list1.count(3)* *4*
To convert a SymPy expression to a function	my_fcn = **lambdify**(x,string_expression) *Ex:* *>>>my_fcn = lambdify(x,'x**2+1')* *>>>my_fcn(2)* *5*

Exception handling	**try:** result = [assignment statement] **except:** [Do something else] *Ex:* ***try:*** *result = **Fraction**(f(g(x1)))* ***except:*** *result="Undefined"* ***print**(result)*
To determine if a numeric value is a real number	Use SymPy method **is_real:** **value.is_real** *Ex:* *>>>num = **simplify**(12)* *>>>num.**is_real*** *True* *>>>num = **simplify(sqrt(−1))*** *>>>num.**is_real*** *False*

PRECALCULUS CONCEPTS

Exponential function	A function in the form $f(x) = a \cdot b^x$, in which a and b are constant real numbers and the exponent contains a variable. *Ex:* $f(x) = 2^x$, $y = 3(4)^x$, $f(x) = 12^x$
Law of exponents	If $a^m = a^n$ then $m = n$. *Ex:* If $3^{x+1} = 3^5$ then $x + 1 = 5 \rightarrow x = 4$.

(Continued)

Exponential growth and decay	If a quantity follows a geometric progression, then its quantity (or population) at time t is: $$P(t) = P_0 e^{kt}$$ where P_0 is the initial quantity (population) and k is a constant called the growth factor (or growth rate). If $k > 0$, the quantity experiences exponential growth. If $k < 0$, the quantity follows exponential decay. *Ex:* $P(t) = 1000e^{0.5t}$ *Has an initial quantity of* $P_0 = 1000$ *and a growth factor of* $k = 0.5$.
Composition of functions	"*f composed with g of x*" $(f \circ g)(x) = f(g(x))$ *Ex:* *If* $f(x) = x^2$ *and* $g(x) = \frac{1}{x}$ *then* $(f \circ g)(x) = \left(\frac{1}{x}\right)^2$ *and* $(g \circ f)(x)) = \frac{1}{x^2}$
Domain of composition of functions	The domain of $(f \circ g)(x)$ is the set of all xs in the domain of g for which $g(x)$ is in the domain of f. *Ex:* *If* $f(x) = \sqrt{x}$ *and* $g(x) = \frac{1}{x+1}$ *The domain of g is* $\{x \mid x \neq -1\}$ *The domain of f is* $\{x \mid x \geq 0\}$ *Thus* $g(x) \geq 0 \rightarrow \frac{1}{x+1} \geq 0 \rightarrow x+1 \geq 0 \rightarrow x \geq -1$. *Combine* $x \neq -1$ *with* $x \geq -1$ *to get:* *Domain of* $f \circ g$ *is* $\{x \mid x > -1\}$.

One-to-one function	A function in which each y is paired with exactly one x. *Ex:* 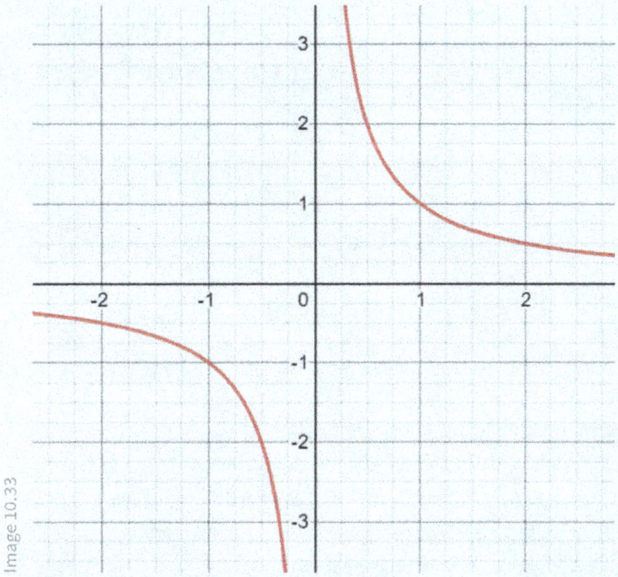
Horizontal line test	If any horizontal line intersects the graph of a function at more than one point, it is NOT one-to-one. *Ex:* *$y = x^2$ is NOT a one-to-one function.* 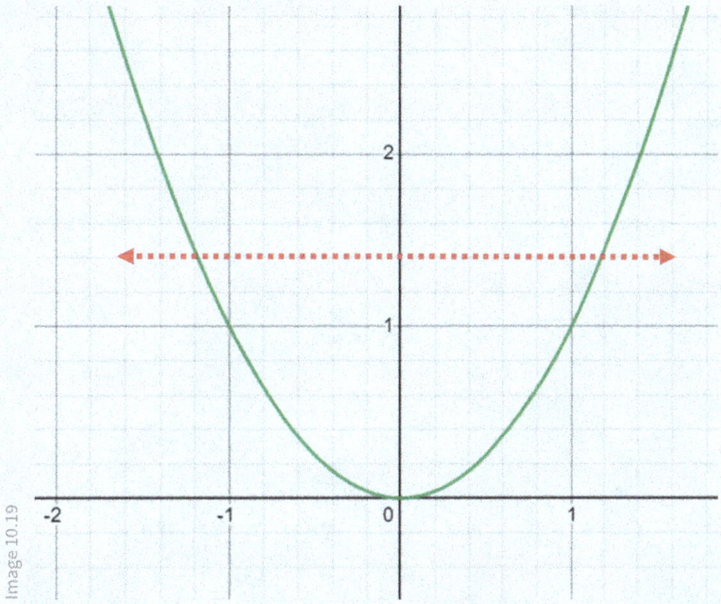

(Continued)

Inverse functions	f and g are inverse functions if their xs and ys are swapped. More formally: $(f \circ g)(x) = (g \circ f)(x) = x$ if and only if f and g are inverse functions. Only one-to-one functions can have inverses. The graphs of inverse functions are symmetric to the line $y = x$. *Ex:* $f(x) = x^3$ *and* $g(x) = \sqrt[3]{x}$ *are inverse functions because* $(f \circ g)(x) = \left(\sqrt[3]{x}\right)^3 = x$ *and* $(g \circ f)(x) = \sqrt[3]{x^3}. = x.$

Credits

CHAPTER 11

Logarithms

11.1 Define and Simplify

In Chapter 10, we learned how to find the inverse of a function using these steps:

1. Replace $f(x)$ with y	Example: $f(x) = \frac{2x}{3x+5}$ $y = \frac{2x}{3x+5}$
2. Swap the x and y	$x = \frac{2y}{3y+5}$
3. Solve for y	$x(3y+5) = 2y$ $3xy + 5x = 2y$ $3xy - 2y = -5x$ $Y(3x - 2) = -5x$ $y = \frac{-5x}{3x-2}$
4. Replace y with $f^{-1}(x)$	$f^{-1}(x) = \frac{-5x}{3x-2}$

However, watch what happens when we try to find the inverse of an exponential function:

1. Replace $f(x)$ with y	Example: $f(x) = 2(3)^{x+2}$ $y = 2(3)^{x+2}$
2. Swap the x and y	$x = 2(3)^{y+2}$
3. Solve for y	$\frac{x}{2} = (3)^{y+2}$ Then what? How do we get y out of the exponent?
4. Replace y with f^{-1}(x)	???

We will soon find out how to get the y out of the exponent. It involves something called a logarithm. First, we are going to define $x = \log_b y$ as the inverse of $y = b^x$. This is called the **logarithm base *b* of *y*.**

INFORMAL DEFINITION OF LOGARITHM

$\log_b y$ is the exponent that b is raised to, to equal y.

$b > 0, b \neq 1, y > 0$

i.e., you can only take the log of a positive number.

Examples:

$\text{Log}_2 8 = 3$ because $2^3 = 8$

$\text{Log}_4 16 = 2$ because $4^2 = 16$

$\text{Log}_2\left(\frac{1}{8}\right) = -3$ because $2^{-3} = \frac{1}{8}$

$\text{Log}_4 2 = \frac{1}{2}$ because $4^{1/2} = 2$

Here's how to convert from a logarithmic statement to an exponential statement and vice versa:

Image 11.1

EXERCISE SET 11.1.1

1. 1.If $\log_b 27 = 3$, what is b?
2. If $\log_{64} 4 = y$, what is y?
3. If $\log_{10} x = 5$, what is x?
4. If $\log_{100} 10 = y$, what is y?
5. Convert $5^3 = 125$ to a logarithmic statement.
6. Convert $\sqrt{81} = 9$ to a logarithmic statement.
7. Convert $11^{-2} = \frac{1}{121}$ to a logarithmic statement.
8. Convert $\sqrt[3]{216} = 6$ to a logarithmic statement.
9. Convert $2 = \log_7 x$ to exponential form.
10. Convert $3 = \log_{10}(x + 8)$ to exponential form.
11. Convert $\log_5 125 = x$ to exponential form.

12. What is $\log_0 5$?
13. What is $\log_{15} 1$?
14. What is $\log_1 15$?
15. Given the domain of $f(x) = \log_b x$ is $(0, \infty)$, what is the domain of $f(x) = \log_3(x + 2)$?

We'll review string manipulation with a simple logarithm conversion program.

Program Goal: Write a program in which the user inputs an exponential statement (such as 5**3). Convert and output the equivalent logarithmic statement (such as $\log_5 125 = 3$).

PROGRAM 11.1.1 CONVERT_TO_LOG.PY

```
## convert_to_log.py
## Input: exponential statement in the form a**b
## Output: equivalent logarithmic statement log base a of y = b
ex_str = input("Exponential a**b? ")
base_exp = ex_str.split('**')        # split the input string at the ** operator
                                     # returns list: ['a','b']
if len(base_exp)!=2:                 # if list does not have exactly two elements
    print("Your exponential statement must include **")

else:                                # list does have exactly two elements
    a = base_exp[0]
    b = base_exp[1]
    y = eval(ex_str)
    print("log base", str(a),"of", y,"=",str(b))
```

Sample Output

Exponential *a****b*? 4**2
log base 4 of 16 = 2
Exponential *a****b*? 64**(1/2)
log base 64 of 8.0 = ½

EXERCISE SET 11.1.2

1. What is the output when the program is run with 27**(2/3) as the input?
2. What is the output for (−3)**4? Is this accurate? If not, modify the program.
3. Does the modified program correctly handle a fractional base such as **(1/2)**(−4)**? If not, fix.
4. Technically, $1^n = 1$ for any value n. However, $\log_1 a$ is considered undefined. Modify the program to give an error when the user inputs an expression with base 1.
5. If y (the result of exponentiation) results in a fraction, output as a fraction rather than a decimal.

Sample Output

Exponential *a****b*? (1/2)**2

log base 1/2 of 1/4 = 2

Formatting the Subscript

Because we will be outputting several log statements over the course of this chapter, our output will look much more professional if we are able to print $\log_b y$ instead of **log base *b* of *y***.

One way to do this is using Unicode representations of subscripts:

\u2080	Subscript 0
\u2081	Subscript 1
\u2082	Subscript 2
etc.	

So to print $\log_2 32$:

print("log\u208232")

Let's modify **convert_to_log.py** to make use of this subscript. (This is the version of the program before any changes in Exercise Set 11.1.2.) Changes are shown in blue.

PROGRAM 11.1.1 CONVERT_TO_LOG.PY V2

```
## convert_to_log.py
## Input: exponential statement in the form a**b
## Output: equivalent logarithmic statement log base a of y = b
subscripts=['\u2080','\u2081', '\u2082', '\u2083', '\u2084', '\u2085', '\u2086',
'\u2087', '\u2088', '\u2089']
ex_str = input("Exponential a**b? ")
base_exp = ex_str.split('**')
if len(base_exp)!=2:
    print("Your exponential statement must include **")
else:
    a = int(base_exp[0])
    b = base_exp[1]
    y = eval(ex_str)    # evaluate input as a mathematical expression
    base = subscripts[a]
    print("log"+base, y, "=", str(b))
```

Sample Output

Exponential *a**b*? 5**4
$\log_5 625 = 4$

Note this program only works if the subscript is a single digit.

EXERCISE SET 11.1.3

Use the modified program to convert the following exponential statements to logarithmic statements if possible. If there is an error message, explain.

1. 4**(–1)
2. 9**(3/2)
3. 5**(–2)
4. 8**(2/3)
5. 7**3
6. Does the program work if the base is a fraction—for example, if the input is (4/5)**(–2)?
7. Modify the program so that float numbers are output as fractions instead of decimals.

Sample Output

Exponential *a***b*? 7**(–2)

$\log_7 1/49 = -2$

8. Write a program that converts a logarithmic expression in the form log(*x*,*b*) (representing $\log_b x$), calculates its value, and outputs a corresponding exponential equation in the form *b***y* = *x*.

Sample Output

logarithmic function log(*x*,*b*)? log(9,3)

3 ** 2.0 = 9

logarithmic function log(*x*,*b*)? log(3,9)

9 ** 0.5 = 3

If you wish your output to display a superscripted exponent such as $3^2 = 9$, use Unicode \u00bN in which you replace the N with the desired digit. This is similar to program **convert_to_log** v2, which uses Unicode to output subscripts.

Properties of Logarithms

Recall the properties of inverse functions:

If $f(x)$ and $f^{-1}(x)$ are inverse functions of each other, then $f(f^{-1}(x)) = x$ and $f^{-1}(f(x)) = x$. Substituting $f(x) = b^x$ and $f^{-1}(x) = \log_b x$, we get:

$f(f^{-1}(x)) = b^{\log_b x} = x$ and $f^{-1}(f(x)) = \log_b b^x = x$ for any $b > 0$.

These two properties and several others, easily proven, are summarized in:

PROPERTIES OF LOGARITHMS

If $b > 0$:

$b^{log_b x} = x$

$log_b b^x = x$

$\log_b 1 = 0$

$\log_b b = 1$

$\log_b 0$ is undefined

EXERCISE SET 11.1.4

1. Simplify $\log_{10} 1$
2. Simplify $\log_5 5^3$
3. Simplify $\log_{12} 12$
4. Simplify $x^{log_x y}$ if $x > 0$.
5. Simplify $\log_5 5 - \log_5 1$
6. Explain why $\log_b 1 = 0$ for any $b > 0$.
7. Explain why $\log_b b = 1$ for any $b > 0$.
8. Explain why $\log_b 0$ is undefined for any $b > 0$ and $b \neq 1$.

11.2 Inverses of Logarithms and Exponentials

Let's return to the example at the beginning of this chapter (which was to find the inverse function of $f(x) = 2(3)^{x+2}$) and fill in the missing steps, now that we know that $y = \log_b x$ is the inverse function of $y = b^x$.

1. Replace $f(x)$ with y	Example: $f(x) = 2(3)^{x+2}$ $y = 2(3)^{x+2}$
2. Swap the x and y	$x = 2(3)^{y+2}$
3. Solve for y	$\frac{x}{2} = (3)^{y+2}$ Convert to a logarithmic statement: $log_3\left(\frac{x}{2}\right) = y + 2$ Subtract 2: $y = log_3\left(\frac{x}{2}\right) - 2$
4. Replace y with $f^{-1}(x)$	$f^{-1}(x) = log_3\left(\frac{x}{2}\right) - 2$

Let's also identify the domain and range of $f^{-1}(x) = log_3\left(\frac{x}{2}\right) - 2$. Because you can only take the log of a positive number, the domain is determined by:

$$\frac{x}{2} > 0 \rightarrow x > 0, \text{ or } (0, \infty).$$

Recall that the range of f is the same as the domain of $f^{-1}(x)$ and vice versa. To find the range of $f^{-1}(x)$, simply find the domain of $f(x) = 2(3)^{x+2}$. This function is defined everywhere, so its domain is $(-\infty, \infty)$. Because the domain of $f(x)$ is $(-\infty, \infty)$, the range of $f^{-1}(x)$ is also $(-\infty, \infty)$.

$$\therefore \text{ Domain of } f^{-1}(x) = log_3\left(\frac{x}{2}\right) - 2 \text{ is } (0, \infty) \text{ and its range is } (-\infty, \infty).$$

Example

Find the inverse function of $f(x) = \log_6(4x) + 3$, also its domain and range.

Solution

1. Replace $f(x)$ with y	$y = \log_6(4x) + 3$
2. Swap the x and y	$x = \log_6(4y) + 3$
3. Solve for y	Subtract 3 to isolate the log: $x - 3 = \log_6(4y)$ Convert to an exponential statement: $4y = 6^{x-3}$ Divide by 4 to isolate the y: $y = \frac{6^{x-3}}{4}$ or $y = \frac{1}{4}6^{x-3}$
4. Replace y with $f^{-1}(x)$	$f^{-1}(x) = \frac{6^{x-3}}{4}$

The domain of $f^{-1}(x)$ is $(-\infty, \infty)$ because it is an exponential function, which is defined for all real numbers.

The range of $f^{-1}(x)$ is the same as the domain of $f(x) = \log_6(4x) + 3$. You can only take the logarithm of a positive number, so set $4x > 0$. Thus $x > 0$, or $(0, \infty)$.

EXERCISE SET 11.2.1

Find the inverse function of the following, also its domain and range.

1. $f(x) = 5^x - 3$
2. $f(x) = -2(5)^x$
3. $f(x) = \log_{\frac{1}{2}}(x + 5)$
4. $f(x) = \log_{10}(-x)$
5. $f(x) = 2 \cdot 3^{3x} - 1$
6. $f(x) = -\ln(1 - 2x) + 1$ (Hint: ln is the same as $\log_e$)
7. Recall from section 10.6 that only one-to-one functions have inverses. Are $y = b^x$ and $y = \log_b x$ one-to-one? What graphical test is used to identify one-to-one functions?
8. Write a program that asks the user to input the domain and range of a function *f*. Output the domain and range of f^{-1}. (Yes, this is a pretty easy program.)

Program Goal: Write a program that (A) asks the user to input a logarithmic function, (B) extracts the argument of the function (expression between the parentheses), and (C) uses the argument of the function to calculate and output the domain of the function.

PROGRAM 11.2.1 LOG_DOMAIN.PY

```
from sympy import *
f = input('logarithmic function? ')
start = f.index('(')
end = f.index(')')
argument = f[start+1:end]
undefs = solve(argument)
undef = undefs[0]
print('x>',undef)
```

Sample Output

logarithmic function? log(7*x+1)
$x > -1/7$

Note: We will find out in Section 11.4 that in Python, a log with no explicitly stated base (such as log(*x*)) is presumed to be $\log_e$ (aka the **natural logarithm** ln). This is different than the [log] key on your calculator and in math textbooks, in which log*x* is presumed to be the **common logarithm**, i.e., $\log_{10}x$. Both ln*x* and $\log_{10}x$ are discussed in section 11.4.

EXERCISE SET 11.2.2

1. What is the output if the user inputs log(7*x−49) ?
2. If the user inputs log(4*x+2), identify the values and types of the following variables: start, end, argument, undefs, undef.
3. What does f.**index**('(') do?
4. What does **solve**(argument) do?
5. What is the output if the user inputs log(49−7*x)? Is this the correct solution? Why or why not?
6. This program is accurate if the argument of the function is a linear expression in the form $ax + b$, in which $a > 0$. Is it accurate if the argument is a linear expression in the form $ax + b$ in which $a < 0$? If not, fix the program.
7. What is the output if the user inputs log(x**2 – 4)?
8. Is it accurate if the argument is a quadratic, such as log(x**2 – 4) ?

11.3 Graphical Features of Logarithms

Because $f^{-1}(x) = \log_b x$ is the inverse function of $f(x) = b^x$, their graphs are symmetric to the line $y = x$ as shown:

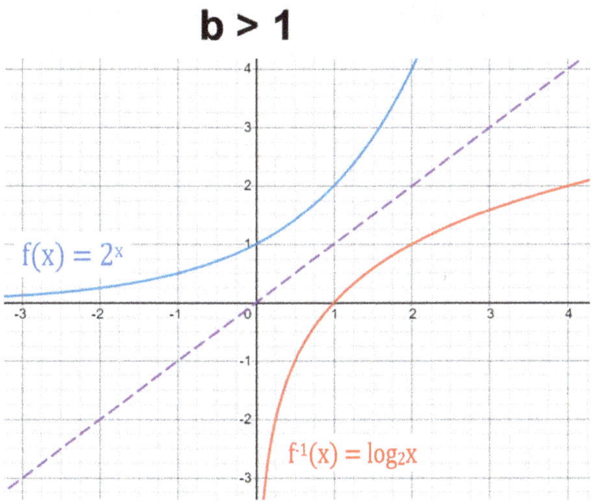

Image 11.2

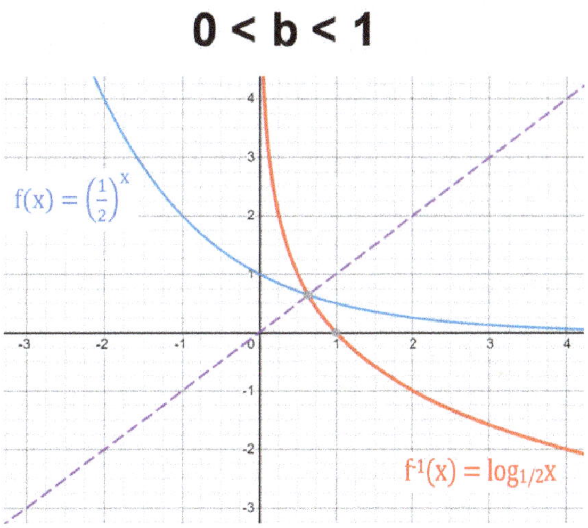

Image 11.3

EXERCISE SET 11.3.1

1. What are the domain and range of $y = \log_b x$?
2. What are the asymptote(s) of $y = \log_b x$?
3. What are the intercept(s) of $y = \log_b x$?

Graph the following transformations of $y = \log_b x$ by hand. Identify the asymptote(s) and intercept(s) on the graph.

4. $f(x) = \log_3 x + 4$
5. $f(x) = \log_3(x + 4)$
6. $f(x) = -\log_{1/2}(x - 5)$
7. $f(x) = \log_2(-x - 3)$
8. Modify program **log_domain** so that rather than identifying the domain of a log function, it identifies the vertical asymptote.

11.4 Special Logarithms

Common Logarithm

If you see the notation $\log(x)$ without a base (such as the [log] key on your calculator or in a math textbook), the base is assumed to be 10, i.e. $\log x = \log_{10} x$. This is called the **common logarithm**. Because $\log x$ is a special case of $\log_b x$, we can adapt its properties:

PROPERTIES OF THE COMMON LOGARITHM LOG(*X*)

$10^{\log x} = x$

$\log 10^x = x$

$\log 1 = 0$

$\log 10 = 1$

$\log 0$ is undefined

Natural Logarithm

Another special logarithm is the **natural logarithm ln*x***. $y = \ln x$ (which is essentially $y = \log_e x$) is the inverse function of $y = e^x$. We recognize it as special because every scientific calculator has a [ln] key. We will soon see that $\ln x$ plays a part in evaluating $\log_b x$ for any $b > 0$.

Both ***e*** and **ln** play important parts in many mathematical formulas including finance and exponential growth and decay.

EXERCISE SET 11.4.1

Assume in these exercises that $\log x$ refers to the mathematical notation for common logarithm and not the Python notation for natural logarithm.

1. Write the five properties of $\ln x$ as adapted from the five properties of $\log_b x$.
2. Simplify $\log 10 + \ln e$
3. Simplify $\log 1000$
4. Simplify $\ln e^3$
5. Simplify $\log \frac{1}{100}$
6. Sketch the graphs of $y = \log x$ and $\ln x$. Identify any asymptotes and intercepts.

The Binary Logarithm

Most precalculus textbooks do not mention that $\log_2 x$ also has a special name: the **binary logarithm**. There are several pretechnology methods for approximating $\log_2 x$.

By definition, $\log_2 10 = x$ means $2^x = 10$. Chapter Project I will ask you to write a program to approximate $\log_2 x$ using three different summation methods.

Python Log Functions

The math module has several useful log functions:

math.**log**(x,b)

returns the value of $\log_b(x)$ where $b > 0$, $b \neq 1$, and $x > 0$. If b is not specified, it is assumed to be e, which means math.**log**(x) is equivalent to the natural logarithm $\ln x$. The return value is type float.

Examples:

math.**log**(8,2) $= \log_2 8 = 3$
math.log(3) $= \ln 3 = 1.0986122886681098$
math.log(math.e) $= 1.0$
math.log(−1) = undefined (math domain error)

Important distinction:

Python Programming	**Math Textbooks**
math.**log**(x) means $\ln x$ (the **natural logarithm**)	$\log x$ means $\log_{10} x$ (the **common logarithm**)

math.**log2**(x)

returns the value of the **binary logarithm** $\log_2 x$ where $x > 0$. Performs the same function as math.**log**(x,2) but is considered more accurate. The return value is type float.

Examples:
math.**log2**(2) = 1.0
math.**log2**(1024) = 10.0

math.**log10**(x)

returns the value of the **common logarithm** $\log_{10} x$ where $x > 0$. Performs the same function as math.**log**(x,10) but is considered more accurate. The return value is type float.

Examples:
math.log10(10) = 1.0
math.log10(1000) = 3.0

EXERCISE SET 11.4.2

Write a program, or use the Python console, to evaluate the following.

1. math.log10(1000000)
2. math.log2(1000000)
3. math.log2(−4)
4. math.log(math.e**2)
5. math.log2(2**15)
6. math.log10(10**36)
7. math.log10(1/100)
8. math.log10(10***n*) for any value *n*
9. The *x*-intercept of the graph of a logarithmic function *f*(*x*) (or any function for that matter) is found by solving *f*(*x*) = 0. Modify program **log_domain** (or write a new one) that finds the *x*-intercept of a logarithmic function.

Sample Output

logarithmic function? log(49−7*x)

The *x*-intercept of log(49−7*x) is (48/7 ,0)

10. Write a program that asks the user to input a log expression in the form log(*x*, *b*). Output the expression using a subscript and the numeric value. If the base is *e*, write ln instead of $\log_e$. Assume the base is a single digit.

Sample Output

Logarithmic function? log(49, 7)

$\log_7 49 = 2$

logarithmic function? log(10,e)

ln 10 = 2.302585092994046

Techniques from programs **convert_to_log** and **log_domain** may be helpful.

11.5 Simplifying Expressions with Logarithms

If you try to use program **find_inverse** (from Section 10.6) to find the inverse of an exponential function, the result might be somewhat inscrutable:

SAMPLE OUTPUT

Function? 5**(x/2)

inverse fcn is g(x) =

[2*log(x)/log(5)]

This corresponds to $f^{-1}(x) = \frac{2logx}{log5}$, but if we were to find the inverse by hand, we would end up with $f^{-1}(x) = log_5 x^2$. Is program **find_inverse** not accurate then?

The program is indeed accurate; it has just provided a different rendering of the function. We will thus introduce the laws of logarithms to help you convert between different logarithmic forms.

LAWS OF LOGARITHMS

If $b > 0$ and $b \neq 1$:

Product Rule: $\log_b(M \cdot N) = \log_b M + \log_b N$

Quotient Rule: $\log_b\left(\frac{M}{N}\right) = \log_b M - \log_b N$

Power Rule: $\log_b M^N = N \cdot \log_b M$

The product rule is commonly mistaken as $\log_b(M \cdot N) = \log_b M \cdot \log_b N$, but which of the following two statements is the correct one?

$$\log_4 64 = \log_4(4 \cdot 16) = \log_4 4 \cdot \log_4 16 = 1 \cdot 2 = 2$$

or

$$\log_4 64 = \log_4(4 \cdot 16) = \log_4 4 + \log_4 16 = 1 + 2 = 3$$

Clearly the second statement is the correct one because $\log_4 64$ = "the power 4 must be raised to, to equal 64" = 3.

Here is a more formal proof of the **product rule**.

Let $x = \log_b M$ and $y = \log_b N$

Then $b^x = M$ and $b^y = N$

So $\log_b M \cdot N = \log_b b^x \cdot b^y = \log_b b^{x+y}$ *(adding exponents)*

$= x + y$ *(using the identity $\log_b b^n = n$)*

$= \log_b M + \log_b N$

$\therefore\ \log_b M \cdot N = \log_b M + \log_b N$

Example 1

Rewrite the expressions as a sum, difference and/or power of logs.

A. $\log_b(x \cdot y^2)$

$= \log_b x + \log_b y^2$	*Product Rule*
$= \log_b x + 2\log_b y$	*Power Rule*

B. $\log_b\left(\frac{x^2\sqrt{y}}{z^5}\right)$

$= \log_b x^2\sqrt{y} - \log_b z^5$	*Quotient Rule*
$= \log_b x^2 + \log_b \sqrt{y} - \log_b z^5$	*Product Rule*
$= \log_b x^2 + \log_b y^{1/2} - \log_b z^5$	*Rewrite $\sqrt{y}$ using fractional exponent*
$= 2\log_b x + \frac{1}{2} \cdot \log_b y - 5\log_b z$	*Power Rule*

Example 2

Rewrite each expression as a single log.

A. $2\log_b x + \frac{1}{2}\log_b(x + 4)$

$= \log_b x^2 + \log_b(x + 4)^{1/2}$	*Power Rule*
$= \log_b[x^2 \cdot (x + 4)^{1/2}]$	*Product Rule*
$= \log_b\left[x^2 \cdot \sqrt{x + 4}\right]$	*Rewrite $(x + 4)^{1/2}$ as a square root*

B. $4\log_b(x+2) - 3\log_b(x-5)$

$= \log_b(x+2)^4 - \log_b(x-5)^3$ *Power Rule*

$= \log_b\left(\frac{(x+2)^4}{(x-5)^3}\right)$ *Quotient Rule*

Example 3

If $r = \log_b 2$, $t = \log_b 3$, $u = \log_b 5$, rewrite each expression in terms of r, t, and/or u.

A. $\log_b 225$

$= \log_b 25 \cdot 9$

$= \log_b 5^2 \cdot 3^2$

$= \log_b 5^2 + \log_b 3^2$

$= 2\log_b 5 + 2\log_b 3$

$= 2u + 2t$

B. $\log_b \sqrt{72}$

$= \log_b \sqrt{2^3 \cdot 3^2}$

$= \frac{1}{2}\log_b(2^3 \cdot 3^2)$

$= \frac{1}{2}\log_b 2^3 + \frac{1}{2}\log_b 3^2$

$= \frac{3}{2}\log_b 2 + \log_b 3$

$= \frac{3}{2}r + t$

EXERCISE SET 11.5.1

For #1–7: Rewrite the expressions as a sum, difference and/or power of logs.

1. $\log 10x$
2. $\log_4 4x^2$
3. $\ln \frac{\sqrt{3x}}{7}$
4. $\ln \frac{xy}{z}$
5. $\log_3 \sqrt[3]{x-2}$

(Continued)

6. $\log_b \dfrac{x^5 z^2}{y^3}$

7. $\ln\left(\dfrac{-b+\sqrt{b^2-4ac}}{2a}\right)$

For #8–13: Rewrite each expression as a single log.

8. $\log 7 - \log x$

9. $\dfrac{1}{2}\log_5 7 - 2\log_5 x$

10. $3\ln x + 2\ln y - 4\ln z$

11. $1 + 3\log_4 x$

12. $\dfrac{3}{2}\ln x^6 - \dfrac{3}{4}\ln x^8$

13. $\dfrac{1}{2}\ln(x^2 + y^2)$

For #14–17: If $r = \log_b 2$, $t = \log_b 3$, $u = \log_b 5$, rewrite each expression in terms of r, t and/or u.

14. $\log_b 45$

15. $\log_b \dfrac{125}{81}$

16. $\log_b \sqrt[3]{225}$

17. $\log_b (12)^{2/3}$

18. Find the inverse function of $f(x) = ½\log 6 + ½\log x$.

19. Prove the quotient rule.

20. Prove the power rule.

If you look at a scientific calculator, it most likely has only has two logarithm keys:

Image 11.4

Using only your calculator (or worse yet, a century-old table of $\log_{10}$ and ln values), the most efficient way to find $\log_b$ for any other base b is the change of base formula:

CHANGE OF BASE FORMULA

$$\log_b M = \frac{\log M}{\log b} \text{ or } \frac{\ln M}{\ln b}$$

if $b > 0$ and $b \neq 1$

Example

Evaluate $\log_9 27$.

Solution

$$\log_9 27 = \frac{\log 27}{\log 9} = \frac{1.431363764}{0.9542425094} = 1.5$$

EXERCISE SET 11.5.2

Rewrite the following logarithms using the change of base formula.

1. $\log_b 81$
2. $\log_2 x$

Review: True or false?

3. $\log_b(x-2) = \log_b x - \log_b x$
4. $\log_b xy = \log_b x \cdot \log_b y$
5. $\log_b \sqrt{x} = \frac{1}{2}\log_b x$
6. $3\ln e = 3$
7. $\frac{\log 10x}{\log 5x} = \log 2$
8. $\log_b(x-2) = \log_b x - 2$
9. $\log_b x + \log_b y = \log_b x \cdot y$
10. $\ln\sqrt{x} = \sqrt{\ln x}$
11. $\frac{\ln 100}{2} = \ln 50$
12. Write a program that asks the user to input a log expression in the form log(*x*,*b*) (representing $\log_b x$) where *x* and *b* are constants. Calculate the value of log(*x*,*b*) using the change of base formula log(*x*) / log(*b*). Assume the base is a single integer digit.

Sample Output

Logarithmic function? log(125,5)

$\log_5 125 = \log(125)/\log(5) = 3.0$

Logarithmic Simplification Functions in the SymPy Library

SymPy has several log simplification functions.

Note that **SymPy** has its own **log** function (to distinguish from math.**log**). One important difference between the two is that sympy.**log**(*x*,*b*) can manipulate **symbols** *x* and *b*. math.**log**(*x*,*b*) treats *x* and *b* as variables that must have predefined values.

Example

What is the output of the following program?

```
1   from sympy import symbols,log,pprint
2   import math
3
4   a,b = 25,5
5   print(math.log(a,b))
6
7   x, y = symbols('x y')
8
9   exprl = 'x**2'
10  expr2 = 'y**2'
11
12  pprint(log(exprl,expr2))
```

Solution

The **print()** function in line 5 will output the value of $\log_5 25$, which is 2.0. (The type is float.)

The **pprint()** function in line 12 will output the equivalent of **SymPy** expression $log_{y^2}\left(x^2\right)$, which is (using the change of base formula) $\frac{log\left(x^2\right)}{log\left(y^2\right)}$.

The **SymPy** library, as opposed to the math library, allows the use of **ln**(x) as an alias for **log**(x).
The log functions (**simplify, expand_log, logcombine**) in the **SymPy** library are best exemplified in the following set of exercises.

EXERCISE SET 11.5.3

For #1–10, use the Python console to print the result of the following expressions. Express answers using both Python notation and mathematical notation (e.g., both log(*x*****a*y/z*) and $ln\left(\frac{x^a \cdot y}{z}\right)$).

*Hint: Pretty Print (**pprint**) might help.*
Begin with:
\>>> **from** sympy **import** *
\>>> *a, x, y, z* = **symbols**('*a x y z*', **positive** = True)

1. math.**log**(8,3)
2. **log**(8,3)
3. **simplify**(log(*x*y*))
4. **expand_log**(log(*x*y*)

5. **simplify**(log(4,*a*)+log(5,*a*))
6. **simplify**(log(*x*,*a*)–log(*y*,*a*)+log(*z*,*a*))
7. **simplify**(*a**log(*x*)+log(*y*)–log(*z*))
8. **simplify**(log(*x**y**2))
9. **expand_log**(log(*x**y**2))
10. **expand_log**(log((*x***2**y*)/*z***3,*a*))
11. **logcombine**(3***log**(*x*))
12. **logcombine**(log(*x*) + 3***log**(*y*))

For #13–17, give the Python statement(s) that will provide the given result, if possible.

13. **log**(*x*,*a*) + **log**(*y*,*a*) simplified to **log**(*x***y*,*a*)
14. **log**(*x*,*a*) – **log**(*y*,*a*)) simplified to **log**(*x*/*y*,*a*)
15. **log**(*x***a*) simplified to *a****log**(*x*)
16. **log(sqrt**(*a***b*)) simplified to **log**(*a***b*)/2.
17. 3***log**(*a*)+4***log**(*b*) simplified to **log**(*a***3**b***4).
18. Type the following lines into your Python console:

    >>>**import** sympy

    >>>**import** math

    >>>**print**(sympy.**ln**(e))

    >>>**print**(math.**ln**(e))

    >>>**print**(math.**log**(e))

Identify the exact output of each of the **print**() functions.

Python will assume, in the absence of **SymPy**, that **log()** refers to the **log()** function in the **math** library.

>>> **log**(8,3)

1.892789260714372

However, once you import the **SymPy** library, **log()** will now be interpreted as the **SymPy log()** function, giving an entirely different result:

>>>**from** sympy **import** *

>>> **log**(8,3)

log(8) / log(3)

Caution: **SymPy** will not always combine expressions into the most succinct and/or readable form.

Conclusions:

1. There are some things humans are better at than computers.
2. There is always room for improvement in existing programming languages.

EXERCISE SET 11.5.4

Write **SymPy** function calls using **simplify**(), **logcombine**(), or **expand_log**(), to simplify each of the following expressions as indicated. Assume your program (or console) begins with the following:

>>>**from** sympy **import** *

>>>*a*, *x*, *y*, *z* = **symbols**('*a x y z*', **positive** = True)

Provide both the function call and result.

1. $log\left(\frac{xy}{z}\right)$ simplified to $\log x + \log y - \log z$
2. $log_y x$ simplified to $\frac{log\ x}{log\ y}$
3. $log\sqrt{xy}$ simplified to $\frac{log\ x}{2} + \frac{log\ y}{2}$
4. $log_z x + 2 \cdot log_z y$ simplified to $\frac{log\left(xy^2\right)}{log\ z}$
5. $ln\, x^2 + ln\sqrt{x}$ simplified to $\frac{5 \cdot logx}{2}$

11.6 Solving Equations with Logarithms

If $\log_2 x = \log_2 5$, then it's fairly obvious that $x = 5$. This is guaranteed by the:

ONE-TO-ONE PROPERTY OF LOGARITHMS

$\log_b M = \log_b N$ if and only if $M = N$.

Note, we are not dividing by $\log_b$ *because* $\log_b M$ *is not a product. It is a function.*

When solving equations involving logarithms, it is important to check your answers to make sure you are only taking the log of a positive number. (Domain of $\log_b x$ is $(0, \infty)$.)

Example 1

Solve for x:

$$\log_2(2x+7) = \log_2 15$$

Solution

$$2x + 7 = 15$$

$$x = 4$$

Check $x = 4$: Is $2(4) + 7 > 0$? Yes.

Example 2

Solve for x:

$$\log_4((x-5)(x-11)) = 2$$

Solution

Convert to an exponential statement:

$$(x-5)(x-11) = 4^2$$

$$x^2 - 16x + 55 = 16$$

$$x^2 - 16x + 39 = 0$$

$$(x-3)(x-13) = 0$$

$$x = 3 \text{ or } x = 13$$

Check $x = 3$: Is $(3-5)(3-11) > 0$? Yes.
Check $x = 13$: Is $(13-5)(13-11) > 0$? No. Reject $x = 13$.
$\therefore$ $x = 3$ is the only solution.

Example 3

Solve for x. Express your answer in terms of e:

$$\ln(4x-1) = 3$$

Solution

Rewrite the logarithmic statement as an exponential statement:

$$4x - 1 = e^3$$

Solve for x:

$$4x = e^3 + 1$$

$$x = \frac{1}{4}\left(e^3 + 1\right)$$

Example 4

Solve for x.

$$3^x - 2 = 12$$

Solution

$$3^x = 14$$

Rewrite as a logarithmic statement:

$$x = \log_3 14$$

Use your calculator to simplify:

$$x \approx 2.4022$$

EXERCISE SET 11.6.1

For #1–7, solve for x. Check your answers.

1. $\log_5(4x + 11) = 2$
2. $\log_2(x + 5) - \log_2(2x - 1) = 5$
3. $\log_8 x + \log_8 (x + 6) = \log_8(5x + 12)$
4. $\log_6 x + \log_6(x - 9) = 2$
5. $\log(x - 2) - \log(2x - 3) = \log 2$
6. $3^{1-x} = 2$
7. $\dfrac{10}{1 + e^{-x}} = 2$

For #8–10, solve and express your answer in terms of e.

8. $\ln(5x) = 2$
9. $\ln(6x - 5) = 3$
10. $\frac{1}{4}\ln(e) = x^2$

For #11–20, write **SymPy** statements that will solve #1–10, if possible. (Remember to use math.**e** in place of e.)

Example: To solve $\log_5(4x + 11) = 2$, use **solve**(**log**(4*x+11,5)−2)

21. Write a program that asks the user to input an exponential or logarithmic equation. Solve the equation.

Sample Output

Enter exponential or logarithmic equation? 3**(1 – x) = 2

[–log(2) / log(3) + 1]

Enter exponential or logarithmic equation? log(x – 2) – log(2*x – 3) = log(2)

[4/3]

If you attempted all the exercises in Exercise Set 11.6.1 above, you can see that **SymPy**'s **solve()** function does not always provide an answer, which is why we need humans to be able to solve these equations to keep check on computers.

In an attempt to simplify #7 above to a form that **SymPy** will solve:

Equation	Solve	Result
$\frac{10}{1+e^{-x}}=2$	solve(10/(1+e**(–x))–2)	[ ]
$10=2(1+e^{-x})$	solve(10–2*(1+e**(–x)))	[ ]
$5=1+e^{-x}$	solve(5–(1+e**(–x)))	[ ]
$4=e^{-x}$	solve(4–e**(–x))	[ ]
$4=1/e^{x}$	solve(4–1/(e**x))	[ ]
$e^{x}=1/4$	solve(e**x–1/4)	[–1.38629436111989]

Finally, we get an answer, but we had to perform virtually all the steps ourselves. This particular flaw is in attempting to raise *e* to a negative exponent, and you will have an opportunity to correct this flaw in Chapter Project IX.

Solving Quadratic Equations in Exponential Form

This is done using an elegant technique called a ***u*-substitution**.

Example 1

Solve for *x*:

$$5^{2x}-5^{x}-12=0$$

Solution

Rewrite 5^{2x} as $(5^x)^2$:

$$(5^{x})^2-5^{x}-12=0$$

Let $u = 5^x$ so it looks more like a quadratic that we're familiar with:

$$u^2 - u - 12 = 0$$

Solve for u:

$$(u + 3)(u - 4) = 0$$

$$u = -3 \text{ or } u = 4$$

But this is not the final answer! You were asked to solve for x, not u. Substitute 5^x back into u:

$$5^x = -3 \text{ or } 5^x = 4$$

Convert each one to a logarithm to get the variable out of the exponent:

$$x = \log_5(-3) \text{ or } x = \log_5(4)$$

Remember you can only take the log of a positive number, so $\log_5(-3)$ is undefined. That leaves $\log_5 4$. Simplify using your calculator:

$$x \approx .8614$$

EXERCISE SET 11.6.2

Solve for x. Express answers in terms of both log and a decimal rounded to 4 places.

1. $e^{2x} - 2e^x = 15$
2. $3^{2x} - 6 \cdot 3^x = -5$
3. $4^x - 2^x - 12 = 0$

11.7 More on Exponential Growth and Decay

In Section 10.3, we looked at the models for exponential growth and decay. Recap:

EXPONENTIAL GROWTH AND DECAY

If a quantity follows a geometric progression, then its quantity (or population) at time t is:

$$P(t) = P_0 e^{kt},$$

where P_0 is the initial quantity (population) and k is a constant called the growth factor (or growth rate). If $k > 0$, the quantity experiences exponential growth. If $k < 0$, the quantity follows exponential decay.

We were able to solve problems that asked us to find P_0 or $P(t)$ but we were not ready to find variables k or t. Why? Because those are the variables in the exponent, and the only way to solve for variables in the exponent is the use of logarithms. In this case, the natural logarithm **ln**.

Example 1

The *Streptococcus* bacteria population P at time t (in months) is given by $P(t) = P_0e^{2t}$. If the initial population was 100, how long does it take for the population to reach one million?

Solution

You are being asked to find variable t, the time. You are given that $P_0 = 100$ and $P(t) = 1{,}000{,}000$. Substitute into $P(t) = P_0e^{2t}$:

$$1{,}000{,}000 = 100e^{2t}$$

$$10{,}000 = e^{2t}$$

Convert from exponential to logarithmic form:

$$2t = \ln 10000$$

$$t = \frac{1}{2}\ln 10000 \text{ (NOT } \ln 5000!)$$

$$t \approx 4.6 \text{ months}$$

Example 2

If \$15,000 is invested in an account that yields 5% interest per year, after how many years will the account be worth \$75,000?

Solution

You are being asked to find t, the time. You are given that $k = .05$, $P_0 = 15{,}000$, and $P(t) = 75{,}000$. Substitute into $P(t) = P_0e^{kt}$:

$$75{,}000 = 15{,}000e^{.05t}$$

$5 = e^{.05t}$ Even though it's money, wait til the last step to do rounding.

$$.05t = \ln 5$$

$t = 20\ln 5$ (Do you see why we multiplied by 20?)

$$t \approx 32.2 \text{ years}$$

Example 3

Iodine-131 is a radioactive material that decays according to the function $P(t) = P_0e^{-0.087t}$ with t measured in days. If a scientist has a sample of 100 g ...

A. How much iodine-131 will be left after 9 days?

B. When will 70 g be left?

C. What is the half-life?

Solutions

A. How much iodine-131 will be left after 9 days?

You are being asked to find $P(9)$. You are given $t = 9$ and $P_0 = 100$.

$$P(9) = 100e^{-0.087(9)}$$

$$\approx 45.7 \text{ g.}$$

B. When will 70 g be left?

Find t when $P(t) = 70$.

$$70 = 100e^{-0.087t}$$

$$0.7 = e^{-0.087t}$$

$$-0.087t = \ln 0.7$$

$$t = -\frac{1}{0.087} \ln 0.7$$

$$\approx 4.1 \text{ days}$$

C. What is the half-life?

Half-life is the amount of time it takes for a substance to decay to half its original amount. Half-life is independent of the initial population. To find half-life, replace $P(t)$ with $\frac{1}{2}\, P_0$:

$$\frac{1}{2}\, P_0 = P_0e^{-0.087t}$$

Cancel out P_0 on each side:

$$\frac{1}{2} = e^{-0.087t}$$

$$-0.087t = \ln .05$$

$$t \approx 8 \text{ days}$$

Alternate Formulas

Given the half-life h of a decaying substance, the constant k in $P(t) = P_0e^{kt}$ is given by:

$$k = \frac{ln(1/2)}{h} \text{ or } \frac{-ln2}{h}$$

The formula for the population at time t is given by:

$$P(t) = P_0(\tfrac{1}{2})^{t/h}$$

EXERCISE SET 11.7.1

1. A bacteria culture starts with 10,000 bacteria and the number doubles every 40 minutes.

   A. Find a formula for the number of bacteria at time *t*.
   B. Find the number of bacteria after one hour.
   C. After how many minutes will there be 50,000 bacteria?

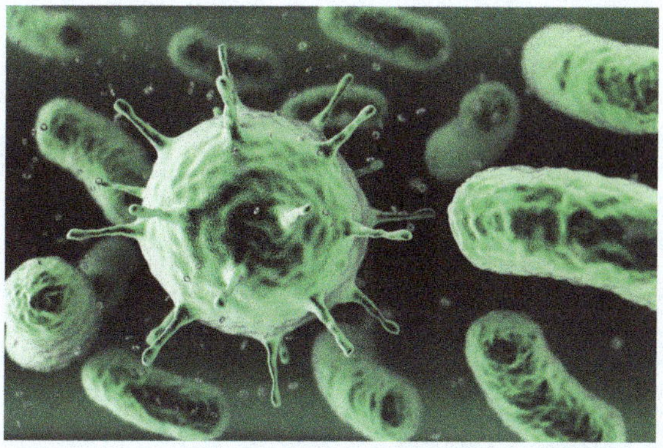

Image 11.5

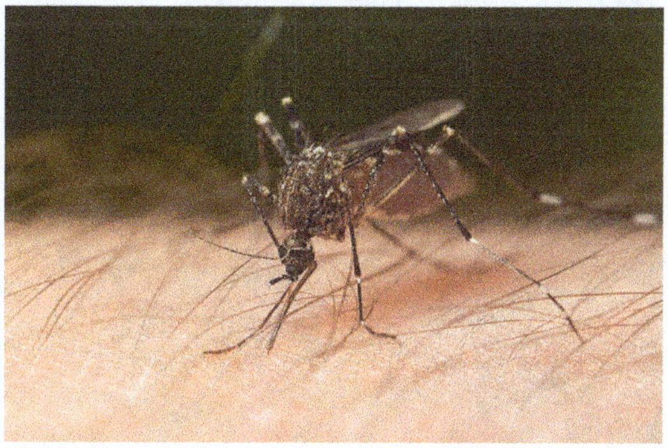

Image 11.6

2. An insect colony starts with 1,000 mosquitos, and there are 1,800 after 1 day.

   A. What is the size of the colony after 3 days?
   B. How long until there are 10,000 insects? Round to the nearest tenth of a day.

(Continued)

3. A fossil contains 70% of its initial amount of carbon-14. The half-life of carbon-14 is 5,730 years.

   A. What is the decay constant k?
   B. How old is the fossil?

Image 11.7

Image 11.8

4. An initial amount of 10g of radium dissipates to 9.797g after 500 years. What is the half life?

5. After the release of radioactive material into the atmosphere from the Chernobyl nuclear power plant in 1986, the hay in Austria was contaminated by iodine-131, which has a half-life of 8 days. It is safe to feed the hay to livestock when 10% of the iodine-131 remains.

   How long did the farmers need to wait to use this hay? Round to the nearest tenth of a day.

Image 11.9

Image 11.10

6. A. The half-life of radioactive potassium is 1.3 billion years. If 10 g is present now, how much will be present in 100 years?
   B. In 1,000 years?

7. How long does it take for a $2,700 investment to double if it is invested at 7% compounded monthly? Use the formula $A(t) = P\left(1 + \frac{r}{n}\right)^{nt}$, where r is the rate (as a decimal, not percent), n is the number of compoundings per year, and P is the initial investment.

   Round to the nearest tenth.

Image 11.11

8. Write a program that will ask the user to input (1) the half-life of a decaying substance, (2) the initial amount, and (3) the amount at time t. Calculate and output the amount of time t corresponding to $P(t)$.

Sample Output

Half-life? 1,690

Initial amount? 50

Ending amount? 38

Time to reach 38 is 669

Congratulations! You can now consider yourself a Python programmer with masterful precalculus skills. Or a mathematician with excellent Python programming skills. It is our hope that you are motivated to pursue both advanced programming and higher mathematics.

Chapter Projects

I. Approximating $\log_2 x$

There are various methods for approximating the binary logarithm $\log_2 x$.

Method 1

If x is in [1,2) and we can express x as $\frac{1+t}{1-t}$, then:

$$log_2 x = log_2\left(\frac{1+t}{1-t}\right) = \frac{2}{ln2}\left(t + \frac{t^3}{3} + \frac{t^5}{5} + \ldots\right)$$

$$If\ x = \frac{1+t}{1-t}\ then\ t = \frac{x-1}{x+1}.$$

Method 2

If x is in (0,2]:

$$log_2(x) = \frac{1}{ln2}\left[(x-1) - \frac{(x-1)^2}{2} + \ldots\right]$$

Write a program that calculates the value of $\log_2 x$ using all applicable methods. Note we are limited to values of x in (0,2]. Because these are infinite series, choose how you will determine when to end the summation. After a given number of terms? After the successive sums, rounded to 4 decimal points, begin to repeat?

Feel free to utilize other series expansions. Keep in mind that $log_2 x = \frac{ln\,x}{ln\,2} = \frac{1}{ln\,2} \cdot ln\,x$, so any series expansion of $\ln x$ can be used if you multiply the result by $\frac{1}{ln2}$.

II. Hyperbolic Sine and Cosine

Two functions defined in terms of the exponential function are **hyperbolic sine** *sinhx* and **hyperbolic cosine** *coshx*:

$$sinhx = \frac{e^x - e^{-x}}{2} = x + \frac{x^3}{3!} + \frac{x^5}{5!} + \frac{x^7}{7!} + \ldots$$

$$coshx = \frac{e^x + e^{-x}}{2} = 1 + \frac{x^2}{2!} + \frac{x^4}{4!} + \frac{x^6}{6!} + \ldots$$

These may seem to bear no resemblance to the trigonometric sine and cosine functions, but their behaviors and relationships are uncannily similar. For example, $\cos^2 x + \sin^2 x = 1$ (as you will prove in Advanced Precalculus) and $\cosh^2 x - \sinh^2 x = 1$.

Write a program that will ask the user to input a value of x. Calculate the values of sinhx and coshx to 4 decimal places using the infinite series definitions. Also, output the value of $\cosh^2 x - \sinh^2 x$ (which should equal 1 or something very close to it).

SAMPLE OUTPUT

x? 10

sinh(10) = 11013.2329

cosh(10) = 11013.2329

cosh**2(10) – sinh**2(10) = 1.0

III. Exponential Quadratic

Write a program that will solve a quadratic equation in the form $ad^{2x} + bd^x + c = 0$. (Similar to the problems in Exercise Set 11.6.2.)

SAMPLE OUTPUT

a, b, c, d? 1,–16,15,e

The solutions of e**(2*x) – 16*e**x + 15 = 0 are:

$x = 2.7081$

$x = 0$

IV. Exponential Regression

Let's say we have a data set of ordered pairs that appear to have an exponential relationship:

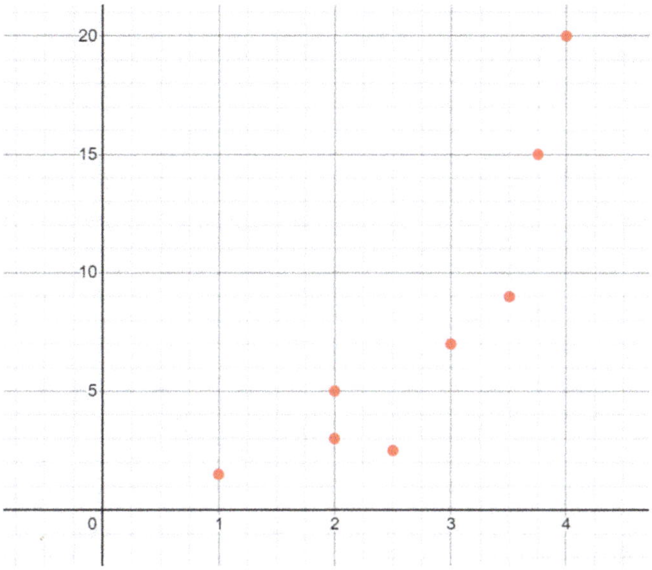

Image 11.12

We'd like to fit an exponential model $y = a \cdot k^x$ to these points. What we'll do is convert the exponential model to a linear model $y = mx + b$ and then use linear regression.

Step 1: Start with $y = a \cdot k^x$ and take the ln of both sides using the one-to-one property of logarithms:

$$\ln y = \ln(a \cdot k^x)$$

$$\ln y = \ln a + \ln k^x$$

$$\ln y = \ln a + x \cdot \ln k$$

Use a u-substitution: $u = \ln y$

$$u = (\ln k)x + \ln a$$

This is a linear equation in which the slope (coefficient of the independent variable) is $\ln k$ and the y-intercept is the constant, which is $\ln a$.

Step 2: Because we will be doing linear regression on (x,u), we need to transform the data using the relationship $u = \ln y$:

Original Data

x	y
1	1.5
2	3
2	5
2.5	2.5
3	7
3.5	9
3.75	15
4	20

Log Transformed Data

x	lny
1	0.405465108
2	1.098612289
2	1.609437912
2.5	0.916290732
3	1.945910149
3.5	2.197224577
3.75	2.708050201
4	1.386294361

Image 11.13

The graph of the transformed data now appears linear:

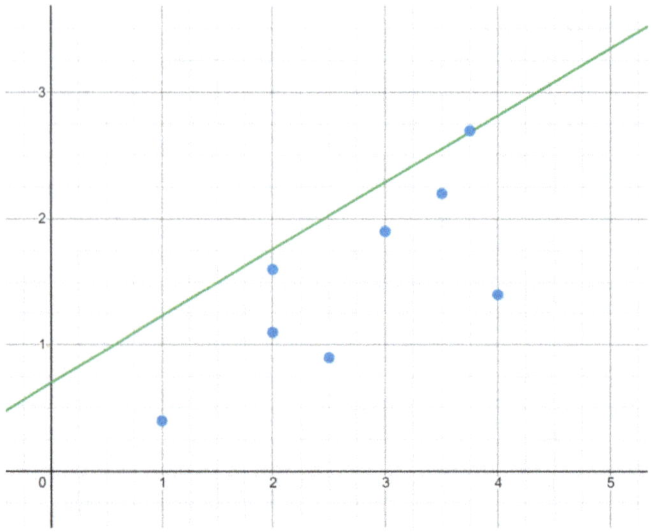

Image 11.14

Step 3: In Exercise Set 3.7.2, you wrote a program **linreg** that took a set of ordered pairs and did linear regression, producing the line $y = mx + b$. Store the log transformed data in two lists ***x*** and ***y***, and run the program, which will provide the constants ***m*** (slope) and ***b*** (y-intercept).
In the given example, linear regression returns $m = 0.54$ and $b = .07$.

Solve $m = \ln k$: $k = e^m$ *In the given example:* $k = e^{0.54} \approx 1.72$	Solve $b = \ln a$: $a = e^b$ *In the given example:* $a = e^{.07} \approx 1.07$

Then substitute k and a into $y = a \cdot k^x$ for your resulting exponential function.

∴ *In the given example, substitute $k = 1.72$ and $a = 1.07$:*

$$f(x) = 1.07 \cdot 1.72^x$$

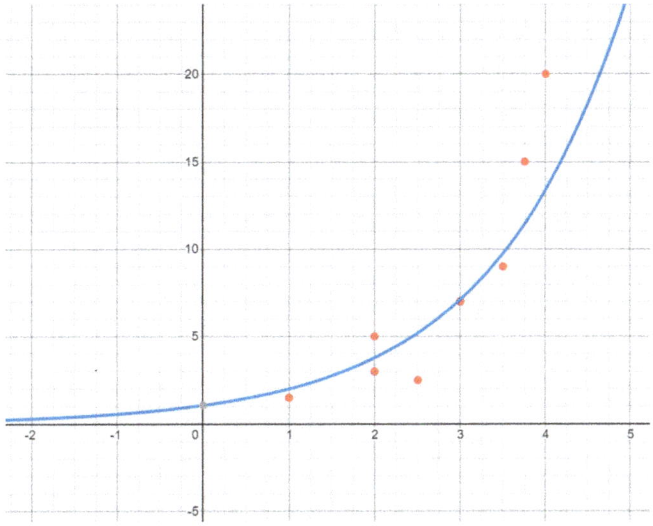

Image 11.15

Write a program that performs exponential regression using this method (thus calculating ***a*** and ***k***) and outputs the exponential model in the form $y = a \cdot k^x$. Plot the points and the approximated exponential graph.

V. Using Prime Factorization to Expand a log

The **SymPy** library will **simplify**(2*log(2)+2*log(3)+log(7)) as log(252), but there does not appear to be a way to do the reverse operation.

Write a program that will expand the natural log of a constant to the sum and difference of the logs of its prime factors.

Example: log(252) – 2*log(2) + 2*log(3) + log(7)

Hint: You wrote a program in Chapter 5 that performs prime factorization.

VI. log Subscripts with More Than One Digit

Program **convert_to_log** v2 is limited to only one digit log subscripts. (For example, it will handle $\log_3$ but not $\log_{13}$.) Modify the program (or design a new one) so that it handles log subscripts of any number of digits.

SAMPLE OUTPUT

Exponential *a****b*? 81**(1/4)

$\log_{81} 3.0 = 1/4$

Exponential *a****b*? 15**3

$\log_{15} 3375 = 3$

VII. Finding the Domain of a Logarithm of a Quadratic

Program 11.2.1 **log_domain** finds the domain of the common logarithm of a linear function, for example, the domain of log(4**x* – 16) is (4, ∞). However, it is not able to accurately find the domain of the common logarithm of a quadratic function; for example, if the user inputs log(*x***2–9), the program will output $x < -3$, which is an incomplete solution.

The complete domain of $f(x) = \log(x^2 - 9)$ is found by solving $x^2 - 9 > 0$. As we saw in Chapter Project IX in Chapter 3, this corresponds to the part of the graph that lies above the *x*-axis, i.e., $x < -3$ and $x > 3$.

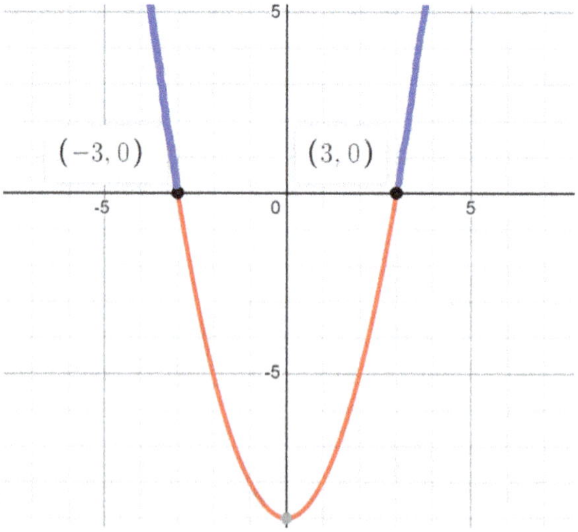

Image 11.16

This corresponds perfectly to the domain of $f(x) = \log(x^2 - 9)$, as evidenced by its graph:

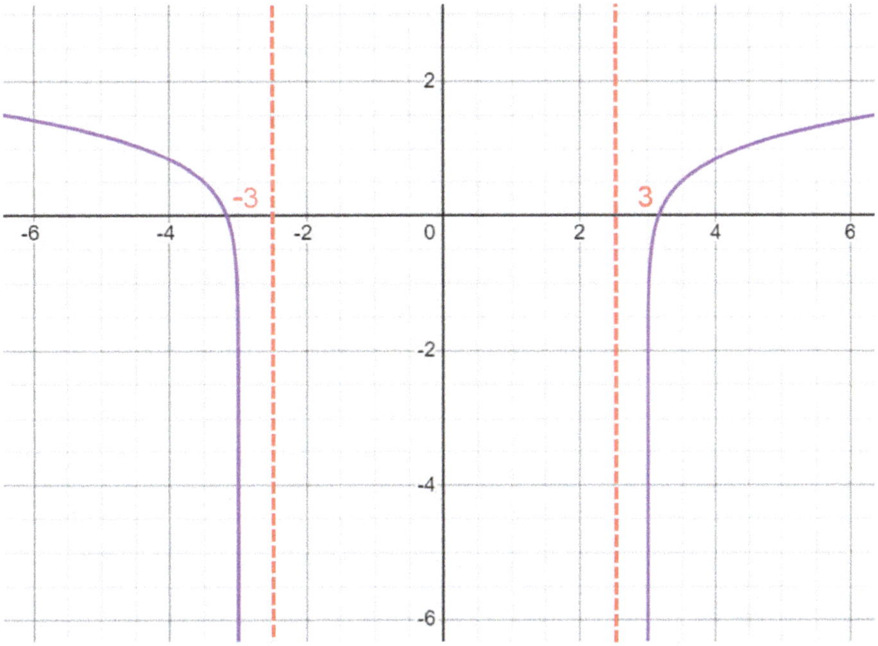

Image 11.17

Write a program that asks the user to input a logarithmic function of a quadratic. Output its domain. Hint: Incorporate Chapter Project IX from Chapter 3 to solve the quadratic inequality formed by setting the quadratic to greater than zero.

SAMPLE OUTPUT

Logarithmic function? log(−x**2 + 5*x)

Domain is (0, 5)

VIII. Numbers to Words

Given a multiple of 10, print out the value in words.

SAMPLE OUTPUT

Multiple of 10? 100

one hundred

Multiple of 10? 10,000,000

ten million

Here is a table of words up to one trillion:

1	One
10	Ten
100	One hundred
1000	One thousand
10,000	Ten thousand
100,000	One hundred thousand
1,000,000	One million
10,000,000	Ten million
100,000,000	One hundred million
1,000,000,000	One billion
10,000,000,000	Ten billion
100,000,000,000	One hundred billion
1,000,000,000,000	One trillion

Hint: Use log10() to figure out how many zeros are in the input. Include checks to make sure the input is indeed a multiple of ten and not larger than one trillion.

If you're curious, there are names for powers of ten all the way up to 10^{100} (which is a **googol**!), but it might be a bit tedious for the user to input that many zeros.

IX. Handling Negative Exponents in solve()

As we saw in Section 11.6, the **SymPy solve**() function is not perfect. (Another reason that *Terminator* and *Westworld* will never come to pass—we need humans to catch programming flaws.) As of this writing, **solve**() has a difficult time with $4 = e^{-x}$. The output is an empty set of solutions, but the actual solution is approximately −1.386.

Write a program that will accept as input an exponential statement in the form $a = e^{-\text{exp}}$ where a is a positive number and exp is any number or expression. Rewrite as a natural log statement (ln) and solve for x.

SAMPLE OUTPUT

Exponential statement $a = e$**(−x)? 4 = e**(−x)

x = ln(1/4) = −1.3863

Exponential statement $a = e$**(−x)? −4 = e**(−x)

not a real number

This may involve a Python string matching technique called **regular expressions,** which you are encouraged to research.

CHAPTER SUMMARY

PYTHON CONCEPTS

Assume x has been declared as a **SymPy** symbol using $x =$ **Symbol**('x')

To output a subscript	'\u208*n*' where *n* is a digit between 0 and 9. *Ex:* *>>> expr = 'log'+'\u2082'* *>>>print(expr)* *log*$_2$
log functions in the math module	math.**log**(x) returns the natural logarithm $\ln x$ math.**log**(x,b) *(logarithm base b)* returns $\log_b x$ when $b > 0$, $b \neq 1$, $x > 0$. (The default value of b is e, as shown above) math.**log2**(x) *(binary logarithm)* returns $\log_2 x$ when $x > 0$ math.**log10**(x) *(common logarithm)* returns $\log x$ when $x > 0$ *Ex:* *>>>math.log(e)* *1.0* *>>>math.log(9,3)* *2.0* *>>>math.log2(512)* *9.0* *>>>math.log10(1000000)* *6.0*

(*Continued*)

log functions in the SymPy library	**simplify**(expression) • Combines logs with the same base into a single log • Applies the change of base formula to express the final answer in terms of ln. **expand_log**(expression) • Applies the three laws of logs to separate the expression into as many logs as possible • Applies the change of base formula to express the final answer in terms of ln **logcombine**(expression) • Applies the three laws of logs to combine the expression into as few logs as possible • Applies the change of base formula to express the final answer in terms of ln *Ex:* *>>>a = **Symbol**('a')* *>>>**simplify(log**(4,a)+**log**(5,a))* *log(20)/log(a)* *>>>x,y = **symbols**('x y',**positive**=True)* *>>>**expand_log(log**(x**2*y**2))* *2*log(x) + 2*log(y)* *>>>**logcombine**(3***log**(x))* *log(x**3)*

PRECALCULUS CONCEPTS

Logarithm	$\log_b y$ is the exponent that b is raised to, to equal y. $b > 0, b \neq 1, y > 0$ *Ex:* *$\log_2 8 = 3$ because 3 is the exponent 2 must be raised to, to equal 8.*
Properties of logarithms	If $b > 0$: $b^{\log_b x} = x$ $\log_b b^x = x$ $\log_b 1 = 0$ $\log_b b = 1$ $\log_b 0$ is undefined *Ex:* *$\log_3 3^5 = 5$*

Common logarithm	Log base 10, written simply **log**. *Ex:* $log1000 = 3$
Natural logarithm	Log base e, written simply **ln**. *Ex:* $ln100 = 4.605170186$ because $e^{4.605170186} = 100$.
Binary logarithm	$\log_2$ *Ex:* $log_2 16 = 4$
Laws of logarithms	If $b > 0$ and $b \neq 1$: **Product Rule**: $\log_b(M \cdot N) = \log_b M + \log_b N$ **Quotient Rule**: $\log_b\left(\frac{M}{N}\right) = \log_b M - \log_b N$ **Power Rule**: $\log_b M^N = N \cdot \log_b M$ *Ex:* $log_5(125) = log_5 5 + log_5 25$ $log_5(100) = log_5 500 - log_5 5$ $log_5 625 = 2 \cdot log_5 25$
Change of base formula	if $b > 0$ and $b \neq 1$: $\log_b M = \frac{logM}{logb}$ or $\frac{lnM}{lnb}$ *Ex:* $log_4 256 = \frac{ln256}{ln4}$
Half-life	Given the half-life h of a decaying substance, the constant k in $P(t) = P_0 e^{kt}$ is given by: $k = \frac{ln(1/2)}{h}$ and the formula for the population at time t is given by: $P(t) = P_0(\,½\,)^{t/h}$ *Ex:* *If the half-life of a substance is 10,000 years then:* $k = \frac{ln(1/2)}{10{,}000}$ $P(t) = P_0(1/2)^{t/10000}$

Credits

APPENDIX A

How to Enter and Run Python Programs

As of this writing (2021), the most current version of Python is Python 3, so this book references commands, functions, and libraries in Python 3. Earlier versions of Python might not be compatible, in particular the **print()** function.

To enter and run a Python program, you need a Python interpreter. As of this writing (still 2021), these are the most popular, easy-to-use Python interpreters. (Keep in mind "popular" and "easy-to-use" are subjective terms.)

In a Browser

This is probably the easiest way because you don't have to install or download anything. One such online Python interpreter is:

https://repl.it/languages/python3

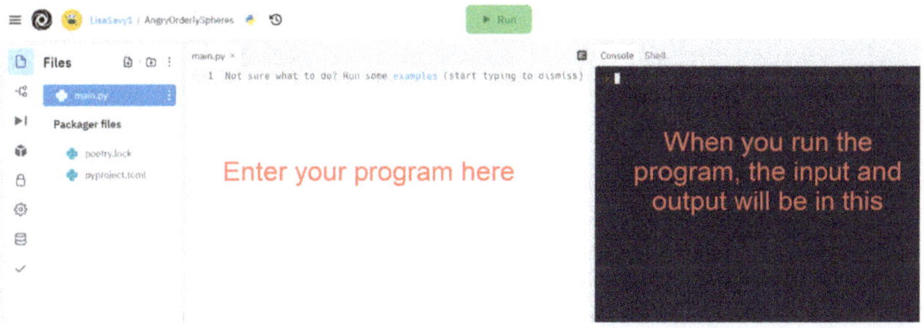

Image A.1

*(The downside with this method is that sometimes it takes a long time to run the program if you are importing libraries like **math** and **SymPy**, and it may decide to not offer certain libraries.)*

https://trinket.io/python

is yet another free and easy Python interpreter:

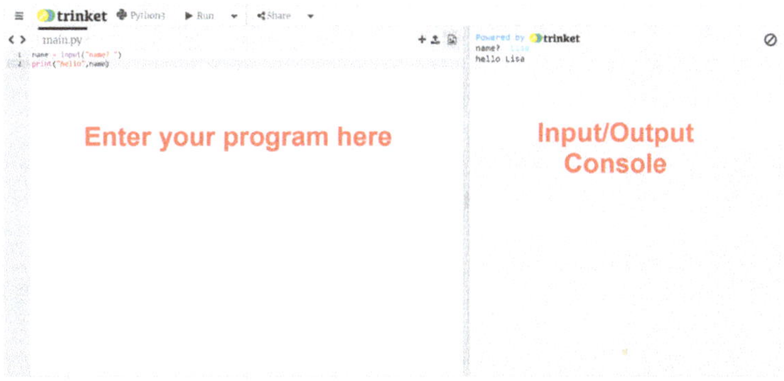

Image A.2

On an iPad

I use Pythonista 3, but it does cost $9.99. There are several other Python interpreters available in the app store.

Image A.3

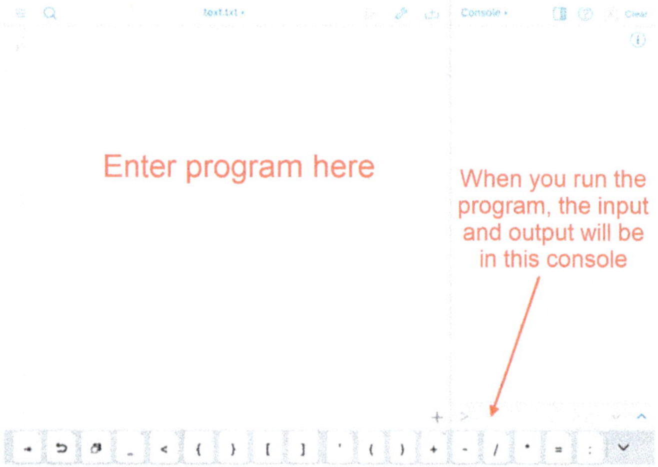

Image A.4

On a Computer

1. Python

Because Python is public domain, its most pure form is free and available for a variety of operating systems: **https://www.python.org/downloads/**

To create and save programs, you need the standard Python development environment called **IDLE** ("**I**ntegrated **D**eve**L**opment **E**nvironment").

IDLE is a shell that you can use either as a console (typing in one Python command at a time) or by accessing a file editor that lets you create and edit existing Python source files.

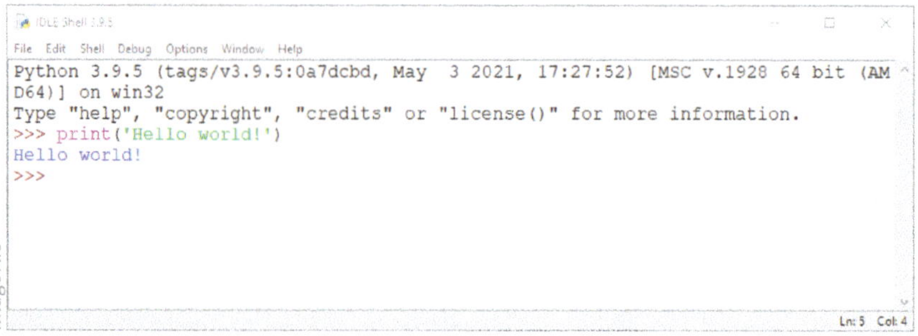

Image A.5

The Python shell might be a bit austere for some beginning programmers, so other organizations have packaged it within more user-friendly interfaces, as follows.

2. Spyder/Anaconda 3

Download the Individual Edition of Anaconda from **https://www.anaconda.com/products/individual.**

To run, select:

Image A.6

Image A.7

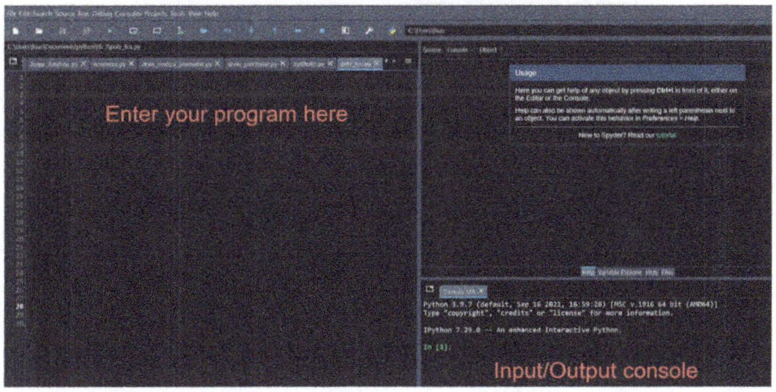

To run the program (after you type it in), use the green arrow button:

Image A.8

3. Pycharm

Free for students at **https://www.jetbrains.com/community/education/#students**. Just register with your student email.

Install the Community Edition using the default installation options. Then select [New Project], then [Create].

Image A 9

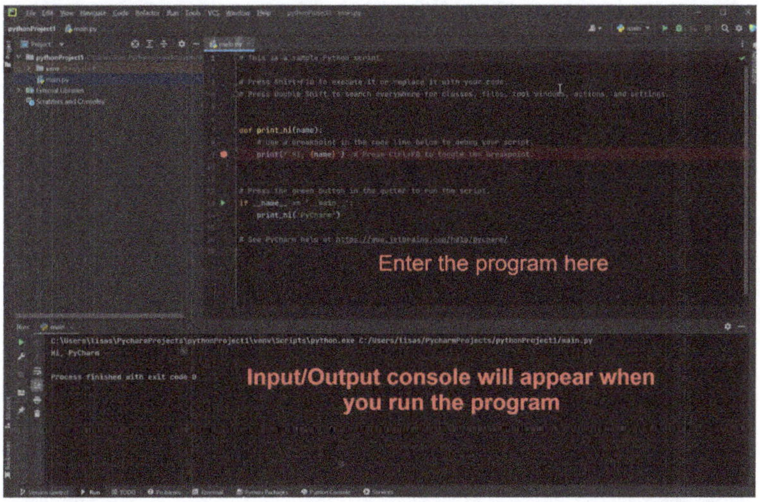

On a Texas Instruments Graphing Calculator

Yes, as of 2021, you can even write and run Python programs on both the TI nSpire and new TI-84 Plus CE graphing calculators. The most recent instructions can be found on the **education.ti.com** website:

TI Codes | Programming in Python with TI technology

These are excellent choices for schools in which instructors wish to have the students program in the classroom but don't have access to a full computer lab. There are grants available to purchase class sets of TI 84 calculators; see the Texas Instruments website for help.

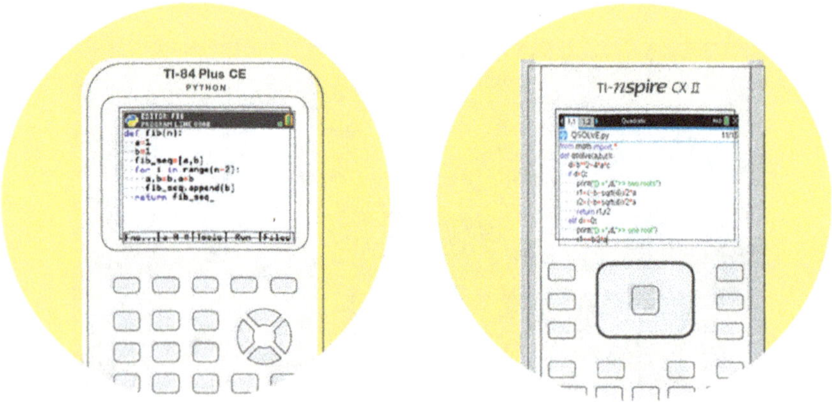

Image A.10

Credits

IMG A.1: Source: http://replit.com.

IMG A.2: Source: http://trinket.io.

IMG A.3: Source: https://apps.apple.com/us/app/pythonista-3/id1085978097?ls=1.

IMG A.4: Source: http://omz-software.com/pythonista/.

IMG A.5: Copyright © by Bernardo Lansing (CC BY-SA 4.0) at https://commons.wikimedia.org/wiki/File:Python_IDLE.png.

IMG A.6: Source: https://www.spyder-ide.org/.

IMG A.7: Source: https://www.anaconda.com/products/individual.

IMG A.8: Source: https://www.spyder-ide.org/.

IMG A.9: Source: https://www.jetbrains.com/pycharm/.

IMG A.10: Source: https://web.archive.org/web/20211231080400/https://education.ti.com/en/activities/ti-codes-overview/choose-your-technology.

APPENDIX B

CSS4 color names (case-insensitive);

Image B.1

Credit

IMG B.1: Matplotlib Development Team, https://matplotlib.org/stable/gallery/color/named_colors.html. Copyright © 2023 by Matplotlib Development Team.

Useful Library Functions

POLY_FCN.PY (CHAPTER 3)

```
# built-in modules
from sympy import sympify
def num_list_to_poly(coeff):
# converts a list of integers to a SymPy polynomial using the integers as coefficients
# example: converts 4,3,2,1 to 4*x**3 + 3*x**2 + 2*x + 1
    degree = len(coeff)−1
    poly_str = " "
    for n in range(0,len(coeff)):
        if n!=0:
        poly_str += "+"
        poly_str += str(coeff[n]) + "*x**" + str(degree)
        degree −= 1
    poly = sympify(poly_str)
    return poly

def poly_to_string(p):
# converts polynomial to string type; replaces ** with ^ and removes *
#  example: converts 4*x**3 + 3*x**2 + 2*x + 1 to 4x^3+3x^2+2x+1     p = str(p)
    p = p.replace("**","^")
    p = p.replace("*","")
    return p
```

MATRIX_FCN.PY (CHAPTER 4)

```
def init_matrix(r,c):
# returns an r × c matrix with all zero entries
    M = [ [ 0 for j in range(c) ] for i in range(r) ]
    return M

def input_matrix(r,c):
# prompts user to input an r × c matrix
```

```
    M = [ ]
    for i in range(r):      # for each row
        row_str = input("Row "+str(i+1)+" separated by commas:")
        row_list = [int(k) for k in row_str.split(",")]
        M.append(row_list)
    return M

def matrix_product(A,B):
# finds the product of matrices A and B
    A_num_rows = len(A)
    A_num_cols = len(A[0])
    B_num_rows = len(B)
    B_num_cols = len(B[0])
    if A_num_cols != B_num_rows:
        print("Sorry, # of A columns must equal # of B rows")
        return [[0]]
    else:
        P = init_matrix(A_num_rows,B_num_cols)
        for i in range(A_num_rows):
            for j in range(B_num_cols):
                for k in range(A_num_cols):
                    P[i][j] += A[i][k]*B[k][j]
    return P

def print_matrix(M):
# output matrix M in rows and columns
    r = len(M)
    c = len(M[0])
    # create format string
    fstring = ''
    for j in range(c):
        fstring += '{:^6n}'
    # print the matrix
    for i in range(r):
        print(fstring.format(*M[i]))      # print row I using format
    return

def matrix_inverse(A):
# returns the inverse of 2 × 2 matrix A as a 2 × 2 matrix of float entries
    if len(A) != 2 or len(A[0]) != 2:
        print("A must be 2x2")
        return [[ 0 ]]
```

(*Continued*)

```
    else:
        I = [[0,0],[0,0]]
        d = A[0][0] *A[1][1]−A[0][1]*A[1][0]
        if d==0:
            print('noninvertible')
            return [[ 0 ]]
        else:
            I[0][0] = round((1/d)*A[1][1],2)
            I[0][1] = round( (1/d)*−1*A[0][1],2)
            I[1][0] = round( (1/d)*−1*A[0][0],2)
            I[1][1] = round( (1/d)*A[0][0],2)
            return I

def determ(M):
    # returns the determinant of 3 × 3 matrix M
    # the following return statement is a single line of code.
    return M[0][0]*(M[1][1]*M[2][2]−M[1][2]*M[2][1]) −M[0][1]*(M[1][0]*M[2][2]−
M[1][2]*M[2][0])+ M[0][2]*(M[1][0]*M[2][1]−M[1][1]*M[2][0])
```

Printed in the USA
CPSIA information can be obtained
at www.ICGtesting.com
CBHW081443101024
15659CB00005B/285